AF581196

Methods in Molecular Biology™

Series Editor
John M. Walker
School of Life Sciences
University of Hertfordshire
Hatfield, Hertfordshire, AL10 9AB, UK

For further volumes:
http://www.springer.com/series/7651

Structure-Based Drug Discovery

Edited by

Leslie W. Tari

Trius Therapeutics, San Diego, CA, USA

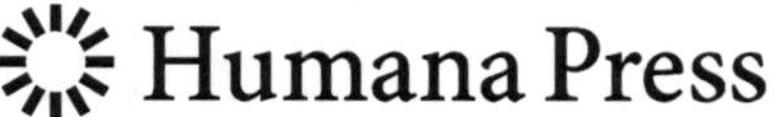

Editor
Leslie W. Tari
Trius Therapeutics
San Diego, CA, USA
ltari@triusrx.com

ISSN 1064-3745 e-ISSN 1940-6029
ISBN 978-1-61779-519-0 e-ISBN 978-1-61779-520-6
DOI 10.1007/978-1-61779-520-6
Springer New York Dordrecht Heidelberg London

Library of Congress Control Number: 2011944430

Printed on acid-free paper

Humana Press is part of Springer Science+Business Media (www.springer.com)

Preface

The potential utility of atomic resolution structures of protein drug targets in drug discovery has long been acknowledged. Without structure, medicinal chemists must rely on the costly, time-consuming endeavor of screening large libraries of compounds for hits, and are often forced to live with high molecular weight, non-ligand-efficient inhibitor scaffolds that must be blindly decorated with thousands of groups to generate SAR, improve potency and properties. With knowledge of the shape and chemical composition of the ligand-binding pocket of the drug target, the de novo design of ligand efficient inhibitor scaffolds is enabled. Also, iterative-structure-guided ligand optimization can be used to rationally improve early leads in a few steps rather than with thousands of analogs. However, despite its promise, structure-based drug design (SBDD) did not live up to expectations in its early days: only a limited range of protein targets were tractable to crystallographic studies, crystal structures took months or years to solve, and limitations in computing power and unrealistic expectations of the capabilities of molecular modeling methods reduced the scope and effectiveness of SBDD.

The last decade has seen the confluence of several enabling technologies that have allowed protein crystallographic methods to live up to their true potential. Off-the-shelf systems exist that allow the rapid cloning, and recombinant expression and isolation of large quantities of protein in a wide range of prokaryotic or eukaryotic hosts. Low-cost nanovolume liquid-handling robotic systems are available for the automated screening of vast arrays of diverse solution conditions to find crystallization conditions for a protein target using minimal quantities of protein. Latest generation synchrotron radiation sources allow for the collection of high-resolution X-ray diffraction data on microcrystals in minutes. Continuing improvements in computing power and advances in crystallographic software have made it possible to go from X-ray dataset to refined crystal structure in less than an hour on a laptop computer. Taken together, these advances have made it possible to tackle difficult biological targets with a high probability of success: intact bacterial ribosomes have been structurally elucidated, as well as eukaryotic trans-membrane proteins like the potassium channel and GPCRs. Of additional importance is the impact the above mentioned advances have had on the throughput of crystallographic structure determinations: it is now possible for medicinal chemists to have access to structural information on their latest small molecule candidates bound to the therapeutic target within days of compound synthesis, allowing structure-guided ligand optimization to occur in "real time." Also, using fragment screening, crystal structures of hundreds of small molecule cores complexed with the protein target can be utilized to construct novel inhibitor scaffolds.

The goal of this book is to provide scientists interested in adding SBDD to their arsenal of drug discovery methods with a practical guide to the methods used to generate crystal structures of biological macromolecules, how to leverage the structural information to design new inhibitor classes de novo, and how to iteratively optimize hits and convert them to leads. Where possible, specific protocols are described. Some examples highlighting the utility of structural biology in the discovery and development of small molecule and protein therapeutic agents are provided in the later chapters.

I am deeply grateful to all contributors who agreed to share their experiences in the development and application of methodologies that support SBDD. I believe their patience and hard work will be rewarded by the impact this volume has on scientists involved in drug discovery. I would like to extend special thanks to John Walker for his guidance, inspiration and patience in the preparation of this volume. Also, I am grateful to Les Tari Sr. for his critical evaluation of this volume and sharp editorial eye.

San Diego, CA, USA ***Leslie W. Tari***

Contents

Contributors

JUAN CARLOS ALMAGRO • *Centocor R&D Inc., Radnor, PA, USA*
JOHN BADGER • *Zenobia Therapeutics, San Diego, CA, USA*
DANIEL C. BENSEN • *Trius Therapeutics, San Diego, CA, USA*
MARK L. CUNNINGHAM • *Trius Therapeutics, San Diego, CA, USA*
JUDIT É. DEBRECZENI • *Structure and Biophysics, Discovery Sciences, AstraZeneca, Alderley Park, Macclesfield, UK*
SUZANNE C. EDAVETTAL • *Centocor R&D Inc., San Diego, CA, USA*
PAUL EMSLEY • *Department of Biochemistry, University of Oxford, Oxford, UK*
ANDREW D. FERGUSON • *Discovery Sciences, AstraZeneca Pharmaceuticals, Waltham, MA, USA*
JOHN FINN • *Trius Therapeutics, San Diego, CA, USA*
GARY L. GILLILAND • *Centocor R&D Inc., Radnor, PA, USA*
ISAAC D. HOFFMAN • *Takeda San Diego, San Diego, CA, USA*
MICHAEL J. HUNTER • *Centocor R&D Inc., San Diego, CA, USA*
ANDY JENNINGS • *Takeda San Diego, San Diego, CA, USA*
FELICE C. LIGHTSTONE • *Lawrence Livermore National Laboratory, Physical and Life Sciences Directorate, Livermore, CA, USA*
JINQUAN LUO • *Centocor R&D Inc., Radnor, PA, USA*
SHAWN P. MADDAFORD • *NeurAxonInc, Mississauga, ON, Canada L5K 1B3*
TOAN B. NGUYEN • *Lawrence Livermore National Laboratory, Physical and Life Sciences Directorate, Livermore, CA, USA*
GYORGY SNELL • *Takeda San Diego, San Diego, CA, USA*
RONALD V. SWANSON • *Centocor R&D Inc., San Diego, CA, USA*
LESLIE W. TARI • *Trius Therapeutics, San Diego, CA, USA*
OMID VAFA • *Centocor R&D Inc., Radnor, PA, USA*
SERGIO E. WONG • *Lawrence Livermore National Laboratory, Physical and Life Sciences Directorate, Livermore, CA, USA*

Chapter 1

The Utility of Structural Biology in Drug Discovery

Leslie W. Tari

Abstract

Access to detailed three-dimensional structural information on protein drug targets can streamline many aspects of drug discovery, from target selection and target product profile determination, to the discovery of novel molecular scaffolds that form the basis of potential drugs, to lead optimization. The information content of X-ray crystal structures, as well as the utility of structural methods in supporting the different phases of the drug discovery process, are described in this chapter.

Key words: X-ray crystallography, Structure-based drug design, Fragment screening, Structural bioinformatics, Lead optimization

1. Introduction

The discovery of new drugs is a time and labor-intensive process. On average, the discovery of a new drug requires the preparation and evaluation of approximately 10,000 compounds over 12 years at a cost of more than $350 million (1). Once in the marketplace, many drugs fail to recover their development costs (as many as 30%, according to data from the 1980s (2)), and many others are ultimately withdrawn from the market. These facts coupled with limits on patent lifetime, escalating global competition, and increasingly stringent government regulations for drug approval have demanded more efficient and accelerated approaches to drug discovery. Conventional "brute force" methods of lead discovery via high-throughput screening (HTS) of proprietary synthetic, combinatorial, or natural product libraries, while effective in many cases, are expensive and have limitations; they require access to large compound libraries (sometimes over 1,000,000 compounds), often yield hits with high molecular weight, poor ligand efficiency,

Leslie W. Tari (ed.), *Structure-Based Drug Discovery*, Methods in Molecular Biology, vol. 841,
DOI 10.1007/978-1-61779-520-6_1, © Springer Science+Business Media, LLC 2012

limited or no potential for optimization, and provide no information to guide ligand optimization.

Advances in crystallographic methods, computational power, molecular biology, and recombinant protein expression systems over the last 30 years have provided researchers with rapid and reliable access to three-dimensional structural information on a wide variety of protein drug targets. Structural information on protein–ligand complexes can eliminate much of the complexity involved in the discovery and optimization of prospective drug leads. Indeed, structure-guided drug design efforts have led to the discovery of high profile drugs in multiple therapeutic areas, including the peptidomimetic HIV protease inhibitors for the treatment of HIV, the neuraminidase inhibitor Tamiflu™ for the treatment of influenza, the carbonic anhydrase inhibitor dorzolamide for the treatment of glaucoma, and the thrombin inhibitor ximelagatran, an oral anticoagulant (3). Access to structural information on the target of interest can streamline all aspects of drug discovery, from target selection to lead discovery and optimization, using methods that are summarized in this chapter.

2. The Information Content of Protein Crystal Structures

Protein crystals, like any crystalline substance, are regular, three dimensionally periodic arrays of identical molecules or molecular complexes (see Fig. 1). A common misconception regarding protein crystal structures is that they are not representative of the protein in solution due to the influence of extensive intermolecular interactions present in the crystalline state. The idea that protein crystal structures are heavily biased by "solid state" artifacts arises from inaccurate comparisons made between protein crystals and crystals of small molecular weight compounds. Crystals of small molecules and proteins differ in ways that extend beyond the properties of their component molecules. Small-molecule crystals typically only comprise the small molecule, while protein crystals contain 25–90% solvent by volume, depending on the protein. The remaining volume in protein crystals is occupied by protein molecules, and is analogous to an ordered gel with large interstitial spaces between protein molecules. By comparison, the number of contacts made in relation to the molecular mass of the protein in protein crystals is smaller by orders of magnitude than it is for small-molecule crystals. This causes the mechanical stability and integrity of protein crystals to be much worse than it is for crystals of small molecules. The high solvent content and tenuous thermodynamic stability of protein crystals complicate the subsequent steps in X-ray diffraction experiments, since these properties result in crystal handling difficulties, susceptibility to temperature changes

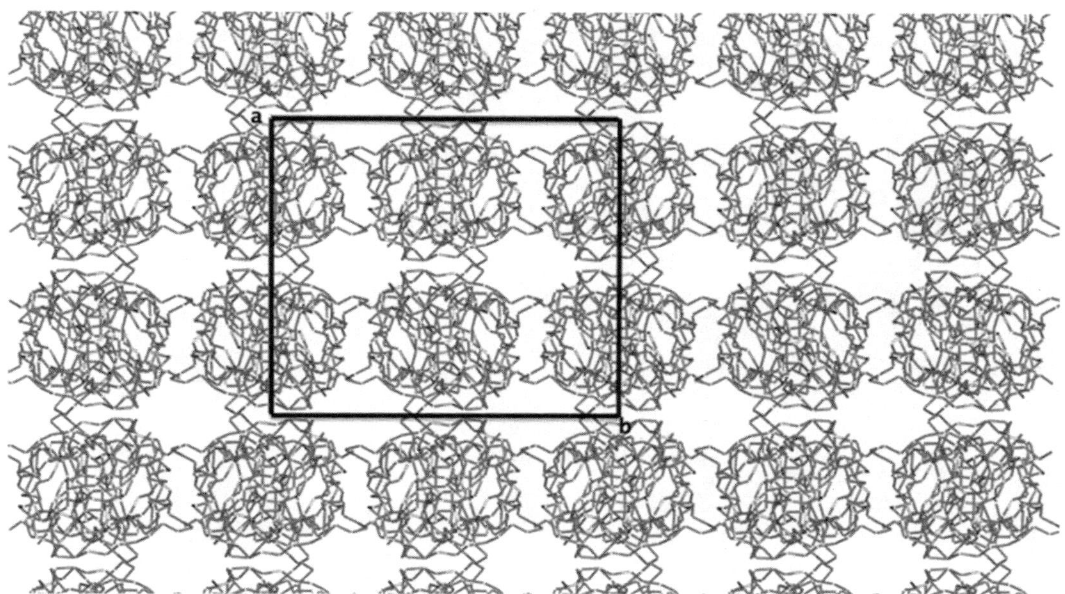

Fig. 1. A view of crystal packing in a *Haemophilus influenzae* dihydrofolate reductase crystal. Boundaries for a single unit cell within the crystal are shown. The view is perpendicular to the *c*-axis of the unit cell. The unit cell is the fundamental building block of the crystal, a translationally periodic substance comprising trillions of unit cells that extend in three dimensions. The unit cell is an arbitrary construction that describes the smallest "box" with the highest metric symmetry.

and dehydration, weaker diffraction, and greater sensitivity to radiation damage. However, the key role played by solvent in protein crystallization is a double-edged sword; while it adversely affects diffraction, it is the very element that makes protein crystal structures valuable. The high solvent content of protein crystals is essential for maintaining the structures of the macromolecules in their solution states. Therefore, to a large extent, proteins in crystals possess the structural, enzymatic, and functional properties of their counterparts in solution. Protein crystal structures must be regarded with care, however. In the hands of the uninformed, the danger exists that crystallographic structural data will be misinterpreted, or overreaching conclusions drawn. An understanding of the parameters derived from crystallographic experiments is essential if structural information from crystallographic experiments is to be used effectively to support drug discovery.

X-ray crystallography and light microscopy share the same basic principle; electromagnetic radiation scattered by the object to be imaged is recombined and focused by a lens to reform the image of the object. Theoretically, the resolving power of any imaging technique is equal to one half of the wavelength of the radiation used for imaging. To resolve the atomic details of protein structures, crystallographic experiments involve the exposure of protein crystals to high-energy monochromatic X-rays (wavelengths on the order of 1 Å). Imaging using X-rays is complicated by the fact

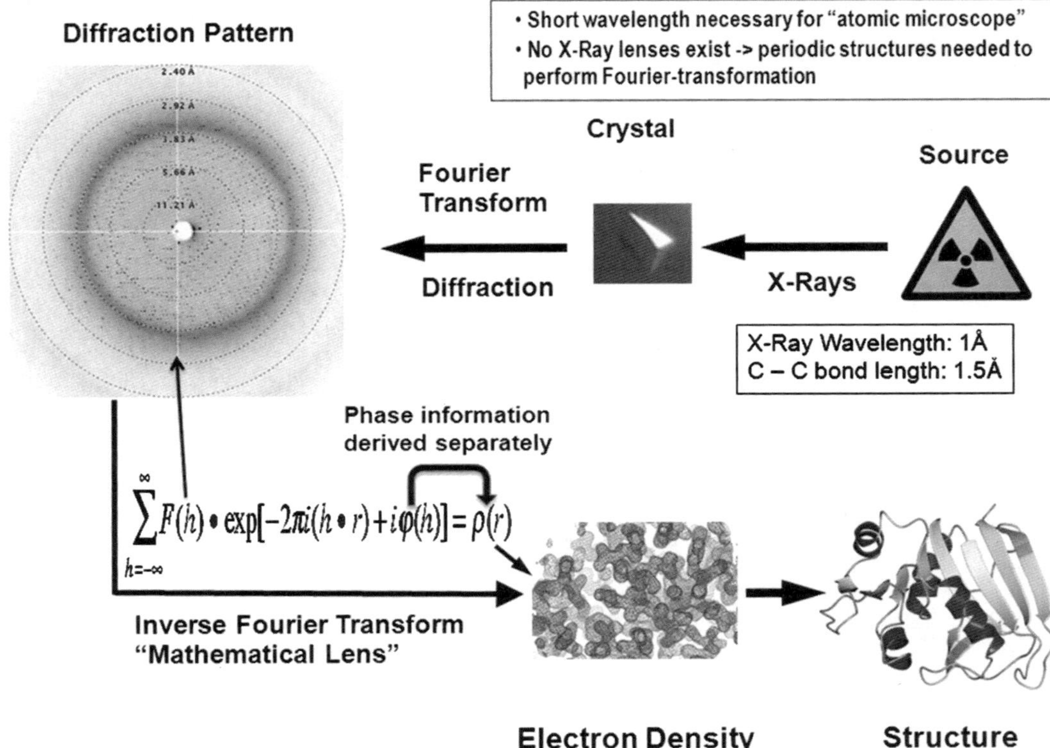

Fig. 2. A schematic outlining the steps in a crystallographic structure determination. Crystals are systematically exposed to monochromatic X-rays in multiple orientations, and the diffraction patterns are captured with electronic detectors. Since crystals are three-dimensionally periodic substances, the diffraction pattern comprises a series of spots rather than a continuous function. Each spot represents a family of diffracted waves that map to discrete spatial periodicities in the unit cell of the crystal. The diffraction pattern is a summation of waves of electromagnetic radiation and can thus be described by a Fourier series, and the diffraction pattern and disposition of the atomic contents of the unit cell are related mathematically by a Fourier transform. An image of the atomic contents of the unit cell of the crystal is derived by applying a mathematical lens (inverse Fourier transform, equation shown on the *lower left*) to the diffracted X-rays. The image reconstruction process is complicated by the fact that only intensities of the diffracted X-rays are measurable ($F(h)$ terms in the equation shown), but not the relative phase shifts between each family of diffracted waves. The missing information is referred to as the crystallographic phase problem. The missing phases are obtained using other experimental or computational methods described in the text. Since the diffraction of X-rays is caused by the interaction of the X-rays with electrons, the resulting image obtained in a crystallographic experiment is of the electron density distribution in the unit cell of the crystal. Interactive model building software is used to build the final atomic model into electron density.

that X-rays interact very weakly with matter, so that no lenses exist which are able to reconstruct the image from the scattered X-rays. Hence, the scattered X-rays from crystals must be captured with electronic detectors and the function of a lens must be simulated mathematically. A schematic describing the steps involved in the solution of a crystal structure is shown in Fig. 2.

Mathematical reconstruction of the structure of the atomic contents of the crystal is complicated by the fact that one of the two key pieces of information describing the diffracted X-ray waves, the relative phase shifts between the different families of diffracted waves, cannot directly be measured (see Fig. 2). Three methods are commonly employed to overcome the phase problem, as summarized below.

(a) *Molecular replacement.* When an approximate model of the unknown crystal structure is available, it can be used to overcome the phase problem. The principle is simple; the model is first oriented and then positioned in the unit cell of the target crystal structure using rotation and translation functions. The correctly oriented model is subsequently used to calculate approximate phases and electron density maps. Alternate cycles of interactive correction and rebuilding of the model into electron density and model refinement are used to improve the quality of the phases and to transform the model structure into the real structure. The success of molecular replacement depends critically on two factors: the fraction of the asymmetric unit for which suitable models exist, and the r.m.s. deviation (after optimal superposition) between the model and target structures. Generally, r.m.s. deviation increases with decreasing sequence identity, or in cases where the target structure undergoes significant conformational changes with respect to the model structure (e.g., movement of protein domains). In the latter case, the model structure can be separated into individual fragments that are sequentially oriented and positioned in the unit cell. Newer maximum-likelihood molecular replacement algorithms, such as those implemented in the program Phaser (4) are more discriminating, and have been successful in solving difficult molecular replacement problems that were previously intractable.

(b) *Isomorphous replacement methods.* This is a classical approach used to solve protein structures with unknown folds. Crystals are soaked in multiple solutions containing salts of heavy atoms such as Hg, Pt, Pb, Au, etc., until conditions are found where a small number of heavy atoms incorporate in well-defined positions on the crystallized protein molecule (without altering the structure of the underlying protein). By analyzing the differences in the intensities of diffraction patterns from the native and heavy atom derivatized protein crystals, it is possible to determine the locations of the heavy atoms in the unit cell and to use the scattering "signal" from the heavy atoms to calculate phases and an electron density map (reviewed in refs. (5–7)).

(c) *Anomalous scattering methods.* For heavier elements, some inner shell electrons have absorption edges in the range of the X-ray wavelengths used in diffraction experiments. The heavy atoms in the protein crystal cause absorption of the impinging radiation, and impart small phase shifts on the radiation scattered from the crystal. This phenomenon is used to determine the positions of the heavy atoms in the unit cell, and subsequently to extract phase information to allow electron density map generation. Anomalous scattering can be used to supplement the phase information obtained from isomorphous heavy atom derivatives, or to independently obtain complete phase information. A very powerful de novo phase determination method utilizes anomalous scattering from proteins that are homogeneously labeled with selenomethionine (incorporated during recombinant expression of the protein in *Escherichia coli*), a derivatized selenium-containing amino acid. Independent diffraction experiments are carried out (on the same crystal, if possible) at multiple X-ray wavelengths on the high and low energy sides of the selenium absorption edge that maximize the anomalous diffraction signal. This method requires a tunable X-ray source, which is present only at synchrotrons (reviewed in refs. (5–7)).

X-ray diffraction is caused by the interaction of the electric field vector of monochromatic X-rays with electrons in a protein crystal. These details, coupled with the fact that crystals are made up of three-dimensionally periodic lattices of molecules, have several important consequences (for excellent reviews see refs. (5–7)): (1) X-ray diffraction experiments generate three-dimensional images of the electron density distribution of the molecular components of the crystal. So heavier atoms generate a proportionally stronger signal, and hydrogen atoms are generally not discernable in protein crystal structures. (2) The short wavelength radiation used in X-ray diffraction experiments allows for the resolution of macromolecular structures at an exquisite level of detail (typical protein crystal structures are determined at resolutions between 1.5 and 3.0 Å resolution). (3) In a crystallographic experiment, the structure of the molecular contents of the unique portion of a crystal (called the asymmetric unit of the unit cell, which is the microscopic building block of the crystal) are obtained, and the resulting crystal can be built by the application of crystallographic symmetry operators to the contents of the asymmetric unit, as shown in Fig. 1. Since the diffraction signal from a crystal arises from constructive interference from trillions of crystallographic asymmetric units, the resulting crystal structure comprises a time- and space-averaged picture of the contents of the copies of asymmetric units that are sampled. Hence, components of the asymmetric unit with a large degree of random spatial heterogeneity, i.e., disordered protein loops or side chains and the bulk solvent occupying the spaces between protein molecules, fade into the background and cannot be modeled. However, in cases where a molecular component of a crystal, such as a protein side chain, occupies a finite number of distinct, low energy conformations in different asymmetric units, it is possible to simultaneously characterize each alternative conformation.

Examination of the equation relating diffracted X-rays to the crystal structure provides insight into the structural parameters that are modeled in a crystallographic experiment (see Eq. 1).

$$F_{hkl} = \sum_{j=1}^{N} f_j e^{-(B\sin^2\theta)/\lambda^2} e^{2\pi i(hx+ky+lz)}. \quad (1)$$

Equation 1 is one of the explicit forms of the structure factor equation (8). Each F_{hkl} term represents a unique family of diffracted X-ray waves from the crystal (diffracted waves from crystals constructively interfere to form patterns of spots, as shown in Fig. 2, which can each be assigned integer indices *h*, *k*, and *l*), which correspond to discrete spatial periodicities in the crystal lattice. The intensity and phase of each family of diffracted waves is derived via a summation of the scattering contributions from all of the atoms in the asymmetric unit of the crystal. The second exponential term in Eq. 1 computes the net phase shift relative to an arbitrary origin of the scattered wave with index *h*, *k*, *l* due to the relative positions of the individual atoms in the unit cell (with fractional coordinates *x*, *y* and *z*). The f_j term corresponds to the scattering factor for each atom in the summation, and is directly proportional to the number of electrons in the atom in question. The first exponential $B\sin^2\theta/\lambda^2$ term (θ is the angle of the scattered radiation with respect to the source X-ray beam, and λ is the wavelength of the X-rays) accounts for the reduction in the intensity of the scattered radiation with scattering angle due to interference between scattered waves from different parts of the electron cloud surrounding each atom. X-ray scattering is attenuated further by smearing of the electron clouds surrounding each atom due to thermal motion of the atoms. Atomic thermal motion is modeled using the extra *B* term in the structure factor equation. As a first approximation it is assumed that the thermal motion of atoms is isotropic (spherically symmetric), with $B = 8\pi^2\mu^2$, where μ is the root mean square amplitude of atomic vibration. Using the calculation above, for a *B*-factor of 15 Å^2, the displacement of an atom from its equilibrium position is approximately 0.44 Å, and it is as much as 0.87 Å for a *B*-factor of 60 Å^2. Thus, analysis of *B*-factors is very important during any structural analysis to provide insight into the dynamics and structural integrity of different regions of a protein molecule. However, one must exercise caution before interpreting *B*-factors too quantitatively. In addition to measuring dynamic disorder caused by temperature dependent vibration of atoms, the *B*-factor is also influenced by subtle structural differences between protein molecules in different unit cells throughout the crystal (which spatially smears the atom positions), steric constraints from intermolecular lattice contacts, and certain systematic experimental errors, such as absorption of the X-ray beam during X-ray data collection. Advanced mathematical models can be used to provide more detailed information on atomic thermal motions. For example, the relative motions of entire protein domains can be characterized using TLS refinement (9). Also, when high-quality X-ray data are available from crystals that diffract to high resolution (typically better than 1.2 Å, rare in protein structure determinations), the isotropic thermal correction can be replaced by a tensor, which corrects not only for the extent of thermal motion of the atoms but also for spatial anisotropy in their motions (10).

Based on the mathematical description of X-ray diffraction provided above, four parameters are optimized in a single crystal X-ray diffraction experiment for each atom in a protein crystal structure: the *x*, *y*, and *z* coordinates of each atom and the *B*-factor describing the thermal motion of each atom. The quality of resulting electron density maps and the accuracy of refined parameters in protein crystal structures are largely dependent on the resolution of the X-ray diffraction data (equivalent to the pixel size of electron density sections). Examples of the effects of diffraction resolution on electron density map quality are shown in Fig. 3. The model is generally manually built (or refit) into electron density by a crystallographer, using two types of electron density maps, $|2F_o - F_c|\alpha_c$ maps, and $|F_o - F_c|\alpha_c$ difference maps, described below.

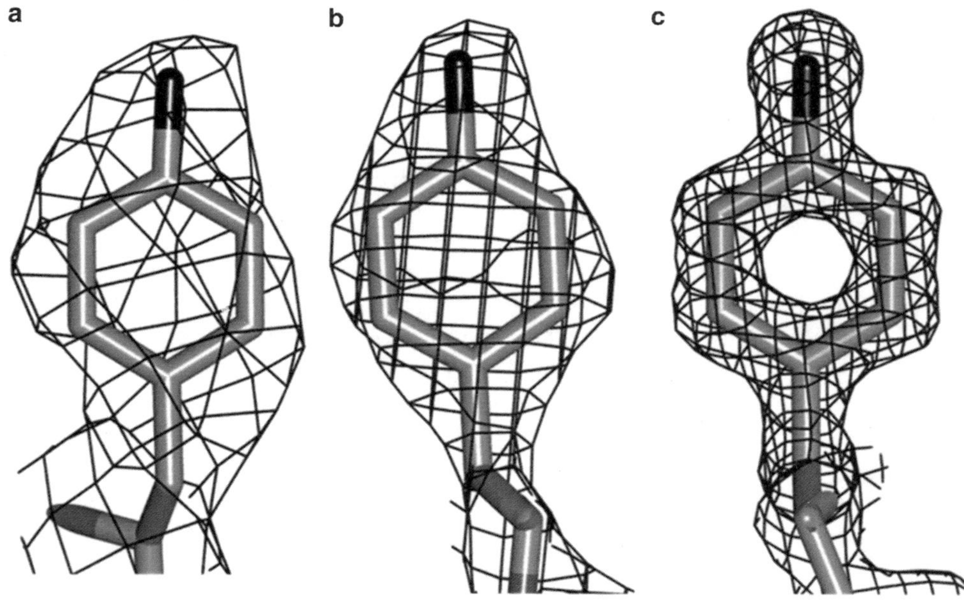

Fig. 3. Representative electron density maps contoured around tyrosine residues (using $|2F_o - F_c|\alpha_c$ coefficients) from three refined crystal structures: (**a**) A 2.8 Å resolution structure of *Francisella tularensis* topoisomerase IV, (**b**) A 2.2 Å structure of *Escherichia coli* topoisomerase IV, and (**c**) A 1.4 Å structure of *Enterococcus faecalis* DNA gyrase B (all from D. Bensen and L. Tari, unpublished results). The electron density maps were contoured using the electron density visualization software COOT (*see* ref. (11), Chapter 6). At better than 3.0 Å resolution, amino-acid side chains can be recognized with the help of protein sequence information, while at better than 2.5 Å resolution solvent molecules can be observed and added to the structural model with some confidence. As the resolution improves to better than 2.0 Å resolution, fitting of individual atoms may be possible and most of the amino-acid side chains can be readily assigned even in the absence of sequence information.

$|F_o - F_c|\alpha_c$ maps. $|F_o - F_c|\alpha_c$ maps, or difference maps, are generated by subtracting the calculated structure factor amplitudes (F_c, from the best current model structure) from the observed structure factor amplitudes (F_o), using phase information (α_c) calculated from the available model structure. To a good approximation, this operation is equivalent to subtracting the electron density calculated from the model from the "real" electron density in the crystal. What is left behind is the electron density for ordered components of the crystal structure that have not been accounted for by the model, or that have not been modeled correctly. Features that are present in the true structure that have not been accounted for in the model structure appear as positive peaks, while atoms that have been incorrectly placed in the model structure (i.e., that do not exist in the real structure) appear as holes or negative peaks. These maps are used to fix improperly modeled side-chains and/or entire polypeptide chains, as well to fit substrates, inhibitors, and ordered solvent molecules into the structure. A special type of difference map called an omit map can be used to confirm the presence of important features in a protein structure. An omit map is calculated by removing the feature of interest (say, an inhibitor) from the model, refining the structure in the absence of that feature, and calculating a new difference map. If the feature of interest is still observed in a difference density map, then it is real, and not an artifact caused by model bias present in the calculated phases. An example of a difference map is shown in Fig. 4.

$|2F_o - F_c|\alpha_c$ maps. $|2F_o - F_c|\alpha_c$ maps are the maps most commonly used for model fitting. They are used instead of $|F_o|\alpha_c$ maps, which suffer from model bias, and tend to show only electron density that is associated with the model. As described above, $|F_o - F_c|\alpha_c$ maps reveal everything in the $|F_o|\alpha_c$ map that has not been modeled. The $|2F_o - F_c|\alpha_c$ map essentially superposes an $|F_o|\alpha_c$ map over an $|F_o - F_c|\alpha_c$ difference map, so that it simultaneously shows both the electron density for the model and the electron density for features that have not been accounted for by the model. Several weighting schemes are used to further diminish the effects of model bias, including figure-of-merit and σ_A weighting schemes (reviewed in refs. (5–7)). An example of a $|2F_o - F_c|\alpha_c$ electron density map is shown in Fig. 4.

In addition to providing a more detailed picture of the electron density, higher resolution X-ray data correlates with a greater number of experimental observations to support structure refinement. For a typical protein structure from a crystal with a solvent content of about 50%, the number of experimental observations and refinement parameters will be about the same at 2.8 Å resolution. The paucity of experimental data compared with the number of parameters that need to be defined make least squares model optimization methods intractable. Additionally, at resolutions lower than 2.8 Å, individual atomic *B*-factors have a very limited

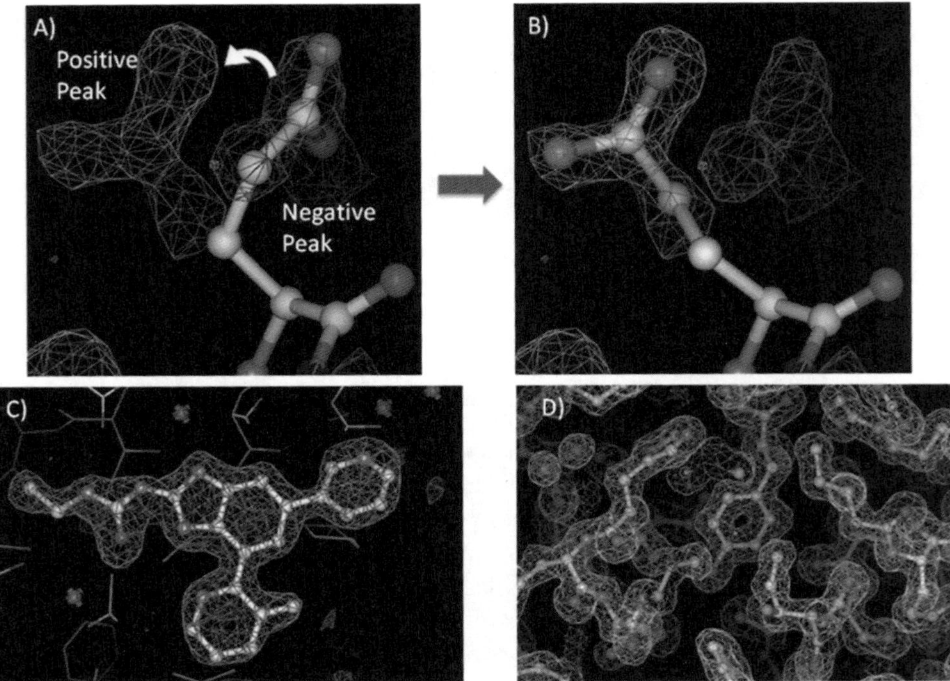

Fig. 4. Examples of $|F_o - F_c|\alpha_c$ and $|2F_o - F_c|\alpha_c$ electron density maps. The electron density maps in all panels are drawn as thin chicken-wire representations. In (**a**) an $|F_o - F_c|\alpha_c$ map contoured at 3σ is used to fit an incorrectly modeled glutamic acid side chain in an *E. faecalis* GyrB crystal structure. In the model structure, part of the side chain is in a negative electron density peak, while a positive difference density peak on the left-hand side of the figure reveals the correct position for the side chain from the experimental data. The correctly positioned glutamic acid side chain is shown in (**b**). In (**c**), an $|F_o - F_c|\alpha_c$ difference electron density map contoured at 3.5σ was used to fit a small-molecule inhibitor into the substrate-binding pocket of *E. faecalis* gyrase B. The difference map was calculated in the absence of the inhibitor, indicating that the difference density shown arises entirely from the experimental X-ray data. Panel (**d**) shows a representative section of a $|2F_o - F_c|\alpha_c$ electron density map contoured at 1σ for an *E. faecalis* GyrB crystal structure. The map displays electron density for both regions of the model that have been correctly fit, as well as regions that have not been accounted for by the model. Because it comprises a superposition of an $|F_o|\alpha_c$ map and a $|F_o - F_c|\alpha_c$ map, $|2F_o - F_c|\alpha_c$ maps are less subject to the effects of model bias than $|F_o|\alpha_c$ maps. During model fitting, crystallographers generally utilize $|2F_o - F_c|\alpha_c$ and $|F_o - F_c|\alpha_c$ simultaneously to trace the polypeptide chain and correct errors in the existing model.

physical meaning. The problem of statistical under determination is overcome by augmenting the X-ray diffraction data with structural parameters of proteins and peptides derived from small-molecule crystallography and spectroscopic data. The resulting function that is minimized in a crystallographic structure refinement incorporates the experimental X-ray data and a molecular mechanics function (which restrains bond lengths, angles, stereochemistry, planarity of peptide bonds and aromatic groups, etc. to reasonable values). The quality of structures refined in this fashion is excellent, even for structures determined at modest resolutions. Properly refined protein crystal structures generated from carefully measured X-ray data yield atomic positions that are precise to within one fifth to one tenth of the stated experimental resolution. Once a structure is fully refined, multiple criteria are used to judge the quality of the model, as described below.

R-factor. The *R-factor* is the averaged error (in percent) between the observed structure-factor amplitudes (the experimentally measured F_{hkl} values) and the calculated structure-factor amplitudes ($F_{hklcalc}$) from the refined model of the contents of the crystal. The ultimate value of the *R-factor* in a well-refined structure depends on a number of variables, including the proportion of the contents of the unit cell that can be correctly modeled, the relative weights assigned to the molecular mechanics restraints vs. the experimental X-ray data during refinement, the experimental resolution of the diffraction experiment and the accuracy and overall quality of the measured experimental X-ray intensities. In protein structures with numerous dynamically disordered loops or domains that cannot be modeled, the *R-factor* will not converge to low values. However, as a general rule of thumb a correctly refined protein structure should have an *R-factor* around 20%.

Free R-factor (R_{free}). The function that is minimized during a protein structure refinement is extremely complex, with multiple false minima. Hence, when not used with care, modern refinement algorithms can converge on convincing *R-factors* for incorrect structures. The R_{free} (12) statistic is an extremely simple and powerful independent validation tool used in modern protein structure refinement. The R_{free} function is identical to the *R-factor*; the only difference is that it is calculated using a small (5–10%) randomly sampled subset of the X-ray diffraction data that is excluded from structure refinement throughout the refinement process. In a correctly refined structure, R_{free} will track with the *R-factor* to within 5–10%. For incorrect structures, R_{free} will remain at a value near the limit observed for random atomic models fit to an X-ray dataset (~57%). In addition to R_{free}, the geometric quality of the refined protein structure should be used to evaluate the model. The averaged bond lengths and angles of the final model should not deviate much from ideal values (r.m.s. deviations from ideality should be within 0.02 Å for bond lengths and 3° for bond angles), and the majority of the protein residues should possess "allowed" combinations of ϕ, ψ main-chain dihedral angles. It is important to note that protein folding can force some residues into disallowed ϕ, ψ values, which can have important functional significance (13). All residues in disallowed regions must be carefully checked to ensure that they are well described by experimental electron density.

Identification and refinement of ordered solvent molecules becomes more reliable when data are available to at least 2.5 Å resolution. Even then, before a water molecule is used in mechanistic or computational analysis, it is always wise to check its *B*-factor and to see if there exists at least one hydrogen bond to hold the water to the protein or a nearby solvent molecule.

Unless the structure has been determined at very high resolution, electron density and refinement do not discriminate between the oxygen and nitrogen atoms of asparagines and glutamines, or the alternative conformations of histidine side chains. In a detailed structural analysis, it is always necessary to check alternative conformations of Asn, Gln, or His side chains to decide which one makes more sense chemically (i.e., by analyzing available H-bonding networks). Also, great care has to be exercised when fitting dynamically disordered protein side chains that are not fully described by electron density. The crystallographer knows they are present from the amino-acid sequence, and incorporates them in conformations commonly observed for that side chain from databases of high-resolution structures. The final refined conformation of the side chain must ultimately be decided using the crystallographer's knowledge of chemistry and side-chain conformational preferences, in conjunction with the refinement program's force field. In many structures, entire loops or even domains are too disordered to show any observable electron density. In such cases, the offending loops/domains are not included in the final model. When analyzing crystal structures, an additional point of caution that must be noted regarding potential artifacts that can arise from contacts between adjacent molecules in a crystal lattice. In the ideal scenario, the protein of interest crystallizes in a lattice that leaves the active-site/receptor pocket solvent exposed, with no lattice contacts preventing the motion of functionally important mobile structural elements surrounding the drug-binding site (i.e., the lattice should not impede ligand-induced conformational changes in the protein). However, protein crystallization does not allow for control of lattice contacts, and the ideal situation does not always exist. Hence, before a new protein crystal form is nominated as a potential candidate for supporting structure-based drug design, a careful analysis of the crystal lattice contacts between neighboring molecules related by crystallographic or noncrystallographic symmetry must be carried out to assess the steric accessibility of the receptor pocket and the solvent space around it, as well as the nature and quantity of lattice contacts in the vicinity. This sort of analysis is particularly important if the crystals are produced for the purpose of ligand soaking experiments to support fragment screening or high throughput structure determination. If multiple crystal forms are available, the crystal forms that approach the ideal criteria should be chosen. Cocrystallization experiments usually circumvent problems related to lattice constraints, since the protein and ligand are mixed in solution, allowing the system to reach a low energy conformational state before crystallization occurs. Additional important parameters to consider when analyzing crystal structures are the solution conditions used in crystallization. Some proteins undergo significant structural changes in different solution conditions. A classic example is ribonuclease A, which undergoes large, pH-dependent conformational changes that have been characterized crystallographically (14).

3. Using Structure in Target Selection and Product Profile Development

In addition to supporting lead discovery and lead optimization, structural information can be used at a very early stage in a drug discovery program to evaluate the viability of a protein as a drug target. Does the protein possess a binding pocket with suitable properties for potent inhibitor development? In a large, structurally related protein family, such as eukaryotic protein kinases, is it possible to develop selective inhibitors against a kinase of interest? More generally, what are the prospects for the development of specific inhibitors against a protein target while avoiding off-target binding? In an antibiotic program, do the protein orthologs encompassed by the proposed target product profile possess sufficient structural homology to allow for the development of a small-molecule agent with the desired spectrum? Careful analysis of the structures of the protein target(s) of interest coupled with structural bioinformatics and molecular modeling can be used to address questions such as those posed above. Such an analysis is important to expose liabilities in target selection or the proposed drug product profile early in a drug discovery program, before a substantial investment of time, money and manpower has been made to pursue a flawed hypothesis.

For example, in the antibacterial arena, the emergence of genomics and proteomics has profoundly changed the approach used for the identification of new targets essential for the survival of bacteria (15). To highlight how this information is used to facilitate target selection, the analysis that led to the selection of bacterial topoisomerases as prospective drug targets at the author's company is summarized below. To pursue a drug discovery program, we sought essential bacterial targets with the following properties: (1) Novel proteins that are not targets of marketed antibiotics, to avoid issues of cross-resistance with existing antibiotics. (2) Targets possessing recessed ligand-binding pockets with mixed polar/lipophilic character, the potential for solvent sheltered "anchoring interactions" and no closely related human counterparts. (3) A high degree of sequence/structure conservation in the ligand-binding pockets of the protein target(s) across bacterial species commonly implicated in bacterial infections. (4) If possible, the option to inhibit multiple bacterial targets with a single therapeutic agent to minimize the threat of resistance emergence. A detailed structural bioinformatics analysis of proteins in several key bacterial pathways revealed the bacterial topoisomerases DNA gyrase and topoisomerase IV as prospective drug targets that met the criteria listed above. DNA gyrase is a type II topoisomerase that plays an essential role in bacterial DNA replication with no direct mammalian counterpart. The enzyme catalyzes the introduction of negative supercoils into DNA using the free energy of

ATP hydrolysis (16). DNA gyrase consists of two subunits, GyrA and GyrB that form a functional heterodimer A_2B_2. GyrA is involved in DNA cleavage and religation, while the GyrB domain contains the ATP-binding site and mediates the passage of the uncut DNA strand through the strand that is cleaved by GyrA (16). A closely related bacterial enzyme from the topoisomerase II family is topoisomerase IV (topo IV), which also forms a heterodimer C_2E_2 consisting of two ParC subunits and two ParE subunits (17). Despite possessing a high degree of sequence identity with DNA gyrase, topo IV is involved in different aspects of DNA replication than gyrase. The two topoisomerase complexes are well established drug targets. Fluoroquinolone antibiotics, such as ciprofloxacin, exert their antimicrobial activity via inhibition of the GyrA and ParC subunits (18). However, no commercial antibiotics have yet reached the market which target the ATP binding domains of the respective topoisomerase complexes (GyrB and ParE), despite the fact that GyrB and/or ParE inhibition has been shown to effectively kill bacteria (19). A sequence alignment of the ATP-binding domains of DNA gyrase and topo IV from key pathogens involved in community acquired pneumonia mapped on to the crystal structure of one of the enzymes (see Fig. 5), suggests that the development

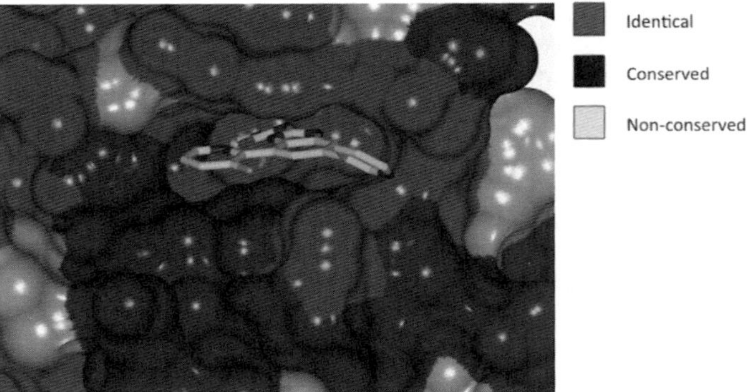

Fig. 5. A solvent accessible surface representation of the ATP-binding pocket of GyrB from the crystal structure of *E. faecalis* GyrB complexed with a benzimidazole inhibitor (D. Bensen and L. Tari, unpublished results). The surface is colored by the degree of sequence conservation observed in the underlying residues for GyrB and ParE enzymes from the major pathogens implicated in community acquired pneumonia. Amino-acid sequences for the relevant proteins were extracted from the KEGG database (20) and sequence alignments were performed with CLUSTALW (21). The high degree of overall sequence conservation (not shown) and the remarkable degree of sequence conservation in the ATP-binding pockets of the selected GyrB and ParE orthologs suggest that the geometries and compositions of the active sites of the enzymes from the different pathogens possess sufficient similarity to allow for the development of dual targeting, broad spectrum inhibitors. Subsequent generation of homology models and crystal structures of several of the orthologs listed on the figure confirmed this hypothesis.

of broad spectrum, dual-targeting inhibitors against these enzymes is feasible. As the above example demonstrates, structural bioinformatics can be an important component in the target selection process and drug product profile determination early in a drug discovery program.

4. Using Crystallographic Methods to Initiate a Drug Discovery Program

The likelihood of success in a small-molecule drug discovery program is greatly enhanced by the availability of multiple molecular scaffolds that bind to and elicit the desired effects on the protein target, while offering prospects for optimization into drug leads. However, the discovery of viable molecular scaffolds for SBDD and medicinal chemistry optimization is not trivial. HTS, when successful, often delivers hits with high molecular weights and poor potential for optimization. The probability of a small-molecule ligand matching the shape and chemistry of a protein target decreases as the complexity and size of the ligand increases, since there exists a greater chance that some part of the ligand will possess features that do not complement those of the protein target. Theoretically, the probability that a small molecule will bind to a protein target decreases exponentially with increasing ligand complexity (22). Thus, there is an advantage to screening for hits using less complex, lower molecular weight compounds (called fragments, with molecular weights ranging from 100 to 250 Da), which interact with only a small number of sites on the protein and possess a greater chance of achieving favorable steric and chemical complimentarity with the protein target. However, the advantage of screening with fragments is offset by the fact that fragments generally bind with much lower affinities than the larger compounds typically screened in HTS. Most biophysical techniques perform poorly at detecting weak binding, limiting their utility in screening fragment libraries. X-ray crystallography, however, is an extremely sensitive technique, capable of detecting compounds with binding constants in the low millimolar range. The extension of crystallographic methods into the high-throughput realm over the past decade has led to the adoption of crystallographic fragment screening in many industrial and academic centers as a drug discovery tool. In this section, the two flavors of crystallographic fragment screening are reviewed: random fragment screening and pharmacophore-based fragment screening.

4.1. Random and Pharmacophore-Based Fragment Screening

The basic premise of crystallographic fragment screening is simple. A protein target is screened against a small library (typically <1,000 molecules) of structurally diverse, highly soluble low molecular weight compounds. The library is screened in one of two ways: pregrown crystals of the protein of interest are soaked with concentrated solutions (in aqueous solution or dimethyl sulfoxide) of individual compounds or mixtures of compounds, or, the protein is crystallized in the presence of compounds/compound mixtures. The latter method has the advantage of allowing the protein to undergo compound induced conformational changes that may be precluded in a preformed crystal lattice. However, cocrystallization generally involves a scan of multiple crystallization conditions to generate usable crystals and can lead to multiple crystal forms, so it is more labor intensive, and requires much larger quantities of protein and fragment solutions for screening. Once putative protein-fragment complex crystals are created, X-ray data collection, crystal structure solution and electron density map interpretation can proceed in a high throughput manner, using high-flux laboratory or synchrotron X-ray sources equipped with sample handling robotics (described in Chapter 5), and automated software for structure solution, refinement and electron density map generation (described in Chapter 6). A schematic representation of random fragment screening and optimization paths from initial fragment hits is shown in Fig. 6. Fragment screening methods are described in more detail in Chapter 7.

An absolute requirement for the application of crystallographic fragment screening is a target protein that is amenable to crystallization (in its apo-form if crystal soaking experiments are used to introduce fragments to the target). Moreover, the crystals must routinely diffract to an adequate resolution (beyond 2.5 Å) to provide a detailed picture of the targeted binding pocket in the protein and the binding modes of bound fragments, and optimally, to resolve ordered waters. When crystal soaking methods are employed, the crystals must possess sufficient mechanical stability to withstand the osmotic pressure generated during exposure to concentrated fragment solutions, and provide unblocked access to the target site. Once these prerequisites are met, crystallographic fragment screening provides insights not offered by other screening techniques. Crystallographic experiments generate electron density maps showing the binding mode of the fragment to the protein target in three dimensions. A detailed knowledge of the

Fig. 6. A schematic representation of crystallographic fragment screening with random fragments. Fragment screening libraries typically contain <1,000 structurally diverse compounds that meet the following criteria: (1) Molecular weights <300 Da, (2) Less than three H–bond donors/acceptors, polar surface area <60 Å^2, less than 3 rotatable bonds, (3) $c\text{Log}\,P < 3$. The libraries can be screened using mixtures of 3–5 compounds, or using individual fragment solutions. (**a**) Two fragments (***the square*** and ***triangle***) are shown binding to distinct regions of a target receptor binding-pocket. Based on crystal structures of the fragment-receptor complexes, several optimization strategies can be employed. Fragments bound to spatially distinct regions of the receptor can be linked as in (**b**) to form a more potent inhibitor, with an inhibition constant (K_i) proportional to the products of the K_{id}s of the individual fragments. Or, as in (**c**), individual fragments can be optimized to improve the steric and chemical fit with the receptor pocket. Additionally, fragments can be elaborated or "grown" into adjacent pockets in the receptor site, as shown in (**d**).

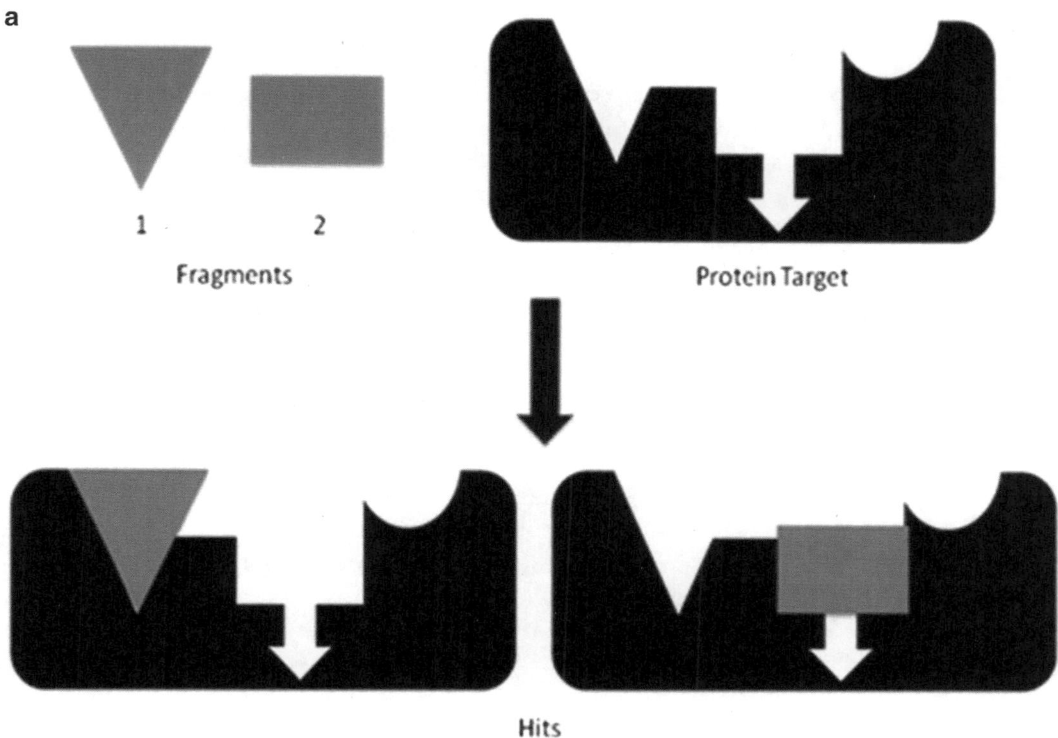
a
1
2
Fragments
Protein Target
Hits

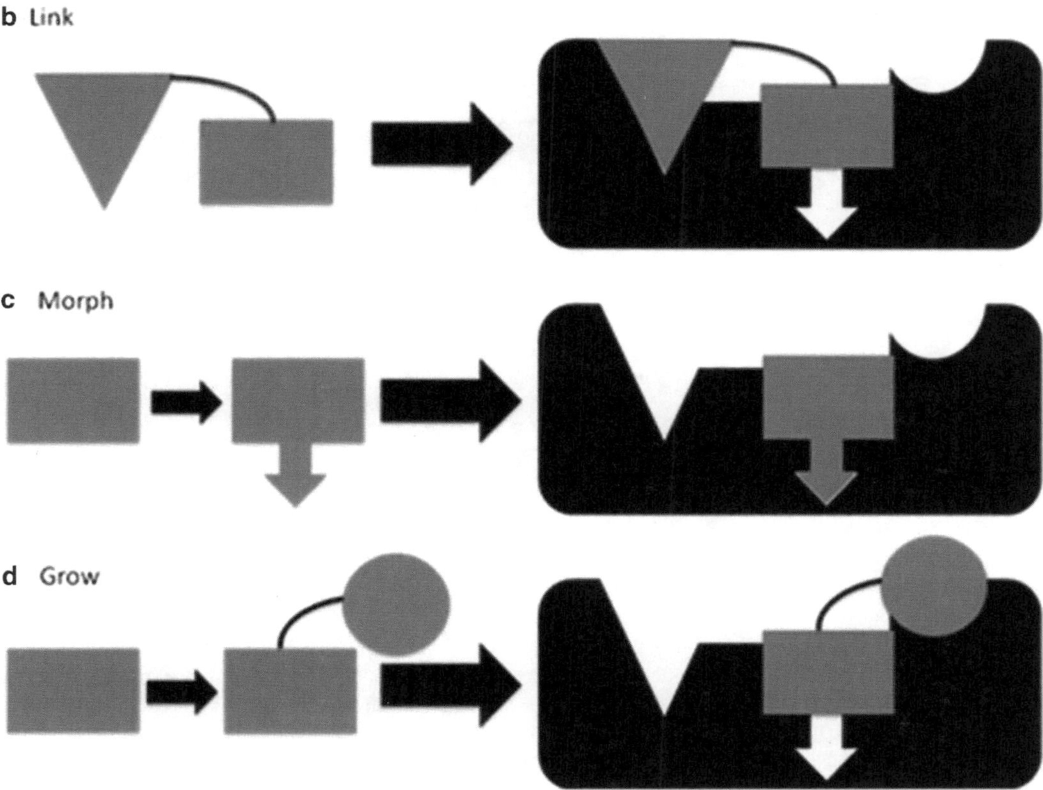
b Link
c Morph
d Grow

key binding interactions, available analoging vectors off the bound fragment(s) and accessible space in the protein defines the spatial and chemical constraints on fragment optimization, streamlining the optimization process. Additionally, crystallographic fragment screening facilitates the identification of false positive hits and fragments that bind to nonproductive sites on the protein target. Conversely, fragment screening can identify novel binding sites that impact protein function with therapeutic development potential. The weakness of X-ray crystallography as a screening method is that it does not provide information on binding affinity and this data must be obtained with a different technique (i.e., a solution based assay, as described in Chapter 8). However, obtaining the binding affinity of initial small fragment hits may be intractable and testing for potency may only become feasible once elaborated follow-on compounds are available.

By utilizing custom designed chemical fragment libraries based on a known target pharmacophore model (as summarized in Fig. 7), pharmacophore-based fragment screening differs philosophically from the use of random chemical fragments in screening. Small molecules are designed or selected from commercial libraries, to key in on specific H-bonding, electrostatic, lipophilic, or π–stacking interactions in the receptor pocket of the target. The same criteria used in random fragment library design for fragment

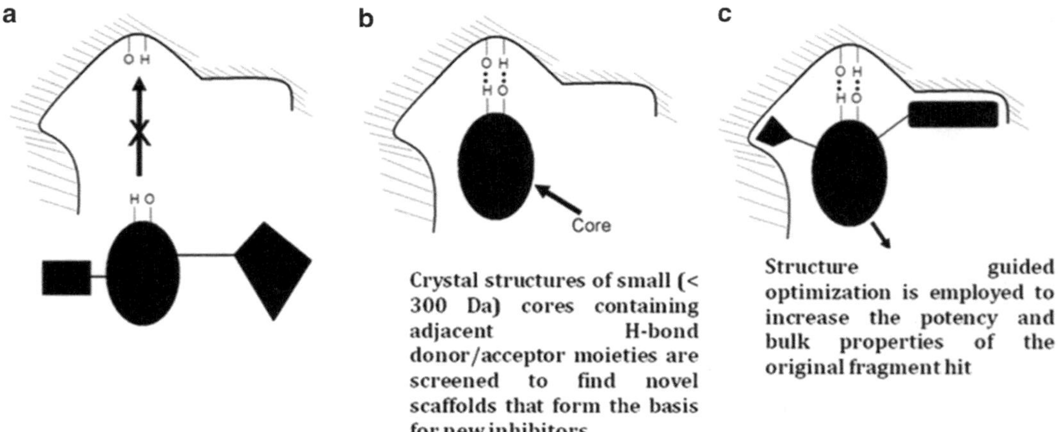

Fig. 7. Schematic outline of pharmacophore-based fragment screening. In this example, a simplified receptor-binding pocket is shown that contains a closely spaced pair of H-bond donor/acceptor moieties comprising the pharmacophore used to guide fragment library design. (**a**) Owing to steric incompatibilities, larger molecules frequently cannot bind to the target, despite containing a potentially useful core. (**b**) To circumvent the problems observed in (**a**), and to find novel inhibitor cores to form the basis for potent inhibitors, pharmacophore-based screening methods are effective. Using cheminformatics software such as MOE™ (23), novel molecules can be designed and synthesized that contain the desired pharmacophore, or commercial libraries can be searched for small-molecule entities with the target pharmacophore. Crystallographic screening methods are then used to screen potential candidates for hits. (**c**) Using the three-dimensional structural information describing the binding modes of fragment hits, the fragments are modified and elaborated with new chemical groups to improve the fit between inhibitor and receptor, and to engage additional pockets in the receptor with potency increasing interactions.

size, solubility, etc. are applied to the custom designed libraries used in pharmacophore-based screening. Starting with small fragment units with known binding interactions, chemical lead series can be rapidly discovered and optimized for drug-like properties. The main advantages of designing fragments around a defined pharmacophore are the creation of highly ligand efficient molecular scaffolds that target the most energetically rewarding interactions (i.e., nonsolvent exposed, available polar interactions) of the target receptor pocket, and the ability to design molecules that achieve the desired selectivity profile by targeting a selected region of the target receptor pocket. For example, if the product of interest is a broad-spectrum antibiotic against a specific bacterial protein target, fragment libraries can be designed to engage only the most conserved regions of the protein target across the different bacterial species.

5. Using X-Ray Crystallography in Lead Optimization

The simplistic view of structure-guided lead optimization is that structural information from crystallographic structures of complexes of lead candidates with the protein target are used as an in vitro assay of sorts, to optimize the potency of the lead against the drug target. However, drug discovery requires optimization of a number of properties, including solubility, intestinal absorption, tissue distribution, metabolic stability, plasma protein binding, elimination, toxicology, and cost of synthesis. To highlight the importance of using structure-based methods in the broader context of a drug discovery program, it is instructive to follow the trajectory of neuraminidase inhibitor development that led to the discovery of the Tamiflu™. Influenza virus neuraminidase has long been recognized as a potential target in the treatment of influenza. Molecular modeling studies based on crystal structures of neuraminidase inhibitor complexes suggested that substitution of the 4-hydroxyl group (structure 1 in Fig. 8) in a compelling lead molecule with a charged basic group would yield a more potent inhibitor (25). Indeed, replacement of the hydroxyl group by a basic guanidine (structure 2 in Fig. 8) resulted in a 5,000-fold increase in potency. Ultimately, this compound (zanamivir) was developed by GlaxoSmithKline and led to the first marketed neuraminidase inhibitor used in the treatment of influenza, Relenza™. However, Relenza™ is not absorbed orally due to its high polarity and basicity, necessitating the development of a dry powder inhaler to topically dose the compound in the lung (26). Substitution of the dihydro-2H-pyran scaffold by cyclohexene and replacement of the polar glycerol and basic guanidinyl moieties with a 1-ethylpropoxy and a primary amine moiety, respectively,

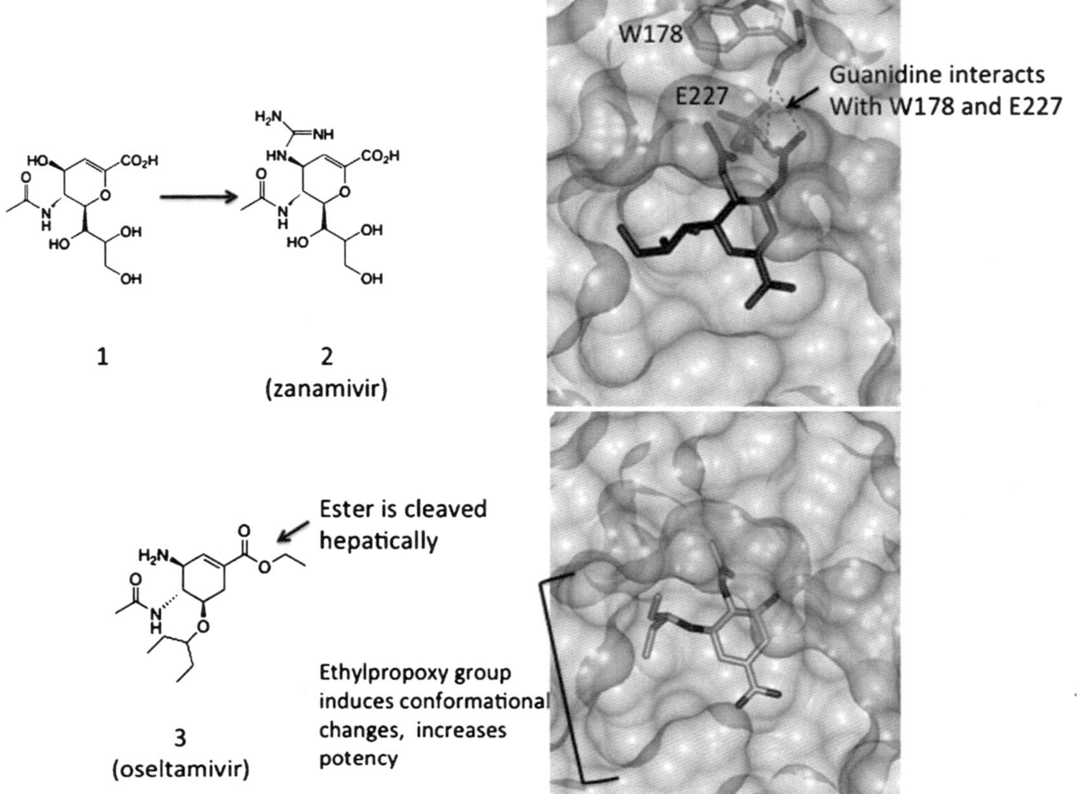

Fig. 8. Structures of neuraminidase inhibitors: (1), lead molecule, (2), Zanamivir (Relenza™), (3), Oseltamivir (Tamiflu™). The ester prodrug is cleaved hepatically to form the carboxylic acid. Semitransparent solvent accessible surface representations of the structures of both drugs bound to H5N1 avian influenza virus neuraminidase (24) (PDB codes 2HTQ and 2HT8) are shown to illustrate the interaction between the zanamivir guanidine and the active-site pocket, and to highlight the conformational changes induced by the carbocyclic scaffold and ethylpropoxy group in oseltamivir upon drug binding to the enzyme.

generate oseltamivir (sold as Tamiflu™, structure 3 in Fig. 8), a smaller, less polar inhibitor than Relenza™ that retains sufficient potency for efficacy. The improved ligand efficiency observed for Tamiflu™ is due in part to the ethylpropoxy group, which induces a conformational change in key active-site pocket residues and participates in lipophilic interactions (24, 27). Conversion of the zwitterionic parent compound to the ethyl ester pro-drug allow the compound to be administered orally, making Tamiflu™ the first neuraminidase inhibitor used as an oral anti-influenza drug. The improved physical property profile of Tamiflu™ vs. Relenza™ translated to considerable commercial success. In 2008, Tamiflu™ outsold Relenza™ by a factor of 5:1 (www.marketresearchmedia.com). This example highlights the difference between good inhibitors and good drugs. When applying structure-based methods, absorption, distribution, metabolism, and elimination (ADME) properties need to be addressed during the quest for potency to avoid complications later in the drug discovery process. In the case of the neuraminidase example above, crystallographic studies revealed a ligand-induced active-site conformation that allowed for the design of a small, moderately polar, less charged molecule with superior drug-like properties to the first generation drug. When used in such a manner, structural information can play a powerful role in drug discovery. X-ray crystallographic methods can provide information about active-site or ligand-binding pocket architecture and its relationship to the functional state of a protein, binding pocket plasticity and small-molecule binding modes that can dramatically streamline lead optimization. Additionally, crystal structures can play a key role in resolving unexpected structure activity relationships (SAR) arising from incorrect small-molecule structure assignments, unanticipated small-molecule binding modes or receptor plasticity. Additional examples highlighting the utility of X-ray crystallography in lead optimization are provided below.

Once an experimental atomic structure of a protein–small molecule lead complex is in hand, available analoging vectors off the lead molecule can be identified by the medicinal chemist and used to guide the synthesis of the next molecule, as described in Fig. 7. Vectors pointing toward empty pockets in the receptor can be filled by complimentary groups to increase potency, while solvent facing vectors can be used to generate analogs with improved bulk properties or metabolic stability. Knowledge of the structural and chemical landscape of the receptor pocket also focuses optimization efforts on analogs that add potency mainly via enthalpic (i.e., polar) interactions vs. analogs that add potency via entropic (i.e., lipophilic) interactions, improving the prospects for selective inhibitor binding and reducing the probability of off-target mediated toxicity. Structure-based methods can also play a key role in more complex systems, where the protein target must be captured in a specific functional and structural state to achieve efficacy or to improve prospects for designing selective inhibitors. For example, stem cell factor receptor, c-Kit, is a receptor protein-tyrosine kinase that initiates cell growth and proliferation signal transduction cascades in response to stem cell factor binding (28). The kinase is activated and transphosphorylates via dimer formation mediated by the binding of stem cell factor dimers to its extracellular domain. Mutations that constitutively activate c-Kit in the absence of the stem cell factor are implicated in several highly malignant human cancers, making it a validated target for the development of anticancer drugs (29). Detailed analysis of the crystal structures of c-Kit in multiple functional states (30), including an autoinhibited form, an activated form, and a drug-bound form, reveal that the kinase adopts discrete structural states when transitioning from an autoinhibited to an activated state (see Fig. 9). The structural results provide a detailed molecular basis for understanding the

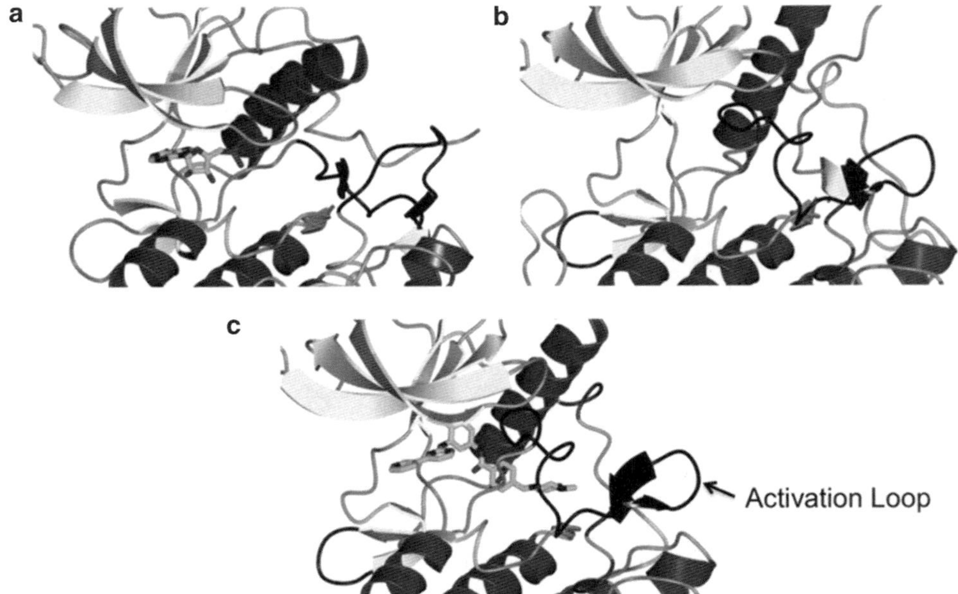

Fig. 9. Ribbon representations of crystal structures of c-Kit kinase in three forms; (**a**) activated, in complex with ADP and Mg^{2+}, (**b**) unphosphorylated, containing the entire juxtamembrane region (autoinhibited state), and (**c**) in complex with the anticancer drug Gleevec™. The mobile activation loop is colored *black* in each panel. Gleevec™ is a fairly selective inhibitor that binds to few kinases, including Abl kinase and platelet-derived growth factor kinase (31, 32). The basis for this selectivity stems from the fact that the inhibitor targets a kinase conformation that resembles the inactive state (compare the activation loop conformations in (**b**) and (**c**)). However, Gleevec™ binding disrupts the fully autoinhibited state by preventing the association of the juxtamembrane domain with the kinase domain. These results demonstrate that selective inhibitors of type III protein-tyrosine kinases can be developed to exploit the unique autoinhibited conformations of these kinases.

mechanism of c-Kit kinase autoinhibition, and snapshots of unique structural states that are exploitable for the structure-based design of specific and potent inhibitors targeting the activated or autoinhibited conformations of c-Kit kinase, as exemplified by the structure of c-Kit bound to the anticancer drug Gleevec™ described in the study.

Many examples exist in the literature that demonstrate the power of crystallographic methods in revealing receptor plasticity in protein drug targets resulting in surprising ligand-induced conformational changes and inhibitor SAR. A case in point is a member of the human histone deacetylase (HDAC) protein family, a series of validated oncology targets. In eukaryotes, HDACs modulate the acetylation of histones and hence play a key role in the regulation of gene expression (33). HDAC deregulation has been linked to several types of cancer, and recently, the HDAC inhibitor suberoylanilide hydroxamic acid (SAHA) was approved for the treatment of cutaneous T-cell lymphoma (34). Crystal structures of several inhibitor bound complexes of human HDAC8 (35) reveal that the surface of the active-site pocket contains flexible

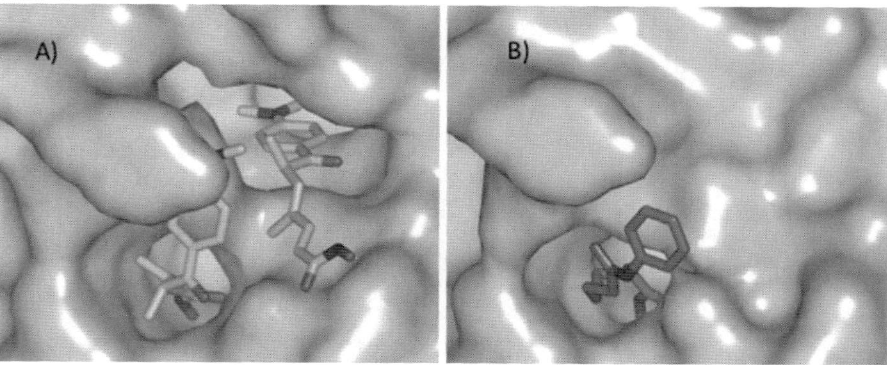

Fig. 10. Solvent accessible surface representations around the active-site pockets of the structures of complexes of HDAC8 with (**a**) trichostatin A (TSA) and (**b**) the anticancer drug suberoylanilide hydroxamic acid (SAHA). TSA induces dramatic conformational changes in several surface elements of HDAC8, creating a second deep pocket adjacent to the active-site pocket. A second TSA molecule occupies the newly formed pocket.

elements that can adopt diverse conformations in response to inhibitor binding. In one of the complexes (see Fig. 10), a loop on the protein surface moves, revealing a deep pocket adjacent to the active-site pocket. This work suggests that HDAC8 inhibitors could be designed with isoform selectivity, despite the highly conserved nature of HDAC active-site pockets (35). The HDAC example highlights the importance of using crystallographic methods for the characterization of novel, low energy conformational states of protein drug targets that can be exploited for the design of selective inhibitors. Such insights would not be possible without the detailed information provided by X-ray crystallography.

In addition to the characterization of receptor plasticity and the correlation of protein functional states with their underlying structures, careful application of crystallographic methods can be used to resolve very detailed questions relating to small-molecule inhibitor structure, binding mode, and, in many cases, ionization state. When high-resolution (typically <2.2 Å) X-ray data are available, cases of mistaken ligand identity can be resolved, or the exact stereochemistry of a protein-bound small molecule (from a mixture of isomers) can be determined unambiguously, revealing the stereochemical preferences of the receptor pocket. In favorable cases, small-molecule ligands can even be fit to electron density without prior knowledge of the structure of the small molecule. Additionally, the experimentally observed conformations of inhibitors bound to protein targets can be subjected to in silico conformational analysis to reveal cases where ligand binding to the target incurs a significant energetic penalty, resulting in reduced inhibitor potency. Based on the results, molecular modeling can be used to design optimized inhibitors that “preorganize” into competent binding conformations in solution, allowing for the development

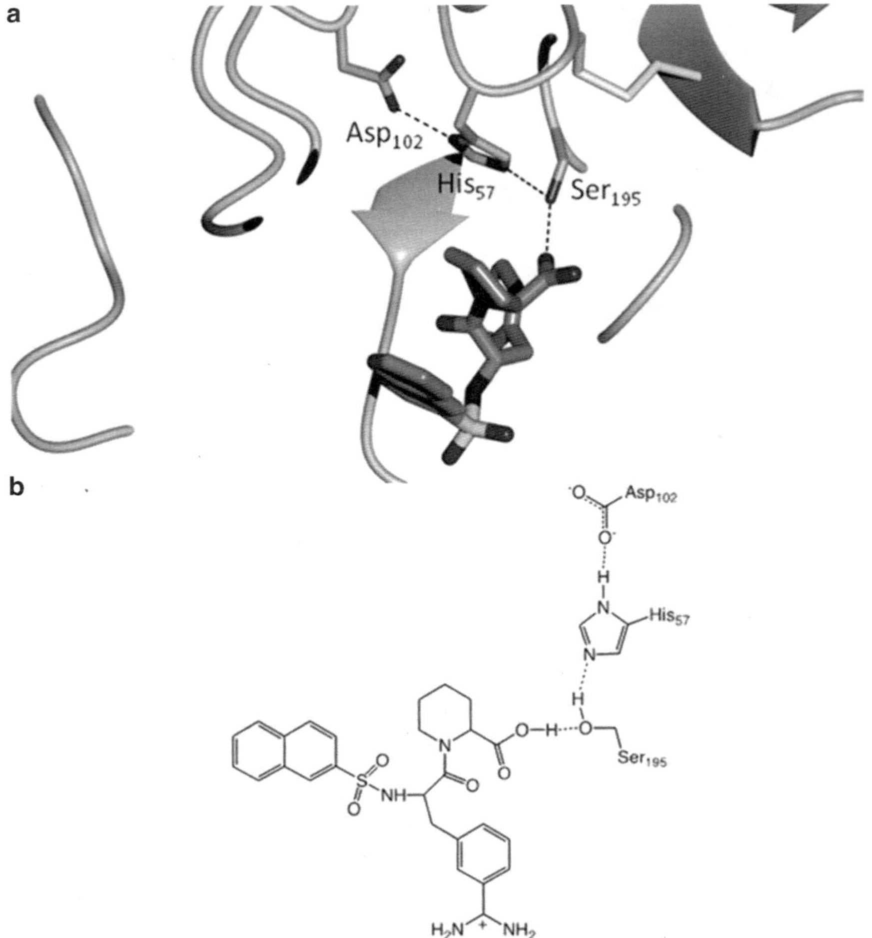

Fig. 11. Demonstration of how crystallographic structural information can be used to deduce the protonation states of ionizable moieties and the hydrogen bonding networks between inhibitors and their protein targets. (**a**) A close-up view of the binding of a napthylsulphonyl-amidino-phenylalanine inhibitor to bovine β-trypsin from the high-resolution crystal structure (36). For clarity, only the interactions between the side chains for the serine protease catalytic triad and the inhibitor are shown. Potential hydrogen bonds are depicted as ***dotted lines***. As shown in (**b**), the carboxylic acid moiety of the inhibitor must be protonated, since the hydroxyl proton of Ser_{195} engages the imidazole ring of His_{57}. The protonation state of His_{57} is locked as shown by its interaction with Asp_{102}.

of inhibitors with improved potency often without increasing molecular weight.

Small-molecule lead optimization is hampered if the protonation states of key acidic and basic amino-acid side chains and/or the tautomeric states of ionizable groups on protein-bound small-molecule inhibitors are not understood. Assuming that the parameters for amino-acid side chains in a folded protein are similar to their counterparts in solution is not always correct; the pK_a values for basic and acidic amino-acid side chains on a protein interior can shift dramatically (>2 units) from their value in aqueous solution, based on the microenvironment created around the residue by the protein structure (36). Furthermore, the pK_a values of amino-acid side chains can change upon complexation by an inhibitor, as can the pK_a values of the ionizable groups on the inhibitor. Although impossible to determine experimentally from a protein crystal structure, the positions (or presence, in the case of groups with exchangeable protons) of hydrogen atoms can usually be inferred from the molecular mechanics force fields used in structure refinement and a careful analysis of the environment surrounding the group in question. An example of how X-ray structural information can be used to deconvolute complex hydrogen-bonding networks and assign protonation states to ionizable groups is shown in Fig. 11. Crystallographic structures and isothermal titration calorimetry were used to map the network of hydrogen bonds of several thrombin and trypsin inhibitors, as well as the protonation states of ionizable inhibitor moieties and active-site side chains (37).

6. Summary

Structure-based drug design is now a staple in the pharmaceutical industry and has contributed to the discovery of many marketed drugs and late-stage clinical candidates. Access to detailed three-dimensional structural information on protein drug targets can streamline many aspects of drug discovery, from target selection and target product profile determination, to the discovery of novel molecular scaffolds that form the basis of potential drugs, to lead optimization. Structural biology is currently in its golden era; the advent of high-throughput methods for all of the steps involved in the generation of protein crystal structures allow empirically derived structural information to drive iterative lead optimization efforts in real time for a wide range of protein targets, avoiding many of the limitations that plague molecular modeling techniques. Crystallographic methods are useful for characterizing the structures correlated with specific functional states of protein targets or alternative conformations of receptor pockets that lead to unique structural states. Such information can be leveraged to develop exquisitely selective small-molecule ligands that target specific proteins, even in closely related protein families. When used carefully in conjunction with ADME data during the lead optimization process, X-ray crystallographic methods are an extremely powerful tool in the drug discovery arsenal that will continue to contribute to the invention of new medicines in diverse therapeutic areas.

References

1. Pharmaceutical Manufacturers Association (1993) 'Facts at a Glance', Washington DC.
2. Grabowski, H. J. G. and Vernon, J. M. (1994) Returns to R&D on new drug introductions in the 1980s. *J. Health Econ.* **13**, 282–406.
3. Gustafsson, D., Byland, R., Antonsson, T., Nilsson, I., Nystrom, J. –E, Eriksson, E., Bredberg, U. and Teger-Nilsson, A. –C. (2004) A new oral anticoagulant: the 50-year challenge. *Nature Rev. Drug. Discov.* **3**, 649–659.
4. McCoy, A. J., Grosse-Kunstleve, R. W., Adams, P. D., Winn, M. D., Storoni, L. C. and Read, R. J. (2007) Phaser crystallographic software. *J. Appl. Cryst.* **40**, 658–674.
5. Blundell, T. L. and Johnson, L. N. (1976) In Protein Crystallography. Academic Press, New York.
6. Stout, G. H. and Jensen, L. H. (1989) In X-ray Structure Determination: A Practical Guide. 2nd ed. Wiley, New York.
7. Drenth J. (1999) In Principles of protein x-ray crystallography. 2nd ed. Springer, New York.
8. Stout, G. H. and Jensen, L. H. (1989) In X-ray Structure Determination: A Practical Guide. 2nd ed. Wiley, New York, Chapters 7–9.
9. Winn, M. D., Murshudov G. N. and Papiz, M. Z. (2003) Macromolecular TLS refinement in REFMAC at moderate resolutions. *Methods Enzymol.* **374**, 300–321.
10. Drenth J. (1999) In Principles of protein x-ray crystallography. 2nd ed. Springer, New York, pp. 89–90.
11. Emsley, P. and Cowtan K. (2004) Coot: model-building tools for molecular graphics *Acta Cryst.* D60, 2126–2132.
12. Brünger, A. T. (1992) Free R value: a novel statistical quantity for assessing the accuracy of crystal structures. *Nature* **355**, 472–475.
13. Jia, Z., Vandonselaar, M., Quail, J. W. and Delbaere, L. T. J. (1993) Active-center torsion-angle strain revealed in 1.6 Å-resolution structure of histidine-containing phosphocarrier protein. *Nature* **361**, 94–97.
14. Bersio, R., Lazmin, V. S., Sica, F., Wilson, K. S., Zagari, A. and Mazzarella, L. (1999) Protein titration in the crystal state. *J. Mol. Biol.* **292**, 845–854.
15. Payne, D. J., Gwynn, M. N., Holmes, D. J. and Pompliano, D. L. (2007) Drugs for bad bugs: confronting the challenges of antibacterial discovery. *Nat. Rev. Drug. Discov.* **6**, 29–40.
16. Champoux, J. J. (2001) DNA topoisomerases: structure, function and mechanism. *Annu. Rev. Biochem.* **70**, 369–413.
17. Peng, H. and Marians, K. J. (1993) Escherichia coli topoisomerase IV. Purification, characterization, subunit structure and subunit interactions. *J. Biol. Chem.* **268**, 24481–24490.
18. Wolfson, J. S. and Hooper, D. C. (1985) The fluoroquinolones: structures, mechanisms of action and resistance, and spectra of activity in vitro. *Antimicrob. Agents Chemother.* **28**, 581–586.
19. Oblak, M., Kotnik, M. and Solmajer, T. (2007) Discovery and Development of ATPase Inhibitors of DNA Gyrase as Antibacterial Agents. *Curr. Med. Chem.* **14**, 2033–2047.
20. Kanehisha, M., Goto, S., Kawashima, S., Okuno, Y. and Hattori, M. (2004) The KEGG resource for deciphering the genome. *Nucleic Acids Res.* **32**, 277–280.
21. Thompson, J. D., Higgins, D. G. and Gibson, T. J. (1994) CLUSTALW: improving the sensitivity of progressive multiple sequence alignments through sequence weighting, position specific gap penalties and weight matrix choice. *Nucleic Acids Res.* **22**, 4673–4680.
22. Hann, M.M., Leach, A.R. and Harper, G. (2001) Molecular complexity and its impact on the probability of finding leads for drug discovery. *J.Chem.Inf.Comput.Sci.* **41**, 856–864.
23. Labute, P. and Clark A. M. (2007) 2D Depiction of Protein-Ligand Complexes. *J. Chem. Inf. Model* **47**, 1933–1944.
24. Russell, R. J., Haire, L. F., Stevens, D. J., Collins, P. J., Lin, Y. P., Blackburn, G. M., Hay, A. J., Gamblin, S. J. and Skehel, J. J. (2006) The structure of avian flu neuraminidase suggests new opportunities for drug design. *Nature.* **443**, 45–49.
25. von-Itzstein, M., Wu, W. Y., Kok, G. B., Pegg, M. S., Dyason, J. C., Jin, B., Van Phan, T., Smythe, M. L., White, H. F., Oliver, S. W., Colman, P. M., Varghese, J. N., Ryan, D. M., Woods, R. C., Bethell, R. C., Hotham, V. J., Cameron, J. M and Penn, C. R. (1993) Rational design of potent sialidase-based inhibitors of influenza virus replication. *Nature.* **363**, 418–423.
26. (2001) In Physicians' Desk Reference. 55th ed. Medical Economics Company Inc. Montvale, NJ, p.1454.
27. Kim, C. U. Lew, W., Williams, M. A., Liu, H., Zhang, L., Swaminathan, S., Bischofberger, N., Chen, M. S., Mendel, D. B., Tai, C. Y., Laver, W. G. and Stevens, R. C. (1997) Influenza neuraminidase inhibitors possessing a novel hydrophobic interaction in the enzyme active-site: design, synthesis, and structural

analysis of carbocyclic sialic acid analogues with potent anti-influenza activity. *J. Am. Chem. Soc.* **119**, 681–690.

28. Linnekin, D. (1999) Early signaling pathways activated by c-Kit in hematopoietic cells. *Int. J. of Biochem. Cell Biol.* **31**, 1053–1074.
29. Hirota, S., Isozaki, K., Moriyama, Y., Hashimoto, K., Nishida, T., Ishiguro, S., Kawano, K., Hanada, M., Kurata, A., Takeda, M., Tunio, G. M., Matsuzawa, Y., Kanakura, Y., Shinomura, Y. and Kitamura, Y. (1998) Gain of function mutations of c-Kit in human gastrointestinal stromal tumors. *Science.* **279**, 577–580.
30. Mol, C. D., Dougan, D. R., Schneider, T. R., Skene, R. J., Krause, M. L., Schiebe, D. N., Snell, G. P., Zou, H., Sang, B. –C. and Wilson, K. P. (2004) Structural Basis for the autoinhibition and ST-571 inhibition of c-Kit tyrosine kinase. *J. Biol. Chem.* **279**, 31655–31663.
31. O'Dwyer, M. E., Mauro, M. J., and Druker, B. J. (2003) STI571 as a targeted therapy for CML. *Cancer Investig.* **3**, 429–438.
32. Buchdunger, E., Cioffi, C. L., Law, N., Stover, D., Ohno-Jones, S., Druker, B. J. and Lydon, N. B. (2000) Abl protein-tyrosine kinase inhibitor STI571 inhibits in vitro signal transduction mediated by c-kit and platelet-derived growth factor receptors. *J. Pharmacol. Exp. Ther.* **295**, 139–145.
33. Khochbin, S., Verdel, A., Lemercier, C. and Seigneurin-Berny, D. (2001) Functional significance of histone deacetylase diversity. *Curr. Opin. Genet. Dev.* **11**, 162–166.
34. Marks, P. A. and Xu, W. S. (2009) Histone-deacetylase inhibitors: potential in cancer therapy. *J. Cell Biochem.* **107**, 600–608.
35. Somoza, J. R., Skene, R. J., Katz, B. A., Mol, C. D., Ho, J. D., Jennings, A. J., Luong, C., Arvai, A., Buggy, J. J., Chi, E., Tang, J., Sang, B. C., Verner, E., Wynands, R., Leahy, E. M., Dougan, D. R., Snell, G., Navre, M., Knuth, M. W., Swanson, R. V., McRee, D. E. and Tari, L. W. (2004) Structural snapshots of human HDAC8 provide insights into the class I histone deacetylases. *Structure* **12**, 1–20.
36. Harris, T. K. and Turner, G. J. (2002) Structural basis of perturbed pKa values of catalytic groups in enzyme active sites. *IUBMB Life* **53**, 85–98.
37. Dullweber, F., Stubbs, M. T., Musil, D., Sturzebecher, J. and Klebe, G. (2001) Factorising ligand affinity: A combined thermodynamic and crystallographic study of trypsin and thrombin inhibition. *J. Mol. Biol.* **313**, 593–614.

Chapter 2

Genetic Construct Design and Recombinant Protein Expression for Structural Biology

Suzanne C. Edavettal, Michael J. Hunter, and Ronald V. Swanson

Abstract

Obtaining diffraction quality crystals is frequently an iterative process which traditionally has involved screening large numbers of crystallization conditions. Due to advances in high-throughput gene engineering, recombinant expression, and purification, the protein of interest has now become one of the many variables routinely investigated during crystallization trials. As such, construct design is a critical step in the path toward successful crystallization. In this chapter will we address construct design strategies frequently employed to improve the solution and crystallization behavior of proteins. Topics covered include choosing a recombinant expression system and reducing disorder through truncations and surface mutagenesis. Also covered are strategies to reduce heterogeneity from posttranslational modifications, impurities, and aggregation.

Key words: Protein Expression Constructs, Recombinant Protein Expression, X-ray crystallography, protein crystallization

1. Introduction

A protein crystal represents a homogeneous population of three-dimensionally arrayed protein molecules. To maximize the likelihood of crystallizing a protein from a solution, it is important to minimize the heterogeneity of the sample. The common misconception is that homogeneity is synonymous with purity at the protein contaminant level. For protein crystallization, this idea that purity and homogeneity are synonymous is an oversimplification. While purity is fundamentally important, the advent of recombinant methods for overexpression, coupled with affinity chromatography tags, has made achieving highly pure protein samples less problematic than in the past. In addition, there are other, perhaps

Leslie W. Tari (ed.), *Structure-Based Drug Discovery*, Methods in Molecular Biology, vol. 841,
DOI 10.1007/978-1-61779-520-6_2, © Springer Science+Business Media, LLC 2012

underappreciated, protein characteristics where heterogeneity arises. At the conformational level, loops and/or the termini of the protein may exhibit disorder, and in larger multi-domain proteins the individual domains may be flexible relative to one another leading to many conformational isomers of the protein in solution and representing a barrier to crystallization due to structural heterogeneity. Poor crystal formation can also arise from aggregation or changes in monodispersity representing higher order solution state heterogeneity. Heterogeneity may also be present at the chemical level of the protein from posttranslational modifications, such as phosphorylation or glycosylation, which are often incomplete or heterogeneous. Proteolysis either during purification or expression can also introduce heterogeneity. In addition, nonenzymatic chemical modifications such as oxidation can occur, creating subspecies of closely related, difficult to distinguish, contaminants. These sources of heterogeneity can be addressed by choosing an appropriate expression system, designing proper construct boundaries, advantageous mutational changes and an effective purification strategy, the latter of which will be addressed in Chapter 3. Frequently, generating a truly homogeneous crystallizable protein solution will require exploring several constructs and variables. In this chapter we explore the considerations used in determining construct design as well as vector–host combinations aimed at minimizing heterogeneity and generating homogenous protein solutions that will lead to well-diffracting protein crystals.

2. Construct Design

Protein engineering is a powerful tool for improving protein physiochemical properties leading to proteins that are more stable, soluble, and have a higher propensity to crystallize. The concept of the protein as a variable in protein crystallization (1) is important and has been enabled by modern molecular biology techniques. Securing downstream success hinges primarily on rational construct design and can be undertaken independent of vector/host choice. The overall goal in construct design is to produce large, homogenous quantities of soluble proteins with a high likelihood to crystallize. An important decision in this effort is the choice of expression boundaries as all other changes take place within these confines. The degree of difficulty in determining domain boundaries depends on the degree of similarity of the protein of interest to other proteins and, in particular, to proteins of known structure. Side chain, loop, or termini flexibility (i.e., changes in entropy) can also be addressed in construct design. Alterations of surface exposed residues, both hydrophilic and hydrophobic, can lead to better behaved protein solutions and protein crystals. Posttranslational modifications, both the addition and subtraction of, can have dramatic results with respect to protein stability and good crystal formation. Even the isolation of protein domains out of the context of the larger protein can lead to better crystal formation when inherent inter-domain flexibility hinders crystallization efforts. In this section, we will address these considerations individually and give the reader a better sense of how the researcher addresses, and ultimately strives to overcome, these issues.

3. Boundaries

Boundary determination often represents the crucial choice for producing sufficient quantities of soluble, high-quality protein for crystallographic purposes. For small prokaryotic proteins the native termini are often ideal. However, for more complex eukaryotic proteins, crystallization of truncated protein or of individual domains for multi-domain proteins is often more appropriate. It is difficult to predict the boundaries of a domain of interest or the proper truncation of a protein's termini from the primary sequence in isolation. Analysis of the similarity of the protein of interest to other family members and in particular to proteins of known structure provides the template for proper boundary choices. The first step is to query the sequence of interest against the Protein Databank (http://www.pdb.org/pdb/home/home.do). This allows a quick assessment of the closest homologs with published structure. If reasonable hits are found, a close approximation of proper boundaries is already at hand through design of a construct with similar termini. Refinement of boundary choice can be completed by analyzing whether the termini used in the database structures are ordered. Multiple sequence alignments also represent an important analytical approach, as primary sequence similarity is a strong predictor of structural similarity. From a structural biology viewpoint, their utility comes in the ability to visually illustrate conserved and variable sites within a protein family that typically correspond to structurally important and dispensable features, respectively. Commonly used alignment tools include algorithms such as ClustalW2, Tree-based Consistency Objective Function for alignment Evaluation (T-Coffee), or multiple sequence comparison by log-expectation (MUSCLE) (2, 3).

A more sophisticated approach entails homology or comparative protein structural modeling (4). This method produces an all-atom model of a sequence based on its alignment to one or more related protein structures. Either sequential or simultaneous modeling of the core of the protein as well as loops and side chains can greatly facilitate the identification of not only end terminal boundaries, but also the identification of surface exposed side chain residues and potential flexible loops. Although it should be noted that many loops will be readily predictable from insertion/deletion gaps in the multiple sequence alignments. Templates for comparative model building are often found by sequence alignment methods such as BLAST, PSI-BLAST, FASTA or SALIGN. Once identified, the atomic coordinates of the templates and a short script file are fed into a computer program for comparative protein structure modeling such as MODELLER. This program implements comparative protein structure modeling by the satisfaction of spatial restraints that are input by the user. MODELLER can also perform a number of auxiliary tasks including the calculation of phylogenetic trees, alignment of two protein sequences or their profiles, multiple alignments of protein sequences or their profiles, multiple alignments of protein sequence and/or structures, and de novo modeling of loops in protein structures. This method remains the most reliable method to predict the three dimensional structure of a protein.

Recently, Mooij et al. presented a web-based tool, ProteinCCD (CCD: Crystallographic Construct Design) which consolidates common tools in structural biology into a single platform that enables comparative analysis of the sequence and allows the design of oligonucleotides for PCR amplification of the chosen protein constructs (5). This suite is divided into four groups of sequence analysis tools, predicting secondary structure, disordered regions, structural motifs, and flexible domains. Secondary structure prediction uses primary sequence information to predict stretches of sequences that are likely to be beta-strands or alpha-helices in the three-dimensional structure and, thus, should not be disrupted. The second group of tools employs algorithms that aim to predict disordered regions in the protein primary sequences. Rigidifying or deleting loop structures predicted to be flexible can improve crystal formation. Predicting specific features of a protein sequence including transmembrane topology, signal peptides, or regions of coiled-coils will also aid in construct design by identifying regions which should have structural rigidity and therefore should not be mutated. The Simple Model Architecture Research Tool (SMART) and the domain Linker Predictor are used to analyze domain structure and identify genetically mobile domains. The information output displays a condensed view of all results against the protein sequence where the researcher can analyze the data and choose, interactively, possible construct boundaries.

These computational methods normally yield multiple possible termini for any given protein. Even in the best-case scenario where a crystal structure of a close homolog exists as a guide, the choice of the precise starting or ending residues may be difficult. In some cases, one terminus may be better defined than the other. It is often advisable to bring forward multiple constructs to test different hypotheses. The expression system, the throughput of the lab, and the ambiguity of the alignment all impact the number of clones that should be generated and evaluated. Ideally, high-throughput expression and purification techniques can be employed to efficiently assess the quality of each construct. High-throughput melting temperature analysis has become a method of choice for ranking the propensity for crystallization of such constructs, as thermal stability has been correlated with crystallizability (6). However, expression level very often provides a good surrogate assessment of the behavior of the protein; better expressers often leading to better crystallizers. The number of constructs can usually be limited by focusing on the most aggressively truncated candidates, in most cases less is more.

In instances of truly novel sequences where computational or comparative methods do not provide guidance, empirical approaches may be employed. Boundaries can be based on information from limited proteolysis/mass spectrometry (LPMS) where a time course of digestion with a protease such as chymotrypsin, subtilisin, or endo-Glu-C is followed by mass spectrometry to identify stable proteolytic fragments. This powerful method can identify exposed flexible termini or loop structures that can be problematic for good crystal formation. Information about cleavage sites gained from limited proteolysis can be translated into new construct design to eliminate inherently flexible regions creating a more minimalist rigid structure and increasing the likelihood of successful structure determination.

A second empirical method, based on enhanced hydrogen/deuterium exchange mass spectrometry can be employed when working with novel sequences (7). This method allows one to identify regions of disorder in a protein through the enhanced exchange rate of backbone amide hydrogens. Slower exchange rates would suggest more highly structured regions whereas faster exchange rates would be indicative of domain boundaries, flexible loops, or disordered termini that could be adjusted or deleted in subsequent construct design. Several examples have been published outlining the utility of this approach (8). It is now routinely employed by both the NESG and JCSG (9, 10).

4. Choosing an Expression System

Several expression systems are routinely used to generate recombinant protein suitable for crystallization purposes, including bacteria, insect cells, yeast, and mammalian cells. Initially, the choice of expression system may be a balance of cost, ease of use, or the complexity of the system. Since most cDNAs can be expressed in many different systems, choosing a host is generally based on expressed protein yields, desired posttranslational modifications,

Table 1
Protein classes

Protein class[a]	Protein size (AA)	Expression system (s)
Peptides	<80	*E. coli* (generally as fusion proteins)
Secreted proteins	80–500	All (proven track record in yeast and mammalian)
Large secreted proteins/ cell surface receptors	>500	Mammalian
Non-secreted proteins	>80	All, based on individual nature or protein

[a] Arbitrary classification of proteins

and relevant purity. The choice of expression system begins with arbitrarily assigning the target protein to one of four broad classes (see Table 1) (11). The first class is small proteins and peptides that are less than ~80 amino acids in length. These are generally best expressed in bacteria with fusion partners that are enzymatically removed post-purification. The second class comprises secreted proteins that range in size from approximately 80–500 amino acids. This class of proteins can be expressed in all expression systems but is generally targeted for yeast, insect cell, and mammalian cell expression. The third class is composed of secreted proteins that are large (>500 amino acids). These are best expressed in mammalian cells as these cells contain the complex machinery needed for processing and adding posttranslational modifications. Finally, the fourth class is cytosolic proteins that are generally larger than 80 amino acids. The choice of expression system for this class of protein depends of the nature of the protein to be expressed as will be discussed later where we address the advantages and disadvantages of each system with respect to the specific class of protein. While membrane proteins are not described here in detail, they are rapidly becoming more common targets for crystallization studies. The eukaryotic hosts are most commonly used for this purpose, although several novel techniques have been described to express membrane proteins in bacterial systems (12, 13). Finally, desired posttranslational modifications, or lack thereof, can influence the choice of expression system where the most limiting system is bacteria (see Table 2). However, the primary driver of choice of expression system should always be probability of success.

4.1. Prokaryotic Expression Systems

Several factors may direct one toward a prokaryotic system including target proteins which are cytosolic, prokaryotic in origin, or lacking in relative complexity, as well as the desire for no or limited

Table 2
Common posttranslational modifications in different host systems

Posttranslational modification	*E. coli*	Insect cells	Yeast	Mammalian cells
Disulfide bond formation	Possible[a]	Yes	Yes	Yes
Proteolytic processing	Signal sequence removal	Yes	Yes	Yes
Phosphorylation	Yes	Yes	Yes	Yes
N-linked glycosylation	No	Yes	Yes	Yes
O-linked glycosylation	No	Yes	Yes	Yes
N-terminal methionine removal	Yes	Yes	Yes	Yes

[a] Possible when expressed in host cells with thioredoxin reductase (trxB) and glutathione reductase (gor) mutations

posttranslational modifications. For most research labs, the choice of bacterial cell host for recombinant protein expression is *Escherichia coli*. This system is often chosen for simple economic considerations, ease of use and a large selection of vector/host combinations allowing one to tackle most protein expression situations. Hosts which promote disulfide bond formation in the cytosol and the titration of IPTG for more uniform expression are available as well as hosts that provide rare codon tRNAs for non-codon optimized cDNAs. A variety of expression vectors are available offering different antibiotic resistances, induction protocols, and fusion partners. Popular fusion partners include poly-histidine for immobilized metal affinity chromatography purification, thioredoxin (Trx) for disulfide bond formation in an oxidizing cytosol, signal peptides or disulfide bond isomerase (Dsb) variants for periplasmic expression and disulfide bond formation and glutathione S-transferase (GST), N utilization substance A (NusA), maltose-binding protein (MBP), or small ubiquitin-related modifier (SUMO) for soluble cytosolic expression of proteins and peptides (see Table 3). Also offered, or engineered directly, are many choices for fusion partner removal by proteolytic digestion. These include enterokinase, thrombin, factor Xa, tobacco etch virus protease (TEV), and human rhino virus 3C protease (HRV3C). The choice of which protease to use is often dictated by the desired, mature N-termini; however TEV usually represents a robust choice.

Direct, cytosolic, expression in bacteria is usually the method of choice for a heterologous protein from the first and fourth classes of proteins as long as this target protein does not contain an inordinate number of cysteines involved in native disulfide bonds. The reducing environment of the bacterial cytosol does not allow disulfide bond formation and overexpression often leads to the

Table 3
Fusion tags and partner proteins

Fusion partner	Placement of tag	Approx. size (AA)	Advantages
Poly-histidine	N, C, I	6–10	Affinity purification
Trx	N	110	Disulfide formation
Signal peptide	N	20	Periplasmic expression, native folding
Dsb	N	220	Periplasmic expression, native folding, disulfide formation
SUMO	N	100	Cytoplasmic solubility, native N-termini
GST	N	220	Cytoplasmic solubility
NusA	N, I	500	Cytoplasmic solubility
MBP	N	400	Cytoplasmic solubility

N N-terminus; *C* C-terminus; *I* internal

formation of insoluble inclusion bodies that require solubilization and refolding to yield the desired product. However, as mentioned earlier, there are several vector choices that not only facilitate the expression of a soluble fusion construct but also the formation of disulfide bonds when used in conjunction with a host cell carrying the thrioredoxin reductase (*trxB*) and glutathione reductase (*gor*) mutations that result in an oxidizing cytosol. There are several examples in the literature where these combinations, along with refolding chaperones, led to native disulfide bond formation (14–16) and the mature protein was acquired following enzymatic removal of the fusion partner.

Intracellular expression in *E. coli* yields a protein containing the initiating methionine residue. Endogenous proteins are normally processed by *E. coli* N-terminal methionine amino peptidase (MAP). For highly expressed recombinant proteins, this processing can be rate-limiting leading to N-terminal heterogeneity. However, the "N-end rule" is often a good guide in construct design when looking to enhance protein expression or to optimize, or minimize, N-terminal methionine processing (17, 18). Effective methionine processing has been shown to be directly related to the radius of gyration of the penultimate residue. Methionine cleavage decreases proportionally to the increase of the minimal side chain length. Smaller residues such as Gly, Ala, Ser, or Cys result in more efficient processing, intermediate residues such as Thr, Pro, Val, Gln, or Glu result in less efficient processing and all other residues in the penultimate position result in little to no methionine processing.

4.2. Eukaryotic Expression Systems

4.2.1. Insect Cells

Another common approach to recombinant protein expression is baculovirus-mediated expression (BEVS) using insect cells. This system has proven itself for the rapid production of high levels of heterologous proteins. Insect cells are slower growing than both bacteria and yeast, and the cost of the media is relatively high compared to those two systems. However, recent advances with intermediate-scale suspension systems (e.g., WAVE bioreactors) and the newer baculovirus-mediated vector systems (e.g., FlashBAC (NextGen Sciences) and BacMagic (Novagen)) have transformed BEVS into the eukaryotic expression system of choice for structural biology. The most commonly used virus is *Autographica californica* nuclear polyhedrosis virus (AcMNPV), which is preferred due to the diversity in commercially available transfer plasmids. Several insect cell lines for baculovirus expression are commonly employed including Spodoptera frugiperda (fall armyworm) lines S*f*9 and S*f*21. These cells double in ~20 h in both monolayer and suspension cultures. Another frequent option that supports AcMNPV replication is the *Trichoplusia ni*-derived BTI-Tn-5B1-4 ("High5") cell line. These cells tend to be better for secreted proteins but they do not grow as well in suspension.

Both intracellular and secreted expression is possible with insect cells. The parameters for optimizing heterologous expression have been well studied and include the kinetics of infection (multiplicity of infection (MOI), cell density, and time) (19). This expression system is categorized as transient expression and is capable of providing milligram quantities of heterologous protein with complex posttranslational modifications. These modifications include correct proteolytic processing of signal peptides and internal cleavage sites, phosphorylation, N-terminal blocking, proper folding and assembly, and glycosylation. The higher expression levels and p10 and polh promoter activity in the very late phase of infection can lead to incomplete posttranslational modifications resulting in higher levels of protein heterogeneity. In addition, BEVS has emerged as a useful host for membrane protein expression (20), especially for GPCRs with all currently published structures derived from BEVS expressed GPCRs.

4.2.2. Yeast

For the past 25 years, *Pichia pastoris* has increased in popularity for the production of heterologous proteins. *P. pastoris* is a methyltropic yeast that uses the tightly regulated alcohol oxidase (AOX) promoter enabling high levels of protein expression by the addition of methanol. Early biochemical studies demonstrated that methanol utilization requires a metabolic pathway involving several unique enzymes. The key enzyme, AOX catalyzes the first step in the methanol utilization pathway. Two genes, AOX1 and AOX2, encode AOX in *P. pastoris* where AOX1 is responsible for the majority of AOX activity in the cell. The presence of methanol is essential for the induction of high levels of protein expression.

An attractive feature of this system is the simplicity of techniques for genetic manipulation. *P. pastoris* has also been shown to express foreign proteins at high levels, both intracellularly and extracellularly. This system is also capable of performing many common eukaryotic posttranslational modifications such as disulfide bond formation, proteolytic processing, and glycosylation (see Table 2). However, since *P. pastoris* has no native plasmids, the expression cassette must be integrated into the chromosome and the selection and optimization of the integrated gene can be a time consuming process. An alternative strategy to employing gene integration in *Pichia* is to exploit episomal expression in *Saccharomyces cerevisiae* (21). This expression technique can provide more rapid evaluation of the feasibility of yeast expression for a given protein.

Compared to a bacterial system, yeast generally produces proteins at lower levels. Even so, yeast has proven its ability to produce biologically active and, hence, biologically relevant proteins in ever increasing cases. Yeast cell expression is a system that provides for both intracellular expression as well as extracellular expression. However, an extracellular expression approach remains the path of choice, as yeast cells grown to maximum stationary phase possess cell walls that are notoriously difficult to remove. Expression in a yeast cell system has some very attractive features including rapid cell growth, low cost of media reagents, high expression levels and the generation of proteins with key posttranslational modifications. Yeast has been shown to be a suitable host for the secreted expression of complex proteins like single chain antibody fragments, serine proteases such as DESC1 (22), matriptase (MTSP1) (23) or urokinase plasminogen activator (24) and various growth factors whereas human superoxide dismutase (25), fibroblast growth factors (26) and α-antitrypsin (27) have been successfully expressed in the cytosol.

4.2.3. Mammalian Cells

Complex proteins are best expressed in mammalian cell culture where processing through the secretion pathway yields predominantly biologically active protein. Mammalian cell culture is still the system of choice for complex, multi-domain proteins. Although, not unlike insect cell culture, there are distinct drawbacks. Like insect cells, mammalian cells grow at much slower rates than microbial cells where doubling times are in the rage of 18–24 h. In addition there are several other considerations that must be evaluated before employing a mammalian cell expression approach, such as substantial time and cost investment, lower overall expression levels, and culture sensitivity issues such as shearing, changes in temperature, pH and oxygen levels and metabolites. The accumulated benefits, including secretion into the media, proper folding, stability and comprehensive posttranslational modifications support this choice of expression system when producing proteins from the third protein class. Efficient protein expression from mammalian systems can be approached in several ways.

Lower productivity has been overcome through the use of stirred bioreactors or Wave systems for suspension cultures. Suspension cultures, in combination with fed-batch approaches allow for very high cell densities and overall increases in protein production. Many commercially available vectors contain strong promoters, such as simian vacuolating virus 40 (SV40) or cytomegalovirus (CMV) immediate early promoter, in addition to dominant and recessive selection markers such as neomycin or dihydrofolate reductase (DHFR), respectively. Homologous recombination of the expression cassette into regions of high transcription has been shown to increase the expression levels of recombinant proteins. However, for most glycoproteins, or proteins with extensive posttranslational modifications, the rate-limiting step in protein production is these posttranslational modifications.

5. Strategies to Facilitate Protein Crystallization

Attrition rates in the production of high quality protein crystals remain high when taking into account the entire process from the generation of high quality protein to the determination of a high resolution crystal structure. One of the key prerequisites to generating high quality protein crystals is effectively addressing disorder and protein instability. Disorder can manifest itself in the form of surface residue dynamics where a high degree of side chain conformational entropy, inter-domain flexibility, or unrestrained termini make the formation of high quality crystal contacts prohibitively difficult. Disorder also appears in the form of heterogenic posttranslational modifications such as glycosylation or phosphorylation. Screening of surface mutations to address hydrophobicity, addressing solvent accessible cysteine residues and even abrogating regions of high conformational flexibility can improve protein stability. In this section, we will address strategies to reduce surface entropy, improve solubility and increase the overall propensity to form high quality crystals capable of diffracting to high resolution.

5.1. Identifying Regions of Disorder

The *apparent* driving force of the universe is the disposition to move toward lower enthalpy (H) and higher entropy (S) according to the Gibbs free energy equation. The *actual* driving force is the tendency to move toward greater entropy. Crystallization is a unique process that occurs at a high entropic cost. The conformational entropy of surface residues and loops, which become ordered in the crystallization process, are the main contributors to this entropic loss. This would suggest that rationally engineering out regions of amino acids with higher conformational entropy, as well as loops with lower flexibility, could lead to more thermodynamically favored crystal formation.

An early example of using surface protein engineering to enhance the propensity for crystal formation was work done by Lawson et al. where surface mutations of human H ferritin promoted crystal formation (28). This group used information taken from the crystal structures of other homologues suggesting that the incorporation of a single L86Q substitution was sufficient to create favorable crystal contacts and led to crystals diffracting to 1.9 Å or better and subsequent three-dimensional structure determination. D'Arcy et al. later demonstrated that the surface mutagenesis of DNA gyrase B led to the formation of new crystal forms that diffracted to higher resolution and concluded that single point mutations can have a marked effect on the crystallization properties of proteins resulting in both the number of crystal hits as well as improvements in crystal quality (29).

Recently, Derewenda has addressed the idea of reducing surface entropy by creating "low-entropy" surface patches through rational site directed mutagenesis (30). The idea that most proteins have evolved a "surface-entropy shield" that is comprised predominantly of lysine and glutamate residues was proposed by Doye et al. (31). The hypothesis is that this prevents nonspecific aggregation and precipitation. The author points out that the spontaneous crystallization of proteins in vivo is responsible for several serious diseases. Thus, an effective entropic-shield makes these interactions less than favorable. Several publications by Derewenda and colleagues have described this concept in more detail (32–34). This approach suggests that mutating residues with high conformational mobility, lysine or glutamate, to smaller amino acids, such as alanine or glycine, should be effective in reducing overall surface conformational entropy. The fact that, together, 68% of these residues are found completely exposed while only 6% are found buried supports the idea that these residues play a large part in creating this entropic shield (35). These authors tested this concept through a series of mutational experiments using the globular domain of the human regulatory protein RhoGDI. Although this molecule is relatively small, it has a surface rich in lysine and glutamate residues. The majority of these experiments involved mutating these residues to alanine with dramatic results. Most of the mutations resulted in proteins with a higher propensity to crystallize. As further validation, in most cases the crystal contacts were mediated by the mutated epitopes. Different sets of mutations also led to novel crystal forms that exhibited superior diffraction. Taken together, the authors suggest that it is the nature of the crystal contacts that is the primary determinant in the quality of the crystal formation. It is important to note however, that mutating polar amino acids like lysine and glutamate to hydrophobic amino acids like alanine can occasionally destabilize the target protein (36).

In aqueous solution, the surface exposure of large hydrophobic regions is energetically unfavorable and can promote aggregation, especially at high protein concentrations typical of crystallization experiments. It has been proposed, however, that increased side chain entropy is more detrimental to crystal lattice formation than decreased hydrophobicity (37). Therefore, a balance must be struck between decreasing surface entropy enough to promote crystallization and maintaining sufficient surface hydrophilicity to deter aggregation when engineering surface mutations.

Limited proteolysis, as mentioned previously, is often used to identify flexible regions at the terminal ends of the protein of interest. Limited proteolysis is also used to identify flexible loops in proteins that could have an adverse effect on crystal packing. This method often results in a more rigid form of the molecule that has a higher propensity for crystal formation. Using information from limited proteolysis, constructs can be designed where offending residues are either removed or substituted with residues that can have a stabilizing effect. Rosenbaum et al. identified a poorly structured intracellular loop in β_2-adrenergic receptor that contributed to the relatively unrestricted movement of two transmembrane helices (38). The conformational heterogeneity resulting from the free movement led to crystallization problems. The authors were able to replace the flexible loop with the well-folded protein, T4 lysozyme. The rational engineering effort added a polar surface while restricting the movement of the transmembrane helices. This chimeric protein was efficiently expressed, retained binding affinities and resulted in crystals diffracting to 2.4 Å.

5.2. Reducing Heterogeneity

Heterogeneity and aggregation are frequently the main impediments to the generation of good quality protein crystals. It is critical therefore to ensure that the initial crystallization sample is homogeneous and monodisperse. Heterogeneity can arise from several sources, including contaminating proteins, aggregates, and post-translational modifications. Sample purity, normally assessed by SDS-PAGE, is necessary to ensure the sample is homogenous with minimal contaminating proteins present (normal good practice is >95% purity by SDS-PAGE). Aggregation, a frequent source of sample heterogeneity, is often detected by SEC (size exclusion chromatography) during purification. However, given that samples are typically concentrated after SEC, and prior to crystallization, this method is not an accurate measure of aggregation in the protein sample as used for crystal screening. Dynamic light scattering (DLS) instruments are now available with both high-sensitivity and plate-readers for higher throughput and are becoming a more common technique to assess aggregation. DLS is well suited to determine protein aggregation as it is compatible with both the concentrations and buffer conditions used in crystallization (reviewed in ref. (39)). Posttranslational modifications can provide a barrier to crystallization by introducing heterogeneity and/or increasing surface entropy. The heterogeneity associated with these modifications is normally assessed using mass spectrometry or DLS.

The addition of a ligand can greatly influence the conformational state of a target protein and increase the probability of crystallization. Useful ligands include substrates, substrate analogs, inhibitors and allosteric modulators. While not part of the construct per se ligands should be part of the screening paradigm.

5.3. Sample Purity

Improving recombinant protein expression levels is a straightforward strategy to improve sample purity. As described above, the choice of expression host can greatly impact recombinant expression levels and should be optimized for a given target protein. Expression optimization involving iterative empirical experiments is usually necessary to achieve high level of recombinant protein expression. Numerous variables can be assessed including expression temperature (*E. coli*), length of induction (*E. coli*) or infection (baculovirus), IPTG concentration (*E. coli*), multicopy integration (*Pichia*), MOI (baculovirus), choice of fusion protein (see Table 3), choice of strain (*E. coli*) or cell line (eukaryotic hosts), or choice of secretion signal. Construct design can also facilitate high levels of recombinant protein expression. Optimizing codon usage for a particular expression host can be beneficial. With the advent of modern gene synthesis techniques codon optimization has become a convenient and cost-effective strategy for improving protein expression. Several different computational software packages are available and vary in terms of the statistical methods used to compute the variation and frequency of optimal codon usage for a given host.

As described in Chapter 3, the choice of chromatographic methods can also greatly influence the purity of sample. When designing constructs it is important to incorporate possible purification strategies in the cloning plan. Most frequently, these manifest themselves in the addition of a fusion or epitope tag to permit affinity based purification. To facilitate tag removal following purification linking the purification tag to the gene of interest with a protease cleavage sequence is often desired. As mentioned before, numerous commercially available proteases are suitable for this purpose, including HRV3C, FactorXa, and TEV. Occasionally, the fusion protein may prove refractory to proteolytic digestion, frequently due to steric hindrance present at the cleavage site. This can be resolved by increasing the number of amino acids between the cleavage site and the fusion protein or the target protein or through the addition of chaotropes (urea, sodium thiosulfate, etc.) to the cleavage reaction mixture to relax the protein structure.

5.4. Post-translational Modifications

Glycosylation, which can be critical for protein folding or function, is frequently an impediment to crystallization as it introduces a source of heterogeneity (40). Several strategies can be employed to reduce or eliminate glycosylation. One straightforward strategy is to employ recombinant expression in a prokaryotic host. This strategy can only be successfully employed if the target protein expresses and folds properly in a prokaryotic host (see choosing an expression system). However since most glycosylated targets are secreted or transmembrane eukaryotic proteins, bacteria are not the first host of choice. If the target protein requires expression in a eukaryotic host two strategies can be employed: enzymatic removal during purification or mutagenesis of the glycosylation sites to eliminate them. Enzymatic removal can be advantageous in that it does not require the generation of additional clones. In addition, enzymatic removal may be necessary for proteins that require glycosylation for proper folding. However, enzymatic removal can result in the introduction of heterogeneity (by incomplete digestion) and also necessitates the development of a deglycosylation protocol. Mutagenesis is a straightforward method to eliminate glycosylation. Asparagines, which are the target for N-linked glycosylation have often been replaced with glutamines or aspartates. However this represents a misplaced attempt to be chemically conservative; mutation to alanine is probably a better choice.

Free cysteines can be impediments to crystallization due to either chemical modification or the formation of undesired disulfide bonds. Commonly, cysteine residues can cross-link monomers forming nonnative intermolecular disulfide bonds or non-native intramolecular disulfide bonds. These nonnative disulfide bonds can frequently result in reduced solubility and aggregation. While the addition of reducing agents can prevent this, often it is necessary to mutate the cysteine residues to prevent non-native disulfide bonds. Mutation to serine can result in marked improvement in both solution and crystallization behavior (41). In addition, it has been observed that mutations of cysteine to serine can improve the crystallization of proteins with good solution behavior (bovine gamma-B crystallin). While mutation to serine is considered more conservative, mutation to alanine can be a better choice.

Other posttranslational modifications can also introduce barriers to crystallization, including phosphorylation, sulfation, and lipidation. Like the previously described posttranslational modifications, these alterations can be critical to both folding and function but also frequently introduce structural heterogeneity or surface entropy. Phosphorylation can be introduced in recombinantly expressed proteins from kinases in the eukaryotic hosts or as the result of autophosphorylation. Heterogenous hyperphosphorylation is frequently an impediment to crystallization. Several strategies can be employed to improve the homogeneity of the protein including co-expression with phosphatases, using chromatographic methods to separate the phosphorylated species or enzymatic dephosphorylation in vitro. A mutagenesis strategy can also be employed, similar to those used in salvage pathways for glycosylated proteins, with the phosphorylated residues mutated to other polar amino acids or alanines (42). However, if phosphorylation is critical to function (and therefore important for determining biological relevant crystal structure) the use of phosphomimetics should be employed when the protein of interest is only partially phosphorylated during expression. This is frequently encountered with protein kinases, as phosphorylation of residues in the activation loop can either activate or inhibit kinase function. Commonly Thr to Glu or Ser to Asp mutations are employed as phosphomimetics (43). Sulfation and lipidation can also introduce heterogeneity, which is frequently resolved either chromatographically, by mutatgenesis, or with primary sequence truncations.

Heterogeneous proteolytic processing can also preclude the isolation of protein amenable to crystallization. Proteolysis can often be overcome through the use of protease inhibitors during cell lysis or through mutagenesis. Proteolytic sites can be predicted in silico using several different software packages (e.g., PeptideCutter) or empirically through mass spectrometry or N-terminal analysis. Once identified the labile sites can be mutated to prevent processing. Note however that proteolytic lability can be an important clue to identifying regions of disorder that may be better addressed by truncation or deletion within a susceptible loop. Occasionally truncated protein is isolated not as the result of proteolysis but rather premature translational termination. Premature translational termination is overcome by either changing expression hosts or by optimizing codon usage for the chosen expression host (44).

6. Conclusion

The probability of successfully attaining a crystal structure weighs heavily on producing a highly homogenous protein sample. As outlined in this chapter, protein heterogeneity can be improved through careful attention to detail and thoughtfulness in construct design via boundary choice and the identification regions of disorder and the selection of expression system and by anticipating sources of potential heterogeneity and minimizing or eliminating them. The most powerful tool available for crystallization construct design is sequence comparison to related proteins of known structure. Frequently the result of such sequence comparison is the design and attempted crystallization of multiple constructs. This parallel multi-construct approach greatly improves likelihood of success while also reducing the time from design to structure. Identifying potential sources inherent heterogeneity that lead to poor crystal formation, such as loop or side chain flexibility or hydrophobic surfaces that increase the propensity of aggregation, creates an iterative loop construct design leading to better behaved and more homogenous protein samples. Understanding the need for posttranslational modifications will walk the researcher through several potential expression hosts depending on the need for little or complex posttranslational modifications. Recognizing potential sources of heterogeneity (e.g., proper folding, disulfide bond scrambling, nonuniform glycosylation, non-uniform phosphorylation) directs the researcher to the system that best suits their needs and the protein in hand.

Taken together, the approaches addressed here are meant to provide a starting point with which to increase the probability of producing well behaved, homogenous protein samples. Boundary choice, rational mutagenesis and expression host choice all impact each other and can have a profound effect of reducing protein heterogeneity. However, the rational approaches described here are not a substitute for empirical data with your protein of interest. Expressing, purifying, and crystallizing multiple constructs will often result in observations that can be used to further refine construct design. As such, it is critical to attempt crystallization with proteins which may not fit all "ideal" criteria for crystallization quality protein.

References

1. Dale, G. E., Oefner, C., D'Arcy, A. (2003) The protein as a variable in protein crystallization. *J. Struct. Biol.* **142**, 88–97.
2. Notredame C., Higgins D.G. and Heringa J. (2000) T-Coffee: A novel method for fast and accurate multiple sequence alignment. *J. Mol. Biol.* **302**, 205–217.
3. Edgar R.C. (2004) MUSCLE: multiple sequence alignment with high accuracy and high throughput. *Nucleic Acids Res.* **32**, 1792–1797.
4. Jacobson M. and Sali A. (2004) Comparative protein structure modeling and its applications to drug discovery. *Ann. Reports Med. Chem.* **39**, 259–276.
5. Mooij W.T.M., Mitsiki E. and Perrakis A. (2009) ProteinCCD: enabling the design of protein truncation constructs for expression and crystallization experiment. *Nucleic Acids Research*, **37**, W402–W405.
6. Malawski, G. A., Hillig, R. C., Monteclaro, F., Eberspaecher, U., Schmitz, A. A. P., Crusius, K., Huber, M., Egner, U., Donner, P. and Mueller-Tiemann, B. (2006) Identifying protein construct variants with increased crystallization propensity-A case study. *Prot. Sci.* **15**, 2718–2728.
7. Pantazatos D., Kim J.S., Klock H.E., Stevens R.C., Wilson I.A., Lesley S.A., Woods V. L. Jr. (2004) Rapid refinement of crystallographic protein construct definition employing enhanced hydrogen/deuterium exchange MS. *Proc. Natl. Acad. Sci. USA.* **101**, 751–756.
8. Spraggon, G., D. Pantazatos, D., Klock, H. E., Wilson, I. A., Woods, V. L. and Lesley, S. A. (2004) On the use of DXMS to produce more crystallizable proteins: structures of the *T. maritima* proteins TM0160 and TM1171. *Prot. Sci.* **13**, 3187–3199.
9. Hamuro, Y., Coales, S. J., Southern, M. R. Nemeth-Cawley, J. F., Stranz, D. D. and Griffin, P. R. (2003) Rapid analysis of protein structure and dynamics by hydrogen/deuterium exchange mass spectrometry. *J. Biomol. Tech.* **14**, 171–182.
10. Sharma, S., Zheng, H., Huang, Y. J., Hamuro, Y., Rossi, P., Tejero, R., Acton, T. B., Xiao, R., Jiang, M., Zhao, L., Ma, L. C., Swapna, G. V., Aramini, J. M. and Montelione, G. T. (2009) Construct optimization for protein NMR structure analysis using amide hydrogen/deuterium exchange mass spectrometry *Proteins.* **76**, 882–894.
11. Gray D. and Subramanian S. (2001) Choice of cellular protein expression system. *Curr. Protocol. Prot. Sci.* **5**.16, 1–34.

12. Dvir, H. and Choe, S. (2009) Bacterial expression of a eukaryotic membrane protein in fusion to various Mistic orthologs. *Prot. Expr. Purif.* **68**, 28–33.
13. Leviatan, S., Sawada, K., Moriyama, Y. and Nelson, N. (2010). Combinatorial method for overexpression of membrane proteins in *Escherichia coli. J. Biol. Chem.* **285**, 23548–23556.
14. Ponniah K., Loo T.S., Edwards P.J.B., Pascal S.M., Jameson G.B. and Norris G.E. (2010) *Prot. Exp. Purif.* **70**, 283–289.
15. Brusehaber E., Schwiebs A., Schmidt M., Bottcher D. and Bornscheuer U.T. (2010) *App. Microbiol. Biotechnol.* **86**, 1337–1344.
16. Yan W.K., Goette M., Hofmann G., Zaror I. and Sim J. (2010) *Prot. Exp. Purif.* **70**, 270–276.
17. Bachmair A., Finley D. and Varshavsky A. (1986) *In vivo* half-life of a protein is a function of its amino-terminal residue. *Science.* **234**, 179–186.
18. Dalboge, H., Bayne, S. and Pedersen, J. (1990) *In vivo* processing of N-terminal methionine in *E. coli. FEBS Lett.* **266**, 1–3.
19. Murphy, C. I., Piwnica-Worms, H., Grunwald, S., Romanow, W. G., Francis, N. and Fan, H. Y. (2004) Expression and purification of recombinant proteins using the baculovirus system. *Curr. Prot. Mol. Biol.* **Chapter 16**, Unit 16 11.
20. Ratnala, V. R. (2006) New tools for G-protein coupled receptor (GPCR) drug discovery: combination of baculoviral expression system and solid state NMR. *Biotech. Lett.* **28**, 767–778.
21. Taxis, C. and Knop, M. (2006) System of centromeric, episomal, and integrative vectors based on drug resistance markers for *Saccharomyces cerevisiae. Biotechniques* **40**, 73–78.
22. Kyrieleis O.J.P., Huber R., Ong E., Oehler R., Hunter M., Madison E.L. and Jacob U. (2007) Crystal structure of the catalytic domain of DESC1, a new member of the type II transmembrane serine proteinase family. *FEBS Journal.* **274**, 2148–2160.
23. Friedrich R., Fuentes-Prior P., Ong E., Coombs G., Oehler R., Hunter M., Pierson D., Gonzalez R., Huber R., Bode W. and Madison E.L. (2002) Catalytic domain structures of MT-SP1/Matriptase, a matrix-degrading transmembrane serine proteinase. *J. Biol. Chem.* **277**, 2160–2168.
24. Katz, B. A., Mackman, R., Luong, C., Radika, K., Martelli, A., Sprengeler, P. A., Wang, J., Chan, H. and Wong, L. (2000) Structural basis for selectivity of a small molecule, S1-binding, submicromolar inhibitor of urokinase-type plasminogen activator. *Chem. Biol.* 7, 299–312.
25. Yoo H.Y., Kim S.S. and Rho H.M. (1999) Overexpression and simple purification of human superoxide dismutase (SOD1) in yeast and its resistance to oxidative stress. *J. Biotechnol.* **68**, 29–35.
26. Barr P.J., Cousens L.S., Lee-Ng C.T., Medina-Selby A., Masiarz F.R., Hallewell R.A., Chamberlain S.H., Bradley J.D., Lee D., Steimer K.S., Poulter L., Burlingame A.L., Esch F. and Baird A. (1988) Antigenicity and immunogenicity of domains of the human immunodeficiency virus (HIV) envelope polypeptide expressed in the yeast *Saccharomyces cerevisiae. J. Biol. Chem.* **263**,16471–16478.
27. Cabezon T., de Wilde M., Herion P., Loriau, R. and Bollen A. (1984) Expression of human alpha 1-antitrypsin cDNA in the yeast *Saccharomyces cerevisiae. Proc. Natl. Acad. Sci.* **81**, 6594–6598.
28. Lawson D.M., Artymiuk P.J., Yewdall S.J., Smith J.M.A., Livingstone J.C., Treffry A., Luzzago A., Levi S., Arosio P., Casareni G., Thomas C.D., Shaw W.V. and Harrison P.M. (1991) Solving the structure of human H ferritin by genetically engineering intermolecular crystal contacts. *Nature.* **349**, 541–544.
29. D'Arcy A., Stihle M., Kostrewa D. and Dale, G. (1999) Crystal engineering: a case study using the 24 kDa fragment of the DNA gyrase B subunit from *Escherichia coli. Acta. Cryst.* D**55**, 1623–1625.
30. Derewenda, Z.S. (2004) Rational protein crystallization by mutational surface engineering. *Structure,* **12**, 529–535.
31. Doye J.P.K., Louis A.A. and Vendruscolo M. (2004) Inhibition of protein crystallization by evolutionary negative design. *Phys. Biol.* **1**, P9–P13.
32. Longnecker K.L., Garrard S.M., Sheffield P.J. and Derewenda Z.S. (2001) Protein crystallization by rational mutagenesis of surface residues: Lys to Ala mutations promote crystallization of RhoGDI. *Acta Cryst.* D**57**, 679–688.
33. Derewenda Z.S. (2004) The use of recombinant methods and molecular engineering in protein crystallization. *Methods,* **34**, 354–363.
34. Czepas J., Devedjiev Y., Krowarsch D., Derewenda U., Otlewski J. and Derewenda Z.S. (2004) The impact of Lys→Arg surface mutations on the crystallization of the globular domain of RhoGDI. *Acta Cryst.,* D**60**, 275–280.
35. Baud F. and Karlin S. (1999) Measures of residue density in protein structures. *Proc. Natl. Acad. Sci.* **96**, 12494–12499.

36. Mateja, A., Devedjiev, Y., Krowarsch, D., Longenecker, K., Dauter,. Z., Otlewski, J. and Derewenda, Z. S. (2002) Entropy and surface engineering in protein crystallization. *Acta Cryst.* D**58**, 1983–1991.
37. Price, W. N. II, Chen, Y., Handelman, S. K., Neely, H., Manor, P. *et al.* (2009) Understanding the physical properties controlling protein crystallization based on analysis of large-scale experimental data. *Nat. Biotechnol.* **27**, 51–57.
38. Rosenbaum D. M., Cherezov V., Hanson M. A., Rasmussen S. G. F., Thian F. S., Kobilka T. S., Choi H. -J., Yao X. -J., Weis W. I., Stevens R. C. and Kobilka B. K. (2007) High-Resolution Crystal Structure of an Engineered Human beta 2-Adrenergic G Protein–Coupled Receptor. *Science*, **318**, 1266–1273.
39. Proteau, A., Shi, R. and Cygler M. (2010). *Curr. Prot. Prot. Sci.* **Chapter 17**, Unit 17 10.
40. Davis S. J., Puklavec M. J., Ashford D. A., Harlos K., Jones E. Y., Stuart D. I., and Williams A. F. (1993) Expression of soluble recombinant glycoproteins with predefined glycosylation: application to the crystallization of the T-cell glycoprotein CD2. *Prot. Eng.* **6**, 229–232.
41. Chrencik, J. E., Patny, A., Leung, I. K., Korniski, B., Emmons T. L. *et al.* (2010) Structural and thermodynamic characterization of the TYK2 and JAK3 kinase domains in complex with CP-690550 and CMP-6. *J Mol Biol* **400**, 413–433.
42. Panneerselvam, S., Marx, A., Mandelkow, E. M., and Mandelkow, E. (2006) Structure of the catalytic and ubiquitin-associated domains of the protein kinase MARK/Par-1. *Structure* **14**, 173–183.
43. Marx, A., Nugoor, C., Panneerselvam, S. and Mandelkow, E. (2010) Structure and function of polarity-inducing kinase family MARK/Par-1 within the branch of AMPK/Snf1-related kinases. *FASEB J.* **24**, 1637–1648.
44. Sallach, R. E., Conticello, V. P. and Chalkof, E. L. (2009) Expression of a recombinant elastin-like protein in *Pichia pastoris. Biotechnol. Prog.* **25**, 1810–1818.

Chapter 3

Purification of Proteins for Crystallographic Applications

Daniel C. Bensen

Abstract

One of the most important parameters correlated with success in protein crystallization experiments is sample purity and monodispersity. Heterologous expression systems have allowed investigators to produce engineered proteins in sufficient quantities which simplify the purification process compared with the days when macromolecules had to be extracted from source tissue. Improvements in the areas of chromatographic media and instrumentation have also dramatically improved throughput and protein yields while maintaining analytical resolution. In a drug discovery setting, efforts can be focused on either a single protein or family of proteins. This requires the development and refinement of general purification methods that can be applied to multiple proteins or construct variants until readily crystallizable forms of the target protein are discovered. It is the aim of this chapter to provide a practical introduction to the techniques and methods used to purify proteins for crystallographic applications. Additionally, a protocol describing the expression, purification, and crystallization of the ATP-binding domain of the important cancer target Hsp90 provides the reader with an example of methods that can be adapted to a wider set of crystallographic target proteins.

Key words: Hsp90, Protein purification, IMAC, Gel filtration, Ion exchange chromatography, Crystallography, X-ray crystallography, Protein crystallization

1. Introduction

In a structure-based drug design (SBDD) program, crystallographic data is a key driver of lead discovery and optimization. Once a validated protein target is nominated, multiple scientific disciplines must be engaged to develop their respective parts in the discovery pipeline. Early in the process, the responsibility of the protein chemist is to develop methods for the purification and crystallization of the target protein(s). This usually requires an approach where many variants of the protein, or, orthologous proteins must be pursued. In the case of published crystallographic data, the literature will allow the researcher to get a jump on the project, but enabling

Leslie W. Tari (ed.), *Structure-Based Drug Discovery*, Methods in Molecular Biology, vol. 841,
DOI 10.1007/978-1-61779-520-6_3, © Springer Science+Business Media, LLC 2012

a robust crystallography tool may require multiple crystal forms to accommodate the chemical entities that are being pursued.

Often, the domain of interest must be winnowed from the mature polypeptide to successfully crystallize. This can present protein stability, as well as purification challenges. An example illustrating this point is the anti-infective target, bacterial DNA gyrase. The fully functional protein exists as a heterotetramer made up of two GyrA (97 kDa) subunits and two GyrB (90 kDa) subunits (1). No crystal structure of the holoenzyme has been published, but domains of both subunits have been structurally characterized (2, 3). Sites of proteolytic lability will often suggest which domain boundaries which should be pursued in crystallographic studies. If the domain of interest can ultimately be crystallized and retains wild-type-like structural and/or kinetic character, it can be beneficial to use the same protein for enzymatic or binding assays when measuring compound affinities. Often this is not the case and the domain that produced diffracting crystals must be reconciled against the SAR generated with the wild-type or orthologous proteins.

A purification scheme is usually developed for a given protein with an initial pilot scale run. The goal of the pilot scale study is to learn how the protein behaves as it progresses through initial purification stages through to its final purification steps. These observations can then be applied to a scale-up run to achieve the ultimate goal of generating sufficient quantities of highly pure protein for crystallization trials. For a protein to successfully crystallize, it must be free of contaminating proteins and, of equal importance, nonaggregated.

The following is a general overview of the methods and theory used in the purification of proteins for crystallographic studies. Materials, technical challenges, and troubleshooting approaches will be discussed. Additionally, a detailed protocol for the purification, formulation, and crystallization of the nucleotide-binding domain of the human cancer target Hsp90 will be presented.

1.1. Expression of Proteins in Escherichia coli

E. coli expression systems provide a robust, inexpensive, and flexible platform suitable for the production of recombinant proteins. High-throughput structural biology studies have highlighted the utility of bacterial expression systems in human protein production (4). The most commonly used platform is the T7 RNA polymerase-based pET family of vectors, which are commercially available from Novogen (5–7).

This pET family of vectors facilitates the directional cloning of the target open reading frame with a very high degree of flexibility with regard to choice of affinity tag position as well the ability to maintain the native DNA sequence with minimal restriction enzyme "scaring." It is common for a laboratory to customize a commercially available vector to suit their preferred cloning methods, restriction sites, affinity tag placement, and to standardize the expression of a wide set of proteins.

The strain of *E. coli* typically employed when using a T7-based expression system is BL21(DE3). Its utility is partially derived from a series of chromosomal knockouts in genes which encode endogenous proteases. The strain also houses the λDE3 lysogen within its chromosome, which carefully controls expression from the T7 promoter (7). While BL21(DE3) is usually sufficient for most expressions, other variants have been developed which tightly regulate basal (pre-induction) expression. These strains and modifications have proven quite useful when the target protein is toxic to the cell. Strains containing supplemental tRNAs to help overcome codon usage problems are also widely used and commercially available.

Expression conditions for each construct must be optimized for the maximal production of soluble protein. A typical expression begins from an overnight starter culture, either grown in solution or on plates containing selective media, typically with LB agar plus the appropriate concentration of antibiotic. Expression cultures are inoculated to a starting point of 0.05–0.1 OD600 and grown to 0.5–0.8 OD600 in 1-L baffled shake flasks before induction with IPTG to a final concentration of 0.4–1 mM. The use of a rich medium such as Terrific Broth is standard, and will speed the doubling time in log phase growth as well as support higher cell densities as the culture reaches stationary phase.

The ideal conditions for each protein expression construct must be determined empirically. It is common to begin testing conditions on a pilot scale in shake flasks (25–100 mL) before moving to a production scale (multiple liters). The two main variables to consider are temperature and time of induction. The most common expression conditions employed are 37°C for 2 h, 30°C for 4 h, and 18–20°C overnight to 24 h. It is our experience that the 18°C overnight expression (up to 24 h) consistently produces much more soluble protein, supports very high cell densities, and is always attempted before exploring other conditions.

1.2. Cell Lysis and Pre-chromatography Processing

When purifying a protein from an expression culture, a buffer system must be chosen. At this point, it is advantageous to consider what the ideal *final* buffer would be. Typically, when a protein enters crystal trials, a minimal buffer is desired to minimize the bias on solution conditions that the protein is exposed to during crystallization trials. Final buffers will often be diluted versions of the buffer chosen for lysis. The most common buffers used for proteins heading into crystal trials are Tris–HCl and HEPES. Most proteins are stable in the ranges of pH supported by these buffers, and most crystallization solutions are compatible with these buffers. Phosphate buffers are typically not used because of a propensity for salt crystal formation in the presence of divalent cations commonly used in crystallization solutions.

Table 1
Common lysis buffer additives

Additive	Advantage/rationale
Lysozyme	Degrades cell walls, aiding in bacterial lysis
PMSF, protease inhibitor cocktails	Protease inhibitors used during lysis
Benzonase	Promiscuous nucleic acid degrader
Triton X-100	Non-ionic detergent aids cell lysis and also help in solubilization of insoluble protein
BME, DTT, or TCEP	Reduction of disulfide-containing proteins
Glycerol	Promotes protein stability

A generic lysis buffer for His-tagged proteins used in our laboratories is 50 mM Tris pH 8.0, 200 mM NaCl. The relatively high buffer concentration allows for the maintenance of pH upon dilution when the buffer is mixed with a cell pellet of equal volume. The relatively high ionic strength typically improves the solubility and stability of proteins. The salt concentrations employed depend on and relate directly to the first chromatographic step used in purification; this will be discussed further in the following sections.

Cell lysis can be conducted in many ways including mechanical cell disruption, liquid homogenization, freeze/thaw cycles, and sonication. In our laboratories, a cell pellet is thawed on ice, buffered, and then subjected to sonication, where high frequency sound waves disrupt cell membranes. All procedures should always be performed on ice and it is important to work quickly, especially if the recombinant protein is sensitive to proteolysis. The lysate is then fractionated by ultracentrifugation, separating the soluble material from the cell debris and insoluble protein. The soluble fraction must be clarified using a 0.22 (or minimally 0.45) μm filter if the next step of purification is HPLC based. For bench top batch or gravity flow purification, the sample does not necessarily need to be further processed.

There are several common additives that may be considered when designing a lysis buffer (see Table 1). As previously mentioned, the minimal lysis buffer required to keep the protein in solution and promote maximal binding at each chromatographic step should be used.

1.3. Immobilized Metal Affinity Chromotography

To facilitate the purification of recombinant proteins, an affinity tag is often engineered in as a transcriptional fusion. The hexahistidine $(His)_6$ tag is one of the most commonly used tags for affinity chromatography. Tagged proteins are bound to Ni^{2+} or other transition metals coupled to an agarose support. The method is highly selective and works under both native and denaturing conditions. Bound protein is generally eluted by competition with imidazole or a lowering of eluent pH. The high affinity of the $(His)_6$ tag for the stationary phase allows proteins to be purified to near homogeneity in a single step.

The optimal pH for protein binding is typically at slightly alkaline pH (7.5–8) in the presence of 200–500 mM NaCl. Chelating agents such as EDTA or citrate will strip metal from the column and should be avoided if possible. Reducing agents may be used at varying levels depending on the different linker chemistry tolerances. Each manufacturer generally outlines levels that are acceptable. Imidazole at low concentrations (15–20 mM) is often used in the binding and wash buffers to prevent the capture of host proteins.

Although $(His)_6$ tags generally have little or no effect on function, fold, or activity, they can impede crystallization. The introduction of a conformationally heterogeneous polypeptide to the domain or protein crystallization target adds a new variable which may or may not impede crystallization. Therefore, it may be advantageous to produce proteins with a removable tag. The expression of proteins with an N-terminal $(His)_6$ tag followed by a tobacco etch virus (TEV) protease recognition sequence is widely used in crystallography laboratories (8).

TEV cleavable proteins are purified by two steps of immobilized metal affinity chromatography. First, the protein is bound to immobilized metal affinity chromotography (IMAC) resin and eluted with imidazole. The tag is then removed by treatment with $(His)_6$-tagged TEV protease. The cleaved protein is then re-applied to IMAC resin to remove the protease, cleaved tag, and any remaining uncleaved protein. Cleavage of the protein while dialyzing away imidazole from the first purification step prepares sample for a second round of IMAC purification.

IMAC provides a flexible first chromatographic step which adapts equally well to high-throughput applications, preparative FPLC methods, and bulk processing of multiliter lysates using simple bench top gravity flow columns. Prepacked columns, bottled resins, and 96-well binding plates are all commercially available from multiple sources.

1.4. Evaluation of Protein Purity and Concentration

1.4.1. SDS-PAGE

SDS-PAGE is a simple, high-resolution method used to separate and analyze protein mixtures. Polyacrylamide gel electrophoresis (PAGE) is performed in the presence of sodium dodecyl sulfate (SDS). SDS binds tightly to hydrophobic residues giving them a net negative charge. By denaturing and normalizing the charge of all proteins in a sample, separation is based almost entirely on molecular weight. The method is used to visually assess the purity of a protein throughout a purification series. SDS-PAGE is highly sensitive; a 50-ng band is typically detectable using standard gel stains (9). Numerous vendors supply precast gels with various separation qualities to suit the molecular weight ranges of the proteins of interest. Buffers and protein standards come in kit format for each formulation.

1.4.2. Bradford Assay

The use of Coomassie dye as a colorimetric reagent for the detection and quantitation of total protein was first described by Dr. Marion Bradford in 1976 (10). Coomassie dyes bind to arginine, lysine, and histidine residues. Upon protein binding, Coomassie shifts from a red to blue color which has a maximal absorbance at 595 nm. The assay is fast and flexible and compatible with a wide range of common buffers. Bradford reagents are currently available in kit format, complete with protein standards from multiple vendors. One disadvantage of the method is its incompatibility with detergents commonly used during protein purification. The assay is somewhat subjective because of the selective binding of amino acids, but still remains the standard for protein detection in solution.

1.5. Ion Exchange Chromatography

Ion exchange chromatography (IEX) separates molecules (proteins in this case) based on differences in overall charge. In cation exchange chromatography, positively charged molecules bind to a negatively charged matrix. Anion exchange chromatography, as you would expect, finds negatively charged molecules binding to a positively charged matrix. Generally, protein solutions or crude extracts are loaded on to a column in a buffer of low ionic strength (salt) and eluted stepwise or with an increasing ionic strength gradient. The binding of molecules to the stationary phase is driven by their ionic attraction. By raising the conductivity, molecules with a weaker ionic interaction will elute first.

Calculating the theoretical isoelectric point (pI) of the target protein will help determine which IEX method to choose. There are many web-based tools which will calculate these values from amino acid sequence. If the pI of the protein is less than the pH of the buffer, anion exchange chromatography should be used. For example, if the calculated pI is 5.8 and the protein is in pH 7.5 buffer, anion exchange should be employed. Shifting the pH to 8.0 will, in theory, strengthen the ionic interaction between the protein and anion exchanger.

IEX works well as a first chromatographic step when purifying an untagged protein. However, in crystallography laboratories, it is more often used as an intermediate purification step following IMAC purification. If IEX is used as a first chromatographic step, we have found that an initial bench top gravity flow separation followed by rebinding and a gradient elution on the same resin provides superior results (D. Bensen, unpublished data).

For example, a clarified (by centrifugation and/or filtration) lysate containing an overexpressed protein buffered in 50 mM Tris pH 8.0 and 20 mM NaCl is mixed with pre-equilibrated Q-Sepharose Fast Flow resin (GE Healthcare). The unbound protein is then allowed to pass through a gravity flow column. Once equilibrated (with binding buffer), four isocratic 10-column volume washes, with 50 mM, 100 mM, 500 mM, and 1 M NaCl, respectively (all with 50 mM Tris pH 8.0) are performed and captured. The fractions are then analyzed by SDS-PAGE. Although this method is very crude, it answers several important questions. First, does the protein bind under these conditions? Second, what is the approximate ionic strength (at the fixed pH) that is required to elute the protein from the column? Finally, the investigator is afforded an early look at yield and relative purity of the target protein.

1.6. Size-Exclusion Chromatography

Size-exclusion chromatography (also known as gel filtration) separates larger proteins from smaller ones. In general, large proteins will not enter the gel pores of the stationary phase and will elute quickly from the column, while smaller proteins will continuously enter, leave the pores, and will remain longer in the column. Therefore, proteins will elute in order of decreasing size. In addition to size separation, gel filtration is commonly used as a buffer exchanger and to remove salts, amino acids, and other low molecular weight contaminants. Size-exclusion chromatography is often used to determine the solution subunit composition of multimeric proteins, and to isolate different multimeric forms of proteins.

Commercially available resins are typically classified by their ability to separate different size ranges of soluble (globular) proteins. These resins have upper and lower ranges and a linear separation range where optimal resolution may be achieved between these two extremes. Typically one should not attempt to separate two components with gel filtration unless they differ by at least twofold in molecular weight. Separations are run isocratically and a wide range of buffer systems are tolerated. Additives such as small molecules, ions, or stabilizing peptides may be added to the buffer.

There are a variety of prepacked analytical and preparative SEC columns used in crystallography laboratories. In our experience, Superdex 75 10/300 from GE Healthcare stands out for its superior separation of proteins and peptides that are less than 70 kDa. Running the protein of interest through a sizing column is one of the most important polishing steps prior to crystallization experiments. It not only separates contaminants of differing molecular weights, but also provides the added benefit of removing aggregated target protein that might have been carried through from earlier chromatographic steps. What might appear to be 95% pure protein by SDS-PAGE may prove to be a different story after examining the trace from a gel filtration run. It is often our experience that proteins, especially those that are of marginal stability,

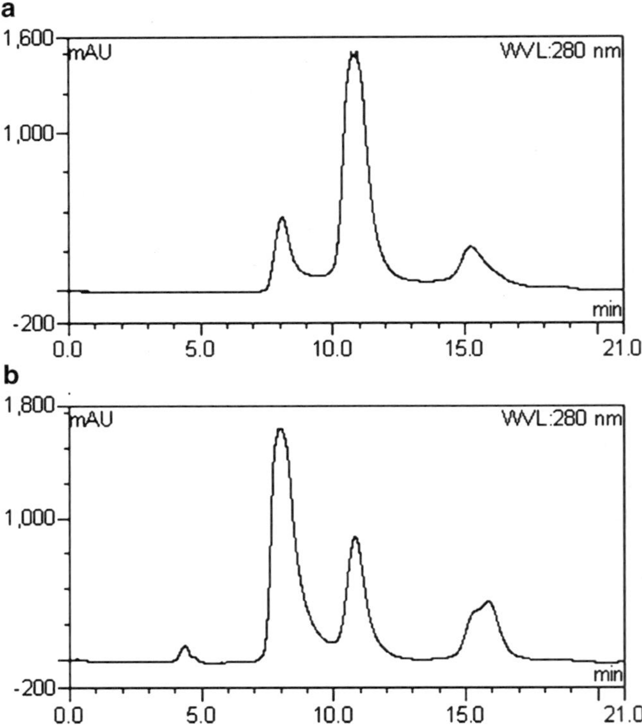

Fig. 1. Gel filtration of purified bacterial cell wall protein *E.coli* MurA (verified by SDS-PAGE) immediately after immobilized metal affinity chromatography (**a**), and gel filtration performed after 24 h (**b**). The peak at approximately 11 min. represents nonaggregated soluble MurA. The 24-h trace shows significant aggregation of the protein. The soluble fractions from multiple runs were pooled and ultimately crystallized. By 48 h, the remaining pre-gel filtration protein contained only soluble aggregate. This example illustrates the utility of gel filtration and the importance of working quickly when purifying proteins destined for crystal trials.

often carry a significant amount of soluble aggregate from early purification stages. Soluble aggregate may accumulate continuously during purification, highlighting the importance of working quickly with a protein to optimize chances for successful crystallization (see Fig. 1).

A typical separation using Superdex 75 10/300 begins with the mobile phase of choice (typically the ideal final buffer), and a "nearly" purified protein at a rough concentration of ~3–5 mg/mL. An injection volume of 1 mL at this concentration is an appropriate starting point (~5% of the column void volume). The run proceeds at 1 mL/min and fractions should be collected every 30–60 s. Fractions containing protein, as judged by UV detection or by a crude Bradford assay should be analyzed by SDS-PAGE and pure fractions should be pooled. Fine-tuning of load volumes and/or protein concentration can be used to optimize resolution. Often, multiple runs will be needed to obtain sufficient quantities of protein for crystallization trials.

1.7. Preparing for Crystal studies

1.7.1. Concentration

Proteins will generally need to be concentrated prior to crystallization trials. This is most commonly achieved by using centrifugal filter units. These single use spin columns can be purchased from a variety of vendors and cover a range of molecular weight cut-offs and sample volumes from 50 μL to 20 mL. It is often strategic to co-concentrate with ligands and/or substrates which will be used in subsequent crystallization trials. The ideal protein concentration needed to grow crystals must be empirically determined for each target protein, but a concentration of 10–20 mg/mL is a reasonable starting point. Detailed information regarding ligand/inhibitor complexing strategies and protein concentration in conjunction with sample preparation for crystallization will be discussed in chapter 4.

1.7.2. Freezing

As a protein chemist working in an SBDD program, considerable time will be spent enabling crystallization targets. Once a robust crystallization condition is discovered, it is highly convenient to simply retrieve an aliquot of crystallization ready protein from the freezer, complex it with new ligands, and crystallize it using predetermined crystallization conditions. Proteins should be snap frozen in small volumes (50–1,000 μL) in a dry ice/ethanol slurry or in liquid nitrogen and stored at –80°C.

Often, it is most practical to freeze 1 mL aliquots of protein at the stage prior to gel filtration. The freezing of protein is often associated with the formation of aggregate or precipitation upon thawing. Running an aliquot over a SEC column will allow one to compare the trace with the pre-frozen chromatogram while polishing your sample for crystallization.

Strategically, freezing a portion of the target protein as soon as it is purified should give the invesitagtor the peace of mind that proteolysis, bacterial carbon scavenging, and other entropic forces will not destroy what could be the “workhorse” protein in your SBDD endeavors.

1.8. The Purification and Crystallization of Human Hsp90

Human Hsp90 is a validated cancer target which functions as a chaperone to help fold, stabilize, and regulate many protein kinases known to be associated with cancers (11). Hsp90 inhibition induces degradation of its client proteins, and has garnered considerable interest as a therapeutic target for anticancer drug discovery (12). Hsp90 contains three domains: an N-terminal ATP-binding domain, a client protein-binding domain, and a carboxy-terminal dimerization domain (13). The ATP-binding domain has been structurally characterized and in recent years has been the focus of intensive crystallography-based fragment screening and SBDD campaigns by numerous research groups (14).

The remainder of this chapter provides a detailed protocol covering the expression, purification, and crystallization of the untagged, Human Hsp90 (residues 1–232) ATP-binding domain.

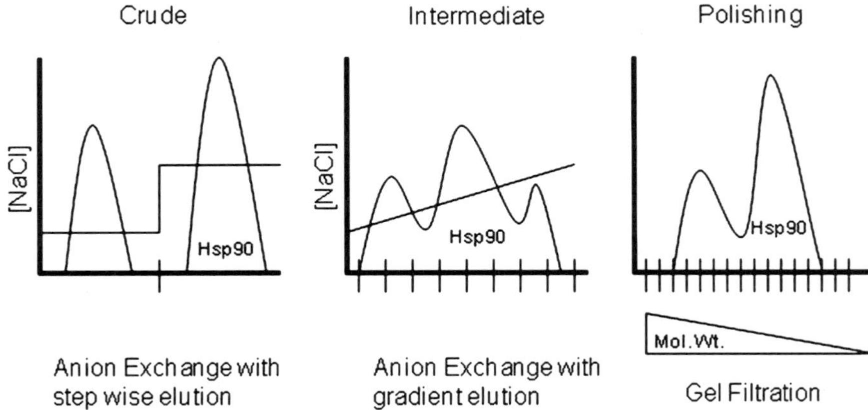

Fig. 2. Hsp90 purification scheme. The tight binding of Hsp90 to Q-Sepharose, allowed two successive ion exchange steps. Protein bound directly from the lysate remained bound at 100 mM NaCl, and could be eluted by 400 mM NaCl. These conditions were leveraged to achieve greater purity in a second gradient elution. Finally, multiple (small volume) runs of Superdex 75 were used to purify the protein to crystallization grade purity.

Initially, $(His)_6$-tagged versions, on both the N- and C-termini were attempted. Although they proved to be fairly robust proteins to work with, reliable, reproducible crystallization of Hsp90 with diverse small molecule ligands has often proved difficult.

The purification method for Hsp90, like most, begins with an initial crude purification step, followed by a more analytical intermediate step, and finally, a high-resolution polishing step (see Fig. 2). Because the estimated pI of the polypeptide is 5.17, binding to Q-Sepharose, a strong anion exchanger is a reasonable first step. SDS-PAGE of the fractions from the crude Q-Sepharose separation indicated that the eluate from IEX Buffer C (400 mM NaCl) contained the bulk of the captured Hsp90. Proteins eluted by IEX Buffer B (100 mM NaCl) contained only cellular contaminants. Because of the tight binding of Hsp90 to Q-Sepharose, a rebinding of the protein to the same resin followed by a shallow elution gradient of 100–400 mM NaCl further improved the purity of the target protein. Finally, gel filtration by Superdex 75 was utilized to purify the protein to homogeneity (>95% as assessed by SDS-PAGE).

2. Materials

2.1. Cell Culture and Lysis

1. Human Hsp90 (residues 1–232) pET3a/BL21(DE3) expression strain.
2. Terrific Broth (TB) (Cellgro) supplemented with 100 μg/mL Carbenicillin (Sigma).

3. 2.5-L Ultra Yield Sterile Flasks and AirOtop seals (Thomson Instrument Company).
4. Isopropyl β-D-1-thiogalactopyranoside (IPTG), 1 M solution can be made up and frozen in 1 mL stocks at –20°C.
5. Hsp90 Cell Lysis Buffer: 50 mM Tris–HCl, pH 8.0, 50 mM NaCl.
6. Complete, EDTA-free; Protease Inhibitor Cocktail Tablets, (Roche).
7. Fast PES Filter Unit, 75 mm diameter membrane, 500 mL 0.2 μm (Nalgene).

2.2. Ion Exchange Chromatography

1. Hsp90 IEX Buffers, Buffer A: 25 mM Tris–HCl pH 8.0, Buffer B: 25 mM Tris–HCl pH 8.0, 100 mM NaCl, Buffer C: 25 mM Tris–HCl pH 8.0, 400 mM NaCl.
2. Q-Sepharose High Performance bulk resin (GE Healthcare).
3. Econo-Pac 14-cm gravity flow columns (Bio-Rad).
4. HiTrap Q-Sepharose Fast Flow 5-mL columns (GE Healthcare).

2.3. SDS-Polyacrylamide Gel Electrophoresis (SDS-PAGE)

1. 1× Tris-Glycine Gel running buffer: 25 mM Tris, 192 mM Glycine, and 0.1% (w/v) SDS pH 8.3. Can be ordered as a 10× Stock from Bio-Rad.
2. Novex® 4–20% Tris-Glycine Precast Gel 1.0 mm, 15 well (Invitrogen).
3. 2× Laemmli Buffer (15) (SDS-PAGE Loading Buffer): 4% SDS, 20% glycerol, 10% 2-mercaptoethanol, 0.004% bromphenol blue 125 mM Tris–HCl pH 6.8.
4. SeeBlue® Plus2 Pre-Stained Molecular Weight Standard (Invitrogen).

2.4. Bradford Assay

1. Coomassie Plus Bradford Assay Reagent (Pierce).
2. Bovine Serum Albumin Standard Ampules, 2 mg/mL (Pierce).

2.5. Size-Exclusion Chromatography (Gel Filtration)

1. Superdex 75 10/300 GL prepacked SEC column (GE Healthcare).
2. Hsp90 gel filtration running buffer: 20 mM Tris pH 8.0, 50 mM NaCl.

2.6. Protein Concentration and Crystallization

1. Amicon Ultracel 10,000 MWCO Centrifugal Filter (Milipore).
2. Crystallization screens: PEGs, PEGs II, pHClear, and pHClear II (Qiagen).
3. CrystalQuick 96-Well Sitting Drop Plate (Greiner).
4. ClearSeal Film (Hampton Research).

3. Methods

3.1. Protein Expression in E. coli

1. Inoculate Hsp90/BL21(DE3) starter cultures from glycerol stock by streaking LB plates containing 100 μg/mL carbenicillin (or ampicillin). Incubate these plates overnight at 37°C.
2. The following day, inoculate two 1 L cultures of TB containing Carbenicillin 100 μg/mL by washing the cells from the starter culture plate (using some of the culture media) into a 2.5-L Ultra Yield Sterile flask (see Note 1). One medium density starter culture should be used for each 1 L flask.
3. Incubate large-scale cultures at 37°C with shaking at 150 rpm. Monitor cell densities in 1 mL cuvettes at 600 nm. Typical starting OD should be ~0.05.
4. Induce expression when OD 600 reaches 0.4–0.8 by adding IPTG to a final concentration of 0.4 mM.
5. Shake for 3 h at 37°C, and harvest cells by centrifugation at 3,500 rpm in 1-L bottles in a swinging bucket rotor.
6. Remove supernatant, freeze pellet at –80°C.

3.2. Cell Lysis and Lysate Processing

1. Thaw both pellets on ice. Add 50 mL Hsp90 Lysis Buffer and one protease inhibitor tablet to each. Use a burette to mix and ultimately reconstitute the pellet.
2. Sonicate the resuspended pellets in a glass beaker for a total of 1 min using 20 s pulses. Allow the lysate to cool/rest on ice between pulses.
3. Transfer lysate to Oak Ridge centrifuge tubes (Nalgene) and centrifuge at maximum speed for 30 min at 4°C (our laboratory uses a Beckman Model J2-21M, rotor JA-30.50).
4. Collect supernatant in 50 mL conicals as soon as the centrifuge stops. The pellet may loosen if left sitting.
5. Filter soluble protein fraction (supernatant) through 0.2 μm filter and return filtrate to ice.

3.3. Gravity Flow Ion Exchange Chromatography

1. Pour Q-Sepharose slurry into two Econo-Pac gravity flow columns so that the bed volume of the resin settles at 7 mL. The lysate will be split over two columns.
2. Chill IEX Buffers on ice.
3. Equilibrate Q resin with 5 column volumes (CV) IEX Buffer C, followed by 7 CV IEX Buffer A.
4. Apply filtered lysate to columns. Chase the lysate with 10 mL of IEX Buffer A. Capture the flow through and keep on ice (see Note 2).
5. Perform 5 CV washes with Buffers B and C. Capture eluates separately. Keep all fractions on ice.

6. All fractions should be analyzed by SDS-PAGE.
7. The eluate from Buffer C should contain the partially purified Hsp90.

3.4. Visualizing Protein Purity by SDS-PAGE Analysis

1. Prepare and load one Novex gel into a XCell SureLock™ Mini-Cell gel apparatus.
2. Prepare 1 L of 1× Tris-Glycine Gel running buffer. Add buffer to the center and forward chambers of the gel box.
3. Prepare protein samples by mixing 5 μL of protein + 5 μL of Laemmli Buffer in a PCR tube. Boil samples in a water bath for 5 min and allow samples to cool to room temperature.
4. Load 5 μL of each sample, include a lane of 5 μL SeeBlue Plus2 gel standard.
5. Set power to 200 V and run gel for approximately 1 h.
6. Stop once dye front reaches the bottom of the gel. The gel can be removed and stained in a solution of 0.1% Coomassie blue, 10% acetic acid, 40% ethanol for 30 min (or longer if needed).
7. Destain in a solution of 10% acetic acid, 40% ethanol (see Note 3).

3.5. Intermediate Purification Step: Ion Exchange Chromatography

1. Dilute the fraction containing Hsp90 (judged by SDS-PAGE) from 400 mM NaCl to 100 mM NaCl using IEX Buffer A (see Note 4).
2. Connect two HiTrap Q Sepharose Fast Flow 5-mL columns in-line and mount on a ÄKTAprime (GE Healthcare) protein purification system.
3. Wash columns with 5 CV of IEX Buffer C and then equilibrate with 10 CV of IEX Buffer B.
4. Load the diluted protein sample onto the column and then equilibrate with 2 CV IEX Buffer B.
5. Finally, elute the protein with a gradient from 100 to 400 mM NaCl over 10 CV. Collect 3 mL fractions and analyze by SDS-PAGE.
6. Pool fractions containing Hsp90 and concentrate to 5 mg/mL (see section 3.8.1).

3.6. Size-Exclusion Chromatography (Gel Filtration)

1. Perform gel filtration with a 20 min, 1 mL/min method using a Superdex 75 10/300 GL column. Collect fractions every 30 s (0.5 mL). Running buffer: 20 mM Tris pH 8.0, 50 mM NaCl.
2. Prepare samples by centrifuging 1 mL aliquots of protein (5 mg/mL) at maximum speed in a refrigerated microfuge for 10 min prior to injection onto the column (see Note 5).
3. Fractions should be analyzed by SDS-PAGE. Fractions containing pure Hsp90 should be combined (see Fig. 3).

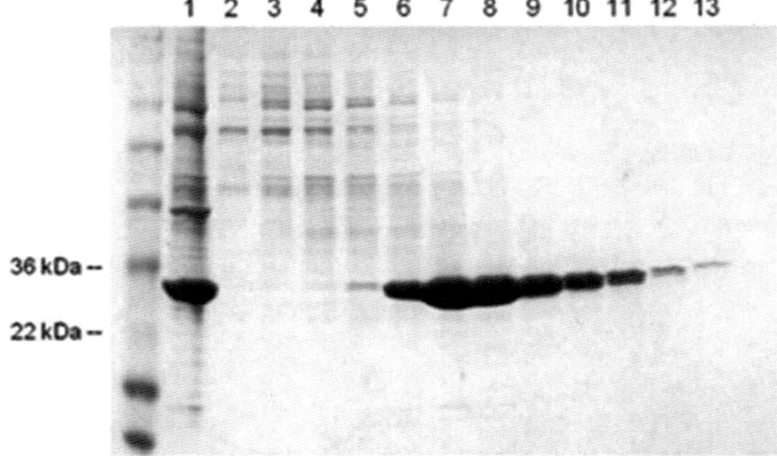

Fig. 3. SDS-PAGE analysis of fractions from Superdex 75 of the 29 kDa ATP-binding domain of Hsp90. Lane 1, Hsp90, pre-gel filtration; lanes 2-13, half-minute fractions (500 μL). Fractions 8–11 were pooled for concentration and crystallization. Fractions 2–7 contain high molecular weight contaminants and were discarded. The absence of Hsp90 in the early fractions from the gel filtration step demonstrates that Hsp90 is highly monodisperse and free from soluble aggregates. Gel filtration also acts as a final buffer exchange step for the protein before entering crystallization trials.

3.7. Determining Protein Concentration Using the Bradford Method

1. Protein standards should be prepared in the same buffer as the samples to be assayed. A convenient standard curve can be made using BSA with concentrations of 0, 0.25, 0.5, 1, 1.5, and 2.0 mg/mL.
2. For both standards and experimental analysis, use 1,000 μL of Bradford reagent and 5 μL of protein solution mixed in a 1.5-mL plastic cuvette.
3. Incubate samples at room temperature for 5 min and read at 595 nm. Bradford assay with BSA standards is typically linear to 1.0 mg/mL.
4. A standard curve can be created by plotting the 595 nm absorbance values (*y*-axis) and the concentration in mg/mL (*x*-axis). Unknown sample concentration can be estimated by using the standard curve.

3.8. Post-Purification Processing

3.8.1. Protein Concentration

1. Proteins may be concentrated by applying the solution to an Amicon Ultracel and centrifuging in a swinging bucket rotor at 4,000 × *g* @ 4°C.
2. Centrifuge for 10–15 min and check reservoir levels. Gently rinse the filter by pipetting the remaining protein solution to keep the protein from becoming locally concentrated near the membrane (see Note 6).
3. Continue concentrating while monitoring protein concentration by Bradford assay.

3.8.2. Ligand Addition and Crystallization

1. Add a two to threefold molar excess of compound (as a 100 mM DMSO stock) directly to pooled fractions from gel filtration. Protein concentration should be approximately 100–200 μM, or ~2.5–5.0 mg/mL (see Note 7).
2. Concentrate Hsp90 to 20–30 mg/mL in a centrifugal concentrator.
3. Prior to crystallization setup, transfer the sample to a microfuge tube and give the protein a "final spin" at maximum speed in a refrigerated microcentrifuge for 10 min. Transfer supernatant to a clean tube, taking care not to transfer any pellet (if any).
4. Set up one plate for each of the following screens in 96-well deep-well block format: PEGs, PEGs II, pHClear, and pHClear II.
5. Set up crystal trials by adding 50 μL of each crystallization solution from the deep-well block to the reservoirs of a CyrstalQuick plate using a multichannel pipette.
6. Working quickly, use a small volume multichannel pipette to lay 0.5 μL of the protein to the stage of the plate.
7. Using the same set of pipette tips, draw 0.5 μL of the corresponding reservoir solution and place it on the protein drop.
8. Quickly cover and seal the plate with a ClearSeal film and incubate at 4°C. Crystals take 2–4 days to appear.

There are two different crystal forms that grow depending on the size of the ligand. The closed form will grow diffraction quality crystals in all four screens. While the open form will typically only appear in the pH Clear screens. The open form always indicates successful ligand binding to the ATP-binding pocket. Closed form unit cell: I222 orthorhombic, $a=67$, $b=91$, $c=100$. Open form unit cell: $P2_1$ monoclinic, $a=53$, $b=45$, $c=54$, $\beta=115°$.

4. Notes

1. Expression cultures are typically inoculated with small volumes of overnight liquid culture. It is our experience that starting from plate-grown overnight cells promotes plasmid stability and better reproducibility.
2. For gravity flow applications, two (or more) lower volume columns will keep the flow rate from being excessively slow. It is important to resist forcing the lysate through the column media at this stage.
3. Once the bulk of the background Coomassie has been removed and the protein bands are visible, the destain solution may be swapped for water. Gels may be stored in water for long periods without overly destaining or damaging them.

4. Diluting proteins, especially when the ionic strength and protein concentration are both adjusted, should be attempted on a small scale first. Incremental adjustments are recommended to keep the shock of the new conditions from precipitating the protein.
5. Centrifuging the sample before gel filtration ensures that no precipitated protein or other insoluble material enters the column. To save time, the entire sample may be filtered using a 0.2-μm filter disk and an appropriate sized syringe. Because a small amount of the solution will be lost to the surface of the filter, this is not recommended when working with a small volume of sample.
6. Centrifugal concentrators are excellent for buffer exchanges when an overnight dialysis or desalting column is not practical (due to protein stability issues). New buffer can be layered onto the protein and iteratively swapped out with short cycles of centrifugation. It should be noted that occasionally swapping out used concentrators for new ones will accelerate the concentration process.
7. Co-concentration of protein and ligand is a useful technique when the target compound is relatively insoluble.

References

1. Peng, H. and Marians, K. J. (1993) Escherichia coli topoisomerase IV. Purification, characterization, subunit structure and subunit interactions. *J. Biol. Chem.* **268**, 24481–24490.
2. Bax BD, Chan PF, Eggleston DS, Fosberry A, Gentry DR, Gorrec F, Giordano I, Hann MM, Hennessy A, Hibbs M, Huang J, Jones E, Jones J, Brown KK, Lewis CJ, May EW, Saunders MR, Singh O, Spitzfaden CE, Shen C, Shillings A, Theobald AJ, Wohlkonig A, Pearson ND, Gwynn MN. (2010) Type IIA topoisomerase inhibition by a new class of antibacterial agents. *Nature.* **466**, 935–940.
3. Tsai FT, Singh OM, Skarzynski T, Wonacott AJ, Weston S, Tucker A, Pauptit RA, Breeze AL, Poyser JP, O'Brien R, Ladbury JE, Wigley DB. (1997) The high-resolution crystal structure of a 24-kDa gyrase B fragment from E. coli complexed with one of the most potent coumarin inhibitors, clorobiocin. *Proteins.* **28**, 41–52.
4. Benita Y, Wise MJ, Lok MC, Humphery-Smith I, Oosting RS.(2006) Analysis of high throughput protein expression in Escherichia coli. *Mol Cell Proteomics.* **9**, 1567–80.
5. Studier FW, Moffatt BA. (1986) Use of bacteriophage T7 RNA polymerase to direct selective high-level expression of cloned genes. *J Mol Biol.* **189**, 113–130.
6. Rosenberg AH, Lade BN, Chui DS, Lin SW, Dunn JJ, Studier FW. (1987) Vectors for selective expression of cloned DNAs by T7 RNA polymerase. *Gene.* **56**, 125–135.
7. Studier FW, Rosenberg AH, Dunn JJ, Dubendorff JW. (1990) Use of T7 RNA polymerase to direct expression of cloned genes. *Methods Enzymol.* **185**, 60–89.
8. Gräslund S, Nordlund P, Weigelt J, Hallberg BM, Bray J, Gileadi O, Knapp S, Oppermann U, Arrowsmith C, Hui R, Ming J, dhe-Paganon S, Park HW, Savchenko A, Yee A, Edwards A, Vincentelli R, Cambillau C, Kim R, Kim SH, Rao Z, Shi Y, Terwilliger TC, Kim CY, Hung LW, Waldo GS, Peleg Y, Albeck S, Unger T, Dym O, Prilusky J, Sussman JL, Stevens RC, Lesley SA, Wilson IA, Joachimiak A, Collart F, Dementieva I, Donnelly MI, Eschenfeldt WH, Kim Y, Stols L, Wu R, Zhou M, Burley SK, Emtage JS, Sauder JM, Thompson D, Bain K, Luz J, Gheyi T, Zhang F, Atwell S, Almo SC, Bonanno JB, Fiser A, Swaminathan S, Studier FW, Chance MR, Sali A, Acton TB, Xiao R, Zhao L, Ma LC, Hunt JF, Tong L, Cunningham K, Inouye M, Anderson S, Janjua H, Shastry R,

Ho CK, Wang D, Wang H, Jiang M, Montelione GT, Stuart DI, Owens RJ, Daenke S, Schütz A, Heinemann U, Yokoyama S, Büssow K, Gunsalus KC. (2008) Protein production and purification. *Nature Methods.* **5**, 135–145.

9. Hempelmann E, Schµlze M, Götze O. (1984) Free SH-groups are important for the polychromatic staining of proteins with silver nitrate. *Neuhof V (ed) Electrophoresis '84, Verlag Chemie Weinheim.* 328–330.
10. Bradford, M. (1976) A rapid and sensitive method for the quantitation of microgram quantities of protein utilizing the principle of protein-dye binding. *Anal. Biochem.* **72**, 248–254.
11. Maloney A, Workman P. (2002) HSP90 as a new therapeutic target for cancer therapy: the story unfolds. *Expert Opin Biol Ther.* **2**, 3–24.
12. Neckers L, Neckers K. (2005) Heat-shock protein 90 inhibitors as novel cancer chemotherapeutics—an update. Expert *Opin Emerg Drugs.* **10**, 137–149.
13. Prodromou C, Pearl LH. (2003) Structure and functional relationships of Hsp90. *Curr Cancer Drug Targets.* **3**, 301–323.
14. Murray CW, Carr MG, Callaghan O, Chessari G, Congreve M, Cowan S, Coyle JE, Downham R, Figueroa E, Frederickson M, Graham B, McMenamin R, O'Brien MA, Patel S, Phillips TR, Williams G, Woodhead AJ, Woolford AJ. (2010) Fragment-based drug discovery applied to Hsp90. Discovery of two lead series with high ligand efficiency. *J Med Chem.* **16**, 5942–5955.
15. Laemmli, U.K. (1970) Cleavage of structural proteins during the assembly of the head of bacteriophage T4. *Nature.* **227**, 680–685.

Chapter 4

Protein Crystallization for Structure-Based Drug Design

Isaac D. Hoffman

Abstract

The crystallization experiment has one main objective: to obtain diffraction quality crystals. This can be achieved through myriad avenues; here the focus will be on crystallization in support of drug discovery. In drug discovery there are two main paradigms for crystallography: high-throughput, and by any means necessary. Each paradigm requires the investigator to formulate strategies based on different priorities. In the high-throughput environment, the emphasis is on rapid prosecution of a large number of protein targets. In the by any means necessary paradigm the target pool is generally smaller and structural information is absolutely necessary for success. The process of growing diffraction quality protein crystals involves deciding on a crystallization method, initial screening, cryoprotection, initial diffraction analysis, and growth optimization. Furthermore, in structure-based drug design it is necessary to obtain crystal structures of protein–ligand complexes.

Key words: Protein crystallization, X-ray crystallography, Industrial protein crystallization

1. Introduction

When undertaking a structure-based drug discovery campaign, one is attempting to leverage the information gleaned from crystal structures with assay data from multiple sources. The trends that develop in comparing the crystal structures of proteins with various ligand bound are known as structure-activity relationships, or SAR. The generation of SAR necessarily relies on the accessibility of protein–ligand complex structures, as well as assay data for the ligands in question. The structures of proteins with ligands bound can be acquired via two basic routes: co-crystallization, where the protein is crystallized in the presence of the ligand; or soaking, where the ligand is added to the crystal after it has been grown. As with all aspects of crystallization, there are many methods for both co-crystallization and the soaking of ligands into crystals.

Leslie W. Tari (ed.), *Structure-Based Drug Discovery*, Methods in Molecular Biology, vol. 841,
DOI 10.1007/978-1-61779-520-6_4, © Springer Science+Business Media, LLC 2012

In order to deconvolute a system with as many variables as macromolecular crystallization, it is important to first take stock of the variables which can be controlled. Once the scope of influence and control is established, the variables must be ranked in order of importance, with respect to the laboratory facilities available. In high-throughput crystallography it is of the utmost importance to have a system that produces abundant protein and easy to grow, diffraction quality co-crystals with small molecule ligands, unless soaking of pre-grown protein crystals with prospective drug candidates is possible. In a *by any means necessary* situation it is important to select the technique, or techniques, best suited for surmounting the obstacles at hand. Every crystallization campaign presents a unique set of challenges, and one must be ready to adapt to the circumstances that present themselves. Above all else, one must never lose sight of the overriding goal of obtaining high quality structural data that will aid in the advancement of the drug discovery process. In drug discovery, crystallization is a means to an end, and that end is the clinic.

1.1. Background

The formation of a macroscopic protein crystal, suitable for X-ray diffraction experiments, requires an assembly of approximately 10^{15} protein molecules to emerge from solution as an ordered, three-dimensional lattice. The process is initiated by the spontaneous association of small numbers of protein molecules into aggregates, where the aggregating molecules dispose themselves spatially in similar relative orientations to those observed in the mature crystal. Once these aggregates reach a critical size, they become stable nuclei, which can support crystal growth via the addition of new protein molecules to the surface of the nuclei. Crystal nucleation and crystal growth only occur from solutions where the protein exceeds its equilibrium solubility value, i.e., where the solution is saturated or supersaturated. The suitable solution parameters for supporting crystal growth can be represented on a simplified two-dimensional phase diagram, as shown in Fig. 1.

Many factors influence macromolecular solubility, such as pH, temperature, salts, detergents, organic additives, and polymers. The most commonly utilized protein precipitating agents are water miscible, hydrophilic polymers like polyethylene glycols, organic alcohols, and salt solutions. Salts are interesting, in that they typically decrease the solubility of proteins at very low- and high concentrations, while increasing a protein's solubility at intermediate concentrations. In addition to influencing macromolecular solubility, the additives to a protein solution can directly influence whether or not a protein crystallizes or precipitates from solution. For example, multivalent cations, in addition to altering protein solubility, often induce the formation of ordered lattices by bridging acidic amino acid side- chains presented on the surfaces of different protein molecules. Throughout the remainder of this section, the

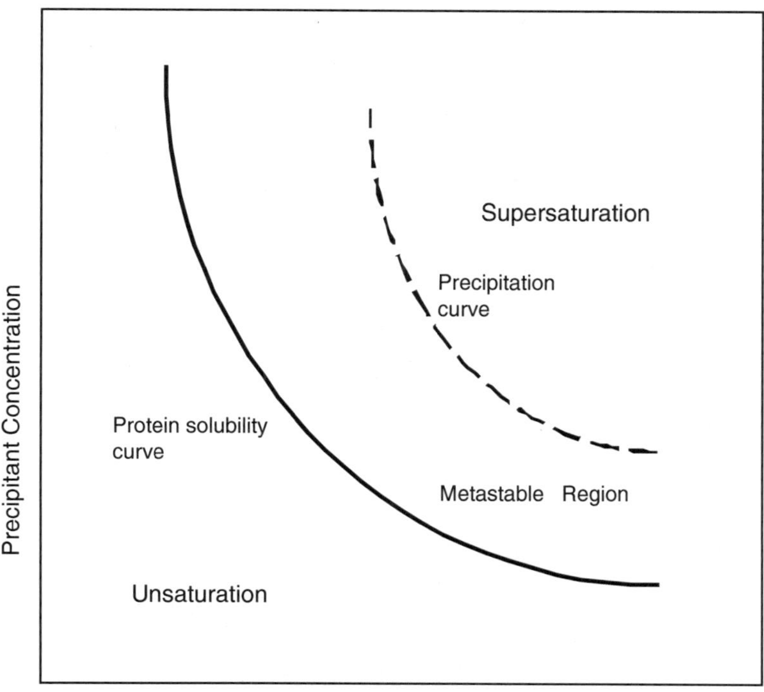

Fig. 1. Representative phase diagram for protein crystallization. As protein concentration increases, the concentration of precipitating agent required to saturate the solution deceases, and vice versa. The plot is divided into three main regions. Below the protein solubility curve is the unsaturated region, where the protein remains in solution and crystals dissolve. Just beyond the protein solubility curve is the metastable region, where the solution is saturated enough to support crystal growth, but not nucleation. As protein and/or precipitant concentration increases, the solution eventually becomes supersaturated (beyond the precipitation curve), and nucleation competes with crystal growth. Successful crystallization generally occurs somewhere in the metastable region.

vapor diffusion method for protein crystallization will be outlined, and phase diagrams will be used to indicate how to manipulate solution conditions to obtain protein crystals.

Crystallization by vapor diffusion is the prevailing method used for the growth of protein crystals for diffraction experiments. The principle is simple; a protein solution is used to dilute a solution that contains a nonvolatile precipitant, such as a salt, nonvolatile alcohol, or hydrophilic polymer. The volume of the crystallization drop is generally on the order of ~100–200 nL when set up by a robot or ~1–2 μL when set up by hand. The ratio of protein to mother liquor in the drop is usually 1:1, however, it is very common to vary this ratio during optimization, and sometimes during screening, in order to alter the path taken through the phase diagram. The protein-containing mixture is then placed in a sealed

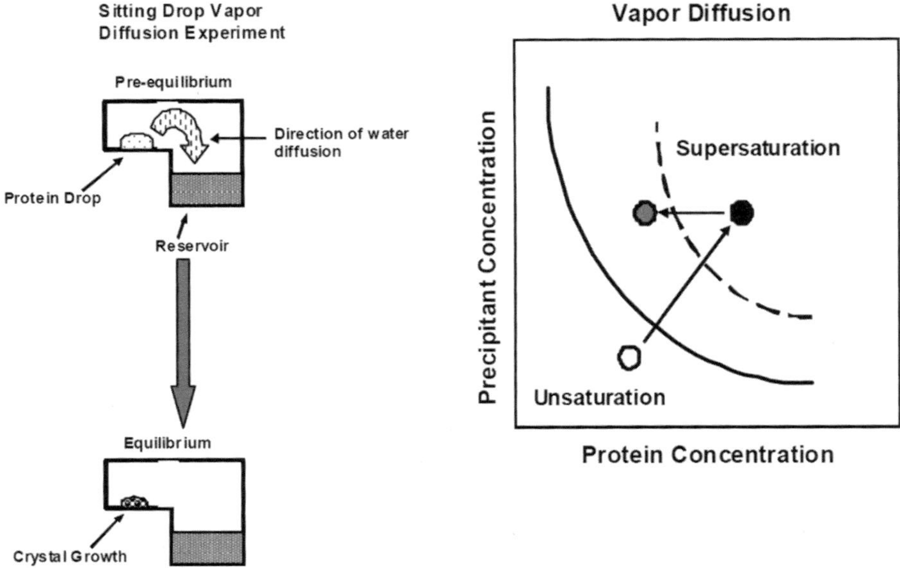

Fig. 2. Crystallization by vapor diffusion. A droplet comprising a mixture of a protein solution with a precipitant containing reservoir solution is dispensed on to a platform over a reservoir solution (sitting drop vapor diffusion, as shown above) or suspended over the reservoir (hanging drop vapor diffusion, not shown). The reservoir solution has a higher concentration of a nonvolatile precipitating agent than the protein drop. At the outset, the protein drop is unsaturated (*open circle*). Over time, water vapor diffuses through the air gap, out of the drop and into the reservoir, slowly concentrating the protein and precipitant in the protein drop, until the solution becomes supersaturated (*black circle*). The protein molecules then start to aggregate out of solution, forming stable crystal nuclei in the process. As the protein leaves the solution phase, the protein concentration decreases into the metastable region, where crystal growth is sustained, but further nucleation does not occur (*gray circle*).

chamber on an isolated platform, where it equilibrates against a large reservoir of the undiluted precipitant containing solution (sometimes referred to as the mother liquor, crystallization cocktail, or crystallization solution). To establish thermodynamic equilibrium, water vapor diffuses through the airspace between the protein-containing drop and the reservoir, concentrating the protein-containing solution to saturation through dehydration. Except in special cases (for example, when crystal seeding is employed), the experiment is left undisturbed until equilibrium is established. A phase diagram charting the path of the protein solution through phase space and representative experimental schematics for vapor diffusion experiments are shown in Fig. 2.

Figure 2 is consistent with an obvious deduction; rapid ascension of a protein solution far into the supersaturated region of the curve results in precipitation or excess nucleation. It seems logical that slowing this ascension should increase the chances of obtaining crystals and/or obtaining fewer, larger protein crystals by keeping the protein solution near the metastable boundary for a longer time period. Methods designed to achieve this end have been demonstrably effectual in reducing precipitation/nucleation density and increasing crystal size in many cases (1). The most facile methods deployed to slow down the kinetics of equilibration are layering mineral oil over the protein drop or reservoir to impede vapor diffusion, increasing the distance between the protein drop and reservoir and/or increasing drop volume to reduce the surface area to drop volume ratio of the protein drop. However, the retardation of equilibration kinetics cannot be regarded as a panacea for at least two important reasons.

Proteins are by nature unstable molecules subject to numerous chemical processes that can lead to their demise in solution over a short time period. Alterations to a crystallization experiment (which can take days to weeks to reach equilibrium) that further extend its lifetime can guarantee failure, particularly with very sensitive protein samples. Recent experiments (2), theoretical considerations and a wealth of empirical evidence reveal that in side-by-side comparisons, higher quality crystals (as determined by superior X-ray diffraction parameters) can be grown from very small volumes (useful crystals can be grown from nL (10^{-9} L) volumes), when compared to crystals of the same proteins derived from conventional experiments, which utilize μL (10^{-6} L) volumes. This result is somewhat surprising in the context of the arguments considered above, since nL volume vapor diffusion experiments can reach equilibrium within hours of experimental setup. In addition to using much less sample, small volume experiments have the benefit of shortening the experimental timescale, allowing proteins to crystallize before deleterious natural processes damage the sample. However, the inverse correlation observed between the volume of a crystallization experiment and crystal quality may be attributable to less obvious factors.

In an ideal crystallization experiment, crystals grow via the diffusion of new protein molecules to the crystal surface. Single protein molecules kinetically diffuse through solution more rapidly than aggregated molecules, so there is a higher probability individual molecules deposit on the surface of a growing crystal first. When crystals emerge from solution, the solution surrounding the protein crystals is temporally depleted of protein molecules, creating a solution density gradient that generates convection currents in the vicinity of the crystal. These convection currents can arrest crystal growth by facilitating the transport of disordered molecular aggregates to the crystal surface, poisoning the crystal lattice. Fluid dynamics calculations on small volumes demonstrate that convection currents are virtually nonexistent in nanoliter drops, so crystallization is diffusion controlled, and equivalent to crystal growth in a microgravity environment (crystals grown in space have also been shown to be of higher quality than their terrestrial counterparts, but at much greater expense than just decreasing the experimental volume!) (2).

To summarize, vapor diffusion is a simple, powerful technique for growing protein crystals for diffraction studies. However, it is impossible to predict in advance whether a given protein crystallization experiment will benefit from a more gradual equilibration or rapid equilibration and the ancillary benefits afforded by small volumes; the experimenter must be willing to test both possibilities.

Although there are several methods of macromolecular crystallization, the focus of this discussion will be on vapor diffusion, as it is the prevailing method in industrial structure-based drug discovery operations. Additionally, the methodologies set forth should be applicable to crystallization endeavors in any format.

2. Materials

The first step in approaching any protein crystallization campaign is the choice of tools; that is, to decide on an experimental format for crystallization, and then of course to procure the required materials. The crystallizer will need crystallization plates, crystallization cocktails (sparse matrix screens) and chemicals (for reproducing and optimizing crystal hits), pipettes/liquid handling robotics, plate-sealing films, a microscope, incubators, crystal freezing materials (including liquid nitrogen and the appropriate dewars and tools), and potentially a shipping dewar certified for liquid nitrogen with a hard shipping case.

In most situations the choice for industrial crystallization is sitting drop vapor diffusion, and due to the popularity of this method there are several choices of hardware for plates, sealing films, and automation/robotics. The most popular screening plates are the Greiner sitting drop plates (CrystalQuick™), but there are many other options from manufacturers such as Art Robbins Instruments, Corning, Emerald, Innovadyne, etc. (see Figs. 3 and 4 for some examples). For sealing films there are two main options: (1) crystal clear tape (regular tacky tape), and (2) pressure seals (Crystal Clear Sealing Film and ClearSeal Film™). These, and most other materials referred to here, are available from Hampton Research as well as many individual manufacturer and distributor websites.

Many other methods for crystallizing macromolecules exist, and they each confer unique advantages and constraints. The choice to use a single method is often driven by the focus on automation and the concomitant constraints a given platform exerts on experimental design. There is no "best" method of macromolecular crystallization. However, for some molecules one method is more effective than others at growing single, diffraction quality crystals; unfortunately, the optimal method can never be determined without experimentation. Thus, it is crucial to keep an open mind and never focus exclusively on one technique. It is not

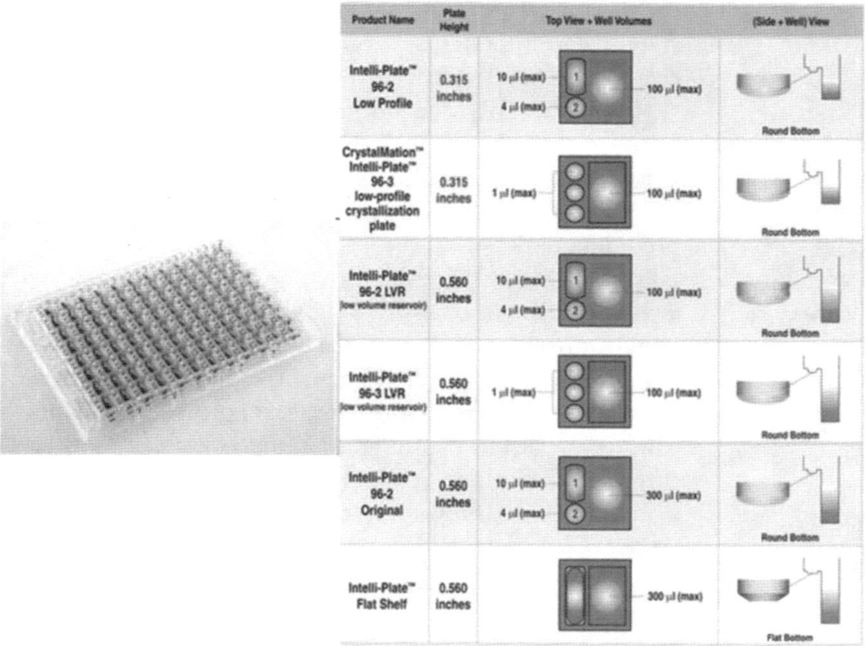

Product Name	Plate Height	Top View + Well Volumes	(Side + Well) View
Intelli-Plate™ 96-2 Low Profile	0.315 inches	10 µl (max), 4 µl (max), 100 µl (max)	Round Bottom
CrystalMation™ Intelli-Plate™ 96-3 low-profile crystallization plate	0.315 inches	1 µl (max), 100 µl (max)	Round Bottom
Intelli-Plate™ 96-2 LVR (low volume reservoir)	0.560 inches	10 µl (max), 4 µl (max), 100 µl (max)	Round Bottom
Intelli-Plate™ 96-3 LVR (low volume reservoir)	0.560 inches	1 µl (max), 100 µl (max)	Round Bottom
Intelli-Plate™ 96-2 Original	0.560 inches	10 µl (max), 4 µl (max), 300 µl (max)	Round Bottom
Intelli-Plate™ Flat Shelf	0.560 inches	300 µl (max)	Flat Bottom

Fig. 3. The various types of Intelli-Plates™ from Art Robbins Instruments, all of which are 96-well SBS-format plates.

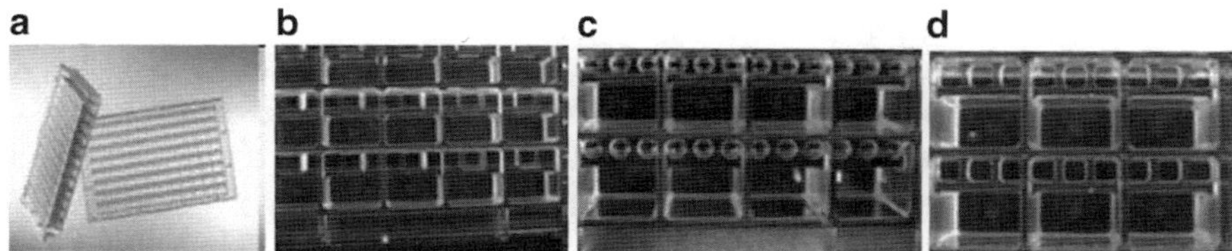

Fig. 4. A few examples of Greiner's CrystalQuick™ crystallization plates. (**a**) Greiner CrystalQuick™ SBS-format standard plate. (**b**) Greiner CrystalQuick™ low profile plate with single, square drop wells. (**c**) Greiner CrystalQuick™ three well, round drop well plate. (**d**) Greiner CrystalQuick™ three well, square drop well plate.

uncommon to find a crystal hit in a sitting drop vapor diffusion screen and, after optimizing the condition via grid screening in both sitting and hanging drops, find that the crystals grow better in the latter. Despite that fact, the most common optimization methods employ screening in sitting drop vapor diffusion plates (because most robots are amenable to this) followed by more exotic methods after exhausting that avenue.

One alternative approach to vapor diffusion crystallization is the free interface diffusion method, employing the Microlytic Crystal Former chips or the Fluidigm Topaz® system. Free interface diffusion explores phase space in a different fashion than vapor diffusion and has been shown to be effective in many situations where vapor diffusion has failed (3, 4).

When implementing laboratory robotics a determination must be made in advance whether to employ a modular system where each process utilizes a stand-alone station, or a fully automated system where plates and protein go in one end and images and crystals (hopefully) exit at the other end.

Many stand-alone liquid handling robots for crystallization setup are available, including the Phoenix (Art Robbins), the Gryphon and the Mosquito (TTP Labtech). In addition to the drop setting machine it is important to have an incubator that can provide a dark, vibration free, fixed temperature area for the crystallization plates to reside in while the crystals grow. This can be anything from a drawer in a temperature regulated room, to a vibration-free incubator with temperature and humidity control and an attached robotic imager. Despite the abundance of stand-alone machines, few options are available offering a fully integrated solution to liquid handling, imaging of experiments and data capture. A few companies like Rigaku and Formulatrix offer production models of fully integrated crystallization workstations, and custom built automation setups are also available. Fully integrated automation platforms are expensive, and generally focus on either speed or flexibility of experimental setup, rarely achieving both in one package.

If budget is an issue it is possible to create a fully functional crystal screening system utilizing hand-operated pipettes. A very effective manual high-throughput setup combines two multichannel pipettes (8 or 12 channels due to the format of 96-well plates) for high and low volumes (one capable of dispensing sub-microliter drops for protein dispensing, and the other capable of 50–100 μL for the reservoirs) with a 96-well sitting drop vapor diffusion plate containing round bottom wells for the crystallization drop (Intelli-Plate™ or CrystalQuick™). The round bottom crystal well simplifies placement of reservoir solution on the protein solution in the finished experiments.

2.1. Plates and Accessories

SBS-format 96-well plates: Intelli-Plate™ (see Fig. 3) from Art Robbins Instruments, CrystalQuick™ (see Fig. 4) from Greiner Bio One GmbH, Clover 384™ from Emerald BioSystems, CrystalEX™ from Corning, and MRC MAXI Plate from Molecular Dimensions Limited.

24-well plates: Cryschem MVD from the Charles Supper Company and the Intelli-Plate™ for sitting drops. For hanging drops there are the EasyXtal™ Tool from Qiagen, the VDX, and the Linbro.

Seals: Crystal Clear Sealing Film and ClearSeal Film™ are both clear polyolefin films with pressure-sensitive silicone-based adhesives sized to seal SBS plates. Also available is Crystal Clear Sealing Tape. The VDX and Linbro plates can be purchased pre-greased and are sealed using siliconized glass coverslides. The EasyXtal™ Tool plates come with screw-caps with silicon gaskets.

2.2. Pipettes and Liquid Handling Robots

Pipettes: Manual and mechanical pipettes can be purchased in single and multichannel formats from companies such as Rainin, Matrix, and Eppendorf, or from suppliers like Fisher and VWR.

Liquid handling robots: Many options are available for liquid handling robotics, even for such a specialized area as macromolecular crystallization. A short list of some suppliers is shown below:

Art Robbins Instruments

Formulatrix

Rigaku

Emerald Biosystems

Tecan

RTS

TTP Labtech

Fluidigm

2.3. Screening Solutions/Crystallization Cocktails

Hampton Research, Jena Bioscience, NeXtal/Qiagen, Emerald Biosystems, and Molecular Dimensions Limited all sell a variety of crystallization screening suites along with the individual component chemical solutions.

2.4. Crystal Mounting Hardware and Microscopes

Hampton Research and Molecular Dimensions Limited are both one stop shops for all crystal mounting tools: mounted loops of various sizes, pin-bases, magnetic wands, cryo-canes, etc.

Microscopes can be purchased from Leica, Olympus, Nikon, etc. The ideal setup is a binocular scope (3D) with decent magnification (40–80×), an articulated head, a large working space on the base and space between the objective lens and the base. Extras that can prove helpful are a polarizing glass window for the base with an analyzer for the lens, and a photo tube for the attachment of a camera or computer. Additionally it is beneficial to use a fiber optic tube and light box because direct lighting tends to heat up the base, whereas an indirect light source does not.

2.5. Cryostorage and Shipping

Hampton Research sells all cryo-storage and transfer equipment except liquid nitrogen. There are however, many alternative sources for liquid nitrogen dewars and shipping cases. A few manufacturers of cryo storage and transfer equipment are Taylor-Wharton, Thermo, and MVE.

3. Methods

The process of growing X-ray diffraction quality crystals for drug discovery is often an iterative process. The initial step is to prepare the protein sample for crystal screening, which often involves the addition of a ligand. Once the sample is ready, it is screened for crystallization. Initial crystal hits should be tested for X-ray diffraction quality, after which it may be necessary to optimize the crystallization conditions, the cryoprotection protocol, or both. Alternative methods for finding and optimizing crystallization conditions may also be necessary if initial screening or optimization efforts prove fruitless. The goal is to find the shortest path to a suitable data set capable of aiding the drug discovery effort.

3.1. Initial Screening

The expanse of chemical space searched during first-pass crystal screening is dependent on the resources and facilities available. Optimally, a high-throughput screening suite is available which sets up unlimited experiments, consumes a modest amount of sample and automatically views experiments on a metered schedule (2). If the proper approach is taken, traditional methods of utilizing hand pipettes, while more labor intensive, can be highly effective. Most importantly, successful crystallization requires careful handling of the protein sample. Ideally, the sample should be kept at 0–4°C throughout purification, and all experiments should be set up in a single day, the same day that the protein purification is completed. Certain proteins, it should be noted, precipitate at cold temperatures and can only be concentrated beyond a certain point if kept at higher temperatures. If possible, the protein should not be frozen, but if it must be frozen it should be snap frozen in liquid nitrogen at low volumes and high concentrations, in thin-walled PCR tubes (5) and stored at −80°C. Emulsification should always be kept to a minimum during all stages of preparation. Just prior to setup the protein should be cleared of all particulate debris by high-speed centrifugation, serially if necessary.

Once a protein formulation has been found that stabilizes the protein of interest and promotes optimal solution monodispersity, the goal is to identify as many crystallization conditions as possible. It is often the case that a protein that crystallizes will do so in multiple conditions, and potentially in different crystal forms. This phenomenon is very important if a given crystal form is intractable due to twinning, poor diffraction characteristics, etc., or if it is not suited for SBDD (e.g., due to unfavorable crystal contacts or an occluded ligand-binding pocket). Before crystallization experiments are initiated, the target protein complexed with the desired ligand(s) must be concentrated to a suitable degree to promote crystallization. Some proteins concentrate in a well-behaved manner without ligand or cofactor, while others require a bound ligand to remain solution stable and to concentrate sufficiently for crystallization (6). Nuclear hormone receptors are a class of proteins that exemplify this phenomenon. Many must be expressed in the presence of a ligand to generate correctly folded protein (6). Ligands with poor solubility should be incubated with dilute solutions of the target protein, and subsequently co-concentrated with the protein.

The central concept behind co-concentration is to find the saturating concentration of the ligand or cofactor of interest and complex it with the protein at an appropriate molar ratio (greater than stoichiometric) before concentrating the protein to the desired level for crystallization (6).

It is often useful to perform a pre-crystallization solubility test to qualitatively assess whether the protein sample of interest is at the proper concentration for crystallization experiments. There are kits available for this purpose (7). Additionally, it is useful to study the behavior of a protein as it is concentrated; the protein concentration should increase in a linear fashion relative to the decrease in sample volume until the protein solution becomes saturated. Many affordable tools are available for assaying protein concentration, however, when a sample contains small molecules with chromophores that absorb 280 nm light (ligands or cofactors as well as some buffer molecules), the method of choice for protein concentration determination is the Bradford method (8).

Once the protein sample has been concentrated, a collection of sparse matrix screens must be chosen for screening. There are two main factors to consider when choosing the initial screens. (1) The diversity of chemical space to sample for crystallization. Factors like pH, different salts, precipitants, organics, and volatiles, are all variables that can be sampled. (2) Redundancy. Unlike most other assays, crystallization has a stochastic component that can be confounding during the screening process. The phenomenon is related to the complexity (and hence probability) of nucleation, and is a function of the propensity of a protein to form viable ordered nucleation complexes that support crystal growth in a given chemical space. In a completely idiosyncratic manner, a protein can display a tendency to nucleate often, just enough, or rarely under certain conditions (9). If a system tends to over-nucleate and generates showers of small crystals, it is usually possible to design conditions to control this. However, if a system rarely nucleates, it is possible (and even likely) to miss fruitful crystallization conditions unless some degree of redundancy is present in the chemical space sampled during the crystallization experiment. To add some degree of redundancy to a crystallization experiment without sacrificing the diversity of chemical space searched, plates with multiple drop positions over each reservoir (Intelli-Plate™ and Greiner multiwell) are advantageous to use. The technique is executed by setting identical drops upon each of the multiple wells associated with each reservoir on one plate, so that greater redundancy can be achieved without the need to set up duplicate plates. This technique can also be adapted to vary the ratio of protein to mother liquor in the wells associated with each reservoir in order to sample different trajectories through phase space without having to set up multiple plates. Additionally, the outcomes of the varied experiments can easily be viewed side by side.

Typically, a minimum of four sparse matrix screens are implemented in the initial screening round, however there is no upper limit and the more screens that are employed the greater the chance for crystallization. Five main companies provide off-the-shelf crystal screening suites; Hampton Research, Jena Bioscience, Qiagen, Molecular Dimensions, and Emerald Biosystems. Each of these companies offer many more screens than are necessary for first-pass crystal screening. They also offer many targeted screening suites for membrane proteins, additive screening, heavy atom soaking, pre-crystallization protein solubility tests, etc. The minimal screening set that is recommended is Hampton Crystal Screen I/II (10), Hampton Index, Hampton SaltRx, and Hampton PEG/Ion. For new systems, screens should always be set up in parallel at two or more temperatures; typically 4°C and room temperature (18–20°C). Nextal/Qiagen has screens with similar compositions to the Hampton screens, but the conditions are organized to make it easier to discern trends by grouping similar conditions spatially. The Jena HT (high-throughput) is a two 96-block screening set that is a good choice when protein sample quantity is limited. Emerald also has a two or four 96-block screen set with further optional blocks. For optimization, most companies supply the chemical cocktails comprising each individual screening condition as well as their component chemicals in solution format.

After setting up the initial screens, and viewing the plates immediately after setup, it is important to allow the screens to sit undisturbed for at least a day or two prior to examining them (unless you are using a robotic imager that minimally disturbs the plates when imaging). Equilibration times will vary depending on the size of the drops, as will the time window of when crystals can still be expected to grow. With 0.5–1 μL drops, equilibrium is generally reached overnight, after which drops continue to dehydrate slowly, with an outside lifetime of approximately 1–2 months depending on the stability of the protein and the temperature of the experiment (11). Experiments should be surveyed at the time of setup, 1 or 2 days afterward, and every 5–7 days after that for 1–3 months. Crystals generally appear within the first 30 days, or not at all.

3.2. Cryoprotection and Diffraction Screening

Once an initial crystal screening hit is obtained, cryoprotection protocols must be developed, diffraction quality checked, crystal growth optimized, and hopefully, the target molecular structure elucidated. For cryoprotection, the components of the crystallization cocktail that yielded the crystal dictate possible cryoprotection strategies.

Almost all modern macromolecular single crystal X-ray diffraction experiments are carried out under cryogenic conditions (100 K) to slow the damaging effects of X-ray bombardment on the crystal. After mounting a crystal in a loop for placement in front of the X-ray beam, it is flash frozen in liquid nitrogen (occasionally the crystals are frozen by other methods, wherein cryoprotection is still necessary). The goal of cryoprotection is to replace enough of the water in the solvent channels of the protein crystal with the cryoprotectant in order to prevent ice crystal formation and destruction of the crystal upon freezing (12). There are numerous methods for achieving cryoprotection. Below, a very general set of strategies and some basic rules.

The two most commonly applied, and most successful cryoprotectants are ethylene glycol (ethane-1,2-diol, $C_2H_6O_2$) and glycerol (1,2,3-propanetriol, $C_3H_5(OH)_3$). A typical cryoprotectant solution is an approximation of the crystallization solution supplemented with 10–30% (v/v) glycerol, ethylene glycol, or other cryoprotectant. Successful cryoprotection incorporates the cryoprotectant while minimizing osmotic shock damage to the crystal. For example, if a crystal hit condition comprises 4% PEG 6000, 0.2M $MgCl_2$, and 0.1M Tris pH 8.5, then a good cryoprotectant solution to try would contain 25% ethylene glycol, 4% PEG 6000, 0.2M $MgCl_2$, and 0.1M Tris pH 8.5. Additionally, if the protein crystallizes in the presence of a ligand or cofactor it is advisable to include it in the cryoprotectant solution at a concentration similar to that used in the protein formulation to avoid diffusion of the small molecule out of the protein. The cryoprotectant can either be added to the drop the crystal was grown in or set down in a separate drop the target crystal can be quickly transferred to. If the initial attempts at cryoprotection damage the crystals it may be necessary to try several cryoprotectant formulations with different cryoprotectants.

There are three main classes of cryoprotectants: First are the organics and sugars (ethylene glycol, glycerol, PEGs, sucrose, xylitol, mannitol, etc.), second are the oils (paratone-N, mineral oil, silicon oil, Al's oil, etc.), and third are high molarity salts (any salt solution near its saturation point). Many crystallization supply companies sell preformulated cryoprotectant kits. The optimal period of time crystals should spend in the cryoprotectant solution must be determined empirically; times range from a few seconds to several minutes, and occasionally hours. To minimize crystal damage from osmotic shock, some crystals must be transferred between solutions where the concentration of the cryoprotectant is incrementally increased to a target value for cryoprotection. There are also many instances when a crystal is damaged when taken out of the drop it was grown in and transferred to the cryoprotection solution; in this case it can be helpful to gradually add the cryoprotectant directly to the crystallization drop. When growing crystals in nano-volumes, the cryoprotectant should always be added directly to the crystallization drop as quickly as possible after breaking the seal on the vapor diffusion chamber due to problems associated with rapid evaporation of small drops. In some instances, crystals grow in cryoprotected crystallization conditions, allowing for direct freezing of crystals.

With very sensitive crystals that eschew all attempts at cryoprotection, it is useful to incrementally introduce different cryoprotectants to the mother liquor as additives during crystallization, starting at about 20% of the concentration necessary for cryoprotection. If the crystals grow in the presence of one of the cryoprotectants, then they will often be amenable to the addition of more of that chemical during cryoprotection.

After the addition of the cryoprotectant to the crystallization drop, or the crystal to a preformulated cryoprotection solution, the crystal must be frozen. A small, open dewar filled with liquid nitrogen is used for this purpose. Before pouring the nitrogen into the dewar, the dewar should be clean, and after pouring the nitrogen the dewar should be covered. The nitrogen should not be allowed to sit exposed to air for long periods in order to prevent ice crystals from forming inside the liquid nitrogen. Ice crystals in the liquid nitrogen can stick to the protein surface, generating ice rings in the X-ray diffraction pattern. However, surface ice can be washed off with liquid nitrogen after the crystal is positioned in the cryostream. It is important when flash freezing to swiftly and completely submerge the entire crystal-pin and cap to ensure rapid cooling and to avoid problems with ice crystal formation during freezing (12, 13).

There are alternative vehicles that are occasionally used for freezing crystals, such as liquid propane, hyperquenching, and "in-the-stream" freezing. Liquid propane freezes the crystals at a higher temperature than liquid nitrogen, but is thought to freeze the crystals faster due to more efficient heat transfer. Freezing "in-the-stream" has the opposite effect. Freezing the crystal on the goniometer in the nitrogen vapor stream occurs at a slower rate, and some dehydration of the crystal occurs in the process, sometimes resulting in superior X-ray diffraction (13). Hyperquenching is a method (14) that involves eliminating the nitrogen vapor layer on the surface of the liquid nitrogen so that the crystal makes the transition from room temperature to 77 K more quickly, thereby preventing the formation of ice crystals. Furthermore, hyperquenching is intended to reduce the amount of cryoprotectant needed for a given system.

Once cryoprotected, the crystals are physically extracted, one by one, from the cryo-solution using a nylon loop approximately the size of the crystal. The nylon loop is attached to the end of an aluminum pin, which in turn is glued into a magnetic base. The proper base/pin combination and height used will depend on the beamline/robotics which will be used in the diffraction experiment. The entire ensemble, with the crystal in the loop, is attached to the magnetic end of a crystal wand for flash freezing in liquid nitrogen prior to cryo storage or transfer to the goniometer.

How one proceeds from this point depends on the X-ray diffractometer setup. There are alternatives to the standard loop and pin method emerging such as in situ diffraction (15), where crystals are exposed directly in the plate in which they are grown, as well as a new class of mounting tools generated by MiTeGen, which are more like scoops.

Once frozen crystals are tested for X-ray diffraction, there are numerous issues that can be diagnosed by analysis of the diffraction pattern. There is no substitute for an experienced eye to determine the nature of the problem in every instance. If one encounters a diffraction pattern comprising small, well-resolved spots to a resolution 3.0 Å or better from a single crystal lattice with no ice rings, this should elicit an audible sigh of relief. However, there are a few common, relatively easy to diagnose conditions that can be addressed by crystal optimization. The five primary adverse results encountered at this point are as follows: no diffraction or weak diffraction, split spots, diffraction patterns from multiple crystals, bad ice rings, and small molecule diffraction. These issues are described in more detail below.

1. *Weak diffraction, broad diffraction spots or no diffraction*: This is generally caused by poor crystal quality, insufficient crystal size or improper cryoprotection. To diagnose and remedy the problem, different cryoprotection protocols (i.e., variable length of exposure of crystals to cryoprotectant, gradual incorporation of cryoprotectant via soaks in solutions which increase the cryoprotectant concentration incrementally) and formulations should be tested. Comparisons of the diffraction quality of crystals frozen without cryoprotection to those frozen with cryoprotection can sometimes be used to diagnose problems with crystal damage from osmotic shock. If possible, diffraction tests should be carried out on capillary mounted crystals at room temperature to assess the effects of cryoprotectants on X-ray diffraction. If the crystals are small, larger crystals should be grown, either by crystallization condition optimization or using seeding. In cases where it is determined that crystals are intrinsically of poor quality, other crystallization conditions should be sought or the protein construct should be modified and rescreened.
2. *Split spots*: The splitting of diffraction spots can be caused by a number of factors, the most common being cracked crystals. Osmotic shock from improper cryoprotection is frequently the culprit. Additionally, growth defects or the presence of multiple lattices (from twinned crystals, see step 3, below) can cause split spots. However, due to the stochastic nature of crystal growth, and variability in freezing protocols, the problem can be frequently overcome by freezing many crystals, and experimenting with different freezing protocols and cryoprotectants.

If possible, at least three crystals of any given type should be frozen using the same protocol and have diffraction tested to ensure that any crystal-to-crystal variability is accounted for (of course this method does not account for variability in the handling of the crystals). In this way one can ensure that the best possible crystal is used for data collection.

3. *Diffraction patterns from multiple crystals*: If the crystals are not cracked, multiple diffraction patterns indicate that more than one crystal is in the X-ray beam. Twinned crystals can grow as optically distinguishable clusters, or as single masses that are only identifiable from diffraction experiments. If the crystals grow in optically distinguishable clusters, then attempts can be made with small tools (like an acupuncture needle) to isolate single crystals from the clusters prior to mounting for further diffraction studies. In any case, crystallization optimization or seeding can be used to attempt to grow more single crystals.
4. *Ice rings*: Ice rings arise from three sources, the crystal, surface ice, or the cryo-stream. Solid ice rings are indicative of internal ice in the crystal, and are a result of insufficient cryoprotection. Some ways to avoid ice rings include using increased concentrations of cryoprotectant during cryoprotection, or soaking the crystals longer in the chosen cryoprotectant. Spotty ice rings are observed if the ice is deposited on the surface of the crystal. Usually surface ice can be washed off with a small splash of liquid nitrogen while the crystal is mounted in the cryo-stream. A misaligned cryo-stream, although not very common, can also create convection currents that deposit ice from atmospheric humidity on the crystal, and can be rectified by proper alignment of the cryo-stream.
5. *Small molecule diffraction*: If there are a small number of strong diffraction spots, and none of them exist close to the origin, then the crystal is probably a small molecule (most often crystallized salts) and one must continue the search for a protein crystal.

3.3. Optimization Methods

Due to the multitude of variables involved in optimizing a crystallization condition, it is imperative to adopt a system of prioritizing variables for optimization. Of primary importance are: temperature, protein concentration, precipitant concentration, pH, buffer, salt concentration, ligand/cofactor concentration (and by proxy carrier solvent concentration, such as DMSO), and drop ratio (protein drop volume: mother liquor drop volume), often in that order.

If the screening is carried out at two temperatures (4°C and ~20°C), the one at which the initial hit was discovered should be used further. Aside from subtle changes, it will be assumed that the proper protein concentration has been established prior to initial screening (caveat: if no hits are observed during initial screening, and very few drops have precipitated, the protein concentration should be increased and initial screens should be repeated). The most important variables then are the precipitant concentration and the pH. A grid that varies the precipitant concentration in one dimension and the pH in the other dimension is generally a prudent initial optimization strategy to employ. If the crystal quality does not improve, more subtle aspects of the formulation should be addressed, such as the other components of the mother liquor, the protein formulation, additive screening, and possibly protein concentration and temperature shifts.

While it is generally believed that no two protein crystal systems react the same way to a given change in a crystallization experiment, there are some basic outcomes that can be expected from some simple changes. First, if a protein crystallizes in a shower of small crystals or clusters of small crystals, then a good first step is to lower the concentration of the precipitant (i.e., the PEG, salt, or any other major component of the mother liquor). Second, if crystals grow in a drop where there is a great deal of precipitate, the protein concentration may be too high, or there are multiple species (or aggregation states) of the protein in the sample; some that precipitate and others that crystallize. Frequently, aggregation or other deleterious processes can be minimized by the addition of (or increase in the concentration of) a reducing agent (16). Reducing agents prevent the formation of intermolecular disulfide bonds, a common cause of soluble protein aggregates, which hinder crystallization. The most common reducing agents are dithiothreitol (also known as DTT or Cleland's reagent, $C_4H_{10}O_2S_2$), 2-mercaptoethanol (also known as β-mercaptoethanol or BME, C_2H_6OS), and tris(2-carboxyethyl) phosphine (also known as TCEP, $C_9H_{15}O_6P$) at approximately 1, 2, and 0.25 mM, respectively. The three reducing agents have different properties, and can act as crystallization additives at the concentrations mentioned, but also at higher concentrations, as they appear in additive screens. Ethylenediaminetetraacetic acid (also known as EDTA, $[CH_2N(CH_2CO_2H)_2]_2$) can have a similar beneficial effect on crystallization on its own or in concert with a reducing agent by chelating trace heavy metals in the crystallization solution. The nature and number of the species of the target protein in the sample can be assessed by mass spectrometry, dynamic light scattering (17), gel filtration chromatography, and SDS/PAGE. Third, when crystals are single but too small it is generally the case that reducing the precipitant concentration, and sometimes reducing the protein concentration, will reduce the number of crystals and hopefully increase their size. Allowing the crystals more time to grow can prove effective, but often crystals reach a point where growth is arrested.

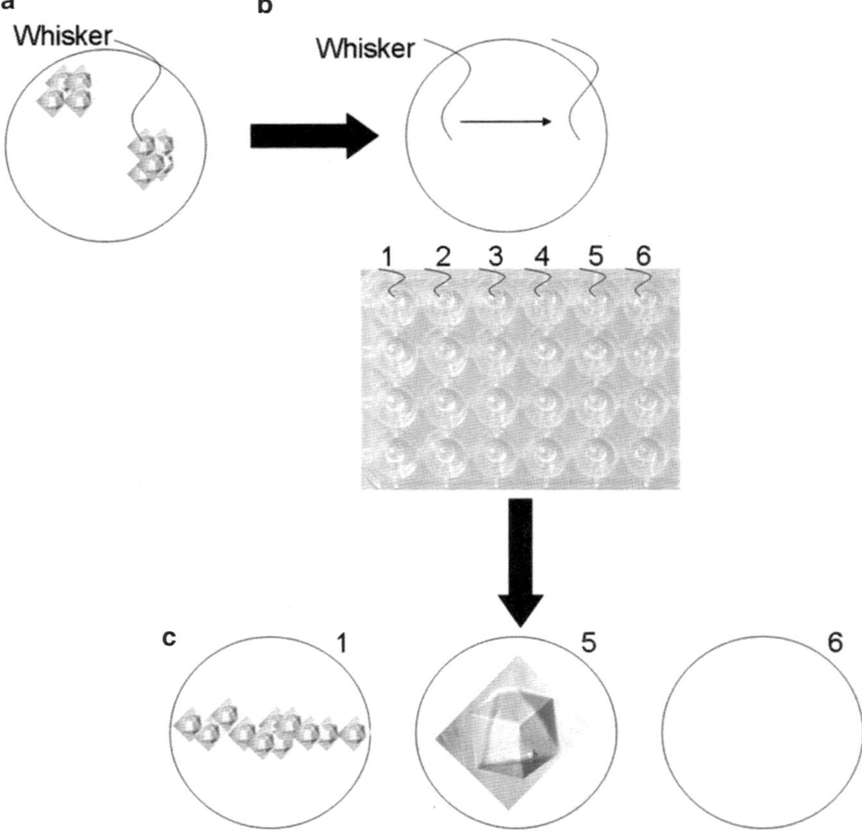

Fig. 5. Touch seeding schematic diagram. (**a**) Source—Touch at least one crystal with the tip of the whisker, microcrystals will be transferred to the whisker from imperceptible friction. (**b**) Pre-equilibrated drop—Drag the tip of the whisker through the pre-equilibrated drops from one side to the other in a swift motion. Streak ~6 drops per touch, in order to vary the number of crystal nuclei deposited in each drop. (**c**) Streak seed results—Crystals will grow where the whisker went through the drop, decreasing in number from the first touched to the last.

If lowering the precipitant or protein concentration results in a lack of nucleation, seeding may produce the desired effect. While there are many different techniques for seeding, the most commonly employed methods are touch seeding and crush seeding (18). For touch seeding, one must first procure either a cat whisker or a horse hair that is very thin at one end (most thin monofilament materials can work, but cat whiskers and horse hairs are the optimal tool for the job). After touching the crystal or cluster of tiny crystals with one end of the hair or whisker, the tip is then streaked through the crystallization drops to be seeded (1–6 per touch) as shown in Fig. 5. All of the conditions streaked after each touch should be the same so that the only variable is the number of nuclei delivered to each drop. Several conditions can, and should, be compared by repeating the touch and streak process for each new variation, ideally varying either the protein concentration or the precipitant concentration. The overall goal is to deliver the

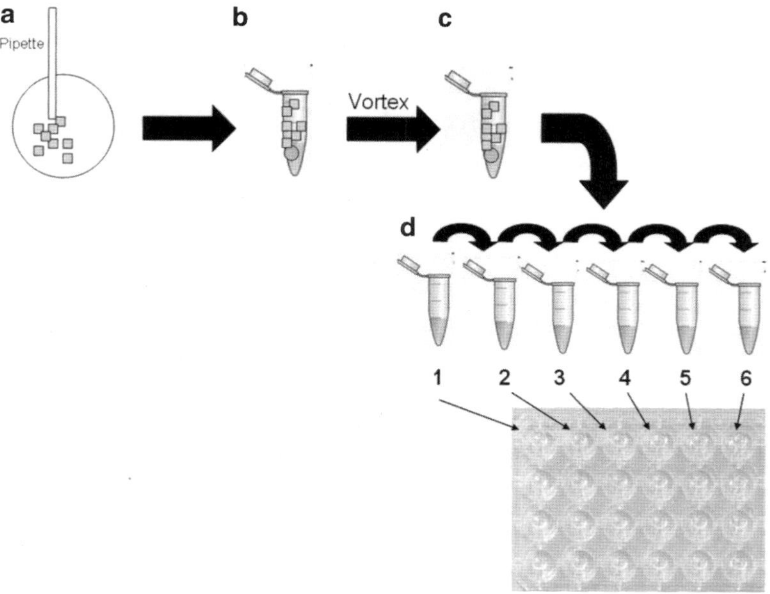

Fig. 6. A schematic of the crush seeding method. Using a pipette (**a**), transfer the crystals from the stock drop into approximately 50 μL of stabilizing solution (**b**), an approximation of the equilibrated drop solution along with a magnetic teflon® bead (like the Hampton Research Seed Bead™). Vortex (**c**) the solution with the bead. Serially dilute (**d**) the seed stock solution (1) into five more tubes of stabilizing solution. After mixing the diluted solutions add ~0.2 μL to equilibrated crystallization drops that have not yet nucleated.

smallest number of nuclei per drop in order to grow the biggest single crystals.

Crush seeding is a technique where crystals for seeding are crushed in a dilution solution of the crystallization condition. The seed stock is further diluted serially, and delivered in a small amount (~0.2 μL) to fresh or equilibrated crystallization drops (~1 μL) that do not contain crystals, as shown in Fig. 6. A unique advantage of crush seeding is that the establishment of a robust combination of seed stock solution and crystallization condition can be determined and the seed stock can be kept stable for a long period of time. Conversely, in the case of streak seeding, there is a necessity to have a fresh crystal drop for seed stock, and the delivery of nuclei to each drop is fairly haphazard.

Additionally, it is fairly common in drug discovery to establish the crystallizability of a target by solving the structure with a ligand of known affinity, only to later experience problems crystallizing the protein with other ligands. There are several reasons this can occur. Most commonly, the binding of a small molecule ligand can result in a change in protein solubility. The change in solubility can be due to a conformational change in the protein, a ligand-induced pH shift in the crystallization solution, the introduction of a ligand carrier solvent into the protein formulation, or a change in the formal charge of the protein upon binding the ligand. In this case, it is generally most efficacious to grid against a fine pH gradient (or a broad pH gradient, if the fine grid fails) by substituting different buffers or using a multi-buffer system (19). The solubility change may only affect the propensity for nucleation, in which case seeding may help (seeding with crystals of a complex of the target protein with a different ligand than the one in the drop to be seeded is not usually problematic).

Co-concentration of the protein and ligand may be necessary when dealing with an insoluble ligand or one with poor affinity for the protein (6). In some cases, the protein does not crystallize due to the fact that the compound is not present in stoichiometric amounts with the protein or because there is too much organic carrier solvent present. Co-concentration can be employed to limit the amount of carrier solvent introduced if one begins the co-concentration at a very low protein concentration (e.g., $[C_{prot}] = 30\ \mu M$, $[C_{lig}] = 100\ \mu M$), a greater than threefold excess of ligand exists, binding occurs, and if the ligand has sufficient affinity for the protein, and a sufficiently long off-rate, then occupancy will remain stoichiometric throughout co-concentration. Another possibility is that the binding of the ligand has induced such a large change in the protein structure that a new crystallization condition will be required for the formation of a new crystal form, necessitating a return to the initial sparse matrix screening step.

Alternatively, if a crystal can be obtained in either an apo-protein or holo-protein form but will not co-crystallize with a particular ligand of interest, it is prudent to attempt to soak the ligand in. Ideally, if one can grow apo-protein crystals, the soaking process is quite straightforward. In order to soak a ligand into an apo-protein crystal it is necessary to create a soaking solution which is a combination of the crystallization cocktail and the ligand of choice, potentially containing some level of cryoprotectant as well. The concentration of the ligand is generally quite high, in the range of 1–10 mM, depending on the solubility of the compound. The crystal is then transferred into the soaking solution and monitored for changes. The crystals will generally appear unchanged, or they will crack or dissolve. Crystals should be extracted at discrete time-points for data collection and analysis to determine if the ligand is bound. The soaking process can take from seconds to hours, or days in extreme cases. The process can be augmented to take the crystal through multiple drops with increasing compound, and/or cryoprotectant, concentration. If apo-crystals cannot be obtained, then it can be possible to soak out a ligand in exchange for another, a process sometimes referred to as back-soaking. This process can be facilitated by multistep soaks and the addition of any chemicals that might expedite the back-soaking process.

Another easy and effective optimization method is to set up the optimization grid across multiple crystallization platforms or plates. Many times crystals grow differently in hanging and sitting drops, as well as between various plate types within the same format (i.e., different sitting drop vapor diffusion plates). Additionally, there are many new plates being produced that enable one to set up crystallization experiments using some alternative methods (i.e., hanging drop, batch, capillary, dialysis, etc.), which were once difficult or messy, in a facile and semi-high-throughput manner. However, it should be noted that conditions for crystallization do not always translate easily from one method or platform to another. Furthermore, when a sample is screened in a high-throughput system and no crystal hits are observed, but the protein seems well behaved, it is prudent to perform a crystallization screen using an alternative crystallization method before attempting more rigorous adjustments to the sample or the protein construct.

3.4. Alternative Optimization Methods

When a direct grid-based approach to optimization is not successful, then a somewhat more random approach must be employed to introduce a further layer of complexity to the crystallization process, and hopefully create order through increasing chemical diversity, or via perturbation of the condition that yielded the original crystal hit. There are a few techniques that can be employed when the options already discussed have been exhausted, and suitable diffraction quality crystals remain elusive.

First is the traditional additive screen (offered by most crystallization reagent suppliers), a collection of low-concentration small molecule solutions that have unique chemical characteristics and sometimes significant effects on protein structure and stability. The idea is to dilute these stock solutions in some constant ratio (usually 1:10) into an array of the optimized crystallization solution. Once the cocktails have been made, the experiments are setup as normal in the plate of choice and monitored for crystal growth. If an improvement in the crystallization is seen with an additive, then experiments can be designed in which the additive is included in the protein formulation and the protein is sparse matrix screened again. At the same time, simple grids can be set up to find the optimal condition using the additive as a new constant or variable.

Second, is the option to either create a customized random additive screen or layer sparse matrices over the successful crystallization condition (20). Some software packages contain a random screen generator (Crystool (21), Rockmaker, etc.) enabling the creation of experiments with elements of randomness incorporated into them. In this way the experiment can be tailored to include a subset of additives at various concentrations or randomly vary the components of the condition to include combinations of components that a human would otherwise not likely try.

The technique of layering sparse matrices over a crystallization condition in the manner of an additive screen as used by the author is known as Iterative Sparse Matrix Additive Layering (ISMAL). ISMAL is a powerful technique that incorporates elements of initial screening, additive screening, and random screening utilizing only pipettes (or multichannel pipettes), commercial sparse matrix screens, and the crystal hit cocktail. The process is similar to the additive screen, except that instead of using an explicit set of single additives, a commercial sparse matrix screen is diluted by some arbitrary ratio (usually 1:10) into a plate containing the crystal hit cocktail. The mixtures are then used to setup new crystallization experiments that can be treated as optimizations or even as a new round of initial screening where multiple experiments can be setup at different temperatures. The one caveat is that incompatible solutions will form precipitates (occasionally small molecule crystals), depending on the initial crystal hit ingredients, of course.

In many instances the purified and ostensibly stable protein is not amenable to crystallization. Sometimes the protein exhibits excellent solution behavior, while in other situations it simply does not concentrate appropriately. When faced with a protein that will not concentrate, there are a number of ways to proceed toward crystallization. First, and easiest, is to attempt to concentrate the protein in the crystallization drop. This is achieved by skewing the ratio of the protein solution to mother liquor in the crystallization experiment in favor of the protein, effectively driving the equilibration point toward a higher protein concentration. Second, attempts can be made to find a stabilizing ligand or buffer formulation. The ligand can be a substrate, substrate analog, cofactor, peptide, or any small molecule which is known to bind to the protein. It is important to select a molecule with a known affinity for the protein target. If no known ligands exist, then a screening assay must be devised to find an appropriate stabilizing chemical or compound. A buffer formulation screen should be assayed at the same time.

There are several methods for determining the binding constants or affinities of ligands for specific protein targets. However, for crystallization the goal is to design an assay that assesses the degree of protein stabilization. The standard method for this purpose is a thermal shift assay and isothermal calorimetry (22, 23). These methods provide a comparative analysis of affinities and melting temperatures of the protein under different conditions. Additionally, one can calculate values for the change in Gibbs free energy, entropy, and enthalpy of small molecule binding. The greater the shift in the melting temperature, the more the protein stability is affected by the variable element. This method can be applied to any soluble, properly folded protein to assess the changes arising from the addition of a ligand or cofactor, additives, buffer variations, or ostensibly viable versions of the protein-containing mutations, construct variations, or posttranslational modifications. One caveat that must be taken into consideration is that a perturbation that will induce crystallization is not always due to stabilization. It is occasionally a result of destabilization. Therefore, when searching for an element to add to your formulation it is sometimes prudent to look at both stabilizing and destabilizing agents.

When no sign of crystallization has presented itself for a given construct, it is a good idea to augment the construct boundaries, or to exercise control over posttranslational modifications, such as phosphorylation, dephosphorylation, methylation, de-glycosylation, etc. Any newly modified form or construct of a given protein should be screened for crystallization as an entirely new protein, and homogeneity of the sample should be of paramount concern. The class and biochemistry of the target protein should dictate which method of modification is appropriate in each case. Ideally, sample homogeneity and monodispersity should be achieved, but this might not be possible. In such circumstances, even if a sample is not 100% pure, homogenous, and monodisperse, it is prudent to screen the protein sample, as it may still crystallize. Although many of the most important variables that promote crystallization are understood, it is not yet possible to predict what will or will not crystallize before it has been tested. Therefore, it is always worth trying. If crystals remain elusive, more new constructs and/or intelligently chosen mutations should be created and screened.

Finally, if a suitable construct cannot be found, but soluble protein is available, limited proteolysis (24) and in situ proteolysis (25) can be performed. When experimenting with limited proteolysis, a protease is added to the protein sample and aliquots are removed and quenched at discrete time-points. If, upon analysis, a discrete protein species is created from the partial digestion, then the product can either be purified and screened, or sequenced and cloned. In situ proteolysis is generally carried out by adding a single protease or several proteases in very small amounts (1:100–1:1,000 mole protease: mole protein target ratios) to the protein sample just prior to setting up the crystal screens. The hope is that a discrete proteolytic event will yield a single, crystallizable product, and as the protease makes its way through the molecule population within the drop, the products will crystallize. If crystals grow using in situ proteolysis, it is important to save some of the crystals to dissolve for analysis by mass spectrometry or N-terminal sequencing.

4. Summary

Although significant advances have occurred in the field of macromolecular crystallization in the past few decades, it remains a bottleneck of structure-based drug design operations and a realm of science shrouded in mystery. Many tools and techniques have been developed to overcome crystallization problems, but one must have some experience to know which methods to employ when. High-throughput laboratories can easily produce hundreds of co-crystal structures of tractable targets, but they can just as easily be stymied by recalcitrant proteins that present insurmountable hurdles for crystallization. What has become clear is that the best strategy for successfully crystallizing new targets is to prosecute as many different constructs as possible up front. The field of macromolecular crystallization remains one dominated by empirical experimentation, and one should always approach it as such.

References

1. Chayen, N. E. (2004) Turning Protein Crystallization from an art into a science. *Curr. Opin. Struct. Biol.* **14**, 577–583.
2. Walter T. S., Diprose J. M., Mayo C. J., Siebold C., Pickford M. G., Carter L., Sutton G. C., Berrow N. S., Brown J., Berry I. M., Stewart-Jones G. B., Grimes J. M., Stammers D. K., Esnouf R. M., Jones E. Y., Owens R. J., Stuart D. I. and Harlos K. (2005) Turning A procedure for setting up high-throughput nanolitre crystallization experiments. Crystallization workflow for initial screening, automated storage, imaging and optimization. *Acta Cryst. D.* **61**, 651–657.
3. Hansen C. L., Skordalakes E., Berger J. M. and Quake S. R. (2002) A robust and scalable microfluidic metering method that allows protein crystal growth by free interface diffusion. *Proc. Natl. Acad. Sci.* **99**, 16531–16536.
4. Emamzadah S., Petty T. J., De Almeida V., Nishimura T., Joly J., Ferrer J. L. and Halazonetis T. D. (2009) Cyclic olefin homopolymer-based microfluidics for protein-crystallization and *in situ* X-ray diffraction. *Acta Cryst. D.* **65**, 913–920.
5. Deng J., Davies D. R., Wisedchaisri G., Wu M., Hol W. G. and Mehlin C. (2004) An improved protocol for rapid freezing of protein samples for long-term storage. *Acta Cryst. D.* **60**, 203–204.
6. Hassell A. M., An G., Bledsoe R. K., Bynum J. M., Carter H. L. 3rd, Deng S. J., Gampe R. T., Grisard T. E., Madauss K. P., Nolte R. T., Rocque W. J., Wang L., Weaver K. L., Williams S. P., Wisely G. B., Xu R. and Shewchuk L. M. (2007) Crystallization of protein-ligand complexes. *Acta Cryst. D.* **63**, 72–79.
7. Watson A. A. and O'Callaghan C. A. (2005) Crystallization and X-ray diffraction analysis of human CLEC-2. *Acta Cryst. F.* **61**, 1094–1096.
8. Noble J. E. and Bailey M. J. (2009) Quantitation of protein. *Methods Enzymol.* **463**, 73–95.
9. Gosavi R. A., Bhamidi V., Varanasi S. and Schall C. A. (2009) Beneficial effect of solubility enhancers on protein crystal nucleation and growth. *Langmuir.* **25**, 4579–4587.
10. Jancarik J. and Kim S. H. (1991) Sparse matrix sampling: a screening method for crystallization of proteins. *J. Appl. Cryst.* **24**, 409–411.
11. Saridakis E., Chayen N. E. (2000) Improving protein crystal quality by decoupling nucleation and growth in vapor diffusion. *Protein Sci.* **9**, 755–757.
12. McFerrin M. B. and Snell E. H. (2002) The development and application of a method to quantify the quality of cryoprotectant solutions using standard area-detector X-ray images. *J. Appl. Cryst.* **35**, 538–545.
13. Edayathumangalam R. S. and Luger K. (2005) The temperature of flash-cooling has dramatic effects on the diffraction quality of nucleosome crystals. *Acta Cryst. D.* **61**, 891–898.
14. Warkentin M. and Thorne R. (2007) *A general method for hyperquenching protein crystals. J. Struct. Funct. Genom.* **8**, 141–144.
15. McPherson A. (2000) *In Situ* X-ray Crystallography. *J. Appl. Cryst.* **33**, 397–400.
16. Dreyfus C., Pignol D. and Arnoux P. (2008) Expression, purification, crystallization and preliminary X-ray analysis of an archaeal protein homologous to plant nicotianamine synthase. *Acta Cryst. F.* **64**, 933–935.
17. Wilson W. W. (2003) Light scattering as a diagnostic for protein crystal growth - a practical approach. *J. Struct. Biol.* **142**, 56–65.
18. Stura, E. A., Wilson, I. A. (1992) "Seeding Techniques" in Crystallization of Nucleic Acids and Proteins: A Practical Approach. Oxford University Press, pp. 99–126.
19. Newman J. (2004) Novel Buffer Systems for Macromolecular Crystallization. *Acta Cryst. D.* **60**, 610–612.
20. Cudney R., et al., (1994) Screening and optimization strategies for macromolecular crystal growth. *Acta Cryst. D.* **50**, 414–423.
21. Rupp B., Segelke B. W., Krupka H. I., Lekin T., Schäfer J., Zemla A., Toppani D., Snell G. and Earnest T. (2002) The TB structural

genomics consortium crystallization facility: towards automation from protein to electron density. *Acta Cryst. D.* **58**, 1514–1518.

22. Ericsson U., Hallberg B., DeTitta G., Dekker N. and Nordlund P. (2006) Thermofluor-based high-throughput stability optimization of proteins for structural studies. *Anal. Chem.* **357**, 289–298.
23. Recht M., Torres F., De Bruyker B., Bell A., Klumpp M. and Bruce R. (2009) Measurement of enzyme kinetics and inhibitor constants using enthalpy arrays. *Anal. Biochem.* **388**, 204–212.
24. Fontana A., Polverino de Laureto P., Spolaore B., Frare E., Picotti P. and Zambonin M. (2004) Probing protein structure by limited proteolysis. *Acta Biochemica Polonica.* **51**, 299–321.
25. Dong A., Xu X. and Edwards A. M. (2007) *In situ* proteolysis for protein crystallization and structure determination. *Nature Methods.* **4**, 1019–1021.

Chapter 5

X-Ray Sources and High-Throughput Data Collection Methods

Gyorgy Snell

Abstract

X-ray diffraction experiments on protein crystals are at the core of the structure determination process. An overview of X-ray sources and data collection methods to support structure-based drug design (SBDD) efforts is presented in this chapter. First, methods of generating and manipulating X-rays for the purpose of protein crystallography, as well as the components of the diffraction experiment setup are discussed. SBDD requires the determination of numerous protein–ligand complex structures in a timely manner, and the second part of this chapter describes how to perform diffraction experiments efficiently on a large number of crystals, including crystal screening and data collection.

Key words: X-ray sources, X-ray generator, Synchrotron radiation, Synchrotron Beamline, Data collection, Data processing

1. Introduction

Protein crystallography is a method to visualize biological molecules on an atomic scale. It can be regarded as a form of microscopy with very large magnification. In optical microscopy, the achievable resolution is similar to the wavelength of the illuminating light source. Visible light falls in the 400–700 nm range and thus the smallest objects viewable are also a few hundred nanometer in size.

The bond lengths between the most common atoms in biological molecules, C, O, N, and H, fall in the range of 100–150 pm. Thus to determine atomic positions the wavelength of the illuminating light source has to be comparable to the distances between atoms. Radiation with the appropriate wavelength is called X-ray radiation.

Leslie W. Tari (ed.), *Structure-Based Drug Discovery*, Methods in Molecular Biology, vol. 841, DOI 10.1007/978-1-61779-520-6_5, © Springer Science+Business Media, LLC 2012

There are two main challenges related to the method of X-ray protein crystallography. First, protein crystals are usually small (10–1,000 μm) and have a high solvent content (around 50%), resulting in very weak diffracting power. Thus, the use of very intense X-rays becomes necessary. Second, no lenses, such as those used in a conventional visible-light microscope for magnification of the specimen, exist for X-rays of 100–150 pm wavelength. This fact alone necessitates the use of the method of crystallography, where diffraction from a periodic structure is used to form an "image" without a lens. The resulting difficulties of obtaining good protein crystals (1) and solving new structures (2) are the main challenges of the field.

The lack of X-ray lenses also complicates the delivery and focusing of the radiation onto the sample. Section 2 of this chapter describes the generation, manipulation and detection of X-rays as they apply to X-ray macromolecular crystallography. The two most common X-ray sources, namely, X-ray generators and synchrotrons, are explained in Section 2.1 followed by a brief overview of X-ray optics in Section 2.2. Components of the diffraction experiment are described in Section 2.3.

The goal of structure-based drug design (SBDD) is to determine the structures of medically relevant proteins in complex with compounds developed during the drug discovery process. These compounds go through numerous iterations as they progress toward the clinic. In an industrial environment, where multiple projects are being worked on simultaneously, with each of them likely having multiple compound series, the number of total complex structures needed can be rather high (hundreds per year). Taking into account an attrition rate of 5–25 between harvested crystals and crystals yielding an actual useful structure, the need for the handling of thousands of samples arises. This can be done efficiently only by using bioinformatics tools for sample tracking and automation for sample handling. The various practical aspects of high-throughput crystallography, including crystal screening, data collection and data processing are described in Section 3.

2. X-Rays

X-rays were discovered in 1895 by the German scientist Wilhelm Conrad Röntgen, who was awarded the first Nobel Prize in Physics in 1901 for his discovery. The term X-ray originates from Röntgen himself because at the time it was an unknown type of radiation. In many languages, including German, X-rays are called Röntgen radiation. X-rays are electromagnetic radiation, just like visible light and radio waves, and this was discovered by the method of crystallography. Max von Laue performed diffraction experiments on crystals using X-rays in 1912 and obtained photographs of diffraction patterns. These patterns are a result of interfering electromagnetic waves with a wavelength similar to the distances between atoms in a crystal. Laue received the 1914 Physics Nobel Prize for this discovery.

Electromagnetic radiation is characterized by its wavelength, λ and X-rays are positioned between vacuum UV radiation and gamma rays with a wavelength in the range of 10 nm to 10 pm. For radiation with a wavelength shorter than visible light, the wavelength is often expressed in terms of the photon energy E, and X-rays fall in the 0.1–100 keV range. Conversion between wavelength and energy can be done using a simple equation: $E\,(\mathrm{eV})\cdot\lambda\,(\mathrm{nm}) = 1{,}239.84$. It is useful to remember that 1 Å[1] wavelength corresponds to 12.4 keV photon energy, since it is a commonly used wavelength for protein crystallography at synchrotron light sources.

For protein crystallography monochromatic radiation is needed. In practice the energy of an X-ray beam is not infinitely well defined, and the term energy bandwidth (ΔE) or energy resolution ($\Delta E/E$) is used to express the energy spread. A monochromator with a resolution of 10^{-4} has a bandwidth of 1 eV at 10 keV photon energy. The bandwidth is usually the full-width half-maximum (FWHM) value of a Gaussian distribution.

In addition to bandwidth, beam intensity is another important property of X-rays. It is measured in photons per second (ph/s) and can be accurately determined by a photodiode or an ionization chamber. Since protein crystals are small, it is desirable to focus the X-rays into a spot of similar size as the crystal (typically around 100 μm). However, strong focusing of an extended light source will result in a highly divergent beam behind the focal point and may impair the separation of diffraction spots on the detector surface. Especially for smaller crystals and/or large unit cells a low beam divergence is advantageous. Based on these practical considerations, a useful measure of beam intensity is the so-called brightness, which defines the beam intensity falling into a unit of solid angle and onto a unit of area (expressed as ph/(s mrad^2 mm^2) at a bandwidth of 0.1%).

In summary, the X-ray beam requirements for protein crystallography experiments are the following: i) high intensity, ii) monochromatic with a narrow bandwidth, iii) tunable energy for multiwavelength anomalous diffraction (MAD) phasing (2), iv) small focal point, v) low divergence, and vi) stability of the light source. Typical parameters of different X-ray sources are listed in Table 1.

2.1. X-Ray Sources

2.1.1. Physics of X-Rays

There are two fundamental physical phenomena that can lead to the emission of X-rays. Acceleration of charged particles generates a continuous spectrum of electromagnetic radiation, such as

[1] Å stands for Ångstrom, which is a widely used unit of X-ray wavelength, 1 Å = 0.1 nm = 100 pm.

Table 1
Typical parameters of synchrotron beamlines in comparison with a high-end beamline (APS ID-23D) (3) and a high-end X-ray generator (Rigaku FR-E + SuperBright™) (4)

	Synchrotron beamline	APS beamline ID-23D	Rotating anode X-ray generator
Energy range (keV)	5–15	3.5–20	5.41, 8.05
Energy resolution ($\Delta E/E$)	10^{-3}–10^{-4}	10^{-4}	
Intensity (ph/s)	10^{10}–10^{13}	1×10^{13}	7×10^{9}
Spot size (μm)	50–200	70 (H) × 20 (V)	208
Divergence (mrad)	1–3	0.14	4.8
Flux density (ph/(s mm²))		4×10^{15}	1.6×10^{11}

Bremsstrahlung and synchrotron radiation. Atomic transitions on the other hand result in the emission of photons with specific wavelengths and narrow bandwidth, such as the characteristic X-rays.

In practice, characteristic X-rays are generated by bombardment of a metal target (e.g. Cu, Al, Mo) with energetic electrons (e.g., 50 keV). These high-energy electrons knock out electrons from the inner-shells of the target atoms thereby ionizing them. This core-hole ionization process leaves the resulting ion in an excited state, which then relaxes to a lower energy state by filling the core-hole with an electron from a higher shell. The energy of the emitted photons is the difference of the electron binding energies of the shells between which the transition occurs. The description of the characteristic X-ray lines is based on using K, L, M, … for the main quantum numbers $n = 1, 2, 3, \ldots$ of atomic shells and α, β, γ, … for the difference $\Delta n = 1, 2, 3$ between shells. Thus, a K_α line denotes an L→K transition and K_β an M→K transition (see Fig. 1, left). In the case of copper, for example, the dominant K_α line has an energy of 8,048 eV (1.541 Å), which is the difference between the binding energies of the K 1*s* and L 2*p* shells.

When the acceleration of charged particles results in a change of the magnitude of their velocity, e.g., when energetic electrons hit a surface and slow down due to repulsive Coulomb interactions with the target atoms' electrons, so-called Bremsstrahlung ("breaking radiation") is generated. The energy distribution of Bremsstrahlung is continuous with the high-energy (short-wavelength) cutoff determined by the energy of the incoming particles (see Fig. 1, left). Radiation is also generated when the acceleration results in a change of the direction of the charged particles' velocity, e.g., when

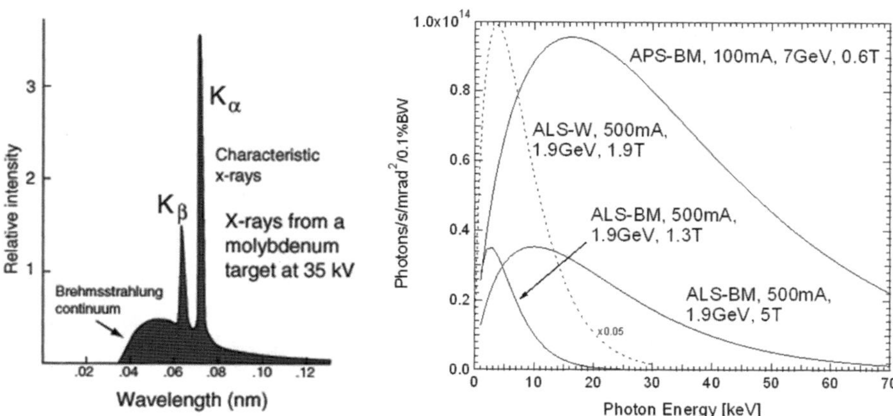

Fig. 1. Spectral distribution of X-rays from an X-ray generator (*left*, reproduced with permission) (5) and from bending magnet (BM) and wiggler (W) synchrotron radiation sources (*right*). The bending magnet and wiggler spectral distribution curves were calculated using (6) and the program SPECTRA (7), respectively.

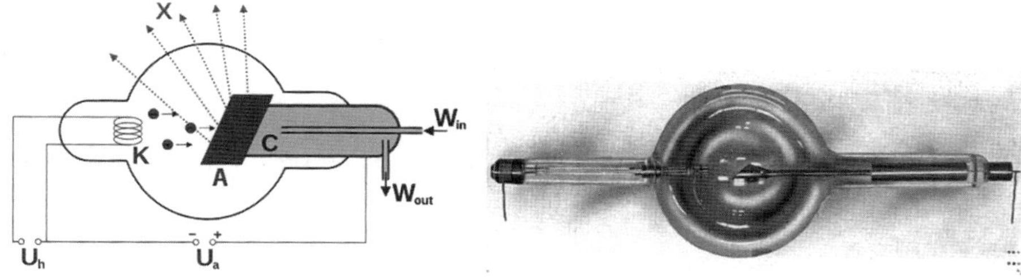

Fig. 2. Schematic of a sealed tube X-ray generator (*left*) and photo of a Coolidge X-ray tube from the early 1900s (*right*) (9).

charged particles are forced onto a circular trajectory by a strong magnetic field. Radiation created this way was first observed in a particle accelerator called a synchrotron in 1947 (8) and was subsequently named synchrotron radiation (SR). SR also has a continuous distribution (see Fig. 1, right), with the high-energy limit related to the energy of the accelerated particles (Equation 1).

2.1.2. X-Ray Generators

The basic layout of an X-ray tube is shown in Fig. 2. In an evacuated vessel usually made out of glass, electrons are emitted from a heated metal filament (e.g., tungsten), the cathode, and accelerated toward the anode by the high voltage (up to 150 kV) applied between the two electrodes. As the fast electrons hit the anode, Bremsstrahlung and characteristic X-rays are generated as described above and their energy will depend on the accelerating voltage and the anode material used. The X-rays exit through the glass wall and can be used for experiments. These devices are also called sealed tubes, since all the components are sealed into a glass tube (see Fig. 2, right).

The production of X-rays is a very inefficient process and most of the electrons' energy (~99%) turns into heat at the anode. This heat load limits the maximum X-ray power possible in a simple sealed tube generator, even if the anode is water-cooled (see Fig. 2). To obtain a small X-ray source size the electrons are focused into a small spot on the anode (e.g., by shaping the cathode appropriately). This focusing increases the local heat load on the anode further.

The heat load problem is much reduced in rotating anode generators, where the anode is a metal disc rotating at high speeds (several thousand revolutions per minute). The target area being hit by the focused electron beam is constantly changing and helps to improve the dissipation of heat. This results in greatly increased performance but the higher technical complexity leads to higher acquisition and maintenance costs for rotating anode systems.

X-ray generators are often characterized by their power rating (watt or kilowatt). This is a measure of the electron beam power hitting the anode (accelerating voltage × current between cathode and anode). Owing to the many factors influencing the usable X-ray power (focal spot size, anode shape, collected solid angle), a higher overall power rating does not necessarily correspond to a more intense X-ray beam at the sample.

X-ray generators fulfill the beam requirements for protein crystallography described above rather well: 1) Highest intensities can be obtained from rotating anode sources and those are indeed the most common devices for protein crystallography in home laboratories. They are well suited for crystals that diffract reasonably well. 2) A characteristic emission line has a specific wavelength with a narrow bandwidth; there are, however, multiple such lines emitted simultaneously, and they are all complemented by the continuous Bremsstrahlung background (Fig. 1, left). To obtain monochromatic radiation for a crystallography experiment, one of those lines has to be selected by using a monochromator or filter. 3) The energy of an X-ray generator is not freely tunable, but it can be changed to values of specific emission lines of different metals by changing the anode material. Copper is the most commonly used anode material (Cu K_α at 1.54 Å), while the chromium K_α line at 2.29 Å is emerging as an excellent source for Se and S SAD phasing in house (10–12). An easy switch between different wavelengths can be achieved by employing stripes of multiple different elements on the same rotating anode (10–12). 4) and 5) The electron beam can be focused into a spot smaller than 100 μm on the anode resulting in a small source size of the emitted radiation. However, the X-rays are emitted in all directions allowed by the anode geometry and have to be collected and focused onto the sample. The amount of focusing needed depends on how much solid angle of the radiation is collected and on the size of the beam in the focal point, where the sample is located. Capturing more solid angle (to increase flux) and/or focusing into a smaller spot (for small crystals) require stronger focusing, which leads to a more divergent beam behind the sample. Increased divergence means that spots closer together will overlap and cannot be resolved on the detector surface. This means there is a trade-off between flux, beam size at the sample, and beam divergence, and those parameters should be optimized for the task at hand. Many modern sealed tube and rotating anode generators are equipped with multilayer optics. These optics are capable of collecting a large solid angle of radiation and efficiently focusing it on a small spot. Additionally, they improve the spectral purity of the X-ray beam. Multilayers are discussed in more detail in section 2.2.2). 6) Beam stability is not an issue with modern generators as they can provide consistent output for extended periods of time (multiple days).

2.1.3. Synchrotrons

Although synchrotron radiation was discovered in the late 1940s, its actual utilization for scientific experiments did not start until the 1960s. Those first experiments were carried out at accelerators built for high-energy physics research and laid the foundation for the very successful expansion of the field. Owing to its broad bandwidth and high intensity (see Table 1), synchrotron radiation is utilized in many different fields of scientific research (physics, chemistry, materials science, biology).

Synchrotron radiation is generated in particle accelerators called synchrotrons (see Fig. 3). All modern light sources are built specifically for the generation of high-brightness synchrotron radiation and are not involved in high-energy physics. In a synchrotron, charged particles travel around a roughly circular path in a metal vacuum tube. The particles are kept on a closed trajectory by strong dipole bending magnets, which generate a magnetic field perpendicular to the plane of the accelerator and deflect the particles due to the Lorentz force (see Fig. 4). Quadrupole and hexapole magnets placed along the trajectory keep the particle beam focused because without focusing the repulsion between the particles with the same charge would "blow-up" the beam instantly. The diameter of the beam in the synchrotron is around 100 μm. The lower the emittance (beam size × beam divergence) of the particle beam in the accelerator the higher the brightness the photon beam will have. The energy of the particles is boosted at every turn by a radio frequency (RF) accelerator cavity. To increase the energy of the particles, the magnetic field in the bending magnets and the frequency of the accelerating voltage in the RF cavity have to be raised in a synchronized manner and that is why this type of accelerator is called a synchrotron. The radius of the particles' trajectory is fixed in a synchrotron, determined by the layout of the vacuum tubes and magnets. This is in contrast to the cyclotron, the circular particle accelerator invented by E. O. Lawrence (1939 Nobel Prize in physics), where the radius changes with increasing energy.

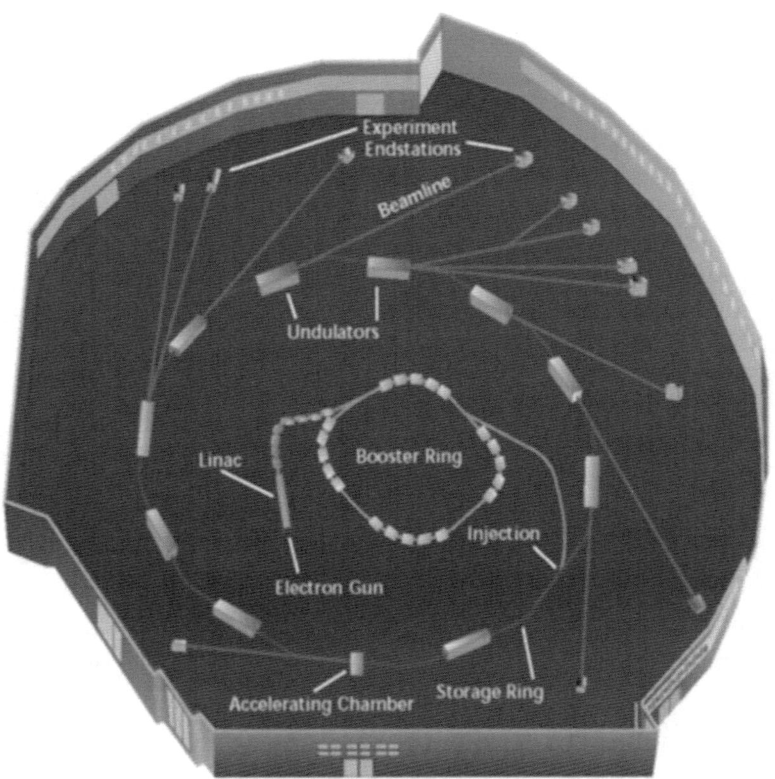

Fig. 3. Schematic representation of a synchrotron radiation facility showing the linac, the booster, and the storage ring. Image courtesy of the Advanced Light Source, Lawrence Berkeley National Lab.

Fig. 4. Dipole bending magnet of the Advanced Light Source Synchrotron (14). The coils of the magnet are made of water-cooled copper tubes.

The emitted power of the synchrotron radiation is proportional to m^{-4} (m—mass of the particle). For high-energy physics experiments, heavy particles (e.g., protons) are used to minimize the energy loss due to synchrotron radiation. Synchrotron light sources on the other hand use the much lighter electrons (or sometimes positrons) to maximize the amount of emitted radiation.

The final, highest electron energy is achieved in multiple steps (see Fig. 3). The electrons originate from the electron gun, which contains a thermionic material that easily releases electrons when heated (e.g., barium aluminate). The first step of acceleration happens in a linear accelerator (LINAC), where microwaves are used to increase the electrons' energy along a straight path to the order of 100 MeV. The electrons are then transferred into the booster synchrotron, which accelerates them to the final, or close to the final, energy, which is typically between 2 and 8 GeV. From the booster the electrons are injected into the much larger storage ring, where electrons are stored and circulated for extended periods of time to generate light for the experiments. The intensity of the radiation is proportional to the number of particles in the storage ring, which is typically in the range of 100–500 mA (depending on the synchrotron). The storage ring is also a synchrotron (it can accelerate electrons within a limited range) and is specifically designed to produce intense photon beams as explained in the next two sections. At energies of a few GeV electrons travel with nearly the speed of light and relativistic theories have to be used for their theoretical description.

Although the vacuum system of the accelerator is kept at a very low pressure (~10^{-10} mbar), there are still enough collisions between the electrons and residual gas molecules such that the beam will decay slowly. Usually, half of the electrons are lost within hours to days (depending on the synchrotron) and have to be replenished by injecting new particles into the ring. To avoid the unwelcome effects of beam decay, such as the constantly changing light intensity, most synchrotron radiation facilities changed over to "top-off mode" operation in recent years. In top-off mode, a small amount of electrons is injected into the storage ring about every minute, keeping the ring current basically constant resulting in a constant photon flux at the sample and a constant thermal load on the beamline optics.

Bending Magnets

As described in Section 2.1.1, acceleration of charged particles results in the emission of electromagnetic radiation. When the electrons circulating in the storage ring pass through a bending magnet they are deflected due to the Lorentz force and emit synchrotron radiation (see Fig. 5, left). The underlying physics is the same as for electrons oscillating in an antenna, thereby creating radio waves for broadcasting or communications.

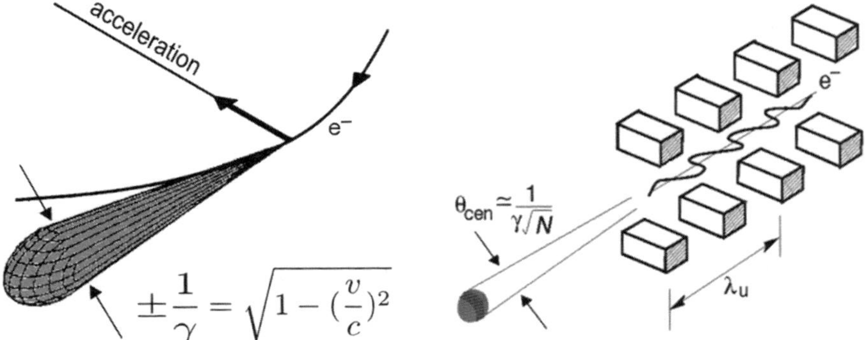

Fig. 5. Synchrotron radiation emitted from a bending magnet (*left*, modified from http://hasylab.desy.de/science/studentsteaching/primers/synchrotron_radiation/index_eng.html) and an undulator (*right*). The opening angle of the cone depends on the ratio of the speed of the electrons *v* and the speed of light *c* for bending magnets. For undulators it is additionally influenced by the number of undulator periods *N* (λ_u—period length of the undulator). The undulator image is courtesy of the Advanced Light Source, Lawrence Berkeley National Lab.

One of the major advantages of synchrotron radiation is its continuous energy spectrum, which covers a wide spectral range from infrared light to hard X-rays (see Fig. 1, right). Using a monochromator, the best photon energy can be selected for a specific experiment (e.g., matching excitation energies of atoms and molecules), providing unmatched flexibility for scientific research. The spectral distribution is described by the critical photon energy ε_c, defined in a way that an equal amount of power is emitted above and below ε_c. The critical energy depends on the radius r of the electron beam in the magnet (or the strength B of the magnetic field) and the particle energy E as follows:

$$\varepsilon_c = 2.218E^3(\mathrm{GeV}^3)\,/\,r(\mathrm{m}) = 0.6651B(\mathrm{T})E^3(\mathrm{GeV}^3). \qquad (1)$$

Raising the critical energy to obtain hard X-rays needed for diffraction experiments can be achieved either by increasing the electron energy in the storage ring, which results in larger and more expensive machines (see Table 2), or by using stronger magnets. The latter, a significantly simpler and cheaper approach, was realized at the ALS, where three of the normal bending magnets with a magnetic field of 1.3 T were upgraded to 5 T superconducting magnets (13), raising ε_c by a factor of 3 (see Table 2; Fig. 1, right).

Another important feature of synchrotron radiation is that it is emitted in a *narrow cone* in the forward direction. This is due to the relativistic nature of the electrons and provides an inherent primary focusing of the photon beam. The angular width of the cone is $1/\gamma$, and it decreases with increasing electron energy (see Fig. 5).

Table 2
Basic parameters of three representative third generation synchrotron radiation facilities

	Advanced Light Source (ALS) (Berkeley, CA)	Stanford Synchrotron Radiation Laboratory (SSRL) (Palo Alto, CA)	Advanced Photon Source (APS) (Chicago, IL)
Electron energy (GeV)	1.9	3.0	7.0
Ring current (mA)	500	500	100
Circumference (m)	197	234	1,104
Critical photon energy (keV)	5.9 (normal), 22.8 (super)	7.6	19.5
Mode of operations	Top-off	Top-off	Top-off

Radiation from a bending magnet is linearly polarized in the plane of the accelerator. Above and below the plane, the radiation is circularly polarized with opposing helicities respectively. The variability of the light polarization is yet another very useful aspect of synchrotron radiation, which is exploited in many experiments.

Insertion Devices (Wigglers and Undulators)

The amount of light emitted from a bending magnet is limited by the electron current in the storage ring and the acceptance angle of the beamline optics (i.e., how much of the radiation cone can be utilized). Since both of these quantities are limited by practical constraints, and because scientists typically need more flux, new devices had to be developed. These devices, called wigglers and undulators, consist of a series of magnetic dipoles with alternating polarity, which cause the electron beam to "wiggle" or "undulate" back and forth around a straight line and emit synchrotron radiation at every deflection (see Fig. 5, right). Wigglers and undulators can be several meters long and are installed in straight sections of the storage ring (see Fig. 3). Their magnetic field is usually perpendicular to the accelerator plane, and thus, the electrons' sinusoidal motion is in the horizontal plane.

A wiggler is a periodic structure of multiple bending magnets with its magnetic field being relatively large resulting in large deflections of the electron beam (as compared to the opening angle of the radiation cone). The emitted light from subsequent poles adds up incoherently and the intensity proportionally increases with the number of periods N. The photon energy distribution is continuous (see Fig. 1, right) and the total emitted power can be significant (several kilowatt). This requires special precautions to avoid damage to the optical components of a beamline.

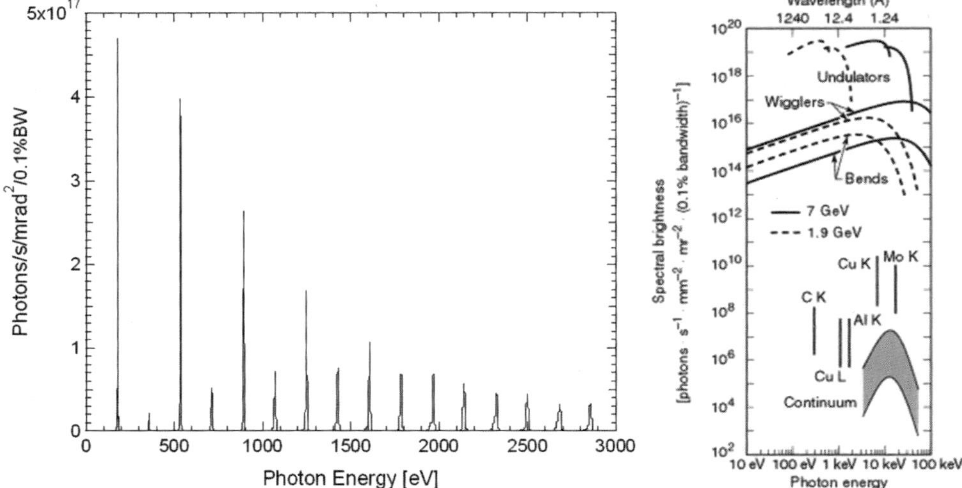

Fig. 6. ***Left***: Spectral distribution of synchrotron radiation from an undulator source (calculated using the SPECTRA program (7) for an ALS U5 undulator). ***Right***: Brightness of bending magnet, wiggler and undulator sources of the ALS and APS storage rings as compared to X-ray generators. Image reproduced from the "X-ray data booklet" (15), with permission.

The magnetic field in an undulator is relatively weak and the deflections of the electron beam are within the opening angle of the radiation cone. The radiation adds up coherently and the resulting interference produces a spectrum where certain wavelengths are amplified while others are suppressed (see Fig. 6, left). The peaks are called harmonics and their intensity scales with N^2. This results in a much higher achievable flux than with bending magnets or wigglers. The central cone of undulator radiation is narrower (see Fig. 5, right) than that of bending magnets and wigglers, which also contributes to their higher brightness. In fact undulators are currently the brightest sources of X-rays available (see Fig. 6, right). The photon energy of the undulator harmonics depends on the electron energy, the period length of the magnetic structure λ_u (see Fig. 5, right) and the strength of the magnetic field. To adjust the energy of the harmonics, the magnetic field strength is changed by mechanically adjusting the gap between the top and the bottom arrays of magnets because most undulators are made of permanent magnets. The photon bandwidth of the first harmonic is given by $\Delta\lambda/\lambda = 1/N$. This bandwidth is not narrow enough for most experiments, and thus, undulator radiation has to be monochromatized further. For large energy changes the monochromator settings and the undulator gap are adjusted together to retain maximum flux.

2.1.4. Future Sources

Naturally, both X-ray generators and synchrotron radiation sources are steadily evolving and improving in performance. Better cooling, higher reliability of operations, lower maintenance, tighter focusing of the electron beam onto the anode, and better focusing of the X-rays are the main areas where advances have been significant for X-ray generators. At synchrotron radiation sources the introduction of the top-off mode resulted in higher brightness photon beams with constant intensity and continuous operations. Stable, high intensity beams a couple of microns in diameter are now possible. In fact, at many modern beamlines the major limiting factor is radiation damage of the samples and not the intensity or other properties of the X-ray beam.

Among the more revolutionary developments in recent years are two new X-ray sources, one shrinking the synchrotron into a room sized device and the other pushing brightness into new territories. Both of these light sources will no doubt have a major impact on X-ray science, albeit in very different ways.

The Compact Light Source ("desktop synchrotron") is being developed by Lyncean Technologies, Inc. in Palo Alto, CA. The basic idea is to replace the conventional undulator with a laser beam (see Fig. 7). The electrons and the laser beam move in opposing directions and the electromagnetic field of the laser acts like an undulator with a very short period length and a large number of periods (20,000). If the laser beam wavelength is 1 μm, 1 Å X-rays can be produced with only 25 MeV electron energy. This dramatic drop in electron energy enables the reduction in size of the accelerator to fit into a home lab. The properties of the emitted radiation are similar to those originating from a regular undulator. The peak photon energy of the undulator

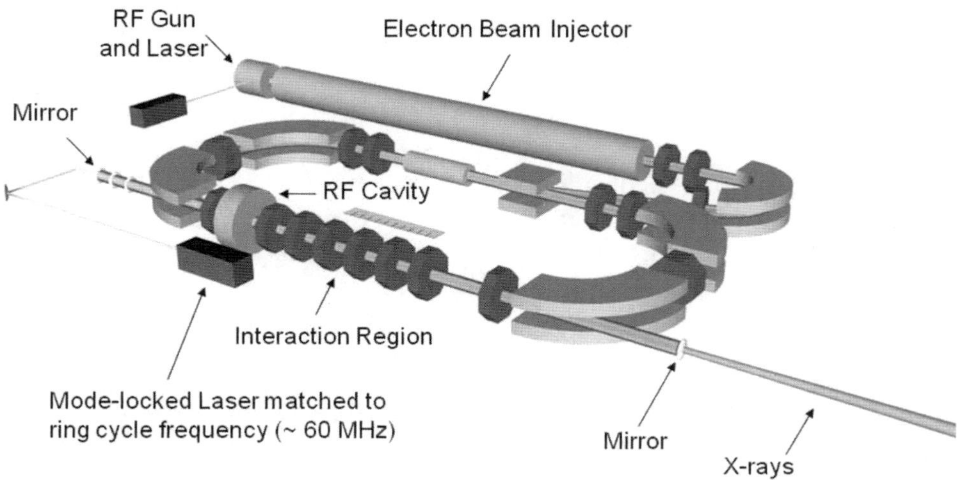

Fig. 7. Schematic drawing of the Compact Light Source (CLS). Image reproduced with permission from Lyncean Technologies (16).

harmonics can be changed by tuning the electron energy in the accelerator (since there are no magnets involved there is no gap to change). The X-ray beam can be further focused and monochromatized with standard beamline components used at large synchrotrons.

The other new development is the X-ray laser. Lasers had a major impact on many facets of technology during the past decades. They produce intense, highly collimated, monochromatic, and coherent radiation from the infrared to the ultraviolet range of the electromagnetic spectrum. Achieving these properties at much shorter wavelengths would provide the ultimate research tool for scientists. In a traditional laser, a resonator cavity is made up of an amplifying medium (gas or solid) enclosed between two mirrors. Light travels back and forth between the mirrors many times and in each pass gets amplified by stimulated emission from the medium. Owing to the absorption of X-rays by matter such a laser setup cannot be realized for photons in the kiloelectronvolt range. It is possible, however, to achieve amplification without excitation of atoms or molecules, by utilizing interactions between free electrons and light. Such a device is called a free electron laser (FEL). Furthermore, the amplification has to happen in one pass, because of the lack of suitable mirrors for X-rays. This can be realized by ensuring a very long interaction region between the electrons and the radiation inside a very long undulator. The technical challenges in building an X-ray FEL are enormous due to the extremely tight tolerances and high precision required of all components (e.g., mechanical alignment of components, electron beam properties, magnetic fields).

The first X-ray FEL to become operational is the Linac Coherent Light Source (LCLS) at the Stanford Linear Accelerator Center (SLAC). A LINAC of 1 km length accelerates electrons to ~14 GeV energy, which then pass through 33 undulators over a 120 m distance (see Table 3) to produce X-rays of ~8 keV energy. These X-rays are coherent, a billion times brighter than any other source and have a short pulse length of ~100 fs. The high brightness might allow the determination of the structure of single molecules without the need for crystals (17). The short pulse duration should help overcome radiation damage by generating an image before damage can occur. If such experiments can be realized in the future, they could shed light on the structure of macromolecules that are difficult or impossible to crystallize.

2.2. X-Ray Optics

Methods of generating X-rays are described in the previous sections. To perform an actual diffraction experiment on a small macromolecular sample the photons have to be delivered and tightly focused onto the crystal while minimizing the loss if intensity. The photon beam also has to be monochromatized because even undulator radiation has too large a bandwidth to be used directly. The devices used to focus and monochromatize X-rays are presented in the following sections.

Table 3
Basic parameters of the ALS wiggler, an APS undulator, and a single undulator of the Linac Coherent Light Source (LCLS) (33 of these are combined for the free electron laser (FEL))

	ALS wiggler	APS undulator A	LCLS
Period length (cm)	11.4	3.3	3.0
Length (m)	3.2	2.4	3.4
Magnetic field (T)	1.83	0.9 (Maximum)	1.33
Number of periods	28	72	113
Gap (mm)	13.7 (Typical)	10.5 (Minimum)	6
Photon energy (keV)	0.1–20	2.9–60 (First, third, fifth harmonics)	8.3

2.2.1. Mirrors

Lenses for visible light are based on refraction: when light passes between materials with different indices of refraction (e.g., between air and glass) their path is deflected according to Snell's law. The index of refraction of all materials is ~1 at X-ray wavelengths and thus a refractive lens thin enough to transmit most X-rays cannot be realized for practical purposes. The index of refraction of vacuum is exactly 1 for all wavelengths, whereas for X-rays it is slightly below 1 for all materials. Thus, if X-rays hit a surface at a small grazing angle (i.e., almost parallel to the surface), they experience a transition to a lower index of refraction and will be reflected. This phenomenon is called "total external reflection" in analogy to total internal reflection for visible light (e.g., when light traveling in water reflects off the water–air boundary). Very high reflectivities can be achieved (see Fig. 8, left) if the grazing angle is sufficiently small and the right material is chosen.

As a consequence of the low grazing angle the mirrors have to be very long. At 0.1° angle an incoming 1 mm^2 diameter parallel beam illuminates a 1×573 mm^2 area on a flat surface. To capture the generally divergent incoming beam, X-ray mirrors are usually a few cm wide and up to 1 m long. Depending on the purpose of the mirror, its shape can be planar to simply deflect the beam, cylindrical to focus in one dimension or toroidal to focus in two dimensions (see Fig. 8, right). There are a number of technical challenges in the manufacturing, metrology and operations of X-ray mirrors. They have to be made of materials that can be machined to very high precision and that have a low thermal expansion (e.g., silicon carbide, Si, fused silica). Cooling channels inside the mirror are often necessary to compensate for the large thermal load from the source, which can distort the mirror surface leading to defocusing of the beam. The surface roughness has to be on the order of a few Ångstrom r.m.s. to maintain high reflectivity.

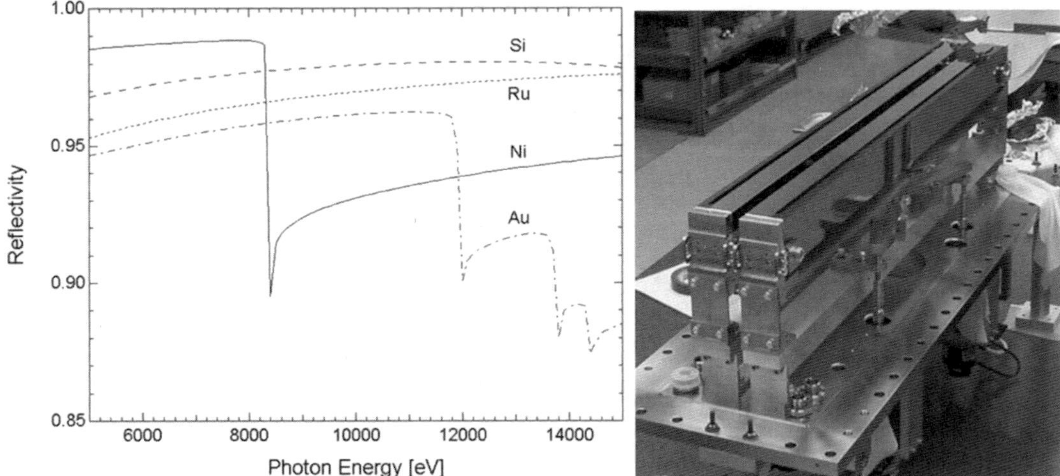

Fig. 8. *Left*: Reflectivity of mirrors at an incidence angle of 0.1° and a surface roughness of 1 nm coated with different elements. The reflectivity curves were calculated using (18). Depending on the photon energy the appropriate coating material is chosen. *Right*: Two X-ray mirrors (each 900 mm long × 51 mm wide) of the ALS sector 5 beamlines 5.0.1 and 5.0.3 (19). The mirrors are made of Si and are coated with Rh/Pt. They are shown before being bent to a cylindrical shape.

A cylindrical or other shape of the mirror along the path of the X-rays is achieved by bending the whole mirror with a mechanical system. The bending radius can be as large as several kilometers and the measurement of such a small deviation from planarity requires special metrology based on laser interferometry. When in operation as part of a beamline, mirrors are mounted inside vacuum vessels and have to be remotely adjustable in position, angle and often bending radius with micrometer precision to align and focus the beam onto the sample.

2.2.2. Monochromators

To avoid marked broadening of the diffraction spots, a monochromatic X-ray beam with a resolution of ~0.1% or better is needed for protein crystallography experiments. However, the X-rays generated by the sources described in Section 2.1 have a broad energy bandwidth for Bremsstrahlung, bending magnet and wiggler radiation, and an energy resolution of a few percent for undulator beams. Some characteristic X-ray lines can be very close to each other, e.g., the Cu $K_{\alpha 1}$ and Cu $K_{\alpha 2}$ lines are 20 eV apart and have widths of 2.3 and 3.3 eV, respectively (20). As a result, the radiation from all these sources has to be monochromatized, either to increase the resolution or to separate nearby emission lines (and also to separate them from the Bremsstrahlung continuum).

Crystal Monochromators

For hard X-rays monochromators are made of crystals with an appropriate lattice constant. Silicon is the most commonly used material for several reasons: very high quality crystals can be obtained in the sizes necessary, it has a low thermal expansion

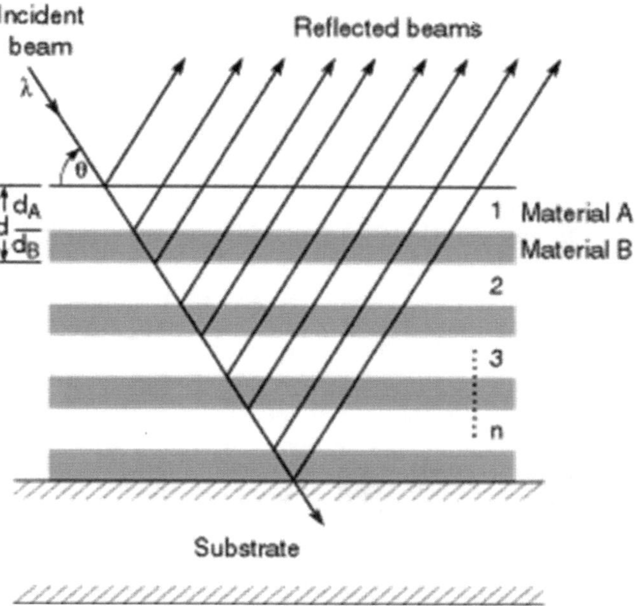

Fig. 9. Reflection of X-rays on a multilayer structure based on Bragg's law (22). Image reproduced from the "X-ray data booklet" (15), with permission.

coefficient and thus deformation is limited under the heat load of radiation, Si(111) and Si(220) have precisely known and suitable lattice constants for the required X-ray wavelengths (Si(111): $2d = 6.27$ Å, Si(220): $2d = 3.84$ Å).

The interference between scattered waves of radiation is the basis for diffraction experiments. Waves that are in phase (with a phase shift of $n2\pi$, where n is an integer) will constructively interfere. This phenomenon is described by Bragg's law:

$$2d\sin\Theta = n\lambda, \quad (2)$$

where λ is the X-ray wavelength, Θ is the angle of incidence, and d is the spacing between atomic layers (see Fig. 9). Based on Bragg's law, the wavelength of the monochromatized radiation can be changed by changing the angle of incidence (in practice by rotating the crystal). In case of a single crystal monochromator, rotation also moves the diffracted beam and the experimental setup has to be moved accordingly. For some applications this might be acceptable, but for synchrotron beamlines such a setup is very cumbersome (although it exists) and thus single crystal monochromators are used primarily for fixed-energy beamlines (see Fig. 10). To avoid the motion of the monochromatized beam described above, a double crystal monochromator can be used (see Fig. 11). In this configuration, the beam is bounced between two parallel crystal surfaces, which are rotated together. The incoming and outgoing

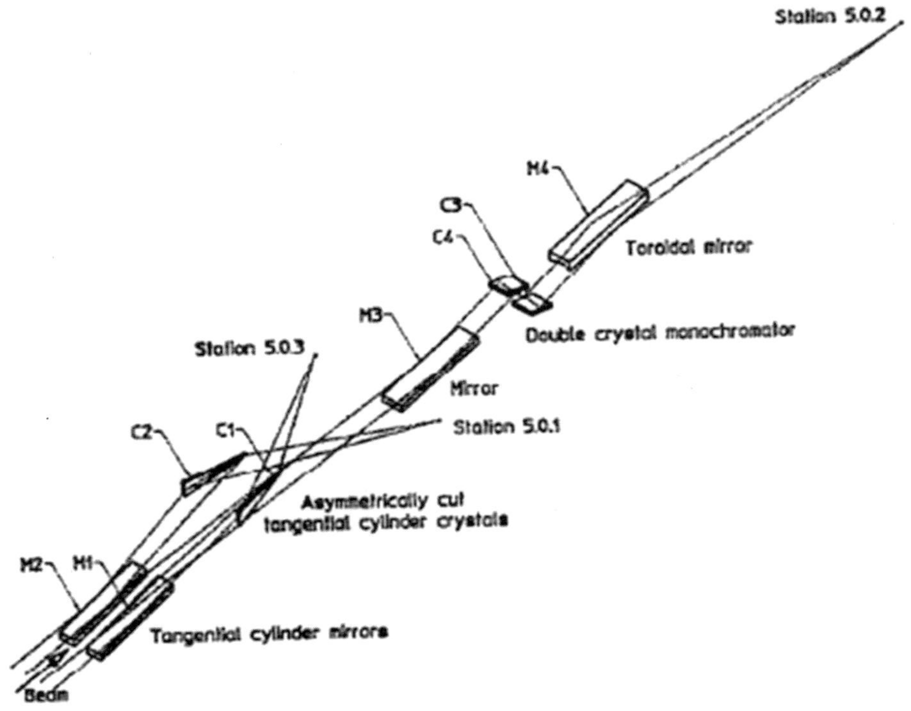

Fig. 10. Schematic layout of the optical components of the three sector 5 beamlines at the ALS. Reprinted with permission from ref. (24). Copyright 1995, American Institute of Physics. The assembly containing mirrors M1 and M2 of the monochromatic side stations 5.0.1 and 5.0.3 is shown in the *right* of Fig. 8.

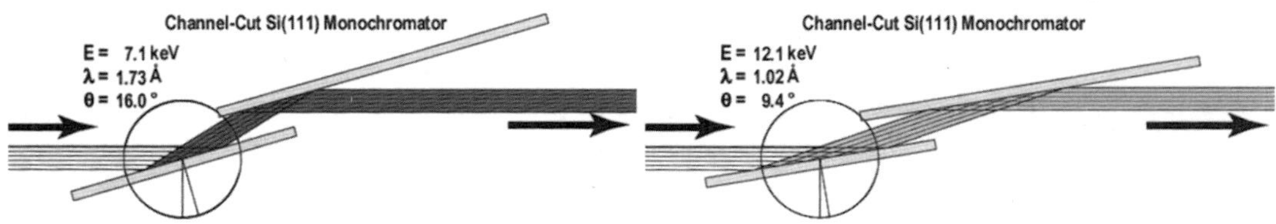

Fig. 11. Channel-cut Si(111) double-crystal monochromator with different incidence angles resulting in monochromatized radiation of different wavelengths (21). Image reproduced with permission.

beams stay parallel during rotation and the outgoing beam stays in the same position. This enables the delivery of a monochromatic beam with tunable wavelength to the same fixed point in space. Parallelity between the two crystals can be ensured either by manufacturing them with high precision from the same block of material (called a channel-cut crystal) or by making the second crystal slightly adjustable. In the so-called sagittally focusing double crystal monochromator, the second crystal is bent perpendicularly to the X-ray beam thereby also focusing the beam. The energy resolution of beamlines based on Si(111) double crystal monochromators is typically $3–8 \times 10^{-4}$.

Multilayers

Both X-ray mirrors and crystal monochromators have some inherent limitations as described above in Section 2.2.1 and "Crystal Monochromators." X-ray mirrors can achieve a high reflectivity only at small grazing angles (~0.1°) resulting in a large reflecting surface. For crystal monochromators, the usable energy range, the incidence angle and also the energy resolution depend on the lattice constant *d* (Equation 2), which is an intrinsic property of the material being used. It would be very beneficial for scientific research if the monochromator/mirror properties could be "tuned" to match the requirements of the experiment at hand. Optical components based on multilayers offer this kind of flexibility for a variety of applications (22).

Multilayers consist of alternating layers (tens to hundreds) of high and low-*Z* materials with thicknesses of a few nm (see Fig. 9). The thickness of a double-layer corresponds to the lattice spacing *d* in a crystal and the reflection of light from a multilayer is also described by Bragg's law (Equation 2). Depending on the materials used, the period length, the ratio of the two materials within a double-layer and the number of periods, multilayers can be manufactured for a wide energy range (13 eV to 21 keV (23)). Multilayers act as mirrors and monochromators at the same time, with a bandwidth in the 0.1–10% range and reaching reflectivities close to 100%. In the X-ray region, the main focus has been the development of high-reflectivity multilayers for Cu K_{α} radiation (~8,050 eV) with tungsten often used as the high-*Z* material and boron carbide (B_4C) or silicon as the low-*Z* material (23). For example, a W/C multilayer with 200 layers of 3 nm thickness has a reflectivity of ~80% and a bandwidth of ~3% at 1.5° grazing angle (23).

New X-ray generators for crystallography are equipped with multilayer optics to efficiently collect and focus a large solid angle of radiation from the anode and to suppress neighboring emission lines and the continuous background at the same time. In this application, multilayers are often further optimized by depositing them on a curved surface for beam focusing and varying the layer thickness either laterally or in depth. At synchrotron beamlines multilayers are used instead of crystal monochromators when high flux is needed and low bandwidth is sufficient (e.g., for small-angle X-ray scattering experiments).

2.3. Beamline

The purpose of a beamline is to deliver a focused and monochromatized X-ray beam to the sample from the source inside the storage ring. The beam properties at the sample are therefore a combination of the source parameters and the beamline specifications. A typical beamline consists of one or more X-ray mirrors, a monochromator, beam diagnostics (such as intensity and position monitors), and beam defining slits. All components are installed in vacuum vessels and are connected by vacuum tubing to minimize absorption of the beam and damage to the optical components.

The beamline is usually separated from the main storage ring vacuum system by a Beryllium window (Be is practically transparent to hard X-rays).

Some technical feats in the construction and operation of a synchrotron beamline are transparent to the end user. All optical components, and the vacuum tanks they are mounted in, have to be installed in the right location with micrometer precision requiring sophisticated metrology. To optimize the beam characteristics once the beamline is operational, the optical components have to be adjusted remotely with high precision, reliability and reproducibility. The deformation of mirrors and monochromators due to the heat load has to be minimized by efficient cooling systems, which sometimes use liquid nitrogen as coolant. Lastly, to provide a stable and optimal beam for an extended period of time to perform the actual diffraction experiment, active feedback systems have to keep the different components in alignment.

Two typical beamline designs are illustrated in Fig. 10. A wiggler source at sector 5 of the ALS provides a wide fan of radiation feeding three separate beamlines (25). One of them (BL5.0.2 in the middle) is an adjustable wavelength MAD beamline consisting of a cylindrically bent prefocusing mirror, a cryogenically cooled Si(111) double crystal monochromator and a toroidal refocusing mirror. This mirror–monochromator–mirror design is common for MAD beamlines, although there are many variations in the details. It gives good performance and adjustability without losing too much beam intensity. The other two beamlines (BL5.0.1 and BL5.0.3) are monochromatic, each consisting of a single cylindrical mirror and a single cylindrically bent Si(220) monochromator crystal. Such a simple design is possible by using both optical elements also for focusing covering both the horizontal and vertical planes. Details of the current performance of the ALS sector 5 beamlines can be found in (26).

All the X-ray beam requirements for protein crystallography listed above are very well fulfilled by modern synchrotron beamlines, as can be seen from the typical parameters of beamlines listed in Table 1. An up-to-date list, including detailed specifications, of structural biology synchrotron beamlines worldwide can be found on the Biosync Web site (27). More details about macromolecular crystallography beamlines can be found elsewhere (28).

2.3.1. Experimental Station (Endstation)

Overview

The section of the experimental setup where the diffraction experiment takes place is called the endstation. A strict separation between the beamline and the endstation is not possible due to the proximity of the components. The main optical elements (mirrors and monochromator) and their supporting systems can be considered part of the beamline, while the components near the sample comprise the endstation.

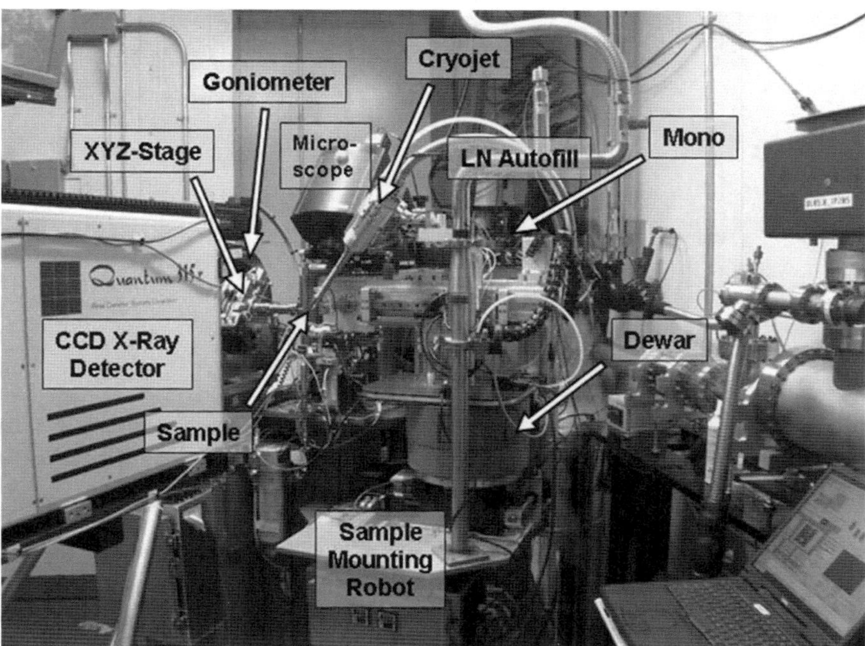

Fig. 12. Overview of the BL5.0.3 endstation at the ALS. The monochromator crystal is mounted in the back of the hutch and the X-rays are moving toward the charge-coupled device (CCD) detector. A detailed description of this endstation can be found in ref. (29).

Owing to the health hazards of hard X-rays and their weak absorption by air, the endstation is located inside a radiation-shielded hutch. The hutch is separated from the rest of the beamline by a beam shutter, which can be opened through a series of interlocks only when nobody is inside. To avoid frequent entries into the hutch, which can be rather time consuming, most of the endstation components can be remotely controlled from the outside. An overview of the ALS beamline 5.0.3 endstation is shown in Fig. 12. A description of the main parts is given in the following sections.

Fast Shutter, Collimator, Slits, Beamstop, Intensity Monitor

The final steps of beam manipulation happen in the endstation. A fast shutter is used to accurately control the exposure time during which the crystal is exposed to X-rays while it is being rotated through the required oscillation angle. It is located upstream of the collimator. A precise coordination of the goniometer rotation and shutter actuation are required to obtain high quality data.

To decrease the beam size beyond the focusing capabilities of the beamline optics, collimators with fixed apertures (e.g., 100 μm diameter) or adjustable slits are used. Fixed diameter apertures are small metal discs with high-precision, laser-drilled holes. A scatter guard is mounted downstream from the collimator to block the X-rays that scatter off the edges of the collimator aperture. The scatter guard is also a metal disc or a metal tube with an opening larger than the collimating aperture (30). The collimator–scatter guard assembly is located close to the sample (~1 cm), leaving just enough room for mounting/dismounting of crystals while not disturbing the flow of the cold stream.

While the collimator decreases the beam size at the sample, the beam size downstream of the sample (i.e., the size of the diffraction spots) can be reduced by decreasing the beam divergence. This might be necessary when the spots overlap on the detector surface due to a large unit cell or high mosaicity. Beam divergence is adjusted with slits upstream of the collimator and typical values are in the 1–3 mrad range (see Table 1).

A beamstop is installed downstream of the sample to block the direct beam from striking the detector. Since the direct beam is orders of magnitude more intense than the diffraction spots, not blocking it (or not entirely blocking it) would generate background radiation that make diffraction measurements impossible and could possibly damage the detector. The beamstop is usually a metal disc with the proper dimensions and absorption characteristics for a given beamline. Some beamstops let a small portion of the direct beam bleed through, which can be helpful to determine the accurate beam position on the detector surface.

To optimize the beam intensity for the diffraction experiment, it has to be measured close to the sample position and downstream of the collimator. A retractable photodiode downstream of the beamstop is often used. Since this requires the removal of the beamstop from the beam and insertion of the diode into the beam, it cannot monitor the intensity during data collection. To overcome this problem, Ellis et al. (31) designed a beamstop with an integrated photodiode, which is shown in Fig. 13.

Goniometer and Crystal Positioning

The goniometer is the endstation component that holds and rotates the crystal in the X-ray beam during data collection. Single-axis goniometers are most common, where the rotation axis is perpendicular to the X-ray beam and usually lies in the horizontal plane. Rotation is provided by an electric motor coupled with a high precision bearing. Modern beamlines use so-called air bearings, where a thin layer of pressurized air keeps the fixed and moving parts separated. As a result, a highly variable rotational speed, high angular resolution, and a low circle of confusion can be achieved (0.01–360°/s, 0.00005°, and <1 μm, respectively, on BL5.0.3 at the Advanced Light Source (29)). The circle of confusion defines how precisely the sample can be kept in one place during rotation and it becomes increasingly more important as crystal and beam sizes decrease.

To remotely center the sample in the X-ray beam, a small motorized positioning stage (*x,y,z* or kinematic) is mounted on the goniometer (see Figs. 12 and 13). This allows movement of the crystal in all three dimensions by a few millimeter. It is controlled

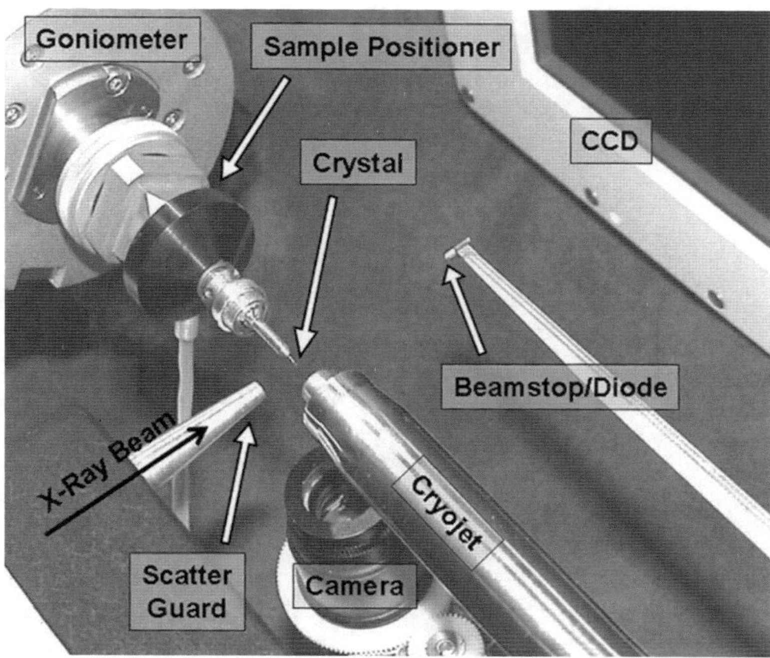

Fig. 13. Sample area on beamline 11-1 of the SSRL. Reprinted with permission from ref. (31). Copyright 2003, International Union of Crystallography.

remotely through a point-and-click graphical user interface, which, in conjunction with the fast goniometer rotation, allows the crystal to be centered within a few seconds. A magnet mounted on the goniometer shaft holds the sample in place during data collection.

Imaging System

Once the sample is mounted on the goniometer it has to be visualized for centering into the X-ray beam and for general inspection (e.g., if there is a crystal in the loop or if the sample is covered with ice). Depending on the beamline, one or more high-magnification video cameras are pointed at the crystal (see Figs. 12 and 13) and their images are displayed on monitors outside the hutch. On newer beamlines in-line viewing is common, so that the path of the incoming X-rays coincides with the video camera viewing direction. Since a camera cannot be mounted in the path of the X-rays, this is accomplished by the use of a small prism or a mirror with a hole for the transmission of the X-rays. To obtain a clear and well-illuminated image, there are usually lights installed both behind and in front of the sample and their intensities can be varied as required.

Sample Cooling

Exposure of macromolecular crystals to X-rays can destroy the crystalline order very rapidly at room temperature. The process of radiation damage can be dramatically reduced by cooling the sample to cryogenic temperatures. For this reason, basically all protein

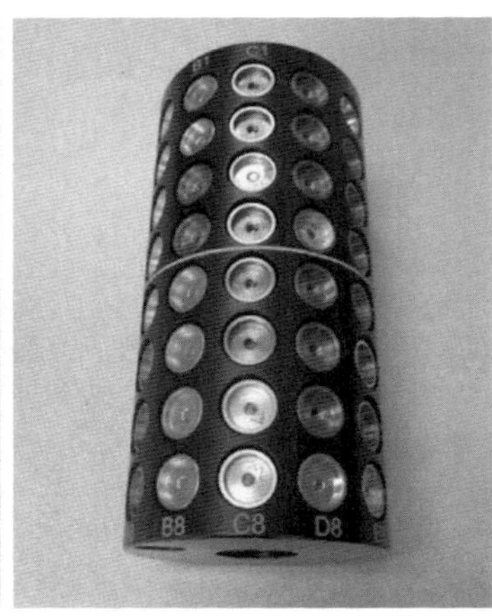

Fig. 14. Sample holder cassettes. The ALS "pucks" (*left*) can hold 16 samples each while the SSRL cassette (*right*) has room for 96 crystals. The Berkeley robot can accommodate six pucks while the Stanford robot can be loaded with three cassettes. Details about these systems can be found in refs. (29, 33).

crystallography experiments today are performed on cryocooled crystals using liquid nitrogen (LN). Owing to its suitable boiling temperature (77.4 K), low cost, and wide availability, LN is well suited for this purpose.

From the time of freezing (usually in the home lab) until exposure to X-rays, crystals must stay cold at all times. This includes sample shipping and handling at the beamline (see Fig. 14). Once the crystal is mounted on the goniometer, it is cooled by a cold nitrogen gas stream with a temperature of 90–100 K. The cold stream is generated by a device called a cryojet or cryostream (see Figs. 12 and 13), which consists of the head containing the nozzle and a separate LN tank. The head is mounted near the sample area (see Fig. 12) with the nozzle pointing at the sample (see Fig. 13). The nitrogen gas is first heated and dried and then cooled down to the required temperature, which can be accurately controlled to better than 1 K and sustained at the sample position (which is 10–15 mm from the tip of the nozzle). To assure an undisturbed laminar flow of the cold nitrogen stream and to avoid sample icing, the cold stream is shrouded in an outer gas stream of dry nitrogen or air. Since this outer stream is not at cryogenic temperature, it is very important that the sample stays well inside the cold stream (ensured by proper alignment).

Sample Mounter

High-throughput SBDD requires the handling of a large number of protein crystals at the beamline and this can be accomplished efficiently only by using automation. The fact that the experiments must be carried out inside a radiation shielded hutch further necessitates the use of automated sample handling. Multiple systems have been developed over the past decade (29, 32, 33) and by now many X-ray generators and most synchrotron beamlines are equipped with one. An up-to-date list of current systems is listed on the Robosync Web site (34).

The main challenge in the development of sample mounting systems was the requirement that the crystal must stay at (or close to) liquid nitrogen temperatures at all times. This means that the sample has to be enclosed in a cold environment while in transition between the sample holding Dewar and goniometer, and the mounting and dismounting processes have to happen within a few seconds to avoid warming up. The Berkeley automounter system (BAM), shown in Fig. 12 as part of the BL5.0.3 endstation setup, can mount or dismount in approximately 5 s by using a sample gripper mounted on pneumatic stages (29). Crystals are stored in a Dewar that can hold 96 samples in liquid nitrogen.

Along with the sample mounters, containers for the easy storage, transport, and handling of a large number of frozen crystals have been developed. Two such systems are shown in Fig. 14. These cassettes fit into common dry shipping containers widely used to send crystals to the synchrotron.

X-Ray Area Detector (CCD)

The purpose of a crystallographic experiment is the accurate determination of the intensities and positions of the diffracted X-rays. Because of the limited availability of synchrotron beamtime and the decay of crystals due to radiation damage the measurements need to be fast, accurate, and efficient. A detector with a large surface area can measure a large number of spots simultaneously and a short readout time can reduce the duration of the experiment. A high enough spatial resolution is required to separate nearby diffraction peaks and to determine the positions accurately. The detector has to have low noise, has to be sensitive enough to measure weak spots, and has to have a large dynamic range to capture weak and strong peaks at the same time.

Charge-coupled device (CCD) based instruments fulfill the above mentioned requirements and currently they are the most common X-ray detectors at structural biology beamlines. The CCD was invented in 1969 by W. Boyle and G. E. Smith (2009 physics Nobel Prize). CCDs are semiconductor based devices containing wells (capacitors) to store charge and have the ability to shift charge between neighboring wells. Originally intended as a memory device ("shift register"), it was soon realized that charge can be generated by light through the photoelectric effect. This discovery laid the foundation for the electronic capture of images, where each well corresponds to a pixel. CCDs can capture light from the near infrared to the ultraviolet and they are widely used in various applications. Once the CCD has been exposed to light, the image is read out by shifting the charge from well to well by rows toward the edge, where each pixel is amplified and digitized. To speed up this sequential readout process, different parts of the CCD can be read out at the same time. Signal-to-noise can be

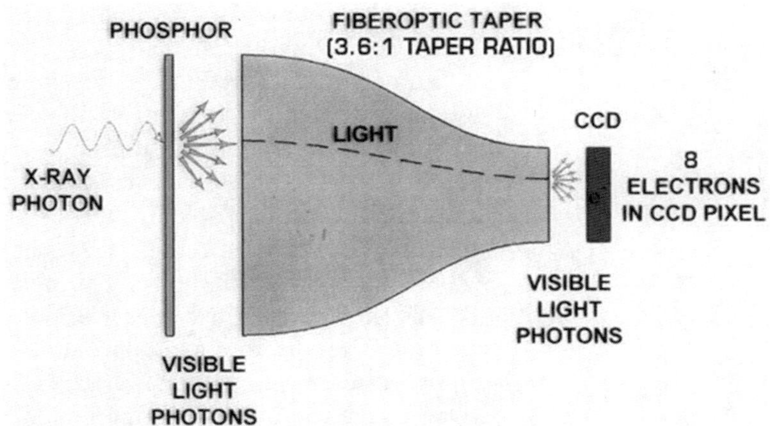

Fig. 15. Schematic overview of a CCD area detector for the measurement of X-rays (35). Image reproduced with permission.

increased by lowering the noise of the sensor, which can be achieved by cooling the CCD. This is especially important in scientific applications; in astronomy CCDs are sometimes cooled with LN, while the detectors for crystallography use thermoelectric cooling (Peltier elements) in conjunction with water cooling to operate at ~−50°C.

For protein crystallography X-rays have to be detected over a large surface area, but CCDs are not sensitive to X-rays and are limited in size to approximately 30–40 mm^2 (tiling 100 chips together would be prohibitively expensive). X-ray detection is achieved by using a phosphor-coated screen, which converts X-ray photons into visible light. The generated image is reduced in size by a factor of 3–4 through a fiberoptic taper, to which the CCD is attached (see Fig. 15). Tiling 4, 9, or 16 such units together in a 2×2, 3×3, or 4×4 array can provide a large sensitive surface with a limited number of CCDs. The specifications of some CCD detectors in use today are listed in Table 4.

Pixel array detectors started to appear recently on protein crystallography beamlines. These devices can detect X-rays directly without the need of conversion to visible light and the active area can be made rather large. They have a very short readout time and large dynamic range; however, their pixel size is rather large and the technology is still new. Specifications of the commercially available Pilatus 6M pixel array detector are also included in Table 4.

Usage of CCD detectors for data collection is described in Section 3.4, along with the problem of overloaded pixels, dark images, and detector binning.

Table 4
Parameters of X-ray area detectors for macromolecular crystallography (35–37)

Manufacturer	ADSC			MAR			DECTRIS
Model	Q210R	Q270	Q315r	CCD225	CCD300	CCD325	Pilatus 6M
Detector type	CCD array (2×2)	CCD array (2×2)	CCD array (3×3)	CCD array (3×3)	CCD array (4×4)	CCD array (4×4)	Silicon diode array (5×12)
Active area (mm)	210×210	270×270	315×315	225×225	300×300	325×325	431×448
No. of pixels	4,096×4,096	4,168×4,168	6,144×6,144	6,144×6,144	8,192×8,192	8,192×8,192	2,462×2,527
Pixel size (μm)	51×51	65×65	51×51	37×37	37×37	40×40	172×172
Spatial resolution (FWHM) (μm)	60	90	60	100	100	100	1 Pixel
Taper ratio	3.7:1	2.7:1	3.7:1	2.7:1	2.7:1	2.9:1	–
Dynamic range	18,100/22,600[a]	41,700	18,100/22,600[a]	65,536	65,536	65,536	1,048,576
Dark current (e/pixel/s)	0.015	0.015	0.015	0.01	0.01	0.01	–
Readout time (s)	0.9/0.25[a]	1.1/0.37[a]	0.9/0.25[a]	1	1	1	0.0036

[a] Full resolution/2×2 hardware binned

3. High-Throughput Data Collection

Many steps of the structure determination process have become increasingly automated in recent years. These developments have been driven by the government funded structural genomics projects (38, 39) and the pharmaceutical/biotechnology industry's desire to implement rational drug design strategies (40). Sample handling for X-ray data collection is no exception and at present many X-ray generators and synchrotron beamlines are equipped with crystal mounting robots and sample tracking systems. The application of these systems in a high-throughput environment is described in the following sections with a focus on obtaining protein–ligand complex structures for SBDD projects at synchrotron beamlines. For the sake of simplicity, it is assumed that the beamlines used are equipped with a single-axis goniometer, a CCD X-ray area detector, and a sample mounting robot.

Takeda San Diego, Inc. (formerly Syrrx, Inc.; "TSD") has been a leader in the field of high-throughput SBDD since its inception in 2000. Its state-of-the-art gene-to-structure platform (41–43) has been employed to determine a number of de novo protein structures that have been implicated in various diseases (44–49) and allowed production of many protein cocrystal structures in complex with drug candidates to guide the SBDD process. TSD's efforts have resulted in multiple clinical candidates (50) and a drug on the market (51).

As part of TSD's protein crystallography platform, all the relevant information on the gene-to-structure process, including the details of crystal harvesting, X-ray screening, data collection, processing, and structure solution, is systematically captured in a centralized database. The author was able to analyze data that has been stored in TSD's database. Throughout Section 3, results of this analysis are used to highlight certain aspects of the topics being described.

The need for sample handling automation and sample tracking can be clearly demonstrated by analyzing TSD's data regarding the attrition rate from crystals harvested into loops to structures solved. At TSD, of a representative selection of approximately 20,000 harvested crystals (100%) ~18,400 were screened (92%) resulting in ~3,000 datasets (13%) and ~800 (co)-crystal structures (4%). This corresponds to a total attrition rate of ~25, underscoring the need to be able to handle a large quantity of crystals efficiently to obtain the desired number of high-quality structures.

3.1. Preparations

The specifications of end-stations and beamlines for protein crystallography can vary considerably (see Table 1), thus the requirements of a specific project have to be examined before deciding which facility to use. MAD experiments require tunable wavelength whereas most cocomplex work can be done at monochromatic sources. For small crystals (<20 μm) a small and intense beam is more advantageous (30). Crystals with large unit cells benefit from a beam with low-divergence, a large detector with 2-Θ capability and a large sample-to-detector distance to separate nearby spots. On the other hand, crystals that diffract well require a small sample-to-detector distance to capture the high-resolution data.

For high-throughput X-ray data collection a sample mounting robot is essential. The different robots found at synchrotron beamlines handle different kinds of storage cassettes (see Fig. 14) and cryoloops and one should bring samples in a compatible format. It is often possible to borrow the cassettes and tools needed for a specific beamline in case the user does not own them.

It is highly recommended to double-check beamline specifications with staff before deciding on a specific facility, as listed/published features can often change. It is also a good idea to confirm the availability of needed features at the beginning of a data collection run to avoid unpleasant surprises "in the middle of the night." Scientists from commercial entities should bear in mind that there are special rules and extra costs for proprietary research at government-run synchrotron sources and those should be checked beforehand.

3.2. Sample Tracking

Automated handling of samples only works efficiently in parallel with some form of crystal tracking system, which can be as simple as a spreadsheet. Most beamlines equipped with an automounter will let the user upload spreadsheets with crystal information into the beamline control system (templates are usually available from the facility's Web site). The minimum information needed for work at the beamline is a list of all samples containing unique crystal identifiers and the locations of the crystals, i.e., information on which cassette and which position in the cassette they are stored.

The bulk of TSD's X-ray data collection work has been performed at ALS beamline 5.0.3. In order to streamline operations and enable fast and accurate information exchange, the TSD database is securely connected over the internet to the MySQL database at BL5.0.3. This has allowed basic crystal information to be sent from TSD to the beamline, and crystal screening and data collection details to be transferred back to TSD. By contrast, when a beamline with no direct connectivity is utilized, spreadsheets or direct entry needs to be used.

Once the cassettes containing the samples have been loaded into the automounter Dewar and the crystal information has been uploaded into the beamline control system, the actual work on crystal screening and data collection can begin.

3.3. Crystal Screening

The purpose of crystal screening is to find the best crystal in terms of diffraction quality from a group of similar crystals, such as crystals of proteins complexed with the same compound.

3.3.1. Crystal Screening Steps

The typical crystal screening workflow comprises the following steps for each sample:

1. Mount crystal from LN dewar on to goniometer.
2. Center crystal in the X-ray beam.
3. Obtain diffraction images, typically two shots 90° apart.
4. Evaluate diffraction.
5. Pause for data collection or further screening.
6. Dismount crystal from goniometer and return it to the dewar.

Mounting and dismounting of the crystals is done by the sample handling robot installed at the endstation. The crystal is usually centered by the user through a "point-and-click" interface but several beamlines also offer automated crystal centering routines. There are different ways to accomplish automated centering, such as centering the cryoloop based on its outline or centering the crystal based on X-ray diffraction or fluorescence (52–54). Manual centering is in most cases faster and more reliable, whereas automated centering may be most efficient when incorporated into a fully automated crystal screening protocol.

The settings for the screening of diffraction images should be similar to typical data collection settings on that beamline. These can be learned from beamline staff. As a rule of thumb, a wavelength around 1 Å, spindle oscillation angle of 1°, exposure time of a few seconds and a detector-to-sample distance of 250 mm (for a detector size in the 300–325 mm range) are, in general, appropriate parameters. Keep these settings the same for a group of similar crystals for accurate comparison. To speed up the screening process, dismounting of a crystal and mounting of the next one should proceed while the snapshots are being evaluated.

Sometimes during screening the diffraction quality of the mounted sample is good enough to collect a dataset without screening more crystals and most screening protocols will pause before dismounting the sample. This also offers the opportunity to collect additional screening images if needed. Large samples should be tested at different areas of the crystal as some regions might diffract better than others. Multiple crystals in the same loop or hardly visible small crystals can require testing multiple areas of the cryoloop. For this purpose, some beamlines are equipped with automated "raster" screening protocols, where a loop is divided into a grid and every element of the grid is tested for diffraction.

3.3.2. Evaluation of Screening Images

When evaluating the screening images, the type of diffraction should be determined first: no diffraction, salt or protein. No diffraction

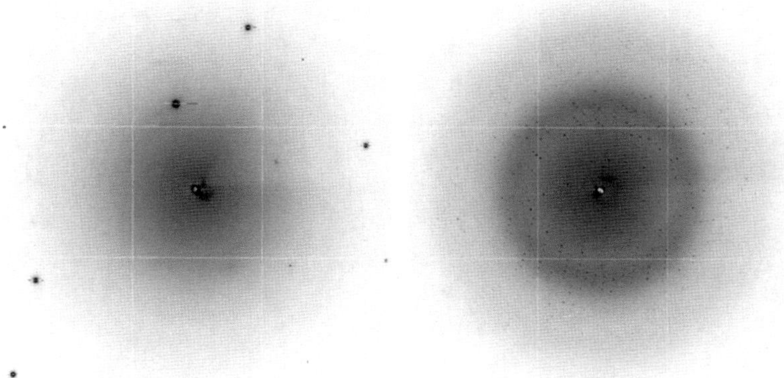

Fig. 16. Diffraction image from a salt (*left*) and a protein crystal (*right*). All diffraction images were recorded by an ADSC Q315r CCD area detector on beamline 5.0.3 at the Advanced Light Source and displayed by the ADXV image viewer program (55).

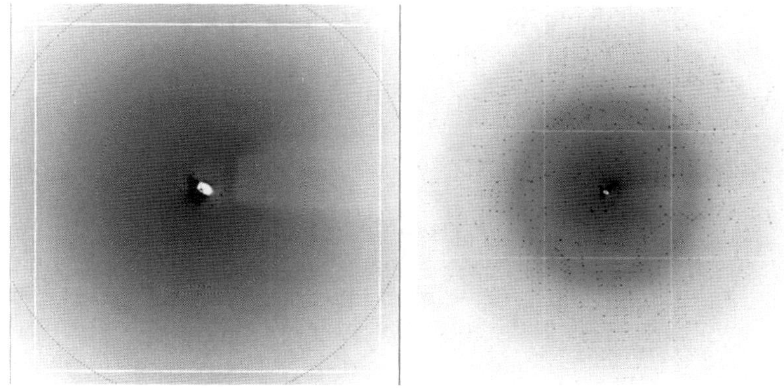

Fig. 17. ***Left***: Low-resolution diffraction spots (30–35 Å) from a protein crystal. ***Right***: Diffraction from a split crystal showing multiple overlapping lattices.

means that there is either no crystal in the loop or that there is one but it is not positioned properly in the X-ray beam. If in any doubt, it is advisable to recenter the crystal and collect more images. Owing to their small unit cells, salt crystals produce few, but very strong diffraction spots (often overloading the detector) and thus are easy to identify in most cases (see Fig. 16, left).

Diffraction from a protein crystal (see Fig. 16, right) should be always carefully analyzed. Even if the diffraction quality is not adequate for data collection, accurate screening results will further aid crystallization trials. Figure 17 shows a few spots arranged in rows close to the beamstop, clearly indicating low-resolution diffraction from a protein crystal and this finding should be noted in the screening results. Typical parameters to be determined from the screening snapshots are resolution limit, mosaicity, diffraction strength, spot shape, anisotropy, and the presence of ice rings.

Diffraction experiments are based on order inside crystalline samples. The higher the order, i.e., the positions of atoms are defined more accurately, as well as more uniformly from unit cell to unit cell in the crystal, diffraction spots with higher *hkl* values will appear. Higher *hkl* values correspond to diffraction spots with a larger deflection angle from the primary beam and thus will be located further away from the beam center on the detector surface. The location of the outermost visible spots is measured in Ångstrom and is called the resolution (or resolution limit) of a particular diffraction pattern. It is best to determine the resolution in the corners of the diffraction image because spot intensities are enhanced along the spindle rotation axis (usually the horizontal) and yield a too optimistic estimate for the resolution.

Protein crystals are not perfect. They can be described as a collection of small perfect crystals (also called mosaic blocks) with each having a slightly different orientation angle. The width of the distribution of these angles defines the mosaic spread, which is measured in degrees. Signs of high mosaicity are a large number of diffraction spots (often suppressing a clear view of the lunes) and tangentially streaky spots (see Fig. 18). At TSD, mosaicity is classified into low, medium, and high categories.

Diffraction strength describes the overall intensity of diffraction spots in the image. Every image consists of a number of pixels corresponding to the resolution of the CCD detector (see Table 4). Every diffraction spot is made up of multiple pixels and the sum of those pixel intensities yields the spot intensity. Image viewer programs (e.g., ADXV (55)) show the average intensity of all pixels, the maximum intensity and the number of overloads. By magnifying a section of the image individual pixels can be also examined. Based on all this information and the overall visual appearance of the diffraction pattern, TSD categorizes diffraction strength into weak, medium, and strong.

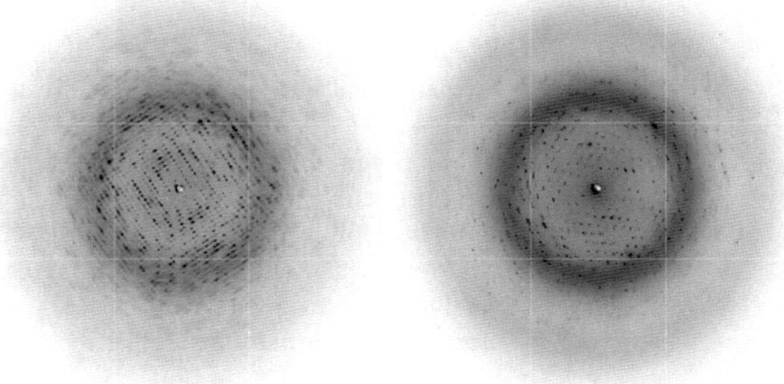

Fig. 18. Examples of diffraction from highly mosaic crystals. Data processing yielded 1.7° mosaic spread at 2.4 Å resolution (*left*) and 1.1° mosaic spread at 1.9 Å resolution (*right*).

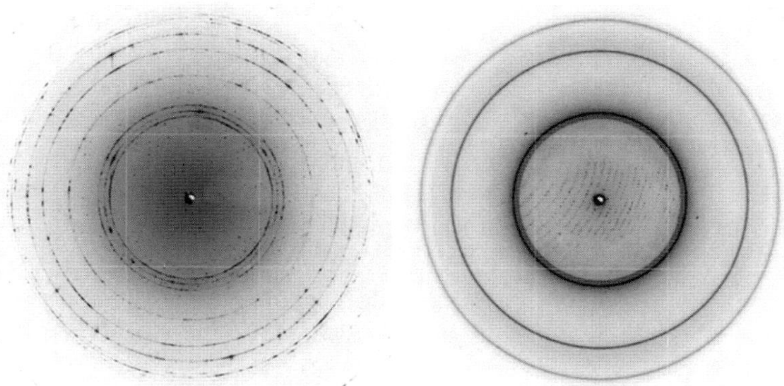

Fig. 19. Examples of ice crystal diffraction rings. The pattern on the *left* is typical of ice located on the surface of the cryoloop.

Spot shape can be evaluated visually, it helps to magnify a region of the diffraction image for this purpose. At TSD, spot shapes are classified into clean, streaked and" split categories. Streaky spots can be an indicator of high mosaicity (see Fig. 18), whereas split spots signal the presence of multiple lattices (see Fig. 17).

Anisotropic diffraction can manifest itself in a single diffraction image, where the spots are roughly within an ellipse and not a circle. It can also mean that the diffraction quality is very different between the images taken 90° apart. All the quality indicators should be noted conservatively in these cases (e.g., a few high-resolution spots in one orientation are not going to result in a complete dataset at that resolution).

Diffraction from ice crystals can be recognized by the strong diffraction rings at 3.9, 3.7, 3.5, and 2.3 Å resolutions (and other, weaker rings) as shown in Fig. 19. Ice rings can originate both from the inside and the surface of the protein crystal. Internal ice is a result of inadequate cryoprotection and can be sometimes recognized by a nonoptically transparent cryoloop. The corresponding ice rings are often broad without any sharp features. Ice buildup on the surface of the cryoloop occurs during crystal handling and is usually clearly visible. The corresponding ice rings are narrow and sharp.

An increasing number of beamlines are equipped with automated crystal analysis software, which can help determine the parameters described above. A popular version is called Web-Ice, which was developed at Stanford Synchrotron Radiation Laboratory (SSRL) and includes the LABELIT (56), and DISTL (57) software packages.

3.3.3. Crystal Screening at Takeda San Diego

As described above, all crystallography related data are stored in a database at TSD. This has enabled a closer inspection of the screening results for a large number of crystals as well as trends within a set of crystals. Of a selection of approximately 17,000 screened crystals, 12% did not diffract and only 2% were salt crystals. These low numbers can be explained by the high percentage of protein/small-molecule cocomplex structures for which the crystallization conditions have been previously optimized. Only 10% of the screened crystals showed diffraction from ice, indicating that crystal freezing and handling is well under control.

To find out how well the manual screening evaluation really describes diffraction quality, the screening results and the processing statistics were compared for those crystals for which a dataset was collected. The average mosaicity (as measured by the averaged angular width of the diffraction spots from a crystal) of all crystals processed (approximately 2,500 of the 17,000) is 0.59°. For those which were defined as low, medium or high mosaicity during screening, data processing yielded measured average mosaicities of 0.49°, 0.81° and 1.05°, respectively. Similar, well correlated behavior was found for the diffraction strength measured as the overall spot intensity-to-background (I/σ) ratio (I/σ being 15, 18 and 22 for weak, medium and strong diffraction respectively).

Comparing the true data resolution with the values from screening reveals a clearly linear dependence (see Fig. 20). It also underscores the notion that during data collection a somewhat better resolution can be achieved than that based on the visible spots on a snapshot (see the equation in Fig. 20). This is mainly due to averaging of symmetry related diffraction spots, which improves the signal-to-noise ratio. The simple evaluation procedure at TSD clearly gives a good assessment of the diffraction quality of the crystals under investigation.

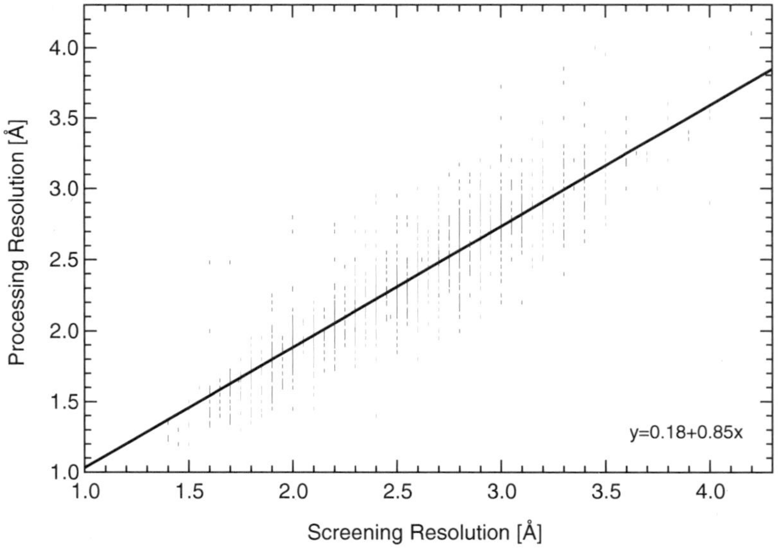

Fig. 20. Comparison between screening and processing resolution values.

3.3.4. Crystal Selection for Data Collection

Before data collection can begin the best crystals have to be selected from the screening process. High-resolution, low mosaicity, strong diffraction, clean spot shape and the lack of ice rings are the prized crystal properties. These conditions are not always all met and some tradeoffs have to be made, e.g., a somewhat higher mosaicity can be accepted (up to approximately 1°) in return for higher resolution. Streaky or split diffraction images should be indexed (if the crystal is otherwise promising), and if a solution is easily obtained, the crystal will most likely yield a useful dataset. Ice contamination should not prevent data collection, unless it is very severe or the resolution limit is close to the ice ring at around 3.5 Å (sometimes ice can be "washed off"). Looking at the impact of ice rings on a number of datasets at TSD, it was found that the overall completeness was reduced only by 1% and the impact on overall R_{sym} was negligible.

If time permits, and the desired structure important enough, data should be collected on as many crystals as possible, even on crystals that are not of the highest quality. At TSD, useful structures for SBDD were obtained from resolutions as poor as 3.5 Å, mosaicity as high as 2.4° and overall completeness as low as 80%.

As a last resort, annealing of a crystal can be tried to improve diffraction quality. During this process, the crystal is warmed up to room temperature (by blocking the cryostream for a few seconds) and cooled down again. In some cases, the lattice will rearrange to a system with superior order and diffraction resolution improves. Occasionally this method works but usually it completely destroys the crystal!

3.4. Data Collection Strategies

There are a number of parameters that have to be set up for each data collection run on the chosen samples; how to optimize those parameters is the subject of this section. The screening results and possible prior knowledge of the project should provide most of the necessary information for this purpose. The goal of data collection is to accurately measure a complete set of diffraction intensities (58). Only monochromatic data collection is discussed here for the purposes of solving structures by the molecular replacement method and to obtain cocomplex structures for SBDD.

With the worldwide proliferation of synchrotron light sources in recent years, availability of beamtime is no longer a limiting factor. For this reason, even in a high-throughput setting, the focus can be shifted from quantity to quality. The best possible datasets should be collected for all samples. This trend is accentuated by the significant performance increase of beamlines, leading to very high intensities and small beam sizes. Interestingly, these developments put another kind of time limit on the data collection process, namely the "survival time" of the crystals before radiation damage occurs (59). Radiation decay, or damage, is the process whereby the energy absorbed by the sample from the impinging X-rays gradually destroys the order in the crystal and thus reduces its diffracting power.

3.4.1. Crystal Symmetry and Unit Cell

Crystal symmetry and unit cell dimensions are crucial pieces of information when determining the correct data collection parameters. For crystals of an ongoing SBDD project, symmetry and unit cell will be already known, and if indexing of the screening images confirms this no more steps are necessary. The situation is similar in the case of a new project, for which a structure is already published, and indexing confirms the published parameters. The crystal information is not known for de novo structures or if the crystal properties change for complexes with different compounds. In these cases finding the proper symmetry is a multi-step process that is described in detail in Section 3.5.5.

3.4.2. Data Collection Parameters

Typical parameters that can be selected by the user for data collection are as follows: X-ray energy, detector-to-sample distance, detector 2-Θ angle, exposure time, oscillation angle, starting rotation angle, rotation angle range, beam size, beam divergence, and detector binning. How to find the best possible settings to obtain an accurate and complete dataset for a given sample is described in detail in the following sections. Optimization is done by collecting a few diffraction images (snapshots) before starting the data collection run and adjusting the parameters.

X-Ray Energy

On monochromatic beamlines the photon energy cannot be changed, but its value should be noted for later reference. On tunable beamlines multiple considerations can be taken into account. The beam intensity usually varies with photon energy, and a value corresponding to the highest flux can be selected. Most CCD area detectors are optimized for maximum sensitivity at 1 Å wavelength. At lower energies there is an increase of X-ray absorption by the sample (resulting in increased radiation damage) and also by the air surrounding the crystal (leading to a higher scattered background and attenuation of the diffracted X-rays). At higher energies radiation damage and air scatter will be lower. In general an energy close to 12.4 keV (1 Å) is a good choice for most practical purposes. When deviating significantly from 1 Å, all the above listed considerations should be taken into account to find the best compromise. One should also make sure that the beamline is optimally tuned for the energy to be used.

Detector-to-Sample Distance, 2-Θ Angle, and Beam Divergence

The two main factors influencing these settings are the expected resolution limit and the spacing of the diffraction spots. The former is known from screening, while the latter is influenced by the size of the unit cell, crystal mosaicity, and the presence of multiple lattices (caused by twinning, split or multiple crystals).

Most beamline control software contains a resolution calculator, which, taking into account the wavelength and the detector size, shows the resolution limits on the edges and the corners of the detector surface for variable detector-to-sample distances (D)

and detector 2-Θ angles. The following steps describe how to find the optimal settings.

1. In the resolution calculator, set 2-Θ = 0 and find *D* for the expected resolution (a few one tenth of an Ångstrom better than the screening resolution) at the edge of the detector surface. Take two snapshots 90° apart (using the anticipated values for all the other settings), and evaluate the images. If the detector surface is filled evenly with diffraction spots to the edge, and the spots do not overlap, data collection can begin (see Fig. 16, right).
2. If the spots do not overlap and the detector surface is filled beyond the edge possibly into the corners or beyond, *D* has to be decreased to capture the high-resolution portion of the data. On the other hand, if the detector surface is not filled with spots to the edge, *D* needs to be increased (see Fig. 18). However, the distance should not be increased too much to fully illuminate the detector surface if the spots are already well separated. A larger than necessary *D* results in additional absorption of the X-rays by air (air absorbs 6% of 1 Å radiation at 200 mm and 14% at 500 mm) and an increase of the spot size due to beam divergence.
3. If the detector surface is filled to the edge, but the spots overlap, either the beam divergence has to be decreased or 2-Θ and *D* have to be increased, or both. Reducing the divergence also decreases the beam intensity, which puts a limit on the minimum usable divergence. To further separate overlapping spots the detector distance has to be increased. To avoid losing high-resolution data, the 2-Θ angle also has to be adjusted at the same time. Additionally, the oscillation angle used during data collection can be decreased to reduce spot overlap (see Section "Starting Spindle Axis Rotation Angle, Rotation Angle Range, and Oscillation Angle" below).

In practice a combination of all those settings will provide the right balance. At the ALS beamline 5.0.3, which is equipped with a Q315r detector and $\lambda = 0.98$ Å, the following settings are used for a crystal with a longest unit cell axis of ~400 Å. Depending on the resolution limit, a detector-to-sample distance in the range of 390–430 mm, a beam divergence of 1.5–1.8 mrad (instead of the standard 2.5 mrad, losing 30–40% flux), and a 2-Θ angle of 0–5° is used.

Exposure Time

Exposure time is the length of time (measured in seconds) for which the shutter is open during the collection of a single diffraction image (during which time the goniometer spindle axis rotates by the chosen oscillation angle). The total data collection time for a dataset is the number of images multiplied by the sum of the exposure times, detector readout time, and beamline overhead. Detector readout time is around 1 s (see Table 4) while beamline overhead is typically in the range of 1–3 s. This means that there is a lower limit on the total data collection time given by the fixed readout and overhead durations. The opening and closing of the shutter is precisely synchronized with the rotation of the spindle axis during the collection of each image. If the exposure time is very short (<~0.5 s), inaccuracies in the synchronization might negatively influence the data quality. For these reasons, even on very intense beamlines the minimum exposure time should be kept in the 0.5–1 s range and the beam should be attenuated if necessary.

Although longer exposure times can be desirable to measure weak reflections accurately, there is an upper limit to exposure time dictated by detector overloads and radiation damage. The dynamic range of the X-ray detector is limited such that there is a maximum intensity that can be stored in every pixel (see Table 4). Once this maximum is reached and passed, the overloaded pixel(s) will not contain useful information about the corresponding diffraction spot(s) and those spots are lost for the structure solution process. A few overloaded pixels (which will most likely be in one or two diffraction spots) are permissible because they will influence low-resolution completeness only slightly and it helps ensure that the full dynamic range of the detector is being used. Overloads are mostly an issue with strongly diffracting crystals.

A rough estimate for the length of data collection before radiation damage occurs on different beamlines can be found in (60). One should keep in mind, however, that radiation damage depends on many factors, such as crystal size, heavy atom content, and X-ray energy (59). Many crystals of the same project (same protein with different compounds) are being collected on in an SBDD environment, and therefore, the properties of these samples will be well known (space group, sensitivity to radiation, typical mosaicity, and resolution). If data are collected under similar conditions on the same or similar beamline (and the proper data collection strategy has been determined), the exposure time can be selected to be near the radiation damage limit. If the sample is less well understood, e.g., the crystal symmetry is not certain yet and the collection of a full 180° dataset is needed, the exposure time should be selected accordingly with an appropriate safety margin. Signs of radiation damage are discussed in the next section under data processing (Section 3.5.4).

On some beamlines data can be collected in “dose mode” where the exposure time of each frame is adjusted to compensate for the decay of the electron current in the storage ring. This is a useful feature for rapidly decaying ring currents, but it is unnecessary for operations in top-off mode.

Starting Spindle Axis Rotation Angle, Rotation Angle Range, and Oscillation Angle

To obtain a complete dataset, the starting rotation angle and the rotation range have to be set up correctly. They are determined by the crystal symmetry (the point group) and the crystal's orientation with respect to the X-ray beam. While the crystal symmetry might be already known (Section 3.4.1), the crystal's exact orientation can be determined by indexing a snapshot (Section 3.5.2). Based on this information, an accurate data collection strategy can be calculated using either dedicated software, such as BEST (61) or Strategy (62) or the built-in predictions in data processing packages (Section 3.5). Data collection should start a few degrees before the recommended value and should always be set up for at least the full 180° rotation range. When the crystal symmetry is ambiguous, collection of the full 180° ensures completeness in any point group. With known symmetry, the proper starting angle will make sure that completeness is reached as soon as possible and if the crystal has not suffered radiation damage (and time permits), more data can be collected to increase the precision of the intensity measurements through the subsequent increased redundancy. There are also other reasons why more data might be needed to reach completeness; some diffraction spots might be rejected during processing due to overlaps with other spots or ice rings; some low-resolution spots might be overloaded in some orientations but not in another; and if a 2-Θ offset is used, high-resolution spots are measured only on one side of the detector surface at a time, and thus, more rotation range is needed.

Special attention needs to be paid to crystals with an anisotropic diffraction pattern, i.e., when diffraction is better in one orientation and worse in another. This can be the case with plate-shaped crystals, for example. In the higher symmetry point groups, there are multiple equivalent choices for starting angles (e.g., every 90°) and one should be chosen where diffraction is best. In this way, the higher quality data are collected first before radiation damage sets in.

As with several other parameters, the choice of the oscillation angle depends on a number of often competing factors. A larger value will result in more diffraction spots on a single image since more lattice points will fulfill the Bragg condition. This might sound enticing, since fewer frames would be needed and thus the data collection time could be shortened (by shortening the total detector readout time). However, higher mosaicity, resolution limit, and beam divergence and a larger unit cell dimension (the one along the beam direction at any given time) will also result in more diffraction spots and thus will limit the maximum allowed oscillation angle for which the spots do not overlap (58). Besides avoiding overlaps, a smaller oscillation angle will have the benefit of an improved signal-to-noise ratio (fine-slicing is not discussed here). Some strategy programs will also recommend what oscillation angle to use and some even allow the simulation of a dataset with different settings to see if overlaps will be a problem. In practice, an oscillation angle in the 0.5–1.0° range works well for most projects.

Beam Size and Detector Binning

In general, the beam size should be set to match the size of the crystal (not the loop!). This will help reduce the scattering background from air and the parts of the loop around the crystal. One has to make sure, however, that the beam covers the crystal in all orientations of the spindle axis. If the goal is to collect multiple datasets on the same crystal (due to radiation damage), a beam size smaller than the crystal should be chosen so different parts of the same sample can be exposed subsequently (see Fig. 21) (30). A properly adjusted beam size can be also used to select a single crystal for data collection when multiple crystals are present in the same loop.

Most modern beamlines are equipped with large CCD area detectors, such as the ones listed in Table 4. Owing to the large number and small size of the pixels standard data collection can be carried out in binned mode, where intensity from four neighboring pixels (in a 2×2 matrix) is combined. Binned mode can speed up detector readout and results in smaller file sizes (e.g., 18 vs. 72 MB for each image of the Q315r).

To reduce the effects of light-independent noise generated in the detector, so called dark images have to be collected regularly. A dark image is taken first, with the exact same settings (exposure time) as the diffraction images, but with the shutter closed. The

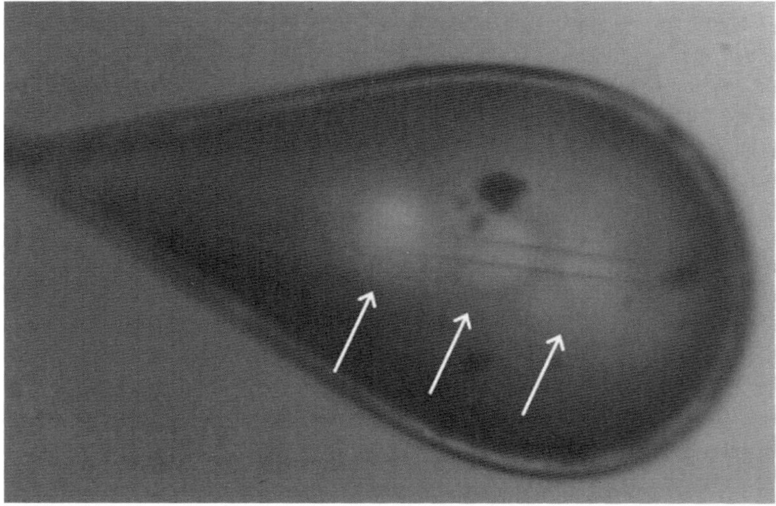

Fig. 21. Three separate data sets were collected on this needle shaped crystal using a 20 μm X-ray beam at the APS ID-23B beamline. Discolorations of the mother liquor surrounding the crystal show where the X-rays hit the loop (marked by *arrows*).

dark image is then subtracted from the subsequent diffraction images by the detector control software. Dark images are handled automatically by the beamline software for the most part (e.g., a new dark is collected before every data collection run), but they can be also requested by the user if needed and it is useful to be aware of this background process.

3.5. Data Processing

During data processing the intensities of all the measured diffraction spots in a data set are accurately determined, combined according to the crystal symmetry and sorted into a final list of *hkl* indices. This process is also called data reduction and involves multiple steps: finding spots, indexing, refinement, integration, and scaling. Some common software packages for the processing of macromolecular diffraction data are HKL2000 (63), MOSFLM (64) and d*Trek (65). Comments throughout this section are HKL2000 specific, since this is the package that has been used at TSD. There is an extensive amount of information available on data processing (manuals, Web sites), thus only a brief overview is given here with special attention to some specific problems.

Data processing should start immediately after data collection has been initiated. This is very important, because problems with the data collection can be caught early and corrected before it is too late. It also enables continuous monitoring of completeness and other data quality indicators that can point to radiation damage or signal if data collection can be stopped.

3.5.1. Finding Spots

To start data processing the dataset has to be loaded into the software. Every diffraction image contains data collection related information (e.g., wavelength, oscillation angle, detector distance) that is automatically read into the program. It is a good idea to briefly double-check if those values are correct and if all the relevant information has been loaded. If not, corrections/additions should be done before proceeding.

Once the first frame is displayed, performing a peak search will find the most intense diffraction spots to be used for autoindexing. On a clean and strong diffraction pattern no more steps are necessary and indexing can begin. Problematic diffraction patterns are discussed below in Section 3.5.3.

3.5.2. Indexing and Refinement

Determination of the possible crystal lattices that correspond to the pattern of diffraction spots found during peak search is called indexing. After indexing is complete, the program provides a list of possible Bravais lattices and corresponding unit cell parameters that match the data (marked green in HKL2000). If the crystal symmetry and unit cell are already known and the right Bravais lattice is shown as an indexing solution, it should be selected and one can move on to refinement.

During refinement a number of crystal and detector parameters are optimized using a least-squares fit to achieve the best possible match between the measured positions of the diffraction spots and their predicted positions based on the crystal symmetry and unit cell parameters. Reliability and accuracy are greatly increased if the known input parameters are provided accurately (e.g., sample-to-detector distance). Select a high-resolution limit beyond the visible spots, refine all the parameters (check "Fit All" in HKL2000) and to also optimize mosaicity using multiple images simultaneously (enter three to five in the 3D window in HKL2000).

To confirm that the right lattice has been picked and everything is in order, the refinement results should be inspected. The χ^2 values (indicating the quality of the fit) should be close to 1 and preferably below 3–4. Unit cell dimensions, sample-to-detector distance and beam position should all be close to their expected values. Visual inspection of the diffraction image should confirm that the predicted spots lie accurately on the measured peaks. If everything is satisfactory, integration can begin. The settings used for refinement (resolution limit, 3D window, etc.) will be also used during integration.

3.5.3. Indexing Problems

Two examples of indexing failures are discussed here. In both situations, it is assumed that Bravais lattice and unit cell parameters are known from previous work.

In the first scenario, the diffraction pattern is clean, there are no obvious signs of a second lattice and the mosaicity is not exceptionally high (below ~1°). Indexing fails, however, with one of the following symptoms: indexing finds a different symmetry and/or unit cell parameters than expected; only the triclinic Bravais lattice matches the data; the right symmetry and unit cell parameters are found, but during refinement the χ^2 values are high and/or some other parameters are far off expected values. In almost all cases the problem is a wrong initial setting for the X and Y positions of the primary X-ray beam. This is easily corrected. Often a visual inspection of a diffraction image can provide good values for the beam position. Occasionally some other setting is off by a large margin: a good candidate to check is the detector 2-Θ offset. Sometimes this is not read into the software and one has to set it manually. A lattice belonging to a large unit cell crystal, i.e., when the spots are close together, is most sensitive to beam position settings. For such samples sometimes the problem does not appear until a later stage of processing: indexing, refinement and integration can all work well but scaling fails (scaling is discussed in the next section). This is due to the fact that the calculated beam center coincides with a low-resolution axial diffraction spot, causing a small indexing error along one axis.

In case of a "messy" diffraction pattern indexing can be difficult and it is an iterative process between finding the right spots, indexing and refinement. If the protein diffraction is contaminated by ice rings, cutting back the high-resolution limit to ~4 Å will eliminate their influence. If the crystal is split and multiple lattices are clearly present (and autoindexing cannot pick one out) manual addition and deletion of spots after the peak search can help. For many difficult cases changing the following settings can help: high-resolution and/or low-resolution limit can be varied; diffraction images from different spindle rotation angles can be tried (ideally a crystallographic zone is visible in some orientation, but in highly mosaic patterns those are hard to recognize); a different number of images can be used at the same time; the size and/or number of peaks during spot finding can be varied (more/less peaks, peak size up/down in HKL2000).

If severe problems with indexing (or integration/scaling) persist, there is either something wrong with the experimental setup or the crystal is not usable. It is a good idea to start a synchrotron run with good crystals, to confirm that there are no problems with the beamline. Processing of these crystals can also reveal how well the beam center and other parameters are set up and those values can be noted and used later.

3.5.4. Integration and Scaling

During integration the diffraction spot intensities are determined for every frame of the dataset. Before starting the integration the anticipated number of frames should be entered if the processing is done during data collection (in HKL2000 on the "Data" tab select "Edit Set(s)"). If indexing was not straightforward, settings for the integration should take that into account, e.g., the integration spot size should be smaller than usual if nearby spots are to be distinguished (e.g., for split crystals).

To take into account small changes in the experiment and the crystal during data collection, crystal and detector parameters are refined for each image. In HKL2000, the refined parameters are shown in a graph and one should keep an eye on their evolution. Small gradual changes are fine, but if any parameter has a sudden change, it could indicate a problem. For example if the crystal orientation changes suddenly (Rot X, Rot Y and Rot Z in HKL2000) it could mean that the crystal slipped on the goniometer. A gradual but steady increase of mosaicity and/or significant changes in the unit cell parameters can be indications of the onset of radiation damage.

In the final steps, the intensities of the images are placed on the same scale (scaling) and the symmetry related intensities are merged together (merging) to generate the final output file of *hkl* indices and the corresponding intensities and their uncertainties (the .sca file in HKL2000). After scaling and merging the overall data quality can be evaluated and the resolution limit determined. This is an iterative process. Resolution limit and other settings (e.g., error model, scaling and B restraints, exclusion of frames) are changed until the final results are satisfactory. The space group is also set during scaling.

The most important data quality indicators are: R_{merge}, completeness, redundancy and the signal-to-noise ratio $I/\sigma(I)$. These parameters and other relevant information can be found in the scaling log file. Overall completeness >95%, redundancy higher than fourfold, $R_{merge} < 0.5$ and $I/\sigma(I) > 2$ are good initial guidelines for stopping data collection or determining the resolution limit. However, a good balance of all these values should be obtained in each specific case. Crystals with low symmetry will have a redundancy lower than four. An $I/\sigma(I)$ lower than two can be allowed if the other parameters are still good (since there is actual signal as long as $I/\sigma(I) > 1$). Similarly, a lower completeness in the high-resolution shell will still contain some real information (in practice one can go as low as 50–60% if $I/\sigma(I)$ is good).

A detailed discussion of R_{merge} is beyond the scope of this review. Owing to the definition of this parameter, its value increases with increasing data redundancy and thus suggests worsening data quality although the precision of the measurement is actually improving. For this reason improved R-factor schemes have been proposed (66, 67) but are still not in widespread use.

3.5.5. Finding the Correct Space Group

The most important rule to follow if the crystal symmetry is unknown or is in doubt is to collect 180° of data. Processing should start with the highest possible symmetry based on indexing. If processing fails during integration (the refinement parameters diverge strongly) or scaling (yielding high R_{merge} and high rejections) the chosen Bravais lattice is incorrect and one should try the next highest one. Once integration and scaling succeeded the correct point group belonging to the chosen Bravais lattice has to be determined (4 or 422 for tetragonal; 3, 321, 312, 6 or 622 for hexagonal; 23 or 432 for cubic). This happens through simple trial and error. If scaling fails the point group is incorrect.

Screw axes in the lattice result in the absence of reflections belonging to certain *hkl* values. These systematic absences can be used to narrow down the list of potentially correct space groups further. If the intensity of reflections that should be absent is larger than zero, the chosen screw axis is incorrect. Some screw axes are indistinguishable based on the diffraction pattern alone (e.g., $P4_1$ and $P4_3$) and can be determined with certainty only during the structure solution process. Even for well established systems the presence of systematic absences should always be checked to potentially detect a change of the space group (e.g., from $P2_12_12_1$ to $P2_12_12$). At the beginning of a data collection run the systematic absences might not be present and one should check again at a later stage.

If finding the correct space group turns out to be a challenge, one can try the following. A solvent content calculation for the crystal can rule out certain space group/unit cell indexing solutions and can even point to the crystallization of the wrong protein. If the diffraction pattern is not clean or there are other problems (e.g., incorrect detector distance), and only the triclinic Bravais lattice appears as an indexing solution, it should be chosen and refinement should be performed (with mosaicity fixed). Sometimes refinement will improve the parameters enough to allow the correct higher symmetry Bravais lattice to appear as a solution.

If time permits it can be a good idea to process a dataset in multiple space groups and to avoid the need to reprocess at a later stage when the experimenter has left the beamline or the data at the beamline is not available.

3.6. Good Practices

3.6.1. Bookkeeping

As with all scientific research, meticulous note taking is essential for protein crystallography experiments. This is especially true at synchrotron beamlines, since the measurements are carried out on equipment owned and operated by someone else and used by many different scientists in close succession. Once an experiment is done, it can be very difficult (or impossible) to find out what certain settings were weeks or months earlier. Even if settings are automatically logged in the background, the scientist's own accurate notes are invaluable. Those notes should contain among others all the relevant settings for every data collection run, details of data processing, and the main results.

3.6.2. Data Backup

Data collection and processing is generally performed on the beamline computers of the synchrotron facility. Since beamlines are utilized by many different user groups and storage is limited, data are usually deleted within 1–2 weeks after an experiment is finished. For this reason it is essential to perform a backup of all the raw and processed data as soon as possible after the experiment is done. The most common forms of backups are DVD's, external hard drives and secure file transfer over the internet. Compressing the raw data after processing can save storage space and reduce backup times. If possible, compressing and data transfer of finished datasets should occur concurrently to data collection.

3.6.3. Beamline "Etiquette"

Synchrotron beamlines are used by many scientists and a changeover between user groups can happen within an hour. It is important to be mindful of all the other users around you and not disturb their work. If you have to stay on to finish data processing or backups after your run has finished, do it in a way that you will not interfere with the next user.

Beamlines are very complex instruments and should be used carefully and thoughtfully according to the instructions of beamline staff. Mechanical and software controls should prevent damage to equipment, but this does not mean that the built-in protections are flawless. If in any doubt, ask somebody knowledgeable.

Safety procedures should be followed, especially regarding the use of liquid nitrogen.

If you are reading this book, you are most likely conducting proprietary research. Keep in mind that beamlines are a rather open environment, where scientists from different institutions and companies work in close proximity. Guard your proprietary information well and do not leave confidential material "lying around" for others to see.

4. Outlook

Over the past decade the field of macromolecular crystallography has enjoyed major advances. Improved synchrotrons, beamlines and X-ray detectors combined with sample handling automation eliminate many of the technical bottlenecks associated with the acquisition of diffraction datasets. Software developments have enabled the quick and robust analysis of data. Overall the structure solution process for routine cases has become significantly more efficient and is being done on an industrial scale in many centers. A natural limit has been reached in certain aspects, which hinders further efficiency improvements (radiation damage, protein crystallizability, human interaction times). In this sense, some of the current developments are evolutionary in nature, improving certain aspects of current systems and upgrading most experimental stations and synchrotrons to similar standards.

However, as new and more challenging projects are being tackled in many laboratories, there will be a strong emphasis on hands-on scientific research that cannot be automated. On the , the technological barriers will be pushed further to overcome other current limitations, such as performing diffraction experiments on crystals as small as a few microns in size.

Acknowledgments

This chapter is dedicated to Peter Boyd of Boyd Technologies, whose invaluable contributions to sample handling automation at synchrotron facilities helped to realize high-throughput SBDD.

The majority of the work described in this chapter has been performed at beamline 5.0.3 of the Advanced Light Source, in Berkeley, CA. The Advanced Light Source is supported by the Director, Office of Science, Office of Basic Energy Sciences, of the US Department of Energy under Contract No. DE-AC02-05CH11231. ALS beamline 5.0.3 has been constructed and is being operated by the Berkeley Center for Structural Biology (BCSB).

Contributions by former and current members of the BCSB are greatly appreciated.

The work presented in this chapter has been carried out as part of the TSD high-throughput SBDD efforts, which were made possible by the members of the structural biology department. Portions of the data have been collected at the Advanced Photon Source (APS) GM/CA CAT beamlines, beamline X6A of National Synchrotron Light Source (NSLS), and the structural biology beamlines of SSRL. Use of the APS was supported by the US Department of Energy, Office of Science, Office of Basic Energy Sciences, under Contract No. DE-AC02-06CH11357. GM/CA CAT has been funded in whole or in part with Federal funds from the National Cancer Institute (Y1-CO-1020) and the National Institute of General Medical Science (Y1-GM-1104). The SSRL is a national user facility operated by Stanford University on behalf of the US Department of Energy, Office of Basic Energy Sciences. The SSRL Structural Molecular Biology Program is supported by the Department of Energy, Office of Biological and Environmental Research, and by the National Institutes of Health, National Center for Research Resources, Biomedical Technology Program, and the National Institute of General Medical Sciences. Beam line X6A is funded by the National Institute of General Medical Sciences, National Institute of Health under agreement GM-0080. Use of the National Synchrotron Light Source, Brookhaven National Laboratory, was supported by the US Department of Energy, Office of Science, Office of Basic Energy Sciences, under Contract No. DE-AC02-98CH10886.

I would like to thank Scott W. Lane (TSD) for careful reading of the manuscript and for the many helpful suggestions. I would also like to thank Simon A. Morton (LBNL) for fruitful discussions and help with the wiggler and undulator spectral distribution calculations.

References

1. Hoffman, I., Protein Crystallization for Structure Based Drug Design, in this book.
2. Drenth, J. (2006) Principles of Protein X-Ray Crystallography. Springer.
3. Ogata, C. (2009) Private communications.
4. http://www.rigaku.com/generators/fre-plus.html
5. http://hyperphysics.phy-astr.gsu.edu/hbase/quantum/xrayc.html; Image reproduced with permission from R. Nave (2010)
6. http://henke.lbl.gov/optical_constants/bend2.html
7. http://radiant.harima.riken.go.jp/spectra/index.html
8. Elder, F. R., Gurewitsch, A. M., Langmuir, R. V., and Pollock, H. C. (1947) Radiation from Electrons in a Synchrotron. *Phys. Rev.* **71**, 829–830.
9. http://en.wikipedia.org/wiki/X-ray_tube
10. Kwiatkowski, W., Noel, J. P., and Choe, S. (2000) Use of Cr K_α radiation to enhance the signal from anomalous scatterers including sulfur. *J. Appl. Cryst.* **33**, 876–881.
11. Watanabe, N. (2006) From phasing to structure refinement in-house: Cr/Cu dual-wavelength system and a loopless free crystal-mounting method. *Acta Cryst.* **D62**, 891–896.

12. http://www.rigaku.com/protein/phasing.html
13. Robin, D. et al (2005) Superbend upgrade on the Advanced Light Source. *Nucl. Instrum. Methods* **A538**, 65–92.
14. Robin, D. (2009) Private communications.
15. X-RAY DATA BOOKLET (2009), Center for X-ray Optics and Advanced Light Source, Lawrence Berkeley National Laboratory. 3rd ed.
16. Ruth, R. (2010) Private communications and http://www.lynceantech.com
17. Miao, J., Ishikawa, T., Shen, Q., and Earnest, T. (2008) Extending X-ray crystallography to allow the imaging of noncrystalline materials, cells, and single protein complexes. *Annu. Rev. Phys. Chem.* **59**, 387–410.
18. Henke B. L., Gullikson E. M., and Davis J. C. (1993). X-ray interactions: photoabsorption, scattering, transmission, and reflection at E = 50-30000 eV, Z = 1–92. *Atomic Data and Nuclear Data Tables* **54,** 181–342, and http://henke.lbl.gov/optical_constants/
19. Morton, S. (2010) Private communications.
20. Deutsch, M., Förster, E., Hölzer, G., Härtwig, J., Hämäläinen, K., Kao, C.-C., Huotari, D., and Diamant, R. (2004) X-Ray Spectrometry of Copper: New Results on an Old Subject. *J. Res. Natl. Inst. Stand. Technol.* **109**, 75–98.
21. Miller, M. (2009) Private communications and http://www.px.nsls.bnl.gov/x12c/CCM-description.html
22. Underwood, J. H. (2009) in X-RAY DATA BOOKLET, Center for X-ray Optics and Advanced Light Source, Lawrence Berkeley National Laboratory. 3rd ed., Section 4–1
23. http://henke.lbl.gov/cgi-bin/mldata.pl
24. Padmore, H. A., Earnest, T., Kim, S.-H., Thompson, A. C., and Robinson, A. L. (1995) A beamline for macromolecular crystallography at the Advanced Light Source. *Rev. Sci. Instrum.* **66**, 1738–1740.
25. Earnest, T., Padmore, H., Cork, C., Behrsing, R., and Kim, S. H. (1996) The macromolecular crystallography facility at the advanced light source. *J. Cryst. Growth* **168**, 248–252.
26. Morton, S. et al (2007) Recent Major Improvements to the ALS Sector 5 Macromolecular Crystallography Beamlines. *Synchrotron Rad. News*, **20**, 2330.
27. http://www.biosync.rcsb.org
28. Helliwell, J. R. (1992) Macromolecular Crystallography with Synchrotron Radiation. Cambridge University Press.
29. Snell, G., Cork, C., Nordmeyer, R., Cornell, E., Meigs, G., Yegian, D., Jaklevic, J., Jin, J., Stevens, R. C., and Earnest, T. (2004) Automated sample mounting and alignment system for biological crystallography at a synchrotron source. *Structure* **12**, 537–45.
30. Fischetti, R. F., Xu, S., Yoder, D. W., Becker, M., Nagarajan, V., Sanishvili, R., Hilgart, M. C., Stepanov, S., Makarov, O., and Smith, J. L. (2009) Mini-beam collimator enables microcrystallography experiments on standard beamlines. *J. Synchrotron Radiat.* **16**, 217–25.
31. Ellis, P. J., Cohen, A. E. & Soltis, S. M. (2003) Beamstop with integrated X-ray sensor. *J. Synchrotron Rad.* **10**, 287–288.
32. Muchmore, S. W., Olson, J., Jones, R., Pan, J., Blum, M., Greer, J.,Merrick, S. M., Magdalinos, P., and Nienaber, V. L. (2000) Automated crystal mounting and data collection for protein crystallography. *Structure* **8**, R243-R246.
33. Cohen, A.E., Ellis, P.J., Miller, M.D., Deacon, A.M., and Phizackerley, R.P. (2002) An automated system to mount cryo-cooled protein crystals on a synchrotron beamline, using compact sample cassettes and a small-scale robot. *J. Appl. Crystallogr.* **35**, 720–726.
34. http://smb.slac.stanford.edu/robosync/
35. http://www.adsc-xray.com
36. http://www.mar-usa.com
37. http://www.dectris.com
38. Rupp, B., Segelke, B. W., Krupka, H. I., Lekin, T.P., Schafer, J., Zemla, A., Toppani, D., Snell, G., and Earnest, T. (2002) The TB structural genomics consortium crystallization facility: toward automation from protein to electron density. *Acta Crystallogr.* **D58**, 1514–1518.
39. Stevens, R.C., Yokoyama, S., and Wilson, I.A. (2001) Global efforts in structural genomics. *Science* **294**, 89–92.
40. Scapin, G., (2006) Structural biology and drug discovery. *Curr. Pharm. Des.* **12**, 2087–97.
41. Santarsiero B. D., Yegian D. T., Lee C. C., Spraggon G., Gu J., Scheibe D., Uber D. C., Cornell E. W., Nordmeyer R. A., Kolbe W. F., Jin J., Jones A. L., Jaklevic J. M., Schultz P. G., and Stevens R. C. (2002) An approach to rapid protein crystallization using nanodroplets. *J. Appl. Cryst.* **35**, 278–281.
42. Hosfield D., Palan J., Hilgers M., Scheibe D., McRee D. E., and Stevens R. C. (2003) A fully integrated protein crystallization platform for small-molecule drug discovery. *J Struct Biol.* **142**, 207–17.
43. Carter, D. C., Rhodes, P., McRee, D. E., Tari, L. W., Dougan, D. R., Snell, G., Abola, E., and Stevens, R. C. (2005) Reduction in diffuso-convective disturbances in nanovolume protein crystallization experiments. *J. Appl. Cryst.* **38**, 87–90.

44. Nowakowski J., Cronin C. N., McRee D. E., Knuth M. W., Nelson C. G., Pavletich N. P., Rogers J., Sang B. C., Scheibe D. N., Swanson R. V. , and Thompson D. A. (2002) Structures of the cancer-related Aurora-A, FAK, and EphA2 protein kinases from nanovolume crystallography. *Structure* **10**, 1659–67.

45. Mol C. D., Dougan D. R., Schneider T. R., Skene R. J., Kraus M. L., Scheibe D. N., Snell G. P., Zou H., Sang B. C., and Wilson K. P. (2004) Structural basis for the autoinhibition and STI-571 inhibition of c-Kit tyrosine kinase. *J. Biol. Chem.* **279**, 31655–63.

46. Aertgeerts K., Ye S., Tennant M. G., Kraus M. L., Rogers J., Sang B. C., Skene R. J., Webb D. R., and Prasad G. S. (2004) Crystal structure of human dipeptidyl peptidase IV in complex with a decapeptide reveals details on substrate specificity and tetrahedral intermediate formation. *Protein Sci.*, **13**, 412–21.

47. Somoza J. R., Skene R. J., Katz B. A., Mol C., Ho J. D., Jennings A. J., Luong C., Arvai A., Buggy J. J., Chi E., Tang J., Sang B. C., Verner E., Wynands R., Leahy E. M., Dougan D. R., Snell G., Navre M., Knuth M. W., Swanson R. V., McRee D. E., and Tari L. W. (2004) Structural snapshots of human HDAC8 provide insights into the class I histone deacetylases. *Structure* **12**, 1325–34.

48. Hosfield D. J., Wu Y., Skene R. J., Hilgers M., Jennings A., Snell G. P., and Aertgeerts K. (2005) Conformational flexibility in crystal structures of human 11beta-hydroxysteroid dehydrogenase type I provide insights into glucocorticoid interconversion and enzyme regulation. *J. Biol. Chem.* **280**, 4639–48.

49. Aertgeerts K., Levin I., Shi L., Snell G. P., Jennings A., Prasad G. S., Zhang Y., Kraus M. L., Salakian S., Sridhar V., Wijnands R., and Tennant M. G. (2005) Structural and kinetic analysis of the substrate specificity of human fibroblast activation protein alpha. *J. Biol. Chem.* **280**, 19441–19444.

50. Fujimoto T., Imaeda Y., Konishi N., Hiroe K., Kawamura M., Textor G. P., Aertgeerts K., and Kubo K. (2010) Discovery of a tetrahydropyrimidin-2(1H)-one derivative (TAK-442) as a potent, selective, and orally active factor Xa inhibitor. *J. Med. Chem.* **53**, 3517–31.

51. Feng J., Zhang Z., Wallace M. B., Stafford J. A. , Kaldor S. W., Kassel D. B., Navre M., Shi L., Skene R. J., Asakawa T., Takeuchi K., Xu R., Webb D. R., and Gwaltney S. L. (2007) Discovery of alogliptin: a potent, selective, bioavailable, and efficacious inhibitor of dipeptidyl peptidase IV. *J Med Chem.* **50**, 2297–300.

52. Karain, W. I., Bourenkov, G. P., Blume, H, and Bartunik, H. D. (2002) Automated mounting, centering and screening of crystals for high-throughput protein crystallography. *Acta Crystallogr.* **D58**, 1519–22.

53. Song, J., Mathew, D., Jacob, S. A., Corbett, L., Moorhead, and P., Soltis, S. M. (2007) Diffraction-based automated crystal centering. *J Synchrotron Radiat.* **14**, 191–195.

54. Jain, A., and Stojanoff, V. (2007) Are you centered? An automatic crystal-centering method for high-throughput macromolecular crystallography. *J Synchrotron Radiat.* **14**, 355–60.

55. http://www.scripps.edu/~arvai/adxv.html

56. Sauter, N. K., Grosse-Kunstleve, R. W., and Adams, P. D. (2004). *J. Appl. Cryst.* **37**, 399–409.

57. Zhang, Z., Sauter, N. K., van den Bedem, H., Snell, G., and Deacon, A. M. (2006). *J. Appl. Cryst.* **39**, 112–119.

58. Dauter, Z., (2005) Efficient use of synchrotron radiation for macromolecular diffraction data collection. *Progess in Biophysics and Molecular Biology* **89**, 153–172.

59. Holton, J. M. (2009) A beginner's guide to radiation damage. J *Synchrotron Radiat.* **16**, 133–42.

60. Holton, J. M. (2008) Expected crystal lifetimes at synchrotron beamlines. http://bl831.als.lbl.gov/damage_rates.pdf

61. Popov, A. N., and Bourenkov, G. P. (2003) Choice of data-collection parameters based on statistic modeling. *Acta Cryst.* **D59**, 1145–1153.

62. Ravelli, R. B. G., Sweet, R. M., Skinner, J. M., Duisenberg, A. J. M., and Kroon, J. (1997) STRATEGY: a program to optimize the starting spindle angle and scan range for X-ray data collection. *J. Appl. Cryst.* **30**, 551–554.

63. Otwinowski, Z., and Minor, W. (1997) Processing of X-ray Diffraction Data Collected in Oscillation Mode. *Methods Enzymol.* **276**, 307–326.

64. Leslie, G. W., (2006) The integration of macromolecular diffraction data. *Acta Cryst.* **D62**, 48–57.

65. Pflugrath, J. W., (1999) The finer things in X-ray diffraction data collection. *Acta Cryst.* **D55**, 1718–1725.

66. Diederich, K., and Karplus, A., (1997) Improved R-factors for diffraction data analysis in macromolecular crystallography. *Nature Structural Biology* **4**, 269–274.

67. Weiss, M.S., Global indicators of X-ray data quality. *J. Appl. Cryst.* **34**, 130–135 (2001).

Chapter 6

The Use of Molecular Graphics in Structure-Based Drug Design

Paul Emsley and Judit É. Debreczeni

Abstract

The use of 3D structures derived from X-ray crystal data in drug development has increased in recent years. Molecular graphics applications are important tools at the end of the data processing pipeline and provide means to build, refine and validate protein models and ligand structures. We describe the requirements on useful data, what such data provide and typical problems in dealing with protein–ligand complexes and how one might address them with an emphasis on the use of Coot.

Key words: X-ray crystallography, Molecular Graphics, Model Building, Ligand fitting, Structure validation

1. Introduction

Structure-based drug design (SBDD) comprises a variety of techniques with one aim: the use of the 3D structure to optimize molecular properties. SBDD represents a step towards an entirely rational design and hence poses a more straightforward path to molecules with desirable properties. SBDD builds on one's ability to capture, organize and interpret a considerable amount of disparate data. These include results of a wide range of experimental techniques, such as enzyme and cell potency, selectivity, pharmacokinetic and physiochemical properties, and also information on chemical tractability and questions around intellectual property. Therefore the substantial number of distinct compounds tested in a particular drug discovery process is the only one aspect of what is essentially a data generation and handling problem.

In SBDD, information derived from a broad variety of independent experimental and virtual techniques on a large chemical

Leslie W. Tari (ed.), *Structure-Based Drug Discovery*, Methods in Molecular Biology, vol. 841,
DOI 10.1007/978-1-61779-520-6_6, © Springer Science+Business Media, LLC 2012

collection is mapped back to the molecular scaffold of the drug molecules or the binding pocket of the protein target, leading to a complex multidimensional picture. This is in turn used to identify starting points and to establish an initial hypothesis for guiding synthetic chemistry efforts moving hits forward into lead series. 3D molecular graphics applications supporting computational chemistry and structural biology play a pivotal role at almost every stage of this design process. Validation and preparation of experimental structures for homology modelling and ligand docking, visualization of hits from High Throughput Screening (HTS) and Virtual Ligand Screening (VLS), pharmacophore mapping and query building are a few examples in the long list of tasks involving the use of molecular graphics applications.

2. Requirements for Molecular Graphics Applications in Drug Discovery

Even though they can be lightweight in terms of the variety of molecular representations, most of the visualization tools used in both computational chemistry and structural biology have a common characteristic: they act primarily as graphical front-ends to the underlying process-driven calculations. In other words, the main requirement for these tools is to be a slick and easy-to-use gateway to the specific tasks to which they are linked. These tools can also handle large datasets, e.g. interfacing compound databases and in-house archives, supporting query building and large-scale visualization (1, 2).

The multidisciplinary nature of the drug design process, together with global discovery models or multi-site development environments, poses an additional set of requirements to modern molecular graphics applications. Over the past couple of years, 3D graphics tools also started to expand into the unfilled niche of structure exploitation by enhancing communication of structural information to non-experts. Several graphics packages now offer various ways to communicate through annotated views of protein and ligand structures, to share them via email-sized state files or embedding into web-pages or "slidepacks" (Molsoft's ICM-Browser (2), OpenEye's VIDA/vivant, AstexViewer (3), proteopedia (4)). Most of these efforts aim at lowering the boundaries between the different disciplines, such as crystallography and computational chemistry. Joint inspection of both the experimental data and the derived model can accelerate the design process and eliminate potential pitfalls and misinterpretations. Chemists can advise crystallographers on the interpretation of the electron density and structure validation. On the other hand, experienced crystallographers are well positioned to spot and highlight relevant structural features (or share their view on the model's ambiguity or inaccuracy) and contribute to the idea generation phase of the design-make-test cycles (5).

In the current competitive pharmaceutical environment, it is essential to ensure that these make-test-cycles are as efficient and fast as possible. The X-ray crystallography process is not exempt from this requirement. Structures solved with short turnaround times generally have higher impact on the direction of synthetic chemistry. As a consequence, crystallography methods in an industrial environment are essentially a set of high-throughput techniques. As the structure of the target protein is known, the main emphasis is placed on fast data processing, quick model completion and efficient interpretation. In practice, this should equate to the swift generation of X-ray structures that are fit for purpose (and not necessarily better) with trustworthy ligand substructures. The computational tools developed for this purpose can range from a simple set of scripts running individual components of the pipeline in a standardized manner up to a fully integrated platform encompassing all steps from data processing to ligand fitting, as pioneered by AutoSolve at Astex (6). Efficient use of model completion and validation applications such as Coot can expedite the process.

3. Requirements for X-Ray Structures for Drug Design

Different stages of the drug discovery process pose different requirements towards experimental structures, but these also vary with the actual drug discovery strategy followed.

At the early stages, e.g. after a HTS, when the main goals are the prioritization of the initial hits and early development of the structure-activity relationships (SAR), the purpose of the first structures is to achieve an understanding on how the hit binds the target of interest. An unambiguous binding mode derived from X-ray structures can tremendously help modelling and subsequent docking of ligands in silico. At this stage, low–medium resolution structures might suffice, provided that the binding mode and the geometry of the ligand is clearly supported by the electron density (see Fig. 1).

As the initial SAR is tested after the synthesis or purchase of follow-up compounds, the accuracy and/or resolution of the experimental structures increases. Docking routines and energy calculations can be imperfect or fail to correct for the static nature of X-ray structures. It is not uncommon that subtle changes in the conformation of the receptor or the importance of ordered solvent molecules are overseen in calculations, but are unambiguously highlighted by high quality experimental structures. At this point, X-ray data are used to test earlier hypotheses or help to explain unexpected activity data. Crystal structures with ligands contradicting early SAR can

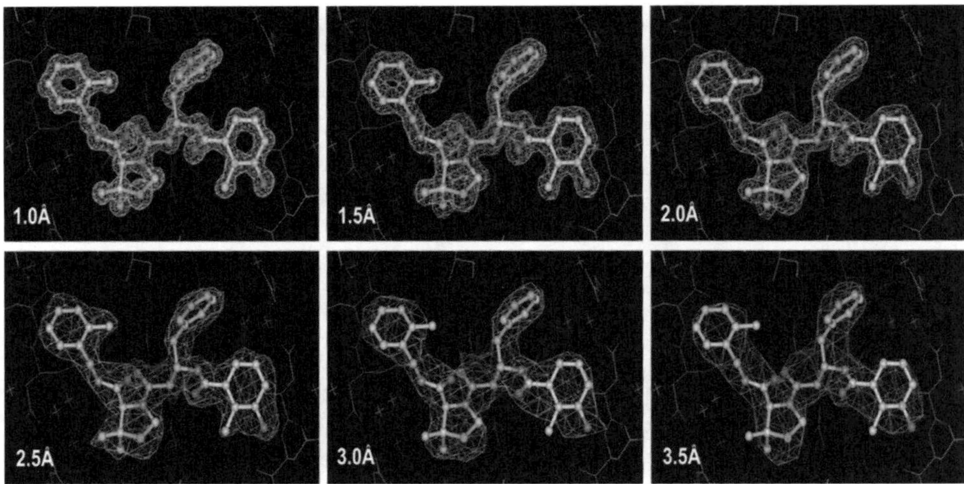

Fig. 1. SigmaA-weighted electron density calculated around the ligand JE2 in the PDB entry 1KZK with different resolution cut-offs. With decreasing resolution, the orientations of the substituents on the phenyl rings on either end of the molecule become less defined.

lead to the discovery of novel binding modes, establishment of alternative lead compounds or identification of new specific interactions.

4. Protein–Ligand Complex Modelling Using Coot

Coot (7) is a modern generation molecular graphics program that is designed particularly for model building of protein and protein–ligand complex structures. Apart from its powerful built-in model building and validation tools, it provides a smooth interface to a variety of external crystallography packages. Coot also has python and scheme interpreters embedded and hundreds of functions are exported to the scripting layer, enabling the integration of Coot into high-throughput structure solution and refinement pipelines or interfacing other in-house or commercial software packages and corporate databases.

After a successful molecular replacement calculation, there are a number of manipulations of the model that need to be undertaken before the model is appropriate for deposition or further use in molecular modelling. Such tasks involve mutations, the building of missing loops and fixing misplaced, partially built or misbuilt residues (particularly at the chain termini). Coot has tools for alignment and mutations. Using the tools in Coot, the model can be aligned, mutated and validated against the target sequence. After the model has been correctly mutated, the rotamers are systematically searched, the candidate conformations are those of the median positions of the rotamers from Molprobity (8). Each of these candidate rotamers undergoes a rigid-body refinement to correct positional errors in the main chain atoms which might propagate to side chain atoms (out of density), preventing recognition of the best rotamer. Optionally (and typically at lower resolutions) one can use a different fitting algorithm, the so-called "Backrub Rotamer" mode based on "Backrub" motions (9) that preserve main chain geometry at the cost of a somewhat inferior fit of each residue to the electron density.

4.1. Non-standard Residues

Non-standard residues occur on occasion in protein structures (e.g., a methylated lysine or phosphorylated tyrosine). Coot provides a convenient tool to replace such residues with their modified form. Given that the atom names of the modified and standard residue may not match (and in the general case the set of matching atoms is not known) then the non-trivial operation is to find which atoms of the modified residues match the atoms in the residue to be replaced. To address this, Coot uses the sub-graph isomorphism method of Krissinel and Henrick (10) which implements graph matching to rapidly find the coincidence of matching atoms. Once the matching atoms are found, a least squares procedure is used to bring the modified residue into the correct position and then replace the standard residue in the internal molecule description.

Additionally, Coot provides a mechanism to "spin-search." A sub-fragment (e.g. the phosphate group in phosphotyrosine) is spun around the bond preceding the P–O bond, so that not only are the standard atoms matched to previously positioned atoms, but also the additional atoms of the modification are fit to density.

4.2. Terminal Residue Addition

By its nature SBDD involves the structure determination of many different, but related protein structures. It is often the case that flexible parts of protein structures can be seen in one complex dataset but not in another. Hence, if one has better quality data, or structures with variable degrees of flexibility it may be possible to add residues to the current model, where the starting model had missing residues.

Residues are added to the end of the protein chain using the fit to density of candidate peptide positions. Candidate positions are generated by randomly selecting phi and psi values from a Ramachandran plot probability distribution. Two torsion angles are needed to position a peptide for N-terminal addition, and additional randomly selected torsion angles are used to provide candidate position for the carbonyl oxygen when adding to the C-terminus. The candidate coordinates are then scored according to the fit to the density and the best one picked after a number of trials. Optionally one can add rigid body refinement of the fragments before scoring.

4.3. Refinement and Regularization

The major feature of Coot is the refinement system. With it one can fit to density a residue/monomer or a selection of residues—the fit to density is improved at the same time the geometry is improved. Coot uses a multi-dimensional gradient minimizer to do this. If the gradients from the density map are excluded from the minimization, then the residue selection simply has its geometry optimized without regard for the electron density. Using an mmCIF (11) dictionary, Coot can be informed about geometry of monomers and how they are linked together. Coot is distributed with the Refmac (12) monomer library (13), which contains monomer descriptions of over 5,000 compounds (including the standard amino acids). Some Refmac dictionary entries are in the form of "minimal descriptions" and when using the CCP4 suite, the program LIBCHECK is used to produce a full description.

4.4. Loop Fitting

A number of approaches to fitting loops in protein structures have been developed; e.g. xpleo (14), loopy (15) and the algorithm implemented in O (16). The mechanisms in Coot are relatively unsophisticated compared to some of these methods. There are two main tools/methods for loop fitting in Coot:

1. Place C-alpha candidate positions using the Baton Build tool, then use the Ca Zone → mainchain tool (an implementation of the algorithm described by Esnouf (17)) and finally mutate, renumber and merge the fragment into the main molecule.
2. Loop fitting directly: a combination of more primitive tools (Ramachandran-weighted terminal residue addition, mutation, rotamer search, rigid-body refinement and real-space refinement). This tool is more suitable for shorter loops, but it is much faster and more convenient than the previous method.

4.5. Dictionary Generation

The drug discovery process by necessity deals with many candidate ligands, representations of which are often stored in a database in a compact form e.g. SDF (18), SMILES (19) or IUPAC's InChI. For use in structural biology, these representations need to be converted into 3D structures with accompanying restraints for crystallographic refinement. Restraints describe the prior knowledge of the chemistry of the ligand in terms of geometrical features, such as bond lengths, bond angles, planar systems, chiral centres and torsions (rotatable or non-rotatable bonds). There are several such compilations of geometry for standard residues and other small molecules and ligands; one of these transferable dictionaries is the mmCIF formatted (11) Refmac dictionary (13) utilized by Coot. For refinement (i.e. geometry optimization in the presence of X-ray data) these individual geometric features need to be associated with a standard deviation. For example, the Refmac dictionary defines the Cα–Cβ bond with an optimal length of 1.541 Å and a standard deviation of 0.033 Å.

The group of "_chem_link" mmCIF items describe the links between various groups or monomer/residue types. For example, the "TRANS" chem link describes a conventional peptide bond between monomers with group type "peptide" and the "SS" chem link describes a disulphide bond between CYS residues. The "_chem_link" system can be used to define bespoke links between the protein and the ligand. Such a link description can be read and used by several refinement programs (including Coot).

Modifications to standard residues can be further represented with the "_chem_mod" formalism that allows addition and deletion of chemical and geometric elements. This provides a transparent way of handling derived or modified residues.

The description of the rotatable bonds is of particular interest to those involved in SBDD. The Refmac dictionary for a torsion angle contains the following: a name for the torsion angle, the names of the four atoms involved in the torsion, an ideal torsion angle, the standard deviation for the torsion angle, and a period (the energy of a torsion angle with standard angle α and period ρ is at a minimum every $\alpha + 360/\rho$ degrees).

However for novel ligands, there will not be a standard dictionary entry, and a dictionary will have to be generated. This can be done in following ways.

4.6. Crystallography Software Suites and Refinement Programmes Tend to have Tools that Create Ligand Description Files in Their Own Particular Format

4.6.1. Phenix.elbow

Phenix.elbow (20) is the ligand builder in the phenix suite (21). Phenix.elbow automates the generation of structures and restraints, although it also has a graphical front-end (REEL). It processes user-defined SMILES strings but also has a built-in library of standard residues. Phenix.elbow can also take Protein Data Bank (PDB) files containing multiple ligands (and protein molecules) as input and write a single mmCIF restraints file with all non-standard residues detected in the input file. It has options to use a simple force-field for energy minimization and also the ability generate better geometries with either a built-in semi-empirical QM method (AM1) or HF/3-21 G using GAMESS.

4.6.2. LIBCHECK and Sketcher

LIBCHECK (13) has been the standard mmCIF dictionary and model generation program in the CCP4 suite for some time. Sketcher is the GUI front-end to this program. JLigand is CCP4's new front-end to LIBCHECK that allows graphical editing of residue modifications and chemical links between residues.

4.6.3. MakeTNT

Similar to phenix and CCP4, there are topology generator, editor and visualizer tools for Buster/TNT (22). Aside from generating restraints used by Buster/TNT, these tools can also be used to create and edit mmCIF style restraints.

4.7. Computational Chemistry and Modelling Tools Contribute to this Field Equally

4.7.1. Corina

Corina (23) is a mature cross-platform 3D structure generator for small organic molecules available from Molecular Networks. It can be used as a tool both to convert 2D structural information (SD files or smiles) to 3D coordinates and to enumerate functional groups e.g. ring conformations. Corina can also generate restraints for refinement in mmCIF format corresponding to the lowest energy conformation. Corina is typically used from the command-line or part of a pipeline, and due to its fast performance it is often used for automatic processing of large datasets.

4.7.2. PRODRG

PRODRG (24) is a small molecule topology generator. It takes input in various formats (including SMILES strings) and creates coordinates and dictionaries suitable for refinement in various packages. PRODRG currently uses gromos 96.1 for energy minimization. It returns output in various formats, including "O", "SHELX" format, mmCIF and PDB. If the input is in PDB format, then the output dictionary and idealized coordinates conveniently have the same atom names as the input. PRODRG is available as a web service or as a command line program via the author's web site [http://davapc1.bioch.dundee.ac.uk/prodrg/] and will be part of the CCP4 Program Suite in due course.

4.7.3. Afitt

OpenEye's Afitt (1) and its command line equivalents, Flynn and Writedict, perform ligand fitting as well as restraint generation for external refinement programs such as Refmac and Coot. It converts a large number of input formats (including SMILES, SD and PDB files) and uses MMFF94 for geometry optimization. Like Phenix.elbow, these tools can detect multiple non-standard or modified residues in their PDB input.

4.8. Geometric Parameters Are also Available via a Number of Web Services and Knowledge Bases

4.8.1. Hetero-compound Information Centre—Uppsala (25)

Hetero-compound Information Centre—Uppsala (HIC-Up) is an information repository for the hetero compounds found in the PDB (26). Although the PDB distributes a chemical component library containing the hetero compounds as both actual (i.e. as in the PDB) and idealized coordinates in mmCIF format, HIC-Up is more useful for crystallographers doing refinement because HIC-Up contains restraint information (although not necessarily for every hetero compound and not in mmCIF format at the time of writing, release 12.1).

4.8.2. PURY

An alternative approach to restraint generation is available in the PURY (27) web server. PURY aims to retrieve geometrical parameters and their standard deviations using a knowledge base derived from the Cambridge Structure Database. This method provides self-consistent parameters that are in agreement with ab-initio calculations and small molecule crystal structures.

There are extensions to Coot that interface with many of the tools listed above. Coot also provides access to restraints in the scripting layer as well as from a restraint editor GUI and allows for manipulation of target values and their estimated standard deviations from target values.

4.9. Ligand Fitting

In general, there are several possible factors that need to be considered when fitting small molecule ligands to electron density in crystallographic experiments:

1. The ligand-binding site can be known or unknown.
2. The ligand can be known, or selected from a choice of several.
3. The structure of the ligand is unknown.
4. The ligand candidate can be static or manipulated by conformer generation.

The ligand fitting algorithm in Coot was initially designed to deal with a known static ligand at an unknown binding site. It has since been extended to choose from a number of ligand candidates, optionally allowing conformation searching at a known binding site.

The "known ligand at a known binding site" is the typical scenario for SBDD. The "one of a limited number" mode is typical of fragment screening, where a cocktail of ligands is used and the best-fitting ligand is selected to fit at each candidate ligand-binding site (28).

If the ligand site is not known, then the whole map is searched for candidate ligand-binding sites. The user chooses an electron density cut-off level, below which density points are not considered to be "in" a ligand density cluster. The map is then masked by the protein "target" molecule so that density points where there are protein atoms are not considered. (It is important therefore that the symmetry of the protein coordinates matches the cell and symmetry of the map.) The masked map is systematically and recursively searched until all map points have been considered; in so doing, each density point can be added to a candidate ligand-binding site cluster. The symmetry of the map is accounted for by mapping points in the electron density grid back into the asymmetric unit (this is a feature of the internal representation of electron density maps provided by Clipper (29)). Having found a list of candidate ligand density clusters, these are filtered so that only those that are bigger than clusters due to waters are considered. The eigenvectors of the density clusters are calculated and similarly the eigenvectors for each of the candidate ligand models are calculated. Then, cluster by cluster, each of the candidate ligand models is rotated and translated to match the eigenvectors of the cluster. There are four orientations to be considered for each eigenvector match. At each of the orientations, the candidate ligand undergoes rigid-body refinement. At each cluster, the best score for each candidate ligand is calculated and the best ligand chosen for each cluster by choosing the ligand that has maximum sum of the electron density at the atom positions after rigid-body refinement.

4.10. Conformation Generation

Ligands can be allowed to vary in conformation before being compared to the electron density clusters. To generate conformers, Coot uses the CIF dictionary for given residue types and varies the rotatable bonds according to the dictionary description of the torsion angles. Only dictionary torsions that do not contain "CONST" or "const" in their ID field are considered as potentially rotatable bonds.

Additionally, rotation angle sampling is not considered for torsion angles that have a standard deviation of less than 11°. This is to increase speed; if the ideal torsion angle is close to the actual torsion angle, then we can get to the correct solution using real-space refinement, rather than search for the correct solution by sampling. The period (D), ideal torsion angle (theta_ideal) and standard deviation (sigma_theta) are used to construct a probability distribution for each rotatable bond (see Fig. 2). A torsion angle is then randomly sampled from the distribution. Each rotatable bond in the dictionary is treated in this manner independently. Of course in real-world molecules the probability/energy distributions for the rotatable bonds are not independent. By treating them as independent we may end up with a chemically unreasonable conformation. Hence, the final stage of conformation generation is a geometry idealization step, where non-bonded contact restraints can relieve close contacts between non-bonded atoms. Currently, tautomeric forms are not generated.

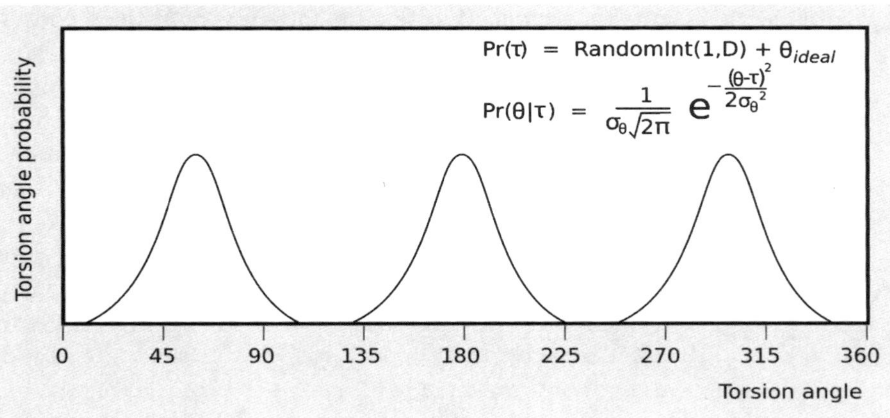

Fig. 2. Representation of a probability distribution for torsion angles. For each rotatable bond in each conformer, a rotation angle is selected from such probability distributions, allowing for the generation of a set of candidate conformers.

4.11. Ligand Overlay

Using the sub-graph isomorphism residue-matching algorithm mentioned above, Coot provides a convenient means to generate an initial fit of the new ligand. Provided that the new ligand is somewhat similar to a ligand of a previously solved complex and has the same binding mode, the new ligand can be automatically positioned based on the position of the previous ligand (and subsequent real-space refinement typically improves the fit).

4.12. Use of Non-crystallographic Symmetry

Coot can comprehend that the input model has non-crystallographic symmetry (NCS). Amongst the NCS tools of Coot is the "NCS Jump", which translates the view between the NCS-related chains, and taking into account the NCS operators so that the relative view remains the same. This provides a means for rapid inspection of NCS-related entities (including NCS-related ligand-binding sites).

Coot can take NCS into account at the ligand fitting stage and perform ligand fitting at NCS-related positions once a ligand has already been fitted at one of the NCS-related protein molecules.

5. Ligand Validation

5.1. Protein-Ligand Interactions

The interaction of the ligand with protein is best (less ambiguously) described if both the ligand and protein model are fully protonated. This is essential for understanding hydrogen-bonding interactions. The Molprobity tool reduce (30) can be used to generate a full-hydrogen model and probe (31) can be applied to find inter-atomic contacts, including polar and van der Waals interactions and hydrogen bonds. These interactions can be used to visually compare binding modes when the resolution of the data is insufficient to uniquely determine the orientation of the small molecule from the density map alone. In due course, the probe interaction (see Fig. 3) will be used to add binding mode information into Coot's ligand position scoring system (which is currently purely density-based). In Coot these can be rendered using Raster 3D (33). It should be noted these black and white screenshots are a poor representation of a typical Coot view, where there are colour, animation and 3-D rotation which improve the 3-D perception.

In the future Coot will have dictionary-based 3-D mark-ups for geometric deviations (e.g. bond lengths, bond angles, torsion angles, chiral centres and planarity deviations).

5.2. Electrostatic Surface Representation

Electrostatics calculations provide a useful tool to analyse molecular recognition processes, whether they involve protein–protein interactions, or protein-ligand interfaces (e.g. (34) and references cited therein). Coot uses the CXXSurface toolkit (35) to solve a form of the Poisson-Boltzmann equation and map the electrostatic potential on the molecular surface.

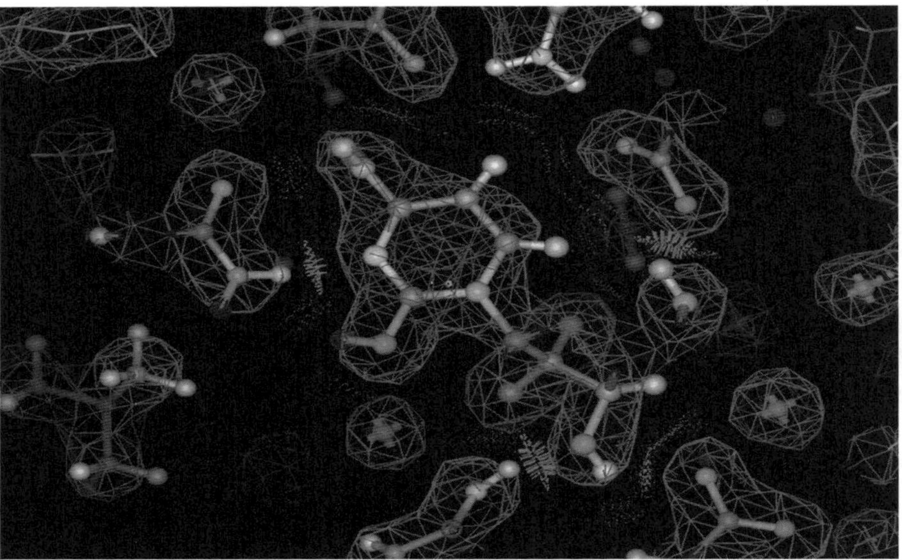

Fig. 3. A representation of Gemcitabine in its binding pocket (32). Probe interactions between the protein and the ligand are displayed, illustrating a number of hydrogen bonds (spiky "pillows") and Van der Waals interactions (e.g. to the hydrogens on the *top right*).

5.3. Ligand Dipoles

As well as providing restraints for monomers, the mmCIF dictionary can provide partial charges for the atoms in a monomer. Using these, Coot can generate the ligand dipole (both residues and residue selections can be used to generate dipoles). In cases where the atom selections are not neutral, the dipole is relative to the centre of the atom selection.

5.4. Assessment of Ligand Geometry

As coordinates delivered by X-ray crystallography feed directly into computational chemistry studies or synthetic chemistry efforts, thorough validation of small molecule substructures should be considered as a key step in structure determination for SBDD. As a consequence of the typical resolution range of macromolecular crystal structures, the final geometry of a small molecule compound is strongly influenced by the restraints applied to it during refinement. Therefore critical assessment of the initial geometric parameters should carry equal weight, particularly in a high-throughput environment.

Validation of chemical structures in the context of macromolecular crystallography can take several directions, depending on the basis of comparison.

Perhaps the most pragmatic approach is to compare the experimental ligand structure to 3D coordinates from quantum mechanical calculations. In this case the geometry can be assessed directly by calculating rms deviations or overall energy difference between the actual and theoretical model. Even though this approach is least prone to experimental and parameterization errors, and probably the best way to obtain initial restraints, it is also the most computationally intensive, so that it might not fit the time constraints of high-throughput crystallography. Additionally, QM calculations are most likely to reflect an in-vacuo state of the ligand (or maybe that in some kind of continuum), therefore subtle conformational or tautomeric changes induced by the protein residues surrounding the ligand may be flagged as errors in the structure.

Alternatively, ligand substructures can be analyzed against those in the PDB, as facilitated by the ValLigURL web server (36). However, it is quite unlikely that the ligand set in the PDB covers any of the novel small molecules of pharmaceutical interest. Consequently, this approach remains to be used mainly in academic research. (It has to be noted that ligand geometries in the PDB leave some room for improvement).

A more appropriate avenue for ligand validation compares the individual structural motifs in the ligand to similar fragments represented in chemical crystallography databases. The Cambridge Crystallographic Data Centre (CCDC) program Mogul (37) is one of the knowledge-bases allowing easy access to preferred bond, angle and torsion values from the Cambridge Structural Database and simple quality indicators (*Z*-scores for bonds and angles, and minimal distance values for torsions). Coot provides a platform to run Mogul as a batch job and visualize its results directly in the context of the electron density and ligand-binding pocket in the protein. This customization to Coot is a rather powerful tool to highlight errors in automatic restraint parameterization, ligand fitting and refinement, and it provides clear guidance for correcting restraint target values and esds. The ability to validate and adjust the ligand geometry within one model building session generally accelerates the iterative process of refinement and validation.

6. Views and Annotations

As we have noted, the optimization of protein–ligand interactions is a multidisciplinary task. It is sometimes desirable to exchange view information with others (i.e. a colleague). You might want to say "go to the active sites in each of the A, B, C chains, contour the map at 0.4 electrons/Å3 and look at the density for the phenyl ring". Coot provides a mechanism to save a set of such views (a Coot "view" currently consists of a centre, a zoom setting, an orientation and an optional annotation). The views can be saved as text files and reloaded (see Figs. 4 and 5). They can be played sequentially, with an animated interpolation

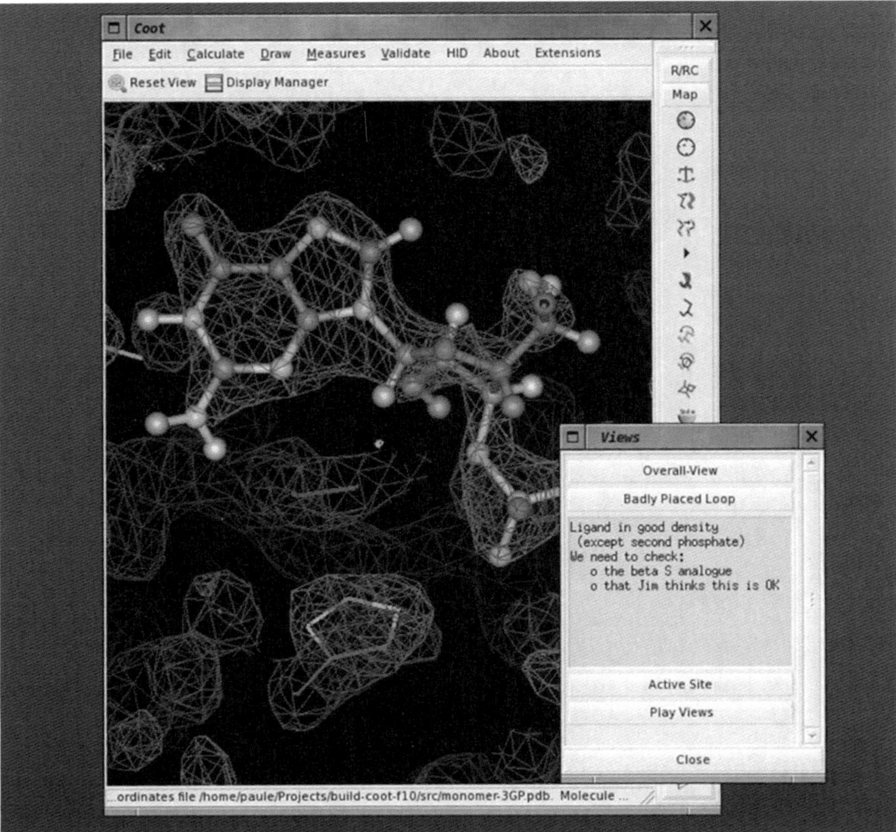

Fig. 4. An example of a Coot "view" and the views panel. The views panel contains a "clickable" list of views, which can be annotated. Smooth Coot transitions between views.

between them. The ability to spin the molecule is an additional view. A rocking view will be added. Coot does not as yet embed representation descriptions or representations of the model or map; these may be added later.

7. The Future of Coot

Coot is still in development, yet to reach version 1.0. As such, there are still several tools useful for SBDD to be added, some of which have been described above. It is often difficult for chemists to purify stereoisomers from a racemic mixture, hence the exact molecular structure of the ligand is often not known, typically being one of two stereoisomers and the determination of the exact configuration is usually done by examination of the electron density. Coot will introduce the ability to select chiral centres for

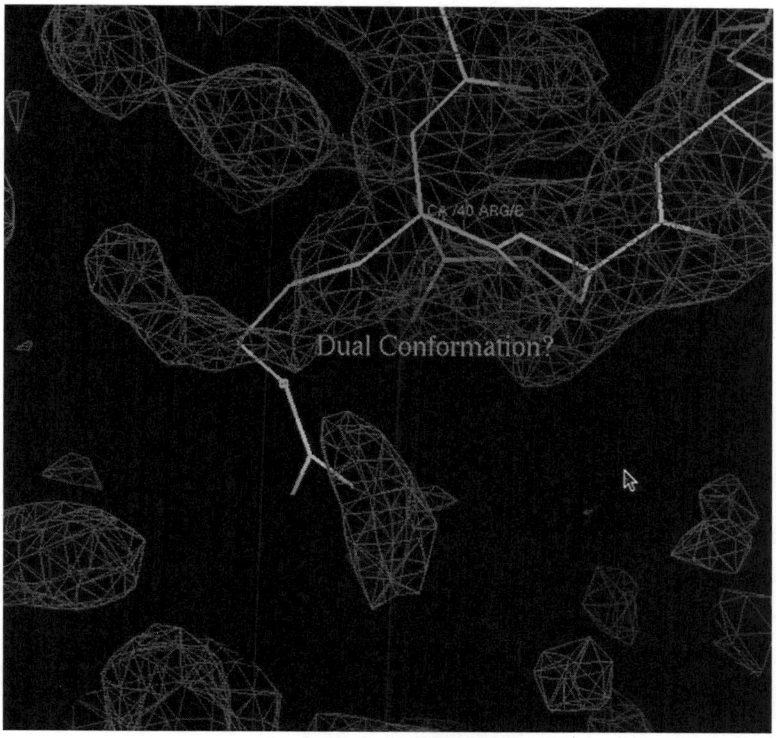

Fig. 5. Coot provides a means to save Annotations in the 3D graphics view. These mark positions in space with arbitrary text, and can be saved and loaded between sessions (e.g. before and after a round of refinement). For example one might leave a tag "the density for the second conformation is poor in refinement round 3".

inversion during conformer generation or interactively. Coot will generate bonds for ligands accordingly using the dictionary definition (e.g. single, double, delocalized), which will assist the analysis of protein-ligand interactions. Nucleotide and carbohydrate model-building and validation tools will be added and improved. We have plans for additional interactions with external validation programs, ligand dictionary-generation programs and web-services.

As DeLano (38) noted, companies engaged in therapeutic discovery have little time to wait for traditional proprietary application vendors to respond to their needs. These companies require modifications and extensions that can be implemented on a day-to-day basis, rather than a lag time of months or years. The distribution and licensing mechanism and configurability of Coot make it well placed for drug discovery.

References

1. Wlodek, S., Skillman, G. A. and Nicholls, A. (2006) Automated ligand placement and refinement with a combined force field and shape potential. *Acta Cryst.* **D62**, 741–749.
2. Abagyan, R., Totrov, M. and Kuznetsov, D. (1994) ICM: a new method for protein modeling and design: applications to docking and structure prediction from the distorted native conformation. *J. Comp. Chem.* **15**, 488–506.
3. Hartshorn, M. J. (2002) AstexViewer(tm): A visualisation aid for structure-based drug design. *Journal of Computer Aided Molecular Design* **16**, 871–881.
4. Hodis, E., Prilusky, J., Martz, E., Silman, I., Molt, J.and Sussman, J. L. (2008) Proteopedia - a scientific 'wiki' bridging the rift between three-dimensional structure and function of biomacromolecules. *Genome Biology* **9**, R121.
5. Davis, A. M., St-Gallay, S. A. and Kleywegt, G. J. (2008) Limitations and lessons in the use of X-ray structural information in drug design. *Drug Discovery Today* **13**, 831–841.
6. Mooij, W. T. M., Hartshorn, M. J., Tickle, I. J., Sharff, A. J., Verdonk, M. L. and Jhoti, H. (2006) Automated Protein-Ligand Crystallography for Structure-Based Drug Design. *Chem. Med. Chem.* **1**, 827–838.
7. Emsley, P. and Cowtan, K. (2004) Coot: model-building tools for molecular graphics. *Acta. Cryst.* D**60**, 2126–2132.
8. Lovell, S. C., Word, M. J., Richardson, J. S. and Richardson, D. C. (2000) The penultimate rotamer library. *Proteins: Structure, Function, and Genetics* **40**, 389–408.
9. Davis, I. W., Arendall, W. B., Richardson, D. C. and Richardson, J. S (2006) The Backrub motion: How Protein Backbone Shrugs When the Sidechain Dances. *Structure* **14**, 265–274.
10. Krissinel, E. and Henrick, K. (2004) Common subgraph isomorphism detection by backtracking search. *Software - Practice and Experience* **34**, 591–607.
11. Bourne, P. E., Berman, H. M., Watenpaugh, K., Westbrook, J. D. and Fitzgerald, P. M. D. (1997) The macromolecular Crystallographic Information File (mmCIF). *Meth. Enzymol.* **277**, 571–590.
12. Murshudov, G. N., Vagin, A. A. and Dodson, E. (1997) Refinement of Macromolecular Structures by the Maximum-Likelihood Method. *Acta Cryst* **D53**, 240–255.
13. Vagin, A. A., Steiner, R. A., Lebedev, A. A., Potterton, L., McNicholas, S., Long, F. and Murshudov, G. N. (2004) REFMAC5 dictionary: organization of prior chemical knowledge and guidelines for its use. *Acta Cryst.* **D60**, 2184–2195.
14. van den Bedem, H., Lotan I., Latombe J. C. and Deacon A. M. (2005) Real-space protein-model completion: an inverse-kinematics approach. *Acta Cryst.* **D61**, 2–13.
15. Joosten, K., Cohen, S.X., Emsley, P., Mooij, W. Lamzin, V. S. and Perrakis, A. (2008) A knowledge-driven approach for crystallographic protein model completion. *Acta Cryst.* **D64**, 416–426.
16. Jones, T. A., Cowan, S., Zou, J.-Y. and Kjeldgaard, M. (1991) Efficient Rebuilding of Protein Structures. *Acta Cryst.* **A47**, 110–119.
17. Esnouf R.M. (1997) Polyalanine Reconstruction from C[alpha] Positions Using the Program CALPHA Can Aid Initial Phasing of Data by Molecular Replacement Procedures. *Acta. Cryst.* D53, 665–672.
18. Dalby, A., Nourse, J. G., Houndshell, W., Gushurst, A. K. I., Grier, D. L., Leland, B. A. and Laufer, J. (1992) Description of Several Chemical Structure File Formats Used by Computer Programs Developed at Molecular Design Limited. *Journal of Chemical Information and Computer Sciences* **32**, 244–255.
19. Weininger, D. (1988) SMILES, a Chemical Language and Information System. 1. Introduction to Methodology and Encoding Rules. *J. Chem. Inf. Comput. Sci.* **28**, 31–36.
20. Moriarty, N. W., Grosse-Kunstleve R. W. and Adams, P. D. (2009) electronic Ligand Builder and Optimization Workbench (eLBOW): a tool for ligand coordinate and restraint generation. *Acta Cryst.* **D65**, 1074–1080.
21. Adams, P. D., Grosse-Kunstleve, R. W., Hung, L., Ioerger, T. R., McCoy, A. J., Moriarty, N. W., Read, R. J., Sacchettini, J. C., Sauter, N. K. and Terwilliger, T. C. (2002) PHENIX: building new software for automated crystallographic structure determination. *Acta Cryst.* **D58**, 1948–1954.
22. Blanc, E., Roversi, P., Vonrhein, C., Flensburg, C., Lea, S. M. and Bricogne, G. (2004) Refinement of severely incomplete structures with maximum likelihood in BUSTER-TNT. *Acta Cryst.* **D60**, 2210–2221.
23. Sadowski, J. & Gasteiger, J. (1993) From atoms and bonds to three-dimensional atomic coordinates: automatic model builders. *Chemical Reviews* **93**, 2567–2581.
24. Schüttelkopf, A. W. and Aalten, D. M. F. (2004) PRODRG: a tool for high-throughtput crystallography of protein-ligand complexes. *Acta Cryst.* **D60**, 1355–1363.

25. Kleywegt, G. J. and Jones, T. A. (1998) Databases in protein crystallography. *Acta Cryst.* **D54**, 1119–1131.
26. Berman, H. M., Westbrook, J., Feng, Z., Gilliland, G., Bhat, T.N., Weissig, H., Shindyalov, I.N. and Bourne, P.E. (2000) The Protein Data Bank. *Nucleic Acids Research* **28**, 235–242.
27. Andrejasic, M., Praznikar, J. and Turk, D. (2008) PURY: a database of geometric restraints of hetero compounds for refinement in complexes with macromolecular structures. *Acta Cryst.* **D64**, 1093–1109.
28. Nienaber, V. L., Richardson, P. L., Klighofer, V., Bouska, J. J., Giranda, V. L. and Greer, J. (2000) Discovering novel ligands for macromolecules using X-ray crystallographic screening. *Nat. Biotech.* **18**, 1105–1108.
29. Cowtan, K. (2003) The Clipper C++ libraries for X-ray crystallography. *IUCr Computing Commission Newsletter* **2**, 4–9.
30. Word, J. M., Lovell, S. C., Richardson, J. S. and Richardson, D. C. (1999) Asparagine and glutamine: using hydrogen atom contacts in the choice of side-chain amide orientation. *J Mol. Biol.* **285**, 1735–1747.
31. Word, J. M., Lovell, S. C., LaBean, T. H., Taylor, H. C., Zalis, M. E., Presley, B. K., Richardson, J. S. and Richardson, D. C. (1999) Visualizing and quantifying molecular goodness-of-fit: small-probe contact dots with explicit hydrogen atoms. *J. Mol. Biol.* **285**, 1711–1733.
32. Sabini, E., Ort, S., Monnerjahn, C. Konrad, M. and Lavie, A. (2003) Structure of human dCK suggest stragegies to improve anticancer and antiretroviral therapy. *Nature Structural Biology* **10**, 513–519.
33. Merritt, E. A. and Bacon, D. J. (1997) Raster3D: Photorealistic Molecular Graphics. *Meth. Enzymol.* **277**, 505–524.
34. Ma, B., Elkayam, T., Wolfson, H. and Nussinov, R. (2003) Protein-protein interactions: structurally conserved residues distinguish between binding sites and exposed protein surfaces. *Proc. Natl. Acad. Sci. USA* **100**, 5772–5777.
35. Gruber, J. Zawaira, A., Saunders, R., Barrett C.P. and Noble M.E.M. (2007) Computational analyses of the surface properties of protein-protein interfaces. *Acta. Cryst.* D**63**, 50–57.
36. Kleywegt, G. J. and Harris, M. R. (2007) ValLigURL: a server for ligand-structure comparison and validation. *Acta Cryst.* **D63**, 935–938.
37. Bruno, I. J., Cole, J. C., Kessler, M., Luo, J., Motherwell, W. D. S., Purkis, L. H., Smith, B. R., Taylor, R., Cooper, R. I., Harris, S. E. and Orpen, A. G. (2004) Retrieval of Crystallographically-Derived Molecular Geometry Information. *J. Chem. Inf. Comput. Sci.* **44**, 2133–2144.
38. DeLano, W (2005) The case for open-source software in drug discovery. *Drug Discovery Today* 10, 213–217.

Chapter 7

Crystallographic Fragment Screening

John Badger

Abstract

Crystallographic fragment screening is a technique for initiating drug discovery in which protein crystals are soaked or grown with high concentrations of small molecule compounds (typically MW 110–250 Da) chosen to represent fragments of potential drugs. Specific binding of these compounds to the protein is subsequently visualized in electron density maps obtained from analysis of X-ray diffraction data collected from these crystals. Theoretical and practical experience indicate that a suitably diverse library of fragment compounds containing only a few hundred compounds may be sufficient to provide a comprehensive screen of the protein target. By soaking crystals in mixtures of 3–10 compounds a fragment screen may be completed within ~100 diffraction data sets. This data collection requirement may be met given reproducible well-diffracting protein crystals and robotic sample handling equipment at a high flux X-ray source. The leading practical issue for most crystallography laboratories that wish to launch a fragment screening project is the design and/or procurement of an appropriate fragment library. Although several off-the-shelf fragment libraries are available from chemical suppliers, the numbers, sizes, and solubility of the compounds in relatively few of these libraries are well-match to the specific needs of the crystallographic screening experiment. Informed consideration of the properties of compounds in the screening library, possibly augmented by additional filtering of available compounds with appropriate search tools, is required to design a successful experiment. The analysis of results from crystallographic fragment screening involves highly repetitive application of routine image data processing and structure refinement calculations from many very similar crystals. Efficient handling of the data applies a high-throughput structure determination methodology that conveniently packages the structure solution calculations into a single process that provides the crystallographer-analyst with ready-to-view maps for evaluating crystals for bound compounds.

Key words: Fragment library, Fragment screening, X-ray Crystallography, Synchrotron Radiation, Crystallography Software

1. Introduction

Fragment-based screening is a lead discovery methodology (FBLD) in which the initial screen of the protein target is performed with low-molecular-weight "fragments of drugs" (typically ~110–250 Da). Bound compound "hits" identified from the fragment screen are

Leslie W. Tari (ed.), *Structure-Based Drug Discovery*, Methods in Molecular Biology, vol. 841,
DOI 10.1007/978-1-61779-520-6_7, © Springer Science+Business Media, LLC 2012

subsequently elaborated into lead molecules. An appealing aspect of FBLD is that it requires relatively small numbers of compounds at the outset and is ready to be enabled as a relatively low-cost "small science" in biotechnology companies and academic laboratories. FBLD activities have the potential to significantly enlarge the research base of researchers involved in the development of early lead compounds. This chapter is written mainly from the perspective of a laboratory performing protein crystallography work that wishes to begin fragment-based screening; it focuses on the more unique aspects of performing the initial fragment screening using protein crystallography as the assay technique.

Theoretical arguments involving matching of compounds to structural features within a protein-binding site predict that screening libraries composed of small compounds will be more efficient than libraries composed of large compounds. The larger compounds will interact with many features on the protein target, and there is a very high chance that some of these features will prevent binding. Conversely, small compounds will interact with only a small number of features and are much more likely to bind to the protein. More precisely, the chance that a compound will bind to protein target is expected to decrease exponentially with increasing compound complexity (1). A great deal of chemical space may be sampled with relatively few compounds provided that the compounds are sufficiently small and screening campaigns using X-ray crystallography typically involve just several hundred compounds. In practice, the hit rate for screening with fragment compounds is ~1–4%, which is 10–1,000 times higher than typical outcomes from high-throughput screening (HTS) using compounds 2–3 times larger.

Offsetting the advantage of requiring only a small library size, fragment compounds generally bind with much lower binding affinities than the larger compounds typically present in HT screening libraries. Special biophysical techniques are required to detect weak binding, and X-ray crystallography is recognized as the most sensitive of currently available techniques, capable of detecting compounds with binding constants in the low millimolar range (see Fig. 1). Protein crystallography applies fragment screening in its most extreme interpretation, by allowing use of smaller and more weakly binding fragments than other techniques. One impetus for the adoption of crystallographic fragment-based screening (2) within the pharmaceutical industry is the realization that protein crystallography has now been reduced to practice to an extent that it allows its application in high-throughput modes (3, 4).

The obvious restriction on the application of crystallographic fragment screening is the requirement that the target protein has been crystallized and that the crystals are relatively robust, routinely diffract to beyond medium resolution (say, 2.5 Å) and provide unblocked access to the target site. However, given these prerequisites, crystallographic screening offers some unique

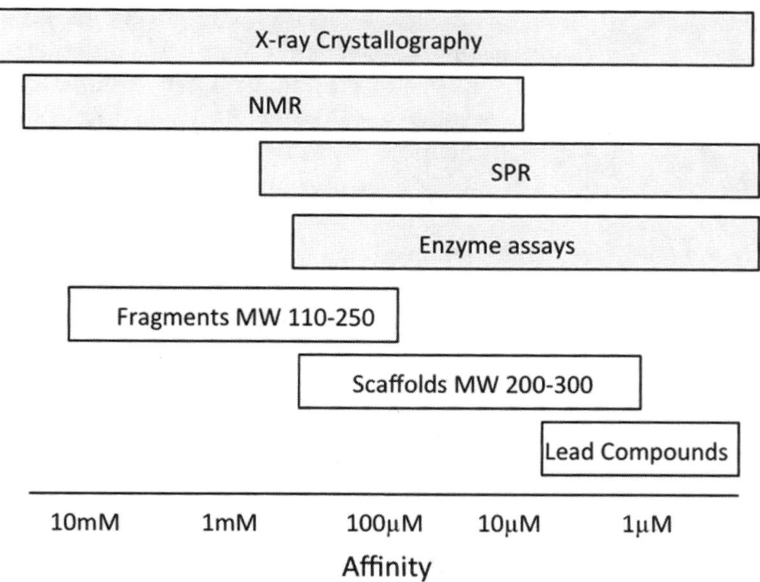

Fig. 1. Fragment molecular weights and biophysical techniques for detection of compound binding across a spectrum of affinities.

advantages over most other techniques. First, because the outcome of the experiment is an image (an electron density map) showing the position and conformation of the fragment on the protein in the crystal, there are few false positive hits in a crystallographic screen. Fragments that bind in nonproductive sites on the protein (i.e., not impacting function) are readily identified. Conversely and usefully, the method may occasional identify unsuspected binding sites as new drug targets. Second, the 3-dimensional structural information on the fragment binding mode in the target site enables the efficient development of a second round of larger follow-up compounds—because the key binding interactions and accessible space in the protein are known, structural constraints on the lead compound development process may be applied to expedite the lead development process. The main shortcoming of X-ray crystallography in terms of providing technology for initiating a lead development process is that the crystallographic identification of a bound molecule does not provide information on binding affinity and this data must be obtained with a different technique. In practice, obtaining the binding affinity of the initial small fragment hit may not be a particularly useful or attainable and testing for potency (most usually by chemical assay or Surface Plasmon Resonance) might begin at the stage of elaborated follow-on compounds.

Due to their small sizes and low potencies, the first set of compounds identified from a FBLD program require significant development before they become "true" lead compounds. However, many practical examples now show the rapid development of lead-like molecules with good binding potency from small weakly binding starting points (5), and this aspect of FBLD does not appear to present an overwhelming challenge. In fact, a key conceptual advance is the view that rather than simple binding affinity, the more sophisticated notion of "ligand efficiency"—the binding energy per non-hydrogen atom (6)—is more central to successful lead development. According to this concept, a relatively weakly binding hit with a small compound would be preferred for lead development over a more potent but larger compound if the ligand efficiency is superior. As compounds developed from fragment screening programs begin to enter clinical trials, there is emerging statistical evidence that growing a lead molecule from a fragment starting point, in which the properties can be monitored and adjusted, may often be a simpler, quicker, and more successful route to a "good" lead compound than attempting to manipulate a larger molecule emerging from an HTS experiment (5). These authors list 47 compounds that have reached significant development milestones with four examples in clinical trials. Two of these compounds appear to have evolved from the fragment-based screen to entry into clinical trials in <2 years. Overall, the mean molecular weight of marketed small molecule drugs is only ~350 Da, perhaps because therapeutic liabilities tend to increase with size, and orally active drugs are generally limited to compounds with molecular weights <500 Da (7). In the CNS therapeutic arena, the inherent biophysical requirements that allow a drug molecule to cross the blood–brain barrier limit the molecular weight, numbers of hydrogen bonds, and polar surface area to compounds that barely extend beyond fragment and scaffold-sized leads (8). Compounds emerging from screening libraries that are heavily populated with relatively high molecular weight compounds may show promising initial potencies but are more likely to encounter therapeutic issues in downstream testing and development.

2. Materials

2.1. Compound Suppliers

1. Pre-plated fragment libraries containing compounds suitable for screening by protein crystallography are distributed by Maybridge at www.maybridge.com and by Zenobia Therapeutics at www.zenobiatherapeutics.com. A list of compounds successfully used for crystallographic fragment screening at the University of Washington is available from those authors upon request (9).
2. Sigma-Aldrich at http://www.sigmaaldrich.com provides many individual low-molecular-weight compounds which may be selected from a substructure search application on the web site.

3. Many chemical vendors provide SD files upon request that contain large collections of screening compounds that may be cherry-picked and purchased. The collections from Maybridge at www.maybridge.com and Chembridge Corporation at http://chembridge.com/index.html contain sufficient low-molecular-weight compounds for a fragment library design.

2.2. Software for Library Design and Analysis

1. A free browser for scrolling through chemical data files in SD format is the *ChemFileBrowser* from Hyleos at http://www.hyleos.net.
2. Free programs for filtering chemical collections presented in SD files according to property criteria are available as part of the *Open Babel* project at http://openbabel.org project. *SDSearch*, which provides a GUI and applies this application code for property filtering and substructure searching, is available from http://deltagtech.com. A point commercial software application for performing these tasks is *FILTER* from OpenEye Scientific Software at http://www.eyesopen.com/.
3. A freeware Chemistry drawing software is *ChemSketch* from ACD at http://www.acdlabs.com/download/chemsketch/.

2.3. Experimental Crystallography

1. A list of synchrotron radiation sources with beam lines for macromolecular crystallography is available from the Biosync website at http://biosync.rcsb.org.

2.4. Software for Protein Crystal Structure Analysis

1. Diffraction image data processing programs that may be run in fully automated (scripted) high-throughput modes are the commercial program *D*TREK* from Rigaku at http://www.rigaku.com and the free *MOSFLM* program from http://www.mrc-lmb.cam.ac.uk/harry/mosflm. *MOSFLM* is also distributed with the *CCP4* suite from http://www.ccp4.ac.uk.
2. Software packages that include programs for diffraction data manipulation, molecular replacement, and structure refinement are the *CCP4* suite from http://www.ccp4.ac.uk, the *CNX/CNS* system which is available for academic users from http://cns-online.org/v1.2 and for commercial users from Accelrys at http://accelrys.com, and the *Phenix* system available from http://www.phenix-online.org. *CCP4* and *CNS* run on operating systems including Windows, Mac, and Linux whereas the *Phenix* system is confined to Linux and Mac.
3. A free interactive model-building and density map display package is *Coot*, available from http://www.ysbl.york.ac.uk/~emsley/coot/ with a Windows version available from http://www.ysbl.york.ac.uk/~lohkamp/coot/wincoot.html. *Coot* is also distributed with the *CCP4* suite from http://www.ccp4.ac.uk. The open source program, *MIFit*, available from http://code.google.com/p/mifit also provides

crystallographic model-building and display capabilities and contains some features specifically designed to assist in fragment screening analysis.

3. Methods

3.1. Selection of Compounds for a Fragment Screening Library

3.1.1. Size of the Library and Molecular Weights of the Library Compounds

A logistical factor for most groups performing crystallographic fragment screening is an expectation that the screen should be accomplished within ~100 crystals. If crystals are soaked in cocktails containing 3–10 compounds, this restriction implies that the fragment library will contain <1,000 compounds. Arguments based on molecular complexity (1) indicate that in order to cover chemical space with a small library the compounds that it contains must also be relatively small. For this reason an appropriate fragment library for crystallographic screening will typically contain compounds with a mean molecular weight of ~160 Da and few compounds with molecular weights >200 Da. Although many chemical vendors now provide "fragment screening libraries," most of these libraries contain large numbers of larger compounds and are not suitable for a crystallographic screening campaign without further filtering. When assessing the suitability of a compound collection for protein crystallography, an initial check should be made on the molecular weight distribution and numbers of compounds within the appropriate size range.

3.1.2. Compound Properties and the Rule of Three

The compounds in fragment libraries are usually selected so that their molecular properties meet the "Rule of three" (Ro3) (10), and this rule places sensible bounds on the nature of compounds in a fragment library.

Ro3 states that the best types of compound to include in the screening library (1) have molecular weights <300 Da, (2) contain no more than 3 hydrogen bond donors and no more than 3 hydrogen bond acceptors, (3) have solubility indices as defined by clogP not greater than 3. Besides the three direct Ro3 rules, other useful limiting property parameters for compound selection (10) are that the number of rotatable bonds should not exceed 3 and that the polar surface area should not exceed 60 Å^2.

The molecular weight bound of 300 Da in the Ro3 is very elastic and would allow inclusion of large compounds that do not fit the "small library, small compound" paradigm applied to crystallographic screening, where it is best to set an upper molecular weight bound of about 200–225 Da. When applying property filters for an in-house fragment library design, it should also be noted that some aspects of Ro3 property calculations are subject to varying definitions and atom typing rules that are potentially problematic for some compound types (rings containing nitrogen atoms).

Depending on property calculation software, an overly strict adherence to the Ro3 might exclude some useful compound types. A recent analysis (11) somewhat undermines the significance of Ro3 properties for determining binding hit rates with well-chosen compound types. The types of compounds most useful for fragment screening are generally small ring-like compounds that resemble fragments of drugs (a useful tabulation is found in ref. (12)) and these chemical types should normally dominate the compound selection for the fragment library.

3.1.3. Compound Solubility

Compound stock solutions are usually created by dissolving dry compound in DMSO. Most crystals will not tolerate changing the composition of the mother liquor beyond addition of more than ~10% DMSO and some crystals might not allow addition of more than ~5% DMSO. In order to allow for this ~10-fold dilution in the crystal soaking mother liquor, and a further 3–10-fold dilution factor from soaking compounds in mixtures, the compound concentration in the original stock solution must be relatively high. Therefore, the compounds that are to be used in mixture soaking experiments need to be soluble at ~200 mM in DMSO in order to achieve final concentrations in the crystal >2 mM. This requirement requires an order of magnitude greater solubility than is typical of most screening experiments. In this author's experience 5–10% of appropriate fragment compounds that are Ro3 compliant (with calculated log P values) fail to meet this level of solubility when tested in the laboratory.

3.1.4. Creating a Compound List

Off-the-shelf fragment libraries suitable for direct use in crystallographic screening without additional compound selection efforts are distributed by Maybridge in sets of 500 and 1,000 compounds and by Zenobia Therapeutics as a set of 352 compounds. These libraries consist of pre-plated compound sets of appropriate chemical classes with suitable molecular properties and experimental solubilities that have been successfully used in crystallographic screening projects.

The majority of screening compounds available from chemical suppliers are at much higher molecular weights (~400 Da) than will be useful for crystallographic fragment screening, and some suppliers stock very few compounds with molecular weight <200 Da. However, it is possible to build an in-house fragment library suitable for crystallographic screening from some of these collections. Two alternative library design pathways seem possible, depending on whether it is more practical for the given vendor to begin the library design by selecting chemical types or begin by filtering on available compound properties.

Approach 1: Chemical substructure search interfaces available through web browsers at vendor web sites may be used to search directly for suitable fragment compounds based on predetermined

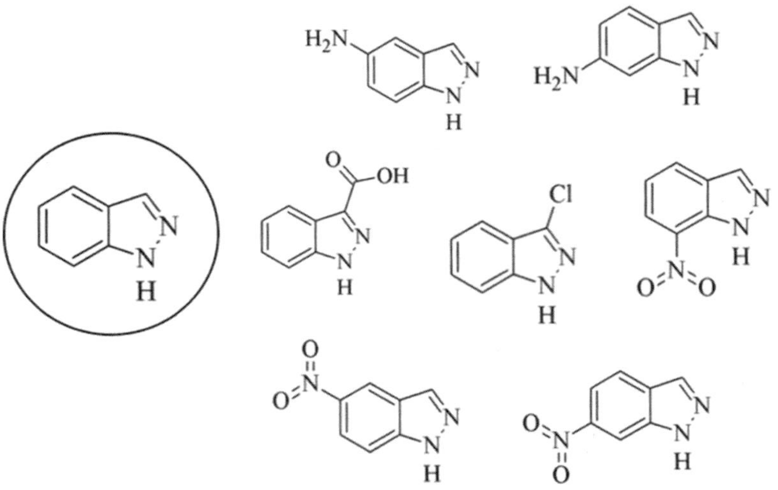

Fig. 2. Substructure searches on an indazole core (*left*) give several possible compounds for inclusion in a fragment library (*right*).

chemical core classes. The working process is to sketch each of a set of designated cores (typically ring structures found in known drugs, see a tabulation in ref. (12)) into the web site structure search page. For each core, those low-molecular-weight compounds that incorporate appropriate side chains and which appear compliant with Ro3 may be purchased (see Fig. 2). Practical factors for acquiring these compounds include immediate availability, relatively low prices, low toxicity, and avoidance of controlled substances. Although it requires some effort to search these individual classes of compounds, the specific attention that is paid to the hand-picked selection helps provide a check on the diversity of compounds across cores and side chain types as well as an implicit check that all of the compounds are suitably similar to fragment of drugs. Several workers have found that Sigma-Aldrich provides sufficient suitable low-molecular-weight compounds for a fragment library containing ~400 compounds and although the library building approach appears unsophisticated, such collections have been the basis of successful screening fragment experiments.

Approach 2: Many chemical vendors provide their available compound data in SD files that typically hold collections of 100,000–500,000 compounds. These data may be filtered using appropriate software to define a subset of compounds for subsequent purchase and inclusion in the fragment library ("cherry-picking"). Suitable suppliers of low-molecular-weight compounds that provide structure data in SD format include the Chembridge Corporation (the Express-Pick™ collection) and Maybridge (the ScreeningCollection). Analysis of these collections at the time of writing shows that they contain ~1,200 and ~1,300 compounds respectively with molecular weight <200 Da that meet the Ro3. Additional analysis of these sets to ensure that they only contain appropriate substructures (for example the drug-like ring systems (12)), with redundancy removal of very closely related compounds, may reduce these lists to several hundred compounds. However, it should be noted that the screening compounds available from reputable vendors should generally be of appropriate types. If available, commercial chemical database software may be used to filter these collections to Ro3 subsets but this software may require a disproportionately large effort and expenditure to install and operate for a single small library design project. An Open Source software suite capable of providing appropriate filters on the Ro3 and other properties from command-line programs is the *OpenBabel* project. Compound properties relating to the Ro3 are usually precalculated and provided as annotations within the SD files supplied by these vendors, simplifying the task of selecting compounds with appropriate ranges of properties. A simple GUI-driven application, *SDSearch*, may be used to facilitate library design by filtering over property annotations with SD files. The main advantage of approaches based on filtering reputable screening collections for Ro3 compliant compounds within a suitable mass range is that it allows very convenient and flexible experimentation with the rules that define the fragment library. A very useful SD file viewer for scrolling through chemical images in order to review the compound selections is the *ChemFileBrowser* from Hyleos.

3.2. Experimental Design with Fragment Mixtures

3.2.1. Factors Determining the Number of Compounds in a Mixture Group

In order to increase the efficiency of the crystallographic fragment screening experiment, it is usual to perform the crystal-soaking experiments with mixtures of 3–10 compounds. Including a relatively large number of compounds in a mixture is superficially attractive in terms of requiring fewer data sets to complete the fragment screen. However, there are potentially negative consequences from having too many compounds in each group. Increasing the number of compounds in a mixture reduces the soaking concentration of each compound within a critical regime of binding constants. A reduced compound concentration could result in failure to achieve a sufficient concentration for detectable binding, effectively limiting the sensitivity of the screen to the most potent examples and losing potential for detecting the weaker binding compounds in the library.

The leading factor controlling the largest possible number of compounds in each mixture is the target concentration of compound in the crystal at the maximum tolerable volume of the DMSO. For example, if the individual compounds are dissolved in DMSO at 200 mM concentration and the crystals tolerate 10% DMSO, then a mixture group of eight compounds will provide a final concentration of 2.5 mM of each compound in the crystal (see Note 1). For crystals less tolerant of DMSO it will be necessary to use smaller groups of compounds in mixtures in order to maintain this compound concentration (see Note 2).

A related demerit of including relatively large numbers of compounds in the mixture is that some compounds may cause crystals to crack or dissolve and the probability that this will occur increases with the number of different compounds in the mixture. This problem creates the nuisance of breaking-up that particular mixture group into smaller sets in order to retain the usable compounds within the experiment while isolating the impact of the problematic compound. If a crystal appears sensitive to this effect, then the number of compounds in each mixture should be kept relatively low.

3.2.2. Creating Shape Diverse Mixtures

The experimental cost of the screening experiment is reduced by arranging the mixture groups so that the compounds are shape-diverse for X-rays (see Fig. 3). A relatively unambiguous identification from the electron density map from the mixture experiments reduces experimental work needed to confirm the compound identity since the most likely candidate can be used first in follow-up soaking experiments containing single compounds. On occasion the higher concentration of compound possible in the single component experiment (or by soaking at the same compound concentration but with proportionately less DMSO) will also provide an improved electron density map of the bound compound.

Unless a second compound within a mixture binds with a free energy within <1 kcal/mol of the most strongly binding compound, the electron density of the first compound will almost completely mask out the second compound. For this reason there is little chance that an electron density feature should be interpreted

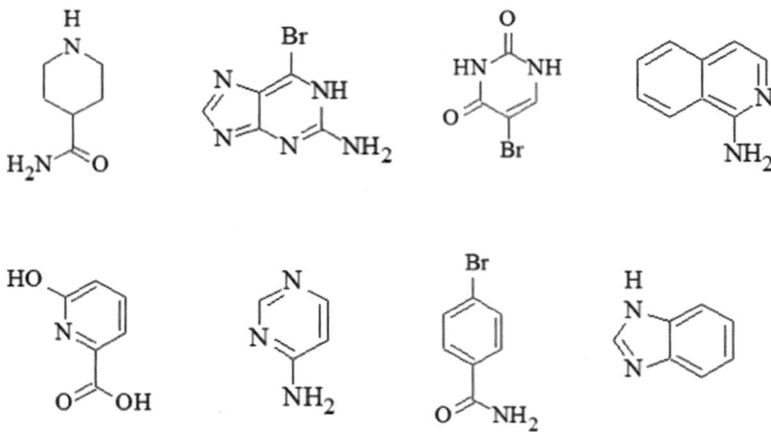

Fig. 3. Mixture group of eight small fragment compounds, differing in ring sizes, R-groups, and inclusion of heavy atoms. A set of four compounds (*top row*) would appear relatively easy to distinguish but more closely related compounds are present after expanding the group to eight compounds.

as a superposition of two different compounds. This effect also means that when a mixture group contains an identified hit, the soaking experiment will need to be repeated with a mixture of the remaining compounds to check for specific binding by any other members of this mixture group. In practice, many experimenters appear to accept or ignore the effective loss of these secondary compounds from the experiment. The loss of compounds will be greater for relatively larger mixture groups.

Although some sophisticated computer methods have been employed to try to objectively maximize the shape diversity of the compound mixtures (12), this is not a critical process and the relatively small size of fragment screening libraries means a simple sorting methodology can be implemented manually.

1. The library may be initially sorted into a few roughly equal-sized sets of compounds that have different sized ring systems and/or other distinguishing features (for example heavy atoms). The number of these sets is made equal to the desired number of compounds in the mixture groups (i.e., 3–10). Mixture groups may then be established by taking one compound for each mixture group from each of these different compound sets. Toward the end of the process some accommodation will be required to account for somewhat different sizes in the initial sets. A visual way to manage this operation on the computer is to create chemical image files for all compounds in the library, apply the *View/thumbnails* (Windows) setting to the folder display, and then drag and sort the images into different folders corresponding to the mixture groups.
2. If the library design was obtained via procedures that place similar compounds next to each other in the final list (for example, if the library design process involved simultaneous selection of compounds with common cores), a simple method for generating reasonable shape diversity is simply to repeatedly count along the compound list into N different mixture groups, i.e., setting group id numbers $1,2,3,\ldots,N,\ldots,1,2,3,\ldots,N,1,2,3,\ldots,N$, along the full compound list ensures that the similar compounds that are close to each other in terms of rank order in the compound list are placed in different mixture groups.

3.3. Requirements for Experimental Protein Crystallography

3.3.1. Data Collection Facilities and Robotics

Most obviously, a project that requires the collection of ~100 data usable sets should be based on the reliable production of crystals that routinely diffract to relatively high resolution (say, to better than 2.5 Å). For fragment screening, it is implicit that most crystals will deliver an adequate data set without requiring special attention in terms of creating individualized data collection strategies, and the goal is to collect as many data sets as possible within a limited amount of data collection time. Soaking of apo crystals in solutions of screening compounds is invariably considered the most economic mechanism for obtaining sufficient numbers of crystals for the screening campaign although co-crystallization might not be completely ruled out. Time spent optimizing the protein purification and crystallization protocols will save a lot of time and cost in the "production" data collection runs.

While not absolutely essential, the availability of robotic crystal mounting and data collection system is a huge advantage for fragment screening experiments. Most fragment screening projects will be most efficiently run if the data collection is performed at a synchrotron source—not only will data collection times be shorter than in the home laboratory but a higher proportion of crystals will yield interpretable map data. With brighter sources, the availability of robotic equipment is increasingly worthwhile since data collection times will be short and overheads in sample changing are to be avoided. Project cost and effort is difficult to anticipate before the reproducibility and typical diffraction characteristics of the crystals are known. However, crystals for workable projects should probably be expected to yield ~2 usable data sets per hour at a third-generation synchrotron X-ray source, and the project will require several shifts to complete. Once the first ~20 data sets have been obtained it should be able to obtain a reasonable idea of the typical crystal characteristics, any special problems with the data processing or map analysis, the hit rate, the overall cost, and the cost-benefit of the project.

3.3.2. Crystal Requirements for Successful Fragment Binding

Before beginning a fragment screening project, it should be established that the crystal form offers prospects for success. Although fragment screening may uncover some interesting and unanticipated binding sites, the main binding (active) sites in the target proteins will often have been already identified. The target binding site(s) should be exposed and available for fragment binding. This requirement may pose a problem for some projects, where crystal structures of ligand-bound forms of the protein are available but the apo form has either resisted crystallization or only diffracts to low resolution. In some "heroic" cases reasonably reliable soak-out protocols have succeeded in enabling a fragment screening project, but this step considerably complicates and increases the amount of crystal handling required and raises questions as to the reliability of the subsequent electron density map interpretation: it may not be clear if a bound entity is a fragment or residual bound ligand.

From an experimental point of view, it should be noted that:

1. Unless some special kinetic event is required for the fragments to access the binding site, fragments are expected to diffuse through a protein crystal extremely rapidly (probably within seconds), and soaking of only a few minutes might be sufficient to achieve fragment binding (see Note 3) (9).

2. It is extremely helpful to be able to identify a bound fragment hit as early as possible in the project because this provides a great deal of confidence in the experimental procedures and particularly in the soaking protocols. Ideally, a control fragment might be available to use in an experimental "work-up" process before beginning data collection production runs. For example, an adenine soak might be tried if a protein target site is known to bind to ADP since this fragment would often be expected to bind at an ADP site (see Note 4).

3.4. Automated Structure Solution and Map Analysis

3.4.1. High-Throughput Structure Solution

In general, the structure solution processes used with fragment screening data are the same as for individual structure determination projects. However, the working philosophy is entirely different from that used for the determination of a new structure. The initial goal of the structure solution in fragment screening work is to achieve an electron density map of sufficient quality to determine whether a fragment is bound or not. Since the protein structure is already known (and fragments will not bind in most crystals), this is usually extremely straightforward but the lack of operational crystallographic difficulty is offset by the need to repeat these calculations in an organized way over ~100 data sets.

Calculations with fragment screening data consist of placing the protein model in the crystal using the method of molecular replacement (using a previously solved parent apo structure as search model) and following up this solution by automated refinement and water-picking. The crystallographic structure solution interfaces from the CCP4 (13) or CNS (14) software are mostly written from the perspective of new structure determination, where successive steps are necessarily somewhat exploratory and there is a separate interface corresponding to each step (i.e., data reduction, molecular replacement, refinement). For fragment screening projects a great deal of labor reduction can be achieved by daisy-chaining these operations so that they are run by a single master script executing the whole process. Many crystallographers develop local shell scripts for combining these operations but the MIFit software contains an interface that runs all structure solution steps (currently consistent with CCP4 6.0 components), and outputs refined models, maps, and summary reports. The input to the automated structure solution process should be the raw image data (or at least the reduced data), and the output should be models and maps in a convenient "ready to load" form for analysis from a crystallographic molecular graphics program. Besides providing an efficient pathway for running the structure solution steps, folding these into a single process also helps maintain consistent and well-defined data naming conventions which are critical for tracking and matching crystal data files with fragment solution. For most projects, the structure solution calculations should not be particularly time-consuming and will require only 5–10 min per data set.

Although automated fragment docking and scoring is achievable (15), the range of possibilities for disorder and the different sizes of the mixture molecules make it difficult to formulate a completely reliable computer-based scoring criteria for deciding which fragment is present in the map (for example, pathological cases include fragments with well-bound and completely disordered portions). Furthermore, given that the rate of detection for specific binding is 1–4% of compounds, the majority of data sets will not show bound compounds. The more critical objective is to be able to easily precompute refined maps and models for easy loading and interpretation.

Key steps in the structure solution process are as follows:

1. Run molecular replacement with an apo search model stripped of waters using a fast program or limited search options. Variability in crystal cells dimensions and protein packing sometimes prevents proper refinement of the protein in the absence of the full molecular replacement search although the parent apo structure will be close to a "drop in" model for some crystals. When the CCP4/MOLREP program is used in a mode where only the top few rotation function solutions are checked, the calculation will typically take less than 1 min. Structure solution with CCP4/PHASER should also be relatively fast since the best solution will stand out above other possibilities (see Note 5).
2. Preliminary automated refinement of the protein model alone (5 internal cycles with CCP4/REFMAC5) followed by three further rounds (5 internal cycles each) of refinement interspersed with rounds of water picking, all using a relatively tight weight on the X-ray term (0.1), is a rapid and sufficient protocol for most cases. Tight weights on the X-ray term are appropriate for structures close to convergence (the case here since the new data is presented with a finished model) and prevent problems with noisy or unexpectedly low-resolution data sets. Since the water-picking step will stud any ligand density with waters, some workers prefer to include an additional step of automatically deleting water molecules from the expected target site and recalculating a maximum-likelihood difference map with an additional refinement run.
3. When bound fragments appear to be present (see Notes 6, 7), the bound molecule may be docked and structure finalized by a few additional cycles of refinement. In some cases, it might be necessary to make trial fits (docking into density) of all of the most promising candidates and perform trial refinements of best-matched candidates. In many cases the bound fragments are not well ordered, and chemical information in the form of putative hydrogen bonds between fragment and protein is used to model the fit to the density (see Note 8) (see Fig. 4).

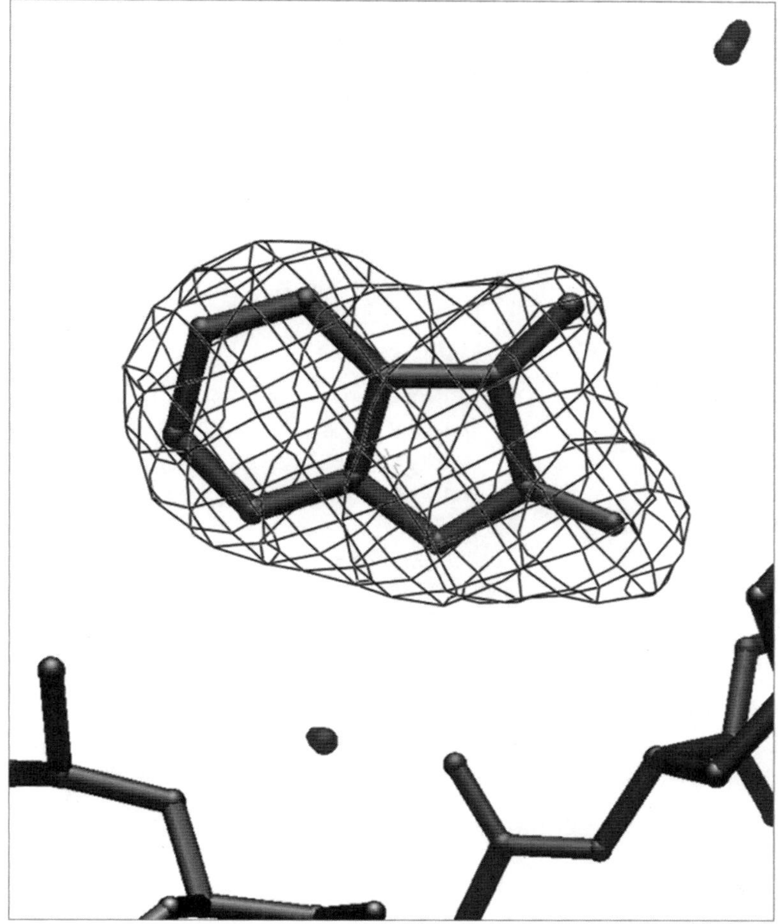

Fig. 4. The electron density of this fragment barely defines its orientation but consideration of interactions between the fragment and protein when atom types are considered would validate the fit.

4. Notes

1. A useful precaution in initial crystal soaking experiments is to include mixture solution in the cryo-protectant at the same concentration as in the crystal. Small molecules can diffuse through the crystal very rapidly and this will prevent bound compounds escaping from the crystal. Once binding has been established for a specific compound, an experiment with the cryo-protectant in absence of that compound can be performed to discover if this precaution is necessary.
2. The maximum concentration of DMSO that the crystal will tolerate gates the achievable concentration of compound in the crystal. A useful preliminary experiment is to collect data

from test crystals containing different concentrations of DMSO (but no compounds) to determine DMSO concentrations and soaking times that can be tolerated while maintaining good quality data.

3. Some crystals will not survive very prolonged soaking in mixture solutions, but binding may be so rapid that relatively "quick dip" (few seconds) exposure may be sufficient. Once binding has been established for a compound, an experimental time series may be performed in which the minimum exposure time required for binding is determined (9).
4. One common pitfall in fragment screening is to find an unwanted small molecule (perhaps a DMSO molecule or a small molecule from the cryo-protectant) bound at the protein target site. Only after a few data sets have been collected, it is fully recognized that rather than a bound fragment, this is a common electron density feature in almost all data sets and a result of some other entity.
5. It is helpful for comparing bound fragments and maps from initial screening if model is placed at the same position in the crystal cell for all cases (i.e., at the same phase origin and assuming a consistent diffraction data indexing). For space groups that permit more than one indexing, the data may be compared with a reference data set and re-indexed for consistency if necessary. When molecular replacement is used for the initial model placement, the option available in CCP4/MOLREP and CCP4/PHASER that places the solution at a consistent position with respect to the crystal cell origin should be used.
6. A useful tactic for electron density map interpretation, especially in the early stages of a project, is to sort the resulting maps into sets that definitely do, definitely do not, and might contain a bound fragment. Only after examining many maps does the appearance of characteristic electron density features at varying resolutions and data quality becomes established in the mind of the interpreter. As part of the interpretation process, it is often useful to be able to view multiple maps at the same time, either in the same or separate viewing canvas.
7. When analyzing maps containing bound fragments from mixture soaking experiments, it is necessary to have images of the chemical structures present in each mixture group for comparison with the density feature. It is very convenient to create a document displaying together the images for all compounds in a particular mixture.
8. Once a bound fragment has been established, independent validation and further information on structural prerequisites for binding may be obtained by obtaining related compounds outside the initial fragment library and collecting data from crystals soaked in them.

References

1. Hann, M.M., Leach, A.R. and Harper, G. (2001) Molecular complexity and its impact on the probability of finding leads for drug discovery *J.Chem.Inf.Comput.Sci.* **41**, 856–864.
2. Rees, D.C., Congreve, M., Murray, C.W. and Carr, R. (2004) Fragment-based lead discovery *Nat. Rev. Drug Disc.* **3**, 660–672.
3. Blundell, T.L., Jhoti, H. and Abell, C. (2002) High-throughput crystallography for lead discovery in drug design *Nat. Rev. Drug Disc.* **1**, 45–54.
4. Nienaber, V.L., Richardson, P.L., Klighofer, V., Bouska, J.J., Giranda, V.L. and Greer, J. (2000) Discovering novel ligands for macromolecules using X-ray crystallographic screening *Nat. Biotechnol.* **18**, 1105–1108.
5. Hajduk, P.J. and Greer, J. (2007) A decade of fragment-based drug design; strategic advances and lessons learned *Nat. Rev. Drug Disc.* **6**, 212–219.
6. Kuntz,I.D., Chen, K., Sharp, P.A. and Kollman, P.A. (1999) The maximal affinity of ligands *Proc.Natl.Acad.Sci.U.S.A.* **96**, 9997–10002.
7. Lipinski, C.A., *Lombardo, B.W. Dominy and P.J. Feeney (1997) Experimental and computational approaches to estimate solubility and permeability in drug discovery and development settings Adv. Drug Del. Rev. 23, 3–25.*
8. Pardridge, W.M. (2005) The Blood-brain Barrier: Bottleneck in Brain Drug Development *The journal of the American Society for Experimental NeuroTherapeutics* **2**, 3–14.
9. Bosch, J., Robien, M.A., Mehlin, C., Boni, E., Riechers, A., Buckner, F.S., Van Voorhis, W.C., Myler, P.J., Worthey, E.A., DeTitta, G., Luft, J.R., Lauricella, A., Gulde, S., Anderson, L.A., Kalyuzhniy, O., Neely, H.M., Ross, J., Earnest, T.N., Soltis, M., Schoenfeld, L., Zucker, F., Merritt, E.A., Fan, E., Verlinde, C.L.M.J. and Hol, W.G.J. (2006) Using Fragment Cocktail Crystallography To Assist Inhibitor Design of *Trypanosoma brucei* Nucleoside 2-Deoxyribosyltransferase *J.Med. Chem.* **49**, 5939–5946.
10. Congreve, M., Carr, R., Murray, C. and Jhoti, H. (2003) A 'rule of three' for fragment-based lead discovery? *Drug Discovery Today* **8**, 876–877.
11. Hubbard, R.E., Davis, B., Chen, I. and Drysdale, M.J. (2007) The SeeDs Approach: Integrating Fragments into Drug Discovery *Current Topics in Medicinal Chemistry* 7, 1–13.
12. Hartshorn, M.J., Murray, C.W., Cleasby, A., Frederickson, M., Tickle, I.J. and Jhoti, H. (2005) Fragment-Based Lead Discovery Using X-ray Crystallography *JACS* **48**, 403–413.
13. Collaborative Computational Project, Number 4 (1994) The CCP4 Suite: Programs for Protein Crystallography *Acta Cryst.* **D50**, 760–763.
14. Brünger, A.T., Adams, P.D., Clore, G.M., DeLano, W.L., Gros,P., Grosse-Kunstleve, R.W., Jiang, J.-S., Kuszewski, J., Nilges, M., Pannu, N.S., Read, R.J., Rice, L.M., Simonson, T. and Warren, G.L. (1998) Crystallography & NMR System: A New Software Suite for Macromolecular Structure Determination *Acta Cryst.* D**54**, 905–921.
15. Mooij, W.T.M., Hartshorn, M.J., Tickle, I.J., Sharff, A.J., Verdonk, M.L. and Jhoti, H. (2006) Automated Protein-Ligand Crystallography for Structure-Based Drug Design *Chem. Med. Chem.* **1**, 827–838.

Chapter 8

The Role of Enzymology in a Structure-Based Drug Discovery Program: Bacterial DNA Gyrase

Mark L. Cunningham

Abstract

The capability to accurately, rapidly, and reproducibly determine the affinity of a ligand for a target protein or enzyme is a vital component for a successful structure-based drug design effort. In order to successfully drive a structure-based drug design (SBDD) project forward, multiple distinct assays, each with particular strengths and weaknesses, need to be employed. Using bacterial DNA gyrase as an example, a range of assays are described that will fully support an SBDD program.

Key words: DNA Gyrase, Malachite green, Enzchek, agarose gel electrophoresis, triplex DNA, Isothermal calorimetry, fluorescence polarization, structure-based drug design, enzyme assay

1. Introduction

Advances in genomics, bioinformatics, and structural biology over recent years are changing the paradigm of modern drug discovery. Increasingly, the screening of hundreds of thousands or millions of compounds, extracts, or combinatorial libraries against panels of potential enzymes or cellular targets is giving way to a more focused approach in which small, pharmacophore-based libraries and specifically designed de novo compounds are screened against a selected enzyme target. The screening of smaller numbers of compounds enriched for attributes tailored for the specific target has a profound effect on enzymatic screening, removing the need for substantial HTS infrastructure, lowering the time and financial cost of mounting a screening campaign, while increasing the hit rate and accelerating the follow-up process.

Leslie W. Tari (ed.), *Structure-Based Drug Discovery*, Methods in Molecular Biology, vol. 841,
DOI 10.1007/978-1-61779-520-6_8, © Springer Science+Business Media, LLC 2012

The structural information used to select both the target and test compounds provides invaluable information as to the proposed mechanism of action, such as a competitive inhibitor of a substrate-binding site. Detailed kinetic analysis of the initial hits can rapidly assess this and serve as a means of discarding unwanted, nonspecific leads. Furthermore, structural knowledge of the target enables otherwise unrelated enzymes with close structural similarities to be identified and if needed counter-screened.

The use of small fragments will almost invariably result in lead compounds with low to moderate potency against the target, often several orders of magnitude away from the desired goal. As such, these leads are unlikely to show adequate activity in cellular assays such as antibacterial MIC or oncology antiproliferative assays. Consequently, during the early stages of most structure-based drug design (SBDD) projects, enzymatic analysis will provide the principal means of assessing the effects of optimization efforts. In concert with structural information obtained from co-crystallization of the ligand with the target protein, accurate measurements of binding affinities will provide the driving force in the iterative process from the initial hit to clinical development candidate.

In this chapter, the questions that will invariably arise from target inception to the routine generation of enzymatic data in SBDD projects will be discussed and methodologies to address these questions in a practical sense exemplified with bacterial DNA gyrase. Typical protocols for key assays will be described and their relative advantages and disadvantages highlighted.

2. Selection of Enzymes for Assay

With the advancement of modern recombinant molecular biology techniques, protein expression, and purification, large quantities of high-quality enzymes are readily available. These facets are covered in other chapters of this publication and close collaboration with the researchers responsible for these areas is essential.

Once a target has been selected for an SBDD project, the question of how to measure ligand potency will arise. Careful selection of the type of protein to be assayed as well as type of assay per se needs to be considered and multiple assays may need to be employed. Bacterial DNA gyrase (and the closely related topoisomerase IV, which will not be discussed here) provides an ideal example of the choices that will arise and potential options available.

DNA topoisomerases control the topological state of DNA, essential in replication and transcription. Bacterial DNA gyrase B is a type II topoisomerase which catalyzes DNA supercoiling. The mechanism involves transient double-stranded cleavage of a segment of DNA, passage of an intact double-stranded segment through this gap and subsequent resealing of the break (1–3). This process is driven by, and dependent on, the free energy of ATP hydrolysis (4, 5). DNA gyrase is a target for numerous antibacterial drugs including the fluoroquinolones such as Ciprofloxacin, and the natural product coumarins such as novobiocin. The former acts primarily through the GyrA subunit, inhibiting the DNA breakage–reunion cycle while the latter acts through the GyrB subunit, inhibiting the ATPase activity (6–8). Uncoupling of the GyrB-encoded ATPase activity abolishes the supercoiling activity of the DNA gyrase A_2B_2 complex. Clinical toxicity and rapid emergence of resistance have limited the use of novobiocin. The design of novel inhibitors of the ATPase activity as a means of disrupting supercoiling activity has thus received considerable interest (2).

DNA gyrase is a heterotetramer, comprising two distinct subunits, GyrA and GyrB, assembled in an A_2B_2 complex (see Fig. 1) (2, 9). In *Escherichia coli*, these subunits are 97 and 90 kDa respectively, resulting in a functional complex approaching 400 kDa. In the absence of gyrase A, the full-length GyrB subunits are able to assemble into homodimers. Moreover, this monomer–dimer transition appears to be driven by ATP (10). These dimers retain the ATPase activities of the A_2B_2 complex but lack the ability to alter the topology of DNA. The ATPase activity of the GyrB homodimer is independent of additional substrates and serves as an excellent means to measure binding of a ligand to the ATP-binding site, and several assays for this will be described below. However, this form of the protein has not been successfully crystallized. Proteins of this size and complexity remain problematic for high-resolution crystallography, where multiple domains, disordered regions, and

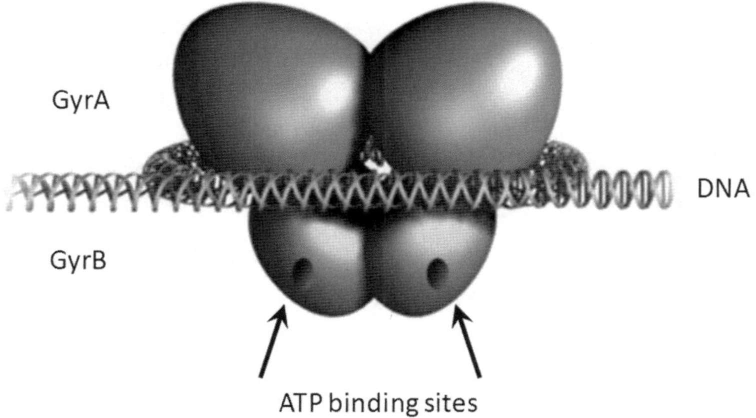

Fig. 1. Schematic representation of bacterial DNA gyrase. The full tetrameric complex comprising two subunits each of DNA gyrase A and DNA gyrase B with DNA substrate. The ATP-binding sites (encompassed entirely within the B subunits) are indicated. The energy released through ATP hydrolysis is an essential step in the gyrase-mediated introduction of negative supercoils into DNA.

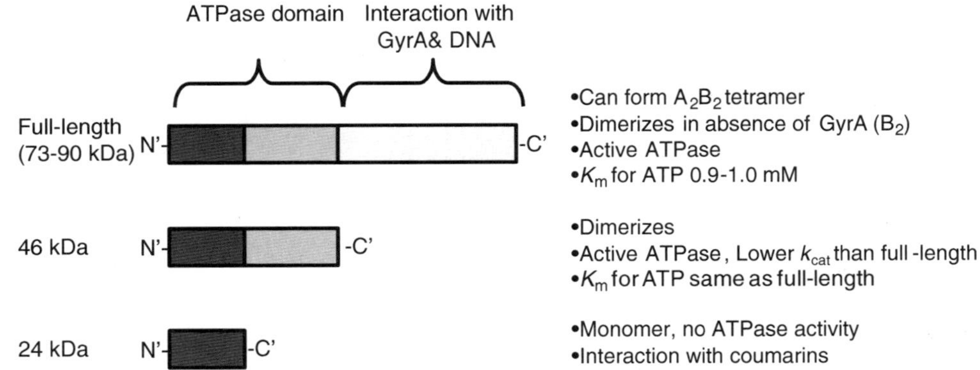

Fig. 2. Summary of DNA gyrase B subdomains and properties.

variable conformations can lead to nonoptimal interactions between protomers, preventing the growth of an ordered crystal lattice. Consequently, truncations, mutation, and deletions of protein domains and residues are often a prerequisite for successful crystallography.

Such dissection of the GyrB protein identified two distinct domains: the N-terminal region (approximately 43–46 kDa) which contains the ATP-binding site (see Fig. 2) (5, 11, 12) and a C-terminal region involved in binding the partner GyrA subunit as well as DNA (13–15). Enzymatic investigation has shown that the 46 kDa fragment fully encompasses the functional ATPase. Kinetics analysis demonstrated that the ATPase activities are not linearly related to enzyme concentration up to 40 μM. The greater than first-order dependence of activity on concentration suggests the active form is an oligomer. Overall, these data are consistent with the active form of the enzyme being a dimer (5, 16).

The N-terminal 46 kDa ATPase domain has successfully been crystallized for multiple bacterial orthologs including *E. coli*, complexed with the nonhydrolyzable ATP analog ADPNP (4, 12), and *Thermus thermophilus*, complexed with the ATP-competitive inhibitor novobiocin (17, 18). These structures are dimeric and in both cases, one ligand is bound per protein subunit. Gel filtration analysis indicated that this fragment exists as a monomer in the absence of nucleotide or presence of ADP + Pi but readily form dimers in the presence of ADPNP. Dimerization appears in part to be driven by polypeptide exchange in which the N-terminal 14 amino acids interact primarily with the partner protomer (4). X-ray crystal structures of the 24 and 46 kDa DNA gyrase proteins are shown in Fig. 3.

Further examination of the 46 kDa X-ray crystal structure revealed that each monomer comprises two distinct subdomains. Notably, the ATP-binding site is encoded by the N-terminal domain (4, 12). The initial 220 amino acids, encoding a protein fragment of approximately 24 kDa, have been shown to bind the

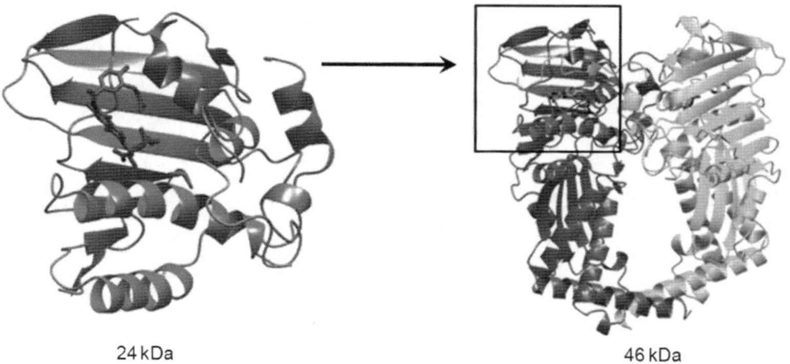

Fig. 3. Ribbon diagram of the X-ray crystal structures of the 24 kDa ATP-binding domain of DNA gyrase B complexed with the inhibitor novobiocin (Bensen, D. and Tari, L.W., unpublished results) (*left*) and the 46 kDa ATP-binding domain of DNA gyrase B from *Escherichia coli* complexed with the substrate analog ADPNP (4) (*right*).

substrate analog ADPNP as well as numerous classes of inhibitors including novobiocin (see ref. (19)). This 24 kDa fragment is present as a monomer and is catalytically inactive. Moreover, the N-terminal 14 amino acids usually associated with dimer formation are invariably disordered in X-ray crystal structures. Removal of this region greatly improves crystallographic properties and enables multiple orthologs to be rapidly and reproducibly optimized for crystallography (Tari, L., personal communication).

This background illustrates the dilemma faced by the enzymologist in an SBDD project; the best crystallographic tool, the 24 kDa ATP-binding domain, is catalytically inactive and least representative of the biologically relevant form of the target, while the most biologically relevant species, the 400 kDa A_2B_2 complex, is unsuitable for X-ray crystallography. However, while not catalytically active, structural studies suggest that the 24 kDa domain possesses all the necessary structural elements needed to bind inhibitors, so that it is a useful tool to drive SBDD. To validate this hypothesis, multiple assays need to be conducted against different subfragments (including the full-length protein) to see if the results from the different assays are well correlated. In the following sections, a suite of available assays will be described that are being used to support and drive forward an SBDD effort targeting DNA gyrase. Strengths and weaknesses of these assays are highlighted in Table 1. The use of these assays in combination will enable a project to screen for and optimize small molecule inhibitors of the ATP-binding site of DNA gyrase B with particular emphasis on assessing potency against fully functional enzymes as well as catalytically inactive truncated crystallographic proteins. In the end, if reliable assays against small target protein subfragments can be developed, they can be utilized in a high-throughput manner to generate data (including thermodynamic parameters) to support SBDD during

Table 1
Comparison of assays used to guide a DNA gyrase SBDD program

Assay	Advantages	Disadvantages
ATPase assays		
Malachite green	No coupling enzymes Amenable to HTS Cheap reagents Ideal for large, primary screen	End point assay, no real time data information Sensitive to minor inorganic phosphate contaminants and commonly used reagents such as DMSO, BSA and glycerol
Enzchek	Highly sensitive, real time assay Amenable to HTS Tolerant of additives such as reductants and DMSO Ideal for follow up assays such as inhibition modality and routine K_i determinations	Dependent on PNP coupling enzyme. Sensitive to minor inorganic phosphate contaminants
Binding assays		
Calorimetry	Does not require catalytically active enzyme Provides insight into underlying components driving potency (enthalpy, entropy, stoichiometry) as well as binding affinity	Low throughput Expensive hardware
Fluorescence polarization	Does not require catalytically active enzyme Amenable to HTS High sensitivity Homogeneous, radiometric assay Ideal for measuring biding affinity to inactive crystallographic proteins	Competition displacement assays requires specific, fluorescently tagged tracer ligand of suitable potency
DNA topology assays		
Agarose gel electrophoresis	Measure ligand efficacy on biologically relevant activity (DNA supercoiling) Ideal in confirming SAR of ATPase inhibition and confirming mechanisms of action Complete kits are commercially available and easily adaptable for non-commercial enzymes	Low throughput Labor intensive and time consuming
DNA triplex formation	High throughput Measure ligand efficacy on biologically relevant activity Complete kits are commercially available	Relatively expensive Fluorescent ligand/highly colored samples may interfere with assay

ligand optimization. More expensive, complex, lower throughput assays against the biologically relevant full-length target protein complexes can be utilized with smaller numbers of advanced lead compounds later in a program as validation tools, or can be tested at an earlier stage to insure that SAR trends observed using protein subfragments translate to the full-length macromolecular complexes that are the drug targets.

3. Materials

3.1. ATPase Assays

Several assays can routinely be employed for the measurement of the DNA gyrase ATPase activity. As described above, the ATPase activity is entirely encoded in the GyrB subunit, consequently GyrA proteins are not required. Both the full-length and 46 kDa truncations are active. However the k_{cat} of the latter is substantially lower such that the full-length protein should be used if available. Two assays are described in detail, the Malachite green and Enzchek assay, both of which follow the hydrolysis of ATP (see Fig. 4). Additional formats such as the pyruvate kinase/lactate dehydrogenase coupled spectrophotometric assay in which Gyrase ATPase activity is coupled ultimately to NADH oxidation (5) and a

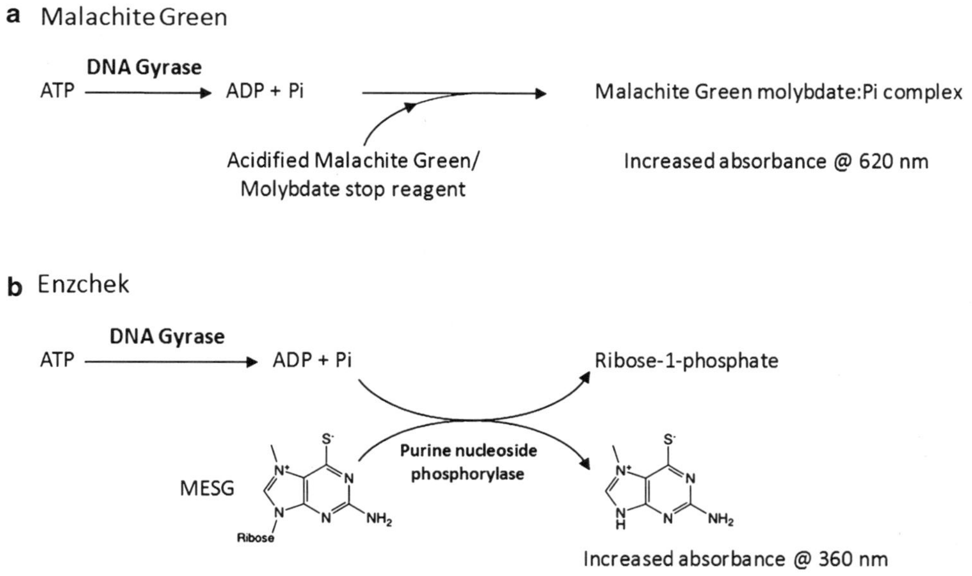

Fig. 4. DNA gyrase ATPase assay schemes. (**a**) Malachite Green assay, (**b**) Enzchek assay. In both cases the DNA Gyrase-mediated hydrolysis of ATP releases stoichiometric amounts of inorganic phosphate. For the Malachite green assay, the reaction is stopped by addition of acidified malachite green/molybdate reagent. In the presence of released inorganic phosphate, a malachite green/molybdate/phosphate complex is formed and is associated with increase in absorbance at 620 nm. In the case of the Enzchek assay, the real-time release of inorganic phosphate is coupled by the enzyme PNP to the substrate MESG. Formation of the purine base product is associated with an increase in absorbance at 360 nm.

radiometric assay in which the gyrase-dependent liberation of ^{32}Pi from [γ-^{32}P]-ATP is determined using charcoal adsorption and scintillation counting can also be employed (4).

3.1.1. Malachite Green

This is a simple end-point colorimetric assay for measuring inorganic free phosphate in aqueous solution. It is based on the quantitation of the complex formed between Malachite Green, molybdate, and free orthophosphate, readily measured using a plate reader at 620 nm (see Fig. 4). This assay has been used extensively for the measurement of phosphate levels, including that generated via enzyme-mediated ATP hydrolysis where stoichiometric levels of Pi are released (20–22).

Materials

1. On the day of use, prepare the malachite green stop reagent by combining the following components in a final ratio of 2:1:1:2. Thoroughly mix until a stable yellow color is attained, typically 2 h.
   (a) 0.0812% (w/v) Malachite green, dissolved in deionized water.
   (b) 2.32% (w/v) polyvinyl alcohol, dissolved in deionized water, may require heating.
   (c) 5.72 % (w/v) Ammonium molybdate, dissolved in 6 M HCl.
   (d) Deionized water.
2. Assay buffer (5×): 250 mM Tris–HCl, pH 7.6, 10 mM $MgCl_2$, 625 mM NaCl. Additives such as EDTA, β-mercaptoethanol, or CHAPS can be added up to 10 mM without effect. By contrast, use of glycerol, DMSO, BSA, DTT, and Triton X-100 can either reduce sensitivity or increase blanks and should be evaluated prior to use and final concentrations kept to a minimum.
3. ATP solution (10×): typically 30 mM, ATP disodium salt is dissolved to approximately 50 mM (27.5 mg/mL in 10 mM NaOH). Concentration is accurately determined using a molar extinction coefficient at 259 nm of 15.4 mM^{-1} cM^{-1} at pH 7.0 and volume adjusted accordingly. Single-use aliquots are stored at –20°C (see Note 1).
4. Numerous 96-well plates are suitable for this assay such as Corning 3641, which has a modified polymer surface which results in a nonionic hydrophilic surface minimizing bio-interactions and enhancing signal-to-noise ratios.
5. DNA gyrase B protein, either full-length or 46 kDa truncations. These should be highly purified, in phosphate-free buffer and of known concentrations.
6. Microplate reader such as Tecan Safire II.

3.1.2. Enzchek Assay

This assay again monitors the stoichiometric release of inorganic phosphate which occurs upon hydrolysis of ATP to ADP. Unlike the Malachite Green assay, detection of the liberated Pi requires the presence of a coupling enzyme/substrate, namely purine nucleoside phosphorylase (PNP) and 7-methyl-6-thioguanosine (MESG). In the presence of Pi, PNP catalyzes the phosphorolysis of the substrate MESG, liberating ribose 1-phosphate and the purine base product (see Fig. 4). The assay is based upon the increased molar extinction coefficient of the latter product relative to MESG (23). This assay is robust, highly sensitive, and tolerant of many commonly required solvents, in particular DMSO. Most notably, it is a real-time assay allowing assay linearity to be monitored and is amenable to high throughput. This assay is regularly used by the author in 384-well format to screen 24 different compounds, each at 15 different concentration in as little as 20 min. As with the malachite green assay, care should be taken to prevent contamination with free phosphate. This is the assay of choice for determining enzyme kinetic parameters such as k_{cat} and K_m using standard methodologies not described here. In addition, this assay is used to accurately measure the enzyme active site concentrations and determine inhibition modalities as described below (24).

Materials

1. Assay buffer (5×): 250 mM Tris–HCl, pH 7.6, 10 mM $MgCl_2$, 625 mM NaCl. Additions including EDTA (typically 0.5 mM), reductants (DTT, TCEP) BSA, or glycerol can be added as needed. This assay is inherently more tolerant of additives than the Malachite green assay. DMSO concentrations approaching 5% (v/v) are acceptable although 2.5% is more commonly employed.
2. MESG solution (10×, or 2 mM). MESG was purchased from Berry and Associates and dissolved in deionized H_2O. Single-use aliquots were stored at –20°C and thawed immediately prior to use (see Note 2).
3. ATP solution (10×): ATP disodium salt is dissolved to approximately 50 mM (27.5 mg/mL in 10 mM NaOH). Concentration is accurately determined using a molar extinction coefficient at 259 nm of 15.4 mM^{-1} cM^{-1} at pH 7.0 and volume adjusted accordingly. Single-use aliquots are stored at –20°C (see Note 1).
4. PNP (100×): enough enzyme is dissolved in 10 mM Tris–HCl pH 7.6, 25 mM KCl to yield a final concentration of 0.1 U/μL, and stored at 4°C.
5. DNA gyrase B protein, either full-length or 46 kDa truncations. These should be highly purified, in phosphate-free buffer and at known concentrations.
6. DNA gyrase dilution buffer: 50 mM Tris–HCl, pH 7.6, 25 mM NaCl.

7. DNA gyrase proteins are stored in 50 mM Tris–HCl, pH 7.6, 25 mM NaCl at −80°C in single-use aliquots.
8. Numerous 96- and 384-well plates are suitable for this assay such as Corning 3640 and 3641, which have a modified polymer surface which results in a nonionic hydrophilic surface minimizing bio-interactions and enhancing signal-to-noise ratios.
9. Microplate reader such as Tecan Safire II.

3.2. Binding Assays

Numerous assays for the measurement of ligand binding to DNA gyrase are available and have proven valuable in the evaluation of ligand potencies against the target. These include analytical ultracentrifugation (5, 25), Surface plasmon resonance (25–28), circular dichroism (29), differential scanning calorimetry (30, 31), and isothermal calorimetry (26, 31).

None of the above methods are classical turnover-based systems such as the ATPase or DNA topology assays described elsewhere in this chapter. This highlights a particular advantage of these methodologies in that the researcher is not confined to using catalytically active enzymes. This is particularly useful is the case of DNA gyrase where, as described above, the most robust crystallographic tool is the catalytically inactive 24 kDa protein which contains the ATP-binding site. These assays allow for the direct comparison of ligand potencies and X-ray crystallography derived co-crystal structures using the same form of the protein. Typical protocols for two distinct binding assays, isothermal titration calorimetry (ITC) and fluorescence polarization, are outlined below.

3.2.1. Isothermal Titration Calorimetry

ITC is an increasingly popular approach to probe binding of a ligand to a target protein. Several texts describing both the theory and practice of this technique are available, and the reader is referred to these for greater detail (32, 33). This section gives a brief overview of the technique and describes a typical protocol used for DNA gyrase (26, 31).

Highly sensitive calorimeters are capable of accurately measuring the endothermic or exothermic enthalpic events that occur when substances bind. A significant advantage of this technique is that all binding parameters, namely the association constant (K_a), enthalpy (ΔH), entropy (ΔS), and binding stoichiometry (n), can be determined simultaneously from a single experiment. This information thus provides not only the inherent potency of a ligand, but also sheds light upon the underlying properties such as hydrophobic interactions, charge–charge interactions, and hydrogen bonds which contribute to and ultimately delineate potency. In a routine experiment, sequential aliquots of a solution of a test ligand are titrated into a solution of target protein, and the heat released upon their interaction (ΔH) is monitored over time. After

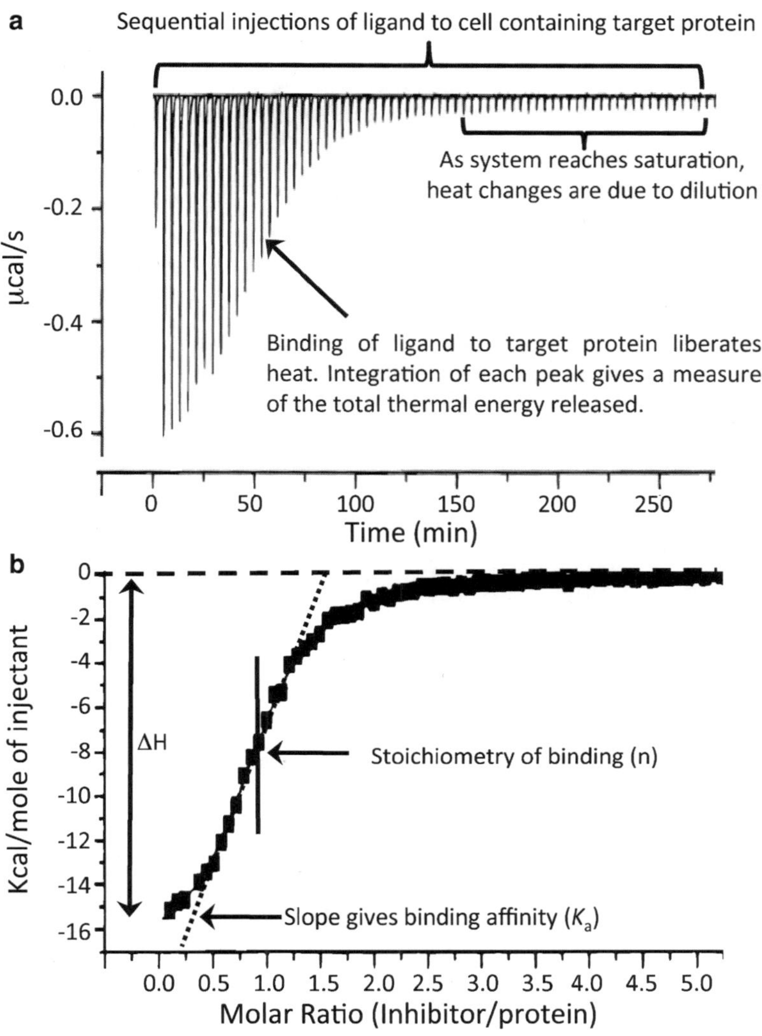

Fig. 5. Thermogram for the binding of an inhibitor to the 24 kDa fragment of DNA gyrase showing both the raw (**a**), and fitted data (**b**). Adapted from Holdgate (32).

each addition, the temperature returns to baseline prior to subsequent additions. The magnitude of ΔH is representative of the degree of ligand binding and decreases as the system reaches equilibrium. Once saturation is reached, the observed changes in ΔH reflect non-binding effects such as dilution. Appropriate analysis of the raw data generates the parameters described above. A typical data set is shown in Fig. 5 (adapted from ref. (34)). Raw data is shown in Fig. 5a and integrated data following appropriate corrections shown in Fig. 5b. The area beneath each injection represents the heat release associated with the binding events which occur following each addition. Integrated data is plotted against the molar ratio of ligand to target, generating a binding isotherm. This data can be analyzed in a number of different ways, most typically using a single saturable binding model from which binding parameters described above can be obtained using the following equation.

$$\Delta G = -RT \ln K_a = \Delta H - T\Delta S$$

where R is the ideal gas constant, T is thermodynamic temperature, and G is Gibbs free energy. Integration and analysis software is typically part of the hardware package and should be used according to manufacturer's instructions. The following protocol is a modification of that used previously for DNA gyrase (26, 31).

Materials

1. Binding parameters are determined using an appropriate calorimeter such as a MicroCal MCS titration calorimeter.
2. Binding buffer (5×): 250 mM Tris–HCl, pH 7.6, 10 mM $MgCl_2$, 625 mM NaCl, 5% DMSO. The use of reductants should be limited as this can affect baseline values. 1× buffer should be prepared and used for dialyzing proteins and preparing ligands.
3. Test ligands should be prepared in the same buffer as the DNA gyrase B protein. Concentrations should be determined as accurately as possible, ideally by enzymatic titration of binding sites or extinction coefficient.

3.2.2. Fluorescence Polarization

Fluorescence polarization (FP) is based upon the fact that when a small fluorescent molecule (such as a fluorescently labeled enzyme inhibitor) in solution is excited by plane-polarized light, the emitted light is depolarized due to the rapid rate of rotation of the molecule during the fluorescent lifetime. By contrast, if this small molecule now binds to a larger molecule (such as a target protein), the rate of rotation is slowed to an extent that the emitted light is now in the same plane as the excitation energy. This assay is ratiometric, comprising measurements of emitted energy parallel and perpendicular to the excitation source, and data is obtained in two, mathematically related forms, anisotropy or polarization, as described below.

$$\text{Anisotropy} = \frac{(\text{Intensity}_{\text{parallel}} - \text{Intensity}_{\text{perpendicular}})}{(\text{Intensity}_{\text{parallel}} + 2 \times \text{Intensity}_{\text{perpendicular}})}$$

$$\text{Polarization} = \frac{(\text{Intensity}_{\text{parallel}} - \text{Intensity}_{\text{perpendicular}})}{(\text{Intensity}_{\text{parallel}} + \text{Intensity}_{\text{perpendicular}})}$$

The degree of polarization reflects the binding of the labeled ligand (often termed tracer ligand) to the target protein and can thus be used to measure affinity. Furthermore, displacement of this tracer ligand by an unlabeled molecule provides a means of measuring potencies of these molecules. Numerous publications are available describing the assay parameters that need to be established for an FP assay prior to use in a displacement assay as well as appropriate methods for determining binding constants from the raw data (35–38).

DNA gyrase B benefits from extensive literature, from which potent inhibitors can be selected and readily derivatized with fluorescent labels such as fluorescein at appropriate solvent-exposed vectors. It is assumed that the fluorescently tagged ligand is fully characterized for suitable properties such as fluorescence intensity, stability, and potency prior to use in the displacement assay. The following protocol describes the use of a highly potent, fluorescein-labeled ligand (MT3-192) with a $K_i < 10$ nM in a displacement assay to measure ligand potency against a 24 kDa DNA gyrase protein (Cunningham, M. L., Bensen, D., Trozss, M., and Tari. L. W., unpublished results).

Materials

1. Assay buffer (5×): 250 mM Tris–HCl, pH 7.6, 10 mM $MgCl_2$, 625 mM NaCl. Additions including EDTA (0.5 mM), reductants (DTT, TCEP) BSA, or glycerol can be added as needed. DMSO concentrations approaching 5% are acceptable although 2.5% is more commonly employed.
2. Microplate reader such as Tecan Safire II equipped with fluorescence polarization module.
3. Numerous 96- and 384-well plates are suitable for this assay such as Corning 3991 or 3575 which have a modified polymer surface which results in a nonionic hydrophilic surface minimizing bio-interactions and enhancing signal-to-noise ratios. The use of black plates also reduces auto fluorescence cross talk with adjacent wells.
4. An appropriately tagged ligand with suitable potency for displacement-based assays. In this case, the ligand used (MT3-192) is a fluorescein-tagged derivative of a known DNA gyrase B inhibitor with $K_i < 10$ nM determined using standard ATPase Enzchek assay. A stock solution of 100 μM prepared in 100% DMSO and stored at −80°C in amber vials (see Note 3).
5. The intact ATP-binding domain of DNA gyrase B (either the 24 or 46 kDa truncations described in Figs. 2 and 3). This should be highly purified and of known concentration.
6. DNA gyrase dilution buffer: 50 mM Tris–HCl, pH 7.6, 25 mM NaCl.
7. DNA gyrase proteins are stored in 50 mM Tris–HCl, pH 7.6, 25 mM NaCl at −80°C in single-use aliquots.

3.3. DNA Topology Assays

As described above, DNA gyrase is responsible for introducing negative supercoils into DNA, this process being dependent upon ATP hydrolysis. The ATPase assays described above provide means of assessing inhibition of the ATPase activity of the GyrB subunit and by implication, disruption of DNA supercoiling activity. However, specific analysis of ligand effects upon gyrase-mediated changes in DNA topology can and should be performed. In this section, two distinct assays using the biologically relevant A_2B_2 complex will be described. The first assay uses agarose gel electrophoresis and exploits the difference in mobility between relaxed (substrate) and negatively supercoiled (product) plasmid DNA. The second is a 96-well-based fluorescent assay which takes advantage of the fact that negatively supercoiled plasmids form intermolecular triplexes more efficiently than relaxed plasmids (39).

3.3.1. Agarose Gel Electrophoresis

Gel-based analysis of DNA mobility is a well established, commonly used method to assess topoisomerase-mediated changes in DNA structure. The ability of DNA gyrase to introduce negative supercoils into relaxed plasmid DNA and the differential mobility of these distinct forms of DNA in agarose gel electrophoresis provide a convenient means of assessing the efficacy of test ligands. Commercially available kits, containing appropriate relaxed and supercoiled plasmid DNA, assay buffer, and most importantly, purified holoenzyme A_2B_2 complexes have greatly simplified this assay. Topogen and Inspiralis offer enzymes from several different bacterial species but the basic protocol can be adapted to other species and protein orthologs rapidly. Moreover, the entire assay can easily be reconstituted from basic reagents for a fraction of the cost if needed. A typical gel is shown in Fig. 6. Supercoiled (product) and relaxed (substrate) DNA are shown as references. A dose-dependent reduction in the quantity of supercoiled DNA is

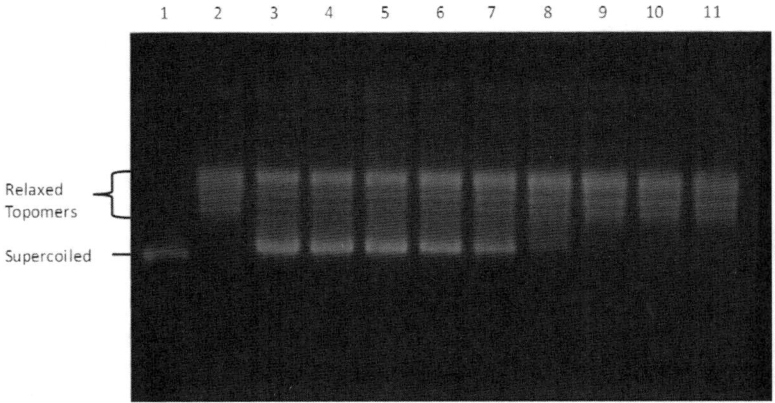

Fig. 6. Agarose gel electrophoresis demonstrating DNA gyrase-dependent changes in DNA topology. Assays were performed as described in Subheadings 3.3.1 and 4.3.1. Supercoiled DNA only (*lane 1*), relaxed DNA only (*lane 2*), relaxed DNA incubated with *E. coli* DNA gyrase (*lane 3*), and relaxed DNA incubated with *E. coli* DNA gyrase plus increasing concentrations of Novobiocin at 0.031, 0.063, 0.125, 0.25, 0.5, 1, 2, and 4 µg/mL (*lanes 4–11*, respectively).

observed as the concentration of novobiocin increases. A standard protocol is outlined below but for more detailed information the reader is referred to the excellent text by Fisher and Pan (40) (see Note 4).

Materials

1. Assay buffer (5×): 175 mM Tris–HCl pH 7.5, 120 mM KCl, 20 mM $MgCl_2$, 10 mM DTT, 9 mM spermidine, 32.5 % (v/v) glycerol, 0.5 mg/mL BSA.
2. ATP solution (20×): ATP disodium salt is dissolved to approximately 25 mM (13.75 mg/mL in 10 mM NaOH). Concentration is accurately determined using a molar extinction coefficient at 259 nm of 15.4 mM^{-1} cM^{-1} at pH 7.0 and volume adjusted accordingly. Single-use aliquots are stored at −20°C (see Note 1). Relaxed plasmid DNA, typically pBR322, is commercially available from Inspiralis or Topogen. Typical stock concentrations are 250 ng/μL. Supercoiled plasmid DNA should also be obtained for use as a gel standard (Inspiralis or Topogen).
3. Molecular biology grade agarose such as Invitrogen Ultrapure Agarose.
4. TBE (10×): 900 mM Tris Base, 900 mM boric acid, 25 mM EDTA pH 8.0.
5. Stop buffer/gel loading dye (5×): 5% SDS or Sarkosyl, 0.125% bromophenol blue, and 25% (v/v) glycerol prepared in deionized water.
6. 10 mg/mL ethidium bromide.
7. 1 mg/mL proteinase K (EMD Biosciences, 600 mAU/mL, note that 1 mAU is equivalent to 1 mg protein).
8. Chloroform: Isoamyl alcohol, 24:1 (EMD Biosciences).
9. Gel documentation system equipped with digital camera and UV transilluminator.

3.3.2. DNA Triplex Formation

The gel electrophoresis protocol described above has several limitations, not least of which is the low throughput. To overcome this limitation, Maxwell and colleagues developed a higher throughput, plate-based assay founded upon the fact that negatively supercoiled plasmids form intermolecular triplexes more efficiently than their relaxed counterparts (39, 41). The authors carried out extensive validation of this assay including inhibition studies which gave similar results to those obtained with the classical gel-based assay. This assay has subsequently been made available from Inspiralis (catalog #TRG01) and can be adapted for use with in-house generated GyrA and GyrB proteins from other bacterial species if needed. The assay measures the formation of a triplex DNA complex between a biotinylated oligonucleotide immobilized on a streptavidin plate and a modified pBR322 plasmid which contains a 20 base-pair insert with triplex-forming potential (pNO1). Relaxed plasmid pNO1 is incubated with DNA gyrase, generating the supercoiled form, which is now capable of forming the triplex DNA complex. Removal of unbound plasmid/reaction mixture and subsequent addition of a fluorescent dye results in increased fluorescent signal in the presence of functional enzyme.

Materials

1. Wash buffer (20×): 400 mM Tris–HCl, pH 7.6, 2.74 M NaCl, 0.2% BSA, 1% Tween 20.
2. Assay buffer (5×): 175 mM Tris–HCl, pH 7.5, 120 mM KCl, 20 mM $MgCl_2$, 10 mM DTT, 9 mM spermidine, 5 mM ATP, 32.5% (v/v) glycerol, 0.5 mg/mL BSA. If necessary, ATP can be omitted and added separately as needed.
3. Dilution buffer: 50 mM Tris–HCl, pH 7.5, 100 mM KCl, 2 mM DTT, 1 mM EDTA, 50% glycerol.
4. TF buffer (20×): 1 M Sodium acetate, 1 M NaCl, 1 M $MgCl_2$.
5. T10 buffer (10×): 200 mM Tris–HCl, pH 8.0, 20 mM EDTA.
6. Relaxed plasmid substrate, pNO1, supplied at 1 mg/mL.
7. DNA gyrase supplied at 5 U/μL in dilution buffer.
8. 96-well black, streptavidin-coated microtiter plate.
9. Fluorescent stain such as SYBR Gold (Invitrogen).

4. Methods

4.1. ATPase Assays

4.1.1. Malachite Green

1. Tests are typically performed in a volume of 50 μL in 96-well, flat bottom plate (such as Corning 3641) but can be scaled proportionately as needed.
2. DNA gyrase solutions, typically 100–1,000 nM final concentration depending on the construct, should be prepared fresh in the assay buffer (an enzyme-free control should also be included). A volume of 40 μL should be added to each well (see Note 5).
3. Test samples are prepared at 10× final concentration in 4% (v/v) DMSO/deionized water. A volume of 5 μL is added to each well and incubated with enzyme for 10 min. DMSO blanks should also be included.
4. The reaction is initiated by addition of 5 μL 10× ATP substrate; final assay volume is 50 μL.
5. Samples are incubated at room temperature for 15–30 min, depending on the enzyme activity. Preliminary experiments should confirm linearity over this period.

6. To the 50 μL sample, add 160 μL of malachite green stop reagent and mix by pipetting 2–3 times.
7. Subsequently add 20 μL of a freshly prepared sodium citrate solution, 34% (w/v) in deionized water, and repeat the mixing step (see Note 6).
8. Leave plates to develop for 15 min prior to reading absorbance at 620 nm. It is important to keep this period constant for accurate comparison across plates.
9. The degree of data analysis is dependent on the question being addressed. In the most basic analysis, yellow indicates inactive enzyme, green indicates active enzyme. This simplistic analysis may be all that is required. However, a more detailed analysis is typically needed; the difference between the average absorbance for the enzyme-independent (blank) and enzyme-containing (uninhibited) wells provides a rapid measure of the assay signal. All other wells, after subtraction of the blank value, can quickly be expressed as a percentage of uninhibited sample and dose responses determined. If a more detailed analysis is required, a standard curve of [Phosphate] vs. absorbance can be plotted and signals converted to nanomoles (nmol) phosphate produced. This assay is most useful for experiments which generate 0.1–1.0 nmol phosphate per well.

4.1.2. Enzchek Assay

1. Tests are typically performed in a volume of 100 μL in 96- or 384-well, flat bottom plate (such as Corning 3641 or 3640) but can be scaled proportionately as needed (see Notes 1 and 5).
2. For a 384-well plate, prepare the following material for 400 assays.
   (a) Deionized water such that final volume is 35,000 μL.
   (b) 8,000 μL 5× assay buffer.
   (c) 4,000 μL 10× MESG.
   (d) 400 μL 100× PNP.
   (e) Finally, DNA gyrase is added to a final active site concentration of typically 20–200 nM. Accurate measurement of active site concentrations should be performed as described below.
   (f) This reaction mixture gives sufficient material for 400 independent reactions of 87.5 μL.
3. Also prepare 16 enzyme-independent blanks as follows:
   (a) 904 μL deionized water.
   (b) 320 μL 5× assay buffer.
   (c) 160 μL 10× MESG.
   (d) 16 μL PNP.

(e) This reaction mixture gives sufficient material for 16 blank samples.

4. Transfer 87.5 μL enzyme containing, and enzyme-independent to 384-well plate as needed.
5. Test samples are prepared at 40× final concentration in 100% (v/v) DMSO. A volume of 2.5 μL is added to each well and incubated with enzyme for 10 min prior to initiation of reaction. DMSO-only blanks should also be included.
6. The reaction is initiated by addition of 10 μL 10× ATP substrate, final assay volume is 100 μL.
7. Samples are incubated at 25°C and the absorbance at 360 nm measured using a microtiter plate reader such as a Tecan Safire II. Assays are typically run for 20–40 min with data collected every 10 s.
8. An initial burst phase is often visible and reflects inorganic phosphate present in the ATP. This initial increase in absorbance should last no longer than 2 min and be low in magnitude. This phase should be excluded from kinetic analysis. After this burst, a linear phase should be observed and fitted to determine the delta OD/min.
9. Again, the degree of data analysis is dependent on the question being addressed. In all cases, analysis should start by determining the enzyme independent (blank) and dependent (uninhibited) rates in the presence of solvent alone. The difference between these values yields the maximal activity. All other rates, after correction for the blank should be normalized to this, percentage often being a convenient measure, although values can be converted to more formal units such as nanomoles product formed per minute.
10. Simple IC_{50} values are readily obtained from this normalized data, typically using as Sigmoidal-dose response curve fitted using non-linear regression analysis software such as GraphPad Prism version 4.00 for Windows (GraphPad Software, San Diego, CA, USA, www.graphpad.com). However, it is well worth the additional effort to extract binding constants from this data. To do so, an accurate measure of substrate K_m is needed as is an understanding of the inhibition modality. For competitive inhibitors, the Cheng and Prusoff (42) equation provides a rapid means of obtaining K_i values:

$$K_i = \frac{IC_{50}}{1 + ([S]/K_m)}$$

11. The Cheng–Prusoff analysis holds while the concentration of inhibitor is in excess of enzyme concentration, such that the inhibitor concentration remains essentially unchanged during

the course of the experiment. As ligand potency increases, the concentrations of inhibitor used decreases, and this assumption no longer holds (see Note 7). Consequently, a more detailed analysis known as the Morrison Tight-binding equation is needed that accounts for ligand depletion (43):

$$\frac{v_i}{v_0} = 1 - \frac{([E]_T + [I]_T + K_i^{app}) - \sqrt{([E]_T + [I]_T + K_i^{app})^2 - 4[E]_T[I]_T}}{2[E]_T}$$

$[E]_T$ is the concentration of catalytically active sites, $[I]_T$ is the concentration of inhibitor, and vi/v_0 is the fractional activity (see Notes 8–10). For a strictly competitive inhibitor, the relationship between K_i and K_i^{app} is:

$$K_i^{app} = K_i(1 + ([S]/K_m))$$

where $[S]$ is the substrate concentration and its associated K_m.

4.2. Binding Assays

4.2.1. Isothermal Titration Calorimetry

1. Proteins should be fully dialyzed against 1× binding buffer, centrifuged (13,000 rpm, 10 min) and degassed immediately prior to use. The concentration of DNA gyrase is typically between 10 and 30 μM.
2. Protein sample is added to the calorimeter cell using a syringe, taking care not to introduce air bubbles. A typical volume of 1.34 mL is used. The reference cell is filled in a similar manner with buffer minus protein.
3. The test ligand is prepared in an identical buffer as the target protein. A useful method for doing this is to dissolve the ligand in the dialysis buffer used to prepare the protein. This may not always be possible due to the need to pre-solubilize the ligand in solvent such as DMSO. Every effort should be made to minimize differences between ligand buffer and protein buffer as these differences can lead to non-binding heat effects upon mixing. Ligand is loaded into a micro-syringe, again taking care to avoid air bubbles, the typical ligand concentration being in the range of 100–500 μM. Sequential injections of 5–10 μL made. Injection parameters such as volume of injection, duration of injection, time between injections, and collection rate are programmed before beginning the experiment and should be adjusted as needed. For standard binding measurements, a final molar ratio of ligand:protein in the range of 3:1 is a reasonable starting point. For a more detailed description of selection of ligand and protein concentrations, the reader is referred to ref. (32).
4. Heat effects observed in the ligand titration are not due solely to the binding event, but also include the heat of ligand

dilution, heat of protein dilution, and heat of mixing (see Fig. 5). These events must be controlled for by titrations of ligand into buffer, buffer into protein, and buffer into buffer, respectively. Once these corrections have been made, raw data from ligand:protein titrations can be corrected and analyzed using associated software (see Note 11).

4.2.2. Fluorescence Polarization

1. All assays are typically performed in a volume of 100 μL in 96- or 384-well, flat bottom plate (such as Corning 3991 or 3575) but can be scaled proportionately as needed (see Note 12).
2. For competition displacement assays in a 96-well plate, prepare the following material for 100 assays.
   (a) 7,150 μL deionized water.
   (b) 2,000 μL 5× assay buffer.
   (c) 100 μL of 500 nM MT3-192. This gives a final concentration of 5 nM.
   (d) Total volume is 9,250 μL, sufficient for 100 assays of 92.5 μL.
3. Remove 925 μL (equivalent to ten assays) and add 50 μL of enzyme dilution buffer, resulting in a volume of 975 μL. Aliquot of 97.5 μL is transferred to eight wells, this will be the unbound ligand control.
4. To the remaining sample (8,325 μL) add 450 μL of 200 nM freshly diluted 24 kDa DNA gyrase B protein; the final enzyme concentration is 10 nM. This gives a volume of 8,775 μL, equivalent to 90 assay of 97.5 μL. Transfer 97.5 μL of this master reaction mixture to the remaining wells of 96-well plate. Enzyme and tracer ligand are incubated at room temperature for 15 min. The concentrations of enzyme and tracer chosen are such that, from a previously determined binding constant, in the region of 80–90% of the tracer ligand will be bound.
5. Test samples are prepared at 40× final concentration in 100% (v/v) DMSO. A volume of 2.5 μL is added to each well and incubated with enzyme. DMSO-only controls should also be included.
6. Samples are incubated at 25°C for 15–30 min in the dark prior to measurement of fluorescence polarization using a Tecan Safire II plate reader using an excitation wavelength of 470 nm, emission wavelength 525 nm (10 nm bandwidth). A typical gain of between 50 and 60 is used and a G-factor previously determined according to manufacturer's instructions.
7. Anisotropies for (1) tracer ligand in the absence of protein, (2) tracer ligand plus protein, and (3) tracer ligand plus protein and increasing concentrations of competing ligand are measured. Simple IC_{50} values are readily obtained from this

normalized data, typically using a Sigmoidal-dose response curve fitted using non-linear regression analysis software such as GraphPad Prism version 4.00 for Windows (GraphPad Software, San Diego, CA, USA, www.graphpad.com).

8. A more sophisticated analysis can be carried out using the following equations. Determination of binding constants is certainly worth the additional effort required (38).

$$F_{SB} = \frac{2\sqrt{(d^2 - 3e)}\cos(\theta/3) - d}{3K_{D1} + 2\sqrt{(d^2 - 3e)}\cos(\theta/3) - d}$$

with

$$F_{SB} = \frac{A_{OBS} - A_F}{A_B - A_F}$$

$$d = K_{D1} + K_{D2} + L_{ST} + L_T - R_T$$

$$e = (L_T - R_T)K_{D1} + (L_{ST} - R_T)K_{D2} + K_{D1}K_{D2}$$

$$f = -K_{D1}K_{D2}R_T$$

$$\theta = \arccos\left(\frac{-2d^3 + 9de - 27f}{2\sqrt{(d^2 - 3e)^3}}\right)$$

where F_{SB} is the fraction of bound fluorescent tracer; A_B and A_F are the anisotropies of the bound and unbound species, respectively; A_{OBS} is the measured anisotropy; L_{ST} is the total concentration of the fluorescent tracer; L_T is the total concentration of the competing ligand; R_T is the total receptor target concentration; K_{D1} is the dissociation constant for the interaction between the target and the tracer ligand; K_{D2} is the dissociation constant for the interaction between the target and competing ligand (38).

4.3. DNA Topology Assays

4.3.1. Agarose Gel Electrophoresis

1. Reactions are carried out in Eppendorf tubes or 96-well PCR plates/strips and each should comprise the following:
   (a) 4 μL of 5× assay buffer.
   (b) 1 μL relaxed plasmid DNA (typically 100–400 ng DNA final).
   (c) 1 μL of 20× ATP, final concentration 1 mM. This can be omitted for a negative control.

(d) Deionized water to 20 μL. This is variable based on the volume of gyrase enzyme included.

(e) The amount of DNA gyrase should be used at the recommended amount (if using commercial kits) or determined empirically. Typically, 1 U of gyrase will supercoil 0.5 ng relaxed plasmid DNA in 30 μL at 37°C in 1 h.

(f) Negative controls, lacking enzyme and/or ATP, should be included as should solvent-only controls. Limit DMSO concentrations to 5% or less. Preincubate the enzyme, DNA, and test ligands for 10 min prior to initiation of the reaction by addition of ATP. Dose responses for test ligands should be examined. Additionally, ATP concentration can be varied and used to probe if inhibition of gyrase function is consistent with ATP-competitive ligands.

2. Reactions are initiated by addition of the ATP, mixed by gentle pipetting, and incubated at 37°C for 15–60 min. Optimal incubation length should be determined empirically for each batch of enzyme. When using individually purified GyrA and GyrB proteins, preliminary experiments fixing the concentration of one, while varying the other, should be performed to find optimal stoichiometry, theoretically 1:1. Slight excess of one subunit over the other should be anticipated.

3. Reactions are stopped by the addition of 5 μL 5× Stop buffer/gel loading dye followed by 1.5 μL of 1 mg/mL Proteinase K. Setting up assays in 96-well format has the advantage of initiating/termination all samples at uniform times. Samples are incubated at 37°C for 30 min.

4. Finally, 20 μL of Chloroform:isoamyl alcohol is added, samples mixed by vortexing (Eppendorf tubes) or pipetting (96-well plates). The organic (lower) and aqueous (blue, upper) phases are allowed to separate.

5. The DNA present in the upper aqueous phase is loaded to a 1% agarose gel in 1× TBE and electrophoresed at 2–5 V/cm. Running gels at increased voltage will result in decreased resolution. It is important not to have ethidium bromide in the gel or buffer at this point as this impacts migration patterns, specifically it intercalates into relaxed DNA, resulting in mobility resembling that of supercoiled DNA. Care should be taken to thoroughly clean lab apparatus such as gel boxes and combs prior to use.

6. After electrophoresis is complete, gels should be stained with 0.5 μg/mL ethidium bromide in 1× TBE for 30–60 min followed by destaining for 1 h in 1× TBE.

7. Gel images should be captured using UV irradiation and a gel documentation system.

8. Data can be analyzed in several ways, such as IC_{50} (the value of compound required to reduce supercoiling by 50%), MNEC (the maximum non-effective concentration, i.e., the highest tested concentration with no visible effect on supercoiling activity), or by titrating to find a dose that causes an equivalent level of inhibition to a known concentration of a reference compound.

4.3.2. DNA Triplex Formation

1. Dilute the supplied wash buffer (20×) to 1× with deionized water, and use this to carefully rehydrate the wells of the supplied 96-well plate with 3×200 μL.
2. The biotinylated oligonucleotide (TFO1), supplied at 10 μM, is diluted 20-fold with 1× wash buffer to a final concentration of 500 nM. 100 μL is added to each well and allowed to immobilize by incubation at room temperature for 5 min. Excess, non-immobilized oligonucleotide is removed and wells washed with 3×20 μL 1× wash buffer.
3. Prepare the following reaction mixture for 100 assays, Each well will be in a final volume of 30 μL.
   (a) 2,025 μL deionized water.
   (b) 600 μL 5× assay buffer, final concentration 1×.
   (c) 75 μL 1 mg/mL pNO1.
   (d) 100 μL of 1.5 U/μL enzyme prepared fresh using dilution buffer.
   (e) Total volume is 2,800 μL. Aliquot 28 μL to each well and immediately add 2 μL DMSO or test ligand in DMSO as needed. Minimize the time between initiation of the assay and addition of potential inhibitors. An alternate approach is to omit ATP from the assay buffer, pre-incubate enzyme, plasmid, and test ligand in the 96-well plate prior to initiating the reaction by addition of ATP. Enzyme-independent controls should also be set up.
4. Incubate the plate at 37°C for 30 min.
5. Prepare 1× TF buffer by dilution of 20× stock with deionized water and add 100 μL to each well and incubate at room temperature for a further 30 min in order for triplex formation to occur.
6. Carefully remove the liquid from each well and discard. Wash each well with 3×200 μL 1× TF buffer to remove any unbound plasmid.
7. Finally, add fluorescent stain; SYBR Gold is diluted to 1× using 1× T10 buffer. Add 200 μL to each well and incubate in the dark for 20 min prior to reading fluorescence using an excitation and emission wavelengths of 495 and 537 nm, respectively (bandwidth 10 nm or less).

8. Simple IC_{50} values are readily obtained by plotting fluorescence against ligand using a Sigmoidal-dose response curve fitted using non-linear regression analysis software such as GraphPad Prism version 4.00 for Windows (GraphPad Software, San Diego California USA, www.graphpad.com).

5. Notes

1. ATP should be prepared fresh and care taken with correct storage to minimize hydrolysis and release of inorganic phosphate that will manifest as high background in this assay.
2. Care should be taken to fully resuspend ATP prior to use.
3. Care should be taken with the storage of fluorescently labeled ligands, particularly limiting their exposure to light.
4. An important control that is often omitted is to examine whether the test ligands affect the mobility of the relaxed plasmid DNA in the absence of enzyme. This may indicate that the compounds are DNA intercalators. A ligand that causes complete inhibition of supercoiling activity but is also a potent DNA intercalator could cause DNA to migrate as if supercoiled, and be mistaken for a target-based inhibitor. A convenient kit to asses this is available from Topogen which should be used as described.
5. Care should be taken throughout this procedure to prevent phosphate contamination. This would manifest as a high signal in the enzyme-independent wells and low signal/background. The most likely suspect if this occurs is phosphate contamination of one or more of the reagents. However, detergents used for cleaning glassware can also be problematic. Make sure that the enzyme is not present in a phosphate-based buffer, commonly used in protein purification.
6. This addition is optional but serves to limit the non-enzymatic hydrolysis of ATP that occurs in acidic solution.
7. Typically, if the concentration of inhibitor required to bind all target sites is 10% or greater of the total inhibitor concentration, then ligand depletion should be accounted for using the Morrison equation.
8. An accurate measure of the active site concentration is vital for this analysis. The use of standard protein estimation techniques such as the Bradford assay should be avoided as this is a measure of total protein content rather than functional active site concentration. Tight-binding inhibitors can be used to accurately determine the number of binding sites as described (24).

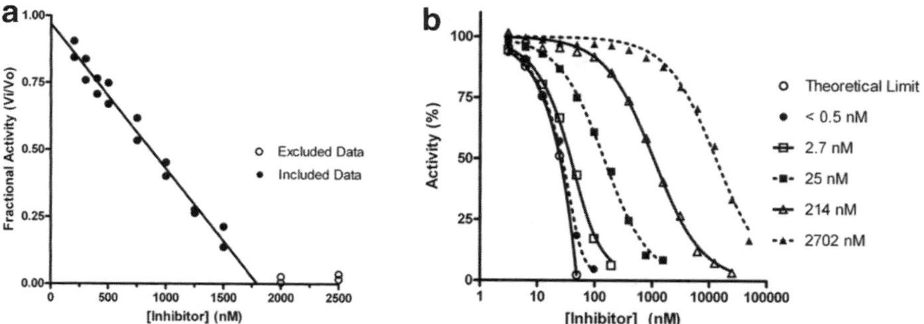

Fig. 7. Enzyme active site titration and inhibition. (**a**) 2 μM sample of enzyme was assayed as described in the methods using the Enzchek ATPase assay in the presence of increasing amounts of a known competitive inhibitor with a K_i of 5 nM. Under these conditions, [*E*] is 200-fold greater than K_i and a simplified form of the Morrison tight binding equation (*see* Subheading 4.1.2) approximates a linear relationship between fractional activity (*y* axis) and inhibitor concentration. Extrapolation of the data (*closed circles*) to the *x* axis gives a measurement of the active site concentration where [*E*] = [*I*]. Data deviating from linearity (*open circles*) are excluded from the analysis. (**b**) A series of inhibition curves obtained using the Enzchek ATPase assay in which 50 nM enzyme (active site concentration) is incubated with a series of inhibitors of varying potency. Inhibitors are tested over a wide concentration range from 50 μM to 3 nM. A theoretical limit of inhibition is shown for reference (*open circles*).

Briefly, if we reconsider the Morrison equation from Subheading 4.1.2 and set up an experiment in which the concentrations of both enzyme ([*E*]) and inhibitor ([*I*]) are much larger than K_i^{app}, then this term can be ignored. Under these conditions, fractional activity decreases essentially linearly as the concentration of inhibitor increases until [*I*] = [*E*]. Experimental design requires the use of higher enzyme concentrations than would typically be employed. Figure 7 shows the active site titration data of a sample estimated to contain 2 μM protein with a competitive inhibitor with a K_i of 5 nM (Cunningham et al. unpublished results). Ideally, the concentration of the enzyme should be at least 200-fold greater than K_i although this is not always achievable given the enzymatic rates measurable. Data in closed circles has been included in the linear extrapolation while the data represented by open circles has been excluded due to deviation from linearity. Extrapolation to the *x* axis gives [*E*] = [*I*] = 1,794 nM.

9. In addition to being necessary for K_i determination, accurate knowledge of the concentration of active sites in an assay can also be used to gauge data quality and the limits of the assay. Figure 7 shows several inhibition curves ranging in potency from μM to nM. In addition, a theoretical limit is also plotted. If we assume the inhibitor is so potent that every molecule added binds to and inhibits an active site, we can generate a theoretical limit of inhibition, i.e., addition of 5, 10, 15, or 19 nM inhibitor to 20 nM active site will maximally cause 25, 50, 75, and 95% inhibition, respectively. By comparing experimentally obtained data with the theoretical limit, data quality

can be assessed. This is particularly important as potency increases. From Fig. 7, one can see that 2.7 nM is clearly distinguishable from the theoretical limit whereas inhibition constants below 0.5 nM are not callable as the data is almost superimposable with the theoretical limit. The operator should be able to define a lower limit for potency. If increased sensitivity is needed, efforts should be made to modulate substrate levels and/or lower enzyme concentration. Data below the theoretical limit likely indicate problems with either the active site concentration estimate, ligand dilution series preparation, or inhibition mechanism.

10. A question that all enzymologists will undoubtedly hear at some point from a medicinal chemist is "Can you increase the sensitivity of the assay?" For a competitive inhibitor, judicially increasing the concentration of the competing substrate and lowering the enzyme concentration are the best tools available, although care has to be taken to maintain adequate signal to noise. We regularly run assay with 20 nM active site concentration of full-length DNA gyrase and are able to measure K_i values as low as 0.2 nM.

11. If the association of the test ligand is too strong to be determined using the standard protocol, a modified methodology is often used in which a weaker binding ligand (L_1) is displaced by a higher affinity ligand (L_2) (31, 32). It is vital that the two ligands compete for binding to the same site and that their binding is mutually exclusive. In this case, binding parameters for the weakly binding ligand (L_1) are first determined in a preliminary experiment as described above. Next, protein and L_1 are mixed in a syringe, degassed, and loaded into the calorimeter cell. The concentration of L_1 is chosen such that it significantly reduces the apparent affinity of the higher affinity ligand L_2. A detailed protocol to estimate an appropriate concentration is described in reference (32). Finally, the higher affinity test ligand (L_2) is titrated into the system in the standard manner and appropriate control titration performed. An apparent binding constant (K_{app}) is obtained corresponding to the displacement of the L_1 by L_2. Knowing the binding constant K_1 for L_1, and the concentration of L_1, the binding constant K_2 for L_2 can be calculated using $K_2 = K_{app}(1 + K_1[L_1])$.

12. Prior to competition assays, an accurate K_d for the fluorescently labeled ligand against the protein target should be obtained. This is done by incubating a fixed concentration of ligand with variable concentration of protein and measuring the resultant anisotropies. Binding constant K_{D1} is calculated using the following equations:

$$F_{SB} = \frac{(K_{D1} + L_{ST} + R_T) - \sqrt{(K_{D1} + L_{ST} + R_T)^2 - 4L_{ST}R_T}}{2L_{ST}}$$

with

$$F_{SB} = \frac{A_{OBS} - A_F}{A_B - A_F}$$

where F_{SB} is the fraction of bound fluorescent tracer; A_B and A_F are the anisotropies of the bound and unbound species, respectively; A_{OBS} is the measured anisotropy; L_{ST} is the total concentration of the fluorescent tracer; R_T is the total receptor target concentration; and K_{D1} is the dissociation constant for the interaction between the target and the tracer ligand (38).

References

1. Kampranis, S. C., Bates, A. D., and Maxwell, A. (1999) A model for the mechanism of strand passage by DNA gyrase. *Proc. Natl. Acad. Sci. USA.* **96**, 8414–9.
2. Reece, R. J., and Maxwell, A. (1991) DNA gyrase: structure and function. *Crit. Rev. Biochem. Mol. Biol.* **26**, 335–75.
3. Wigley, D.B. (1995). Structure and mechanism of DNA gyrase. In *Nucleic acids and molecular biology, Vol. 9* (Eckstein, F., and Lilley, D. M.J., eds.) Berlin, Springer-Verlag. pp 165–176
4. Brino, L., Urzhumtsev, A., Mousli, M., Bronner, C., Mitschler, A., Oudet, P., and Moras, D. (2000) Dimerization of *Escherichia coli* DNA-gyrase B provides a structural mechanism for activating the ATPase catalytic center. *J. Biol. Chem.* **275**, 9468–75.
5. Ali, J.A., Jackson, A. P., Howells, A. J., and Maxwell A. (1993) The 43-kilodalton N-terminal fragment of the DNA gyrase B protein hydrolyzes ATP and binds coumarin drugs. *Biochemistry* **32**, 2717–24.
6. Drlica, K., Coughlin, S. (1989) Inhibitors of DNA gyrase. *Pharmacol. Ther.* **44**, 107–21.
7. Wolfson, J. S., Hooper, D. C. (1985) The fluoroquinolones: structures, mechanisms of action and resistance, and spectra of activity *in vitro*. *Antimicrob. Agents Chemother.* **28**, 581–6.
8. Maxwell, A. (1992) The molecular basis of quinolone action. *J. Antimicrob. Chemother.* **30**, 409–14.
9. Maxwell, A., and Gellert, M. (1986) Mechanistic aspects of DNA topoisomerases. *Adv. Protein Chem.* **38**, 69–107.
10. Tingey, A. P., Maxwell, A. (1996) Probing the role of the ATP-operated clamp in the strand-passage reaction of DNA gyrase. *Nucleic Acids Res.* **24**, 4868–73.
11. Jackson, A. P., Maxwell, A., and Wigley, D. B. (1991) Preliminary crystallographic analysis of the ATP-hydrolysing domain of the *Escherichia coli* DNA gyrase B protein. *J. Mol. Biol.* **217**, 15–7.
12. Wigley, D. B., Davies, G. J., Dodson, E. J., Maxwell, A., and Dodson, G. (1991) Crystal structure of an N-terminal fragment of the DNA gyrase B protein. *Nature* **351**, 624–9.
13. Brown, P., O., Peebles, C.,L., and Cozzarelli, N.,R. (1979) A topoisomerase from Escherichia coli related to DNA gyrase. *Proc. Natl. Acad. Sci. USA.* **76**, 6110–4.
14. Gellert, M., Fisher, L. M, O'Dea, M. H. (1979) DNA gyrase: purification and catalytic properties of a fragment of gyrase B protein. *Proc. Natl. Acad. Sci. USA.* **76**, 6289–93.
15. Adachi, T., Mizuuchi, M., Robinson, E. A., Appella, E., O'Dea, M. H., Gellert, M., and Mizuuchi, K. (1987) DNA sequence of the *E. coli* gyrB gene: application of a new sequencing strategy. *Nucleic Acids Res.* **15**, 771–84.
16. Ali, J. A., Orphanides, G., and Maxwell, A. (1995) Nucleotide binding to the 43-kilodalton N-terminal fragment of the DNA gyrase B protein. *Biochemistry* **34**, 9801–8.
17. Lamour, V., Hoermann, L., Jeltsch, J. M., Oudet, P., and Moras, D. (2002) An open conformation of the *Thermus thermophilus* gyrase B ATP-binding domain. *J. Biol. Chem.* **277**, 18947–53.
18. Sugino, A., Higgins, N. P., Brown, P. O., Peebles, C. L., Cozzarelli, N. R. (1978) Energy coupling in DNA gyrase and the mechanism of action of novobiocin. *Proc. Natl. Acad. Sci. USA.* **75**, 4838–42.

19. Maxwell, A., and Lawson, D. M. (2003) The ATP-binding site of type II topoisomerases as a target for antibacterial drugs. *Curr. Top. Med. Chem.* **3**, 283–303.
20. Anderle, C., Stieger, M., Burrell, M., Reinelt, S., Maxwell, A., Page, M. and Heide, L. (2008) Biological activities of novel gyrase inhibitors of the aminocoumarin class. *Antimicrob Agents Chemother.* **52**, 1982–90.
21. Rowlands, M. G., Newbatt, Y. M., Prodromou, C., Pearl, L. H., Workman, P., and Aherne, W. (2004) High-throughput screening assay for inhibitors of heat-shock protein 90 ATPase activity. *Anal. Biochem.* **327**, 176–83.
22. Chène, P., Rudloff, J., Schoepfer, J., Furet, P., Meier, P., Qian, Z., Schlaeppi, J. M., Schmitz, R., and Radimerski, T. (2009) Catalytic inhibition of topoisomerase II by a novel rationally designed ATP-competitive purine analogue. *BMC Chem. Biol.* **9**, 1.
23. Webb, M.R. (1992) A continuous spectrophotometric assay for inorganic phosphate and for measuring phosphate release kinetics in biological systems. *Proc. Natl. Acad. Sci. USA.* **89**, 4884–7.
24. Copeland, R. A. (2000) Tight binding inhibitors. In *Enzymes: A Practical Introduction to Structure, Mechanism, and Data Analysis* (Copeland, R. A., ed). John Wiley & Sons, New York, NY, pp 305–317.
25. Chatterji, M., Unniraman, S., Maxwell, A., and Nagaraja, V. (2000) The additional 165 amino acids in the B protein of *Escherichia coli* DNA gyrase have an important role in DNA binding. *J. Biol. Chem.* **275**, 22888–94.
26. Flatman, R. H., Howells, A. J., Heide, L., Fiedler, H. P., and Maxwell, A. (2005) Simocyclinone D8, an inhibitor of DNA gyrase with a novel mode of action. *Antimicrob. Agents Chemother.* **49**, 1093–100.
27. Nakanishi, A., Imajoh-Ohmi, S., and Hanaoka, F. (2002) Characterization of the interaction between DNA gyrase inhibitor and DNA gyrase of *Escherichia coli. J. Biol. Chem.* **277**, 8949–54.
28. Boehm, H.J., Boehringer, M., Bur, D., Gmuender, H., Huber, W., Klaus, W., Kostrewa, D., Kuehne, H., Luebbers, T., Meunier-Keller, N., and Mueller, F. (2000) Novel inhibitors of DNA gyrase: 3D structure based biased needle screening, hit validation by biophysical methods, and 3D guided optimization. A promising alternative to random screening. *J. Med. Chem.* **43**, 2664–74.
29. Sissi, C., Vazquez, E., Chemello, A., Mitchenall, L. A., Maxwell, A., and Palumbo, M. (2010) Mapping simocyclinone D8 interaction with DNA gyrase: evidence for a new binding site on GyrB. *Antimicrob. Agents Chemother.* **54**, 213–20.
30. D'Arcy, A., Stihle, M., Kostrewa, D., and Dale, G. (1999) Crystal engineering: a case study using the 24 kDa fragment of the DNA gyrase B subunit from *Escherichia coli. Acta Cryst.* **D55**, 1623–1625
31. Lafitte, D., Lamour, V., Tsvetkov, P. O., Makarov, A. A., Klich, M., Deprez, P., Moras, D., Briand, C., and Gilli, R. (2002) DNA gyrase interaction with coumarin-based inhibitors: the role of the hydroxybenzoate isopentenyl moiety and the 5′-methyl group of the noviose. *Biochemistry* **41**, 7217–23.
32. Holdgate, G. (2009) Isothermal titration calorimetry and differential scanning calorimetry. *Methods Mol. Biol.* **572**, 101–33
33. Lewis, E. A., and Murphy, K. P. (2005) Isothermal titration calorimetry. *Methods Mol. Biol.* **305**, 1–16.
34. Holdgate, G. A., Tunnicliffe, A., Ward, W. H., Weston, S. A., Rosenbrock, G., Barth, P. T., Taylor, I. W., Pauptit, R. A., and Timms, D. (1997) The entropic penalty of ordered water accounts for weaker binding of the antibiotic novobiocin to a resistant mutant of DNA gyrase: a thermodynamic and crystallographic study. *Biochemistry* **36**, 9663–73.
35. Kim, J., Felts, S., Llauger, L., He, H., Huezo, H., Rosen, N., and Chiosis, G. (2004) Development of a fluorescence polarization assay for the molecular chaperone Hsp90. *J. Biomol. Screen.* **9**, 375–81.
36. Onuoha, S. C., Mukund, S. R., Coulstock, E. T., Sengerovà, B., Shaw, J., McLaughlin, S. H., and Jackson, S. E. (2007) Mechanistic studies on Hsp90 inhibition by ansamycin derivatives. *J. Mol. Biol.* **372**, 287–97.
37. Gooljarsingh, L. T, Fernandes, C., Yan, K., Zhang, H., Grooms, M., Johanson, K., Sinnamon, R. H., Kirkpatrick, R. B., Kerrigan, J., Lewis, T., Arnone, M., King, A. J., Lai, Z., Copeland, R. A., and Tummino, P. J. (2006) A biochemical rationale for the anticancer effects of Hsp90 inhibitors: slow, tight binding inhibition by geldanamycin and its analogues. *Proc. Natl. Acad. Sci. USA.* **103**, 7625–30.
38. Roehrl, M. H., Wang, J. Y., and Wagner, G. (2004) A general framework for development and data analysis of competitive high-throughput screens for small-molecule inhibitors of protein-protein interactions by fluorescence polarization. *Biochemistry* **43**, 16056–66.
39. Maxwell, A, Burton, N. P, and O'Hagan, N. (2006) High-throughput assays for DNA gyrase and other topoisomerases. *Nucleic Acids Res.* **34**, e104.

40. Fisher, L. M., and Pan, X.S. (2008) Methods to assay inhibitors of DNA gyrase and topoisomerase IV activities. *Methods Mol. Med.***142**, 11–23.

41. Kawabata, Y., Ooya, T., Lee, W. K., and Yui, N. (2002) Self-assembled plasmid DNA network prepared through both triple-helix formation and streptavidin-biotin interaction. *Macromol. Biosci.* **2**, 195–198.

42. Cheng, Y., and Prusoff, W. H. (1973) Relationship between the inhibition constant (Kl) and the concentration of inhibitor which causes 50 per cent inhibition (I50) of an enzymatic reaction. *Biochem. Pharmacol.* **22**, 3099–108.

43. Morrison, J. F. (1969) Kinetics of the reversible inhibition of enzyme-catalyzed reactions by tight-binding inhibitors. *Biochem. Biophys. Acta.* **185**, 269–86.

Chapter 9

Leveraging Structural Information for the Discovery of New Drugs: Computational Methods

Toan B. Nguyen, Sergio E. Wong, and Felice C. Lightstone

Abstract

Escalating problems with drug resistance continue to compromise the effectiveness of commercial antibiotics, necessitating the search for novel classes of antimicrobial agents. To circumvent problems with resistance, a multitarget single-pharmacophore approach has been employed to discover inhibitors that possess balanced activity against multiple target enzymes. In this chapter, we examine the application of computational techniques, in particular, structure-based drug design approaches, to design new dual-targeting antibacterial agents against bacterial topoisomerases.

Key words: Structure-based drug design, Structure-activity relationships, Multi-target single pharmacophore, Virtual screening, Docking, DNA gyrase

1. Introduction

Almost 70 years ago, penicillin was used to save the life of an infected patient suffering from *Streptococcal* sepsis (1). Since then, many new classes of antibiotics have been discovered. However, bacteria, the champions of evolution, have adapted and developed resistance against most frontline antibiotics, such as vancomycin, methicillin, fluoroquinolones, and macrolides (2–7). A growing clinical concern is that bacteria are becoming increasingly multi-drug-resistant where even new classes of antibiotics, such as linezolid (8) and daptomycin (9), are encountering significant resistance. Consequently, there is renewed interest in the discovery and development of new classes of antibiotics (10) with novel mechanisms of action (11–13). Pharmaceutical researchers are pursuing a multitarget single-pharmacophore approach (14) to fight antibiotic resistance. This new strategy focuses on the design of inhibitors that possess balanced activity against multiple target

Leslie W. Tari (ed.), *Structure-Based Drug Discovery*, Methods in Molecular Biology, vol. 841,
DOI 10.1007/978-1-61779-520-6_9, © Springer Science+Business Media, LLC 2012

enzymes from multiple pathways, such as bacterial topoisomerase IV and DNA gyrase, dramatically reducing the probability of resistance incidence (15).

The utilization of microbial genomics (16) for target identification, advances in X-ray crystallography for 3D target–ligand interactions, and exploitation of computational approaches for designing and developing (17, 18) novel small molecules should accelerate the discovery of new antibacterial agents with broad-spectrum efficacy and low potential for resistance development.

This chapter discusses the application of computational structure-based drug design (SBDD) methods in the discovery of novel small-molecule inhibitors against well-validated targets involved in bacterial DNA replication (19, 20) and cell division: the type IIA topoisomerases, gyrase B (GyrB) and topoisomerase IV (ParE). The focus is on specific ligand–protein interactions observed in publically available crystallographic structures of novobiocin and/or adenylyl-imidodiphosphate (ADPNP) bound to the *Escherichia coli* GyrB and ParE subunits as the initial guide for inhibitor discovery and optimization.

2. The Design of Antibacterial Agents That Inhibit the Function of Type IIA Topoisomerases—GyrB and ParE

2.1. Background

There are many classes of antibiotics that target different aspects of bacterial function. Two important classes of antibiotics, including the coumarins (e.g., novobiocin) and fluoroquinolones (e.g., ciprofloxacin), target proteins involved in DNA replication: the type IIA family of topoisomerases, DNA gyrase and topoisomerase IV (19, 20). These enzymes are structurally homologous proteins that exist as heterotetramers in vivo, comprising 2 GyrA and 2 GyrB subunits for DNA gyrase and 2 ParC and 2 ParE subunits for topoisomerase IV. Type II topoisomerases possess both DNA cleavage and ligation functionalities and an ATP-dependent clamp that works in a concerted fashion to resolve DNA catenates and supercoils during replication. The GyrA and ParC subunits catalyze DNA cleavage and religation via an intermediate where the 5′-ends of the DNA chain are covalently bound to conserved tyrosine residues. The GyrB and ParE subunits are responsible for passage of a separate DNA strand through the cut DNA strands in an ATP-dependent manner. At the cleavage site, DNA becomes single stranded due to a four base-pair staggered break within the binding cavity, allowing the fluoroquinolone inhibitors to bind to the DNA-topoisomerase complex. The drug–DNA–enzyme ternary complex locks the tetrameric complex in a catalytically noncompetent state (21), arresting DNA replication. The accumulation of single- and double-stranded breaks irreversibly damage the bacterial chromosome (22, 23).

Although dual-targeting in principle, fluoroquinolones appear to kill via single targets in different bacterial classes. In Gram-negative bacteria such as *E. coli*, fluoroquinolones kill mainly via inhibition of DNA gyrase, while in Gram-positive organisms such as *Staphylococcus aureus*, fluoroquinolones appear to operate primarily via inhibition Topoisomerase IV (24). Coumarin antibiotics, which target the ATP-binding domains of the GyrB and ParE subunits, are poor dual targeting agents that operate mainly via the inhibition of GyrB. However, the high degree of similarity at the sequence and structural level between GyrB and ParE suggests that the design of potent dual targeting agents against the GyrB and ParE domains should be possible (25, 26).

2.2. Structural Features of the Type IIA Topoisomerases

A number of X-ray structures (27) of the ATPase domain of DNA-gyrase complexed with phosphoaminophosphonic acid-adenylate ester (ADPNP) and novobiocin are available (see Fig. 1). The detailed analysis of these structures demonstrates that the ligands bind using a similar pharmacophore to a key aspartic acid side chain and a conserved water molecule in the binding cavity. In both cases, each ligand donates a hydrogen bond to the aspartate side chain and accepts a hydrogen bond from the conserved water molecule. Figure 2 illustrates the hydrogen-bond network observed for the adenine of the ADPNP and the carbamate moiety of novobiocin. Novobiocin is an ATP-competitive inhibitor that overlaps partially with the adenine moiety of ATP (27). This overlapping competitive binding area is the focus of novel inhibitor design.

Recently, Bellon et al. from Vertex (27) have reported the structure of the ATP-binding domain of *E. coli* ParE complexed with ADPNP at 2.0 Å resolution as well as a ParE novobiocin complex at 2.1 Å resolution (see Fig. 3). A comparison of the GyrB and ParE structures demonstrate a striking similarity within the ATP-binding sites. Superposition of the *E. coli* ParE and GyrB structures shows the amino acids in the active site overlap closely. The ParE

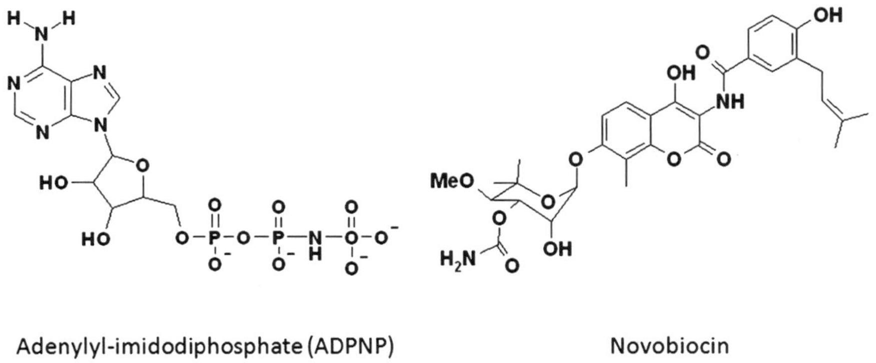

Adenylyl-imidodiphosphate (ADPNP)

Novobiocin

Fig. 1. Chemical structures of a substrate analog and a natural product inhibitor of type IIA topoisomerases; adenylyl-imidodiphosphate (ADPNP) and novobiocin.

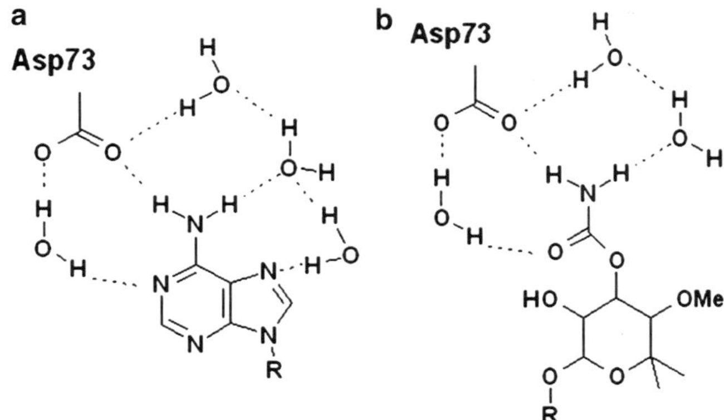

Fig. 2. The "anchoring" hydrogen-bond network used in GyrB and ParE to bind (**a**) the adenine moiety of ADPNP and (**b**) the carbamate moiety of novobiocin.

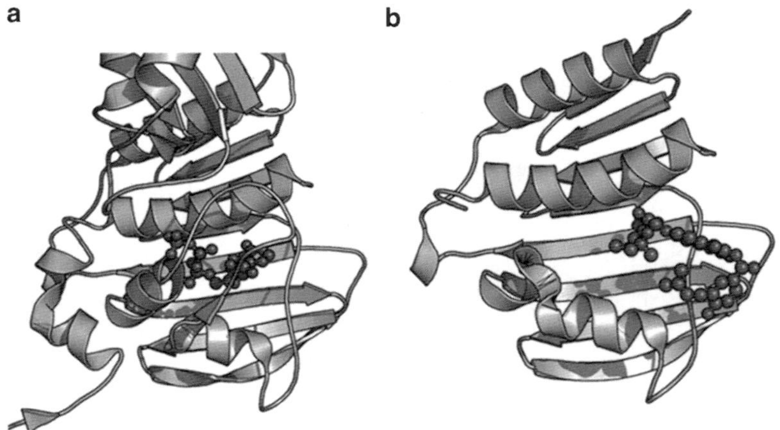

Fig. 3. Crystal structures of *Escherichia coli* ParE. (**a**) A 43-kDa fragment (PDB code 1S16) complexed with ADPNP at 2.0 Å resolution. (**b**) A 24-kDa fragment (PDB code 1S14) complexed with novobiocin at 2.1 Å resolution (27).

and the corresponding *E. coli* GyrB residues (in parenthesis) are as follows when ADPNP is bound (see Fig. 4): Y5 (Y5), E38 (E42), N42 (N46), E46 (E50), D69 (D73), M74 (I78), K99 (K103), Y105 (Y109), T163 (T165), Q332 (Q335), and K334 (K337). The authors also point out that the specific functional roles of GyrB vs. ParE residues in the ATP active site are transferable. Similarly, the key residues of both proteins that interact with novobiocin (see Fig. 5) are as follows: E46 (E50), D69 (D73), R72 (R76), M74 (I78), D77 (D81), I90 (I94), R132 (R136), and T163 (T165). Moreover, Figs. 4 and 5 illustrate two important features: (1) the hydrogen-bond interactions with D1069 observed for the adenine scaffold of the ADPNP also engage the carbamate moiety of novobiocin and (2) the orientation of these ligands with respect to the

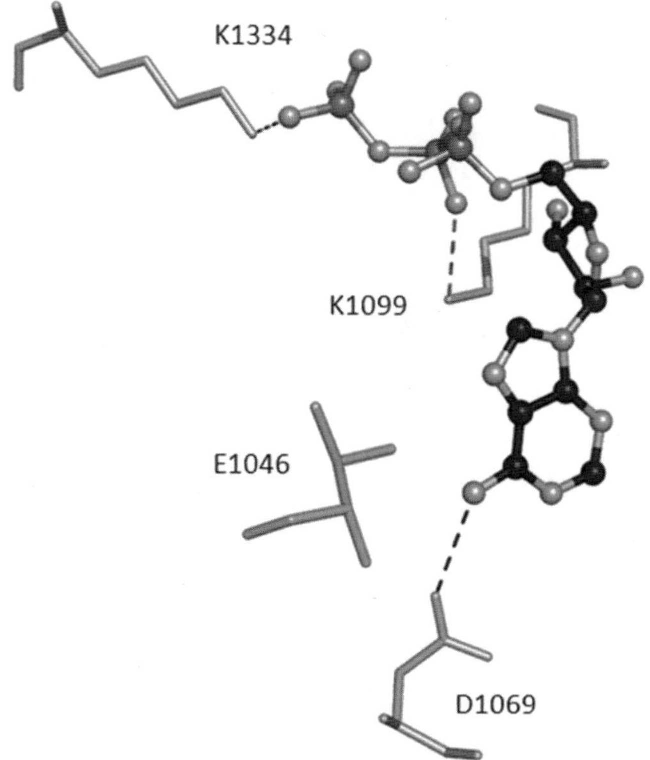

Fig. 4. The ATP-binding site of *E. coli* ParE with ADPNP drawn in ball-and-stick, and the protein shown in cylinders with key side-chains labeled. The illustration shows the hydrogen-bond interactions between the adenine of ADPNP and key amino acid D1069 and its orientation with respect to the binding environment—the polar phosphate group from the tail end of the ADPNP hydrogen-bonds with K1099 and K1334.

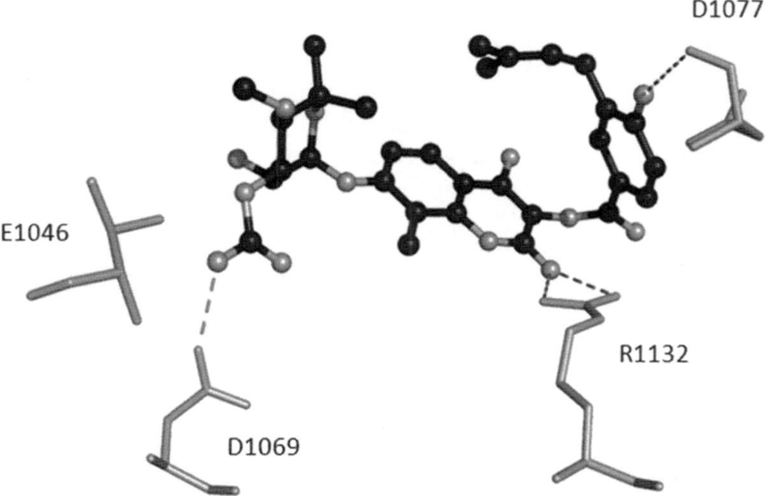

Fig. 5. The novobiocin-binding mode from the crystal structure of *E. coli* ParE with novobiocin shown in ball-and-stick and key side-chains labeled. The illustration shows the hydrogen-bond interactions between the carbamate moiety of novobiocin and the key amino acid D1069 and its orientation with respect to the binding environment—the carbonyl-ester of the middle ring and the hydroxyl of the tail end of novobiocin hydrogen-bonds with R1132 and D1077, respectively.

binding site environment—the polar phosphate group from the tail end of the ADPNP hydrogen bonds with K1099 and K1334, while the carbonyl-ester of the middle ring and the hydroxyl of the tail end of novobiocin hydrogen bonds with R1132 and D1077, respectively.

To maintain dual targeting activity, the authors report that the key structural difference between GyrB and ParE is a single amino acid change from Ile78 to Met74, respectively. Charifson et al. also report (28) that the binding site of ParE is narrower than that of GyrB, thus imposing a coplanarity requirement on ligands that bind in the cavity. Moreover, structures determined by our collaborators (L. Tari and D. Bensen, unpublished results) for both targets complexed with different ligands suggest that the coplanarity requirement of the ligands is not solely dictated by Met74 of ParE or Ile78 of GyrB but also by the key aspartate side chain and conserved crystallographic water molecule in the binding site. Further analysis of the complexes shows the distance from the Cβ of the Met or the Ile residues to the centroid of our ligand is 3.489 Å for ParE and 3.523 Å for GyrB. The resultant 0.034 Å narrowing in the ParE binding cavity seems small but is consistent with the selectivity profile of ParE with respect to the ligand's chemotype. This agrees well with Charifson's observation.

2.3. Identification of the "Hot Spots" in the GyrB Binding Cavity

Once a binding cavity is characterized, interaction "hot spots" within the binding cavity need to be identified. These "hot spots" are specific locations within the binding cavity that would potentially be exploited to interact with the inhibitor pharmacophore. Thus, the "hot spots" can be used for fragment-based screening and lead identification. One method to identify these "hot spots" is to use a variation of virtual screening (see Subheading 2.4). Schechner et al. (29) screened the binding site of GyrB with small fragments. Combinations of these fragments are then used as the basis for the design of novel ligands. They used the Multiple Copy Simultaneous Search (MCSS) technique developed by Miranker and Karplus (30), where all chemical fragment conformations are generated and randomly distributed in the binding cavity, removing fragments that occupy the same positions. The fragment library contained 23 functional groups, representing various chemical characteristics, such as charged, polar, hydrophobic, aromatic and aliphatic groups. Multiple structural states of the 24 kDa subdomain of GyrB identified crystallographically (to encompass active-site flexibility) were used to collect unbiased results. Overall, the protein conformations did not affect the functional group binding. However, binding energies and specific orientations of small functional groups were affected. The tightest binding pharmacophore among the groups tested was a phenol, forming a hydrogen bond with the key Asp73 in the binding cavity. Other functional groups were identified to target the phosphate-binding site. A deep hydrophobic binding pocket near the ATP binding site was also identified and suggested a path for optimization of the inhibitor scaffolds identified in the pharmacophore search.

2.4. Computationally Aided Structure-Based Drug Design

SBDD takes advantage of structural information about a protein target to identify and optimize lead compounds. The protein target structure is often derived from X-ray diffraction data, but may also be an NMR-based structure or based on a homology model (31–33). SBDD still relies on all the available information about the target and its biological role. Experimental data help to guide the computational chemist as to the most relevant calculations to drive optimization. This section aims to broadly describe the role of computational approaches in SBDD for lead identification and optimization.

At the lead compound identification stage, virtual screening searches large chemical databases to eliminate nonbinders. When large compound libraries are screened, i.e., in the range of 100 K or larger, the speed of the calculations is paramount, and this consideration leads to using relatively simple and fast docking algorithms. Each compound is assigned a score that is used to produce a sorted list with the most promising predicted compounds at the top. Success at this stage is not focused on predicting accurate binding affinities, but rather on enrichment of strongly binding compounds ("hits"). Docking programs are sometimes evaluated for robustness by seeding libraries with known binding compounds to see if they appear at the top of the ranked list. Zhou et al. (34) used enrichment to compare the performance of Glide (35), DOCK (36), and GOLD (37).

After initial virtual screening of a large database, top-scored compounds are reordered using more accurate scoring methods. Rescoring aims to correct some of the deficiencies in docking scores ignored in the initial stage. Effects to consider for rescoring may include solvation (32) and accounting for flexibility of the protein target upon ligand binding. Two common rescoring methods are Molecular Mechanics Generalized Born Surface Area (MM-GBSA) (38) and Linear Interaction Energy (LIE) (39). For some well studied cases, such as kinases (33) and DHFR (40), the protein has larger motions that respond to the ligand binding, making flexible conformations of the protein necessary for more accurate binding predictions.

2.4.1. Structure-Based Virtual Screening

Virtual screening is a process where chemical structures are evaluated in silico against the target protein structure via docking and/or a 3D pharmacophore model. This approach enables the identification of the biologically relevant molecules more efficiently when the virtual library is large (>5 million) than a random experimental high throughput screening. Figure 6 illustrates a typical workflow for a virtual screen with a protein receptor as a target, from which

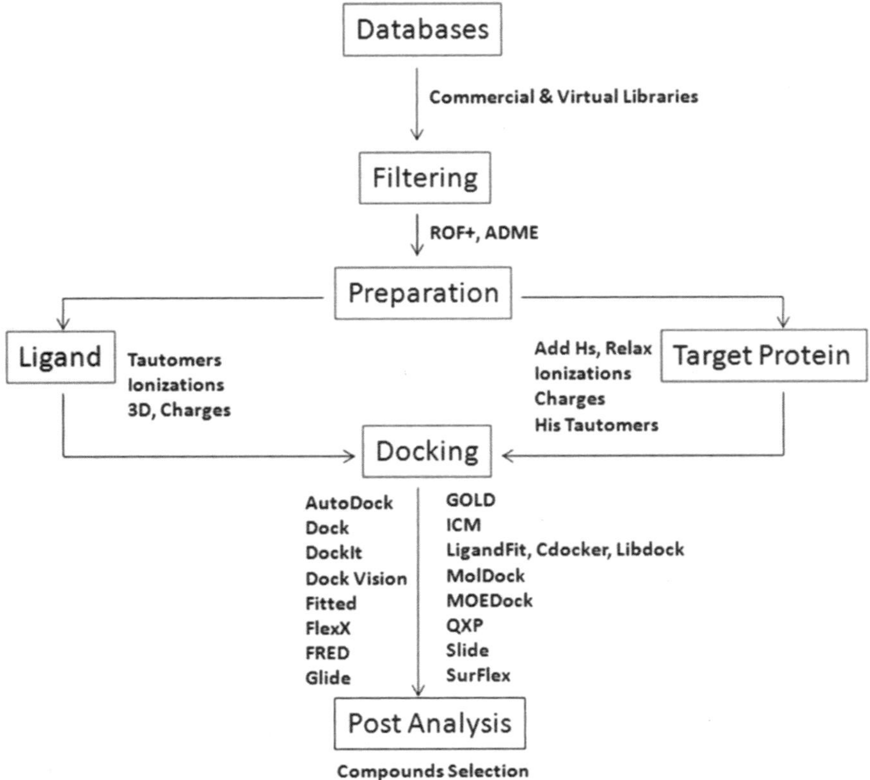

Fig. 6. An illustration of a typical workflow used for in silico screening.

a library of compounds is screened for a subset of compounds for further analysis. This section surveys a limited number of docking programs for virtual screening and their performance for studying topoisomerase inhibitors against GyrB and/or ParE with emphasis on the identification of novel leads and improvement of these leads. The related in silico techniques such as target, ligand preparation, scoring functions, and molecular dynamics are herein broadly discussed.

Target Preparation

Several factors should be considered when preparing protein targets for molecular docking. Frequently, the structures used for the protein targets are derived from X-ray crystallography experiments. Within these structures, small molecules, such as water molecules, ions, cofactors, and crystallization agents can appear in the binding site. In other cases, regions of the protein may be unresolved or disordered. Also, because of the experimental limitations, hydrogen atoms are not usually resolved, meaning that protonation states of titratable groups must be assigned. Each of these issues needs to be addressed in the context of the relevant biology and the desired goals for the molecular docking. For example, conserved water molecules may play a key role in mediating contacts between a ligand and its protein receptor (41, 42). Certainly, this is the case for novobiocin (27). Thus, including key crystallographic water molecules can improve docking results (43, 44). However, the water positions may be different for each putative ligand, and in some cases displacing key waters with a ligand (45) may be more favorable. Other molecules to consider as necessary in the binding site include cofactors, metals, and ligands of metals so that a more biologically relevant protein environment is represented. Most importantly, each of the residues in the binding site must be completely represented. If portions of the binding site are missing in the original crystal structure, then homology modeling or loop prediction methods need to be used to complete the binding site structure.

Fundamentals of Molecular Docking

Proteins are well known to be flexible and conformationally dynamic. Typically, in molecular docking the protein targets are held fixed during virtual screening to minimize the computational cost. The key routine for docking programs is to (1) generate poses for the ligand within the binding site and (2) score each ligand pose as the predicted binding mode for that ligand.

In addition to the internal conformational space, the ligand rotational and transitional degrees of freedom are explored to generate a set of poses. The number of low-energy conformers increases exponentially with the number of rotatable bonds, so efficiency in the search algorithm is imperative. To this end, several algorithms have been developed with modified protocols including the following: incremental construction of the ligand (DOCK (36), FlexX (46)), genetic algorithms that optimize the conformation and orientation of the ligand simultaneously in the binding pocket (AutoDock (47, 48), Fitted (49)), Monte Carlo-based algorithms (MCDOCK (50), Ligandfit (51), QXP (52)), pregeneration of low-energy ligand conformers before introducing them into the binding pocket (FLOG (53), FRED (54)), high-temperature molecular dynamics simulations to generate random ligand conformations that are rigidly translated into the binding site (Cdocker (51), ICM (55)), matching ligand functional properties with hot spots in the binding site (Libdock (51)), calculated distance geometry for intraligand and ligand–sphere interactions (DockIt (56)), or exhaustive search algorithms (Glide (35)). Many codes combine a number of the algorithms to meet the user's needs (Dock Vision (57, 58), MolDock (59), SurFlex (60), MoeDock (61)). The major approaches are mentioned with common docking methods as examples only. Neither the details of available methods nor the completeness of the list of docking codes is the intention for this survey. For a more complete discussion on pose generation, the reader is referred to the review by Brooijmans and Kuntz (62).

The next step is to assign a score for each pose to (1) select the best pose for each compound and (2) rank ligands based on their scores. Key challenges for the scoring functions are to account for the conformational entropy penalty of the ligand, the ligand desolvation, and the intramolecular interaction upon binding. While in principle we formally understand the theoretical basis to calculate the free energy of binding (63), in practice calculating free energy remains a research problem that requires massive (hundreds of cpu hours) computer power per ligand (see for example the work by Wang et al. (64) and Mobley et al. (65)). Owing to the large number of ligands that need scores in a typical virtual screening campaign, simplified scoring functions assign pose scores. The scoring functions are a compromise between accuracy and computational efficiency.

Three broad categories of scoring functions exist: physics (force-field) based, knowledge-based, and empirical scoring functions. Force-field based scoring functions base their scores on a function that explicitly lists physical interactions such as electrostatics and van der Waals. For example, the scoring function in AutoDock (47) originally took its parameters from the Amber force field (66).

Another approach is to consider a database of ligand–protein complexes and calculate the likelihood of forming specific contacts between ligands and proteins. A score based on this "knowledge" is assigned by summing the likelihood of each contact a ligand pose has with a protein. This type of scoring is called knowledge-based scoring (67, 68).

In the case of an empirical scoring function, the strategy is to fit the scoring function to reproduce binding data from a set of training data. The development of this type of function often starts with either a force-field or knowledge-based scoring function. Each contribution is weighed to reproduce binding affinity data (69, 70). For example, the later versions of Autodock (3.0 and above) introduced a scoring function that weighed the Amber force field parameters to better reproduce binding data.

Strategies exist to make the molecular docking results more reliable or at least represent experimental data better. Tirado-Rives and Jorgensen (71) have recently introduced a conformer-focusing contribution to the uncertainty of predicting affinities. This contribution was used in most available docking programs such as Discovery Studio (51), MOE (61), and Glide (35) to account for the penalty of conformational change from the unbound state to the bound state of the ligand.

A number of studies (72–79) have evaluated several docking programs and scoring functions for different major targets such as kinases, metalloenzymes, serine proteases, nuclear receptors and DNA gyrase. Warren et al. (80) from GSK reported that these programs performed well for some targets but poorly for others, in terms of the correct identification of actives or enrichment. The report showed that the best-to-least performers were FlexX,

GOLD, LigandFit, QXP, Dockit, FRED, DOCK, Glide, and MoeDock for GyrB. Schulz-Gasch and Stahl (81) at Hoffman LaRoche reported their findings using FlexX, Glide, and FRED toward GyrB, such that FlexX and Glide predicted the inhibitors well, while FRED's performance was the least among three programs in virtual screening. Generally, the docking programs—Glide, FlexX, DOCK, and GOLD—were the most popular for using in lead identification by pharmaceutical companies, and their performance is not consistent across target classes. For example, the top performing programs for kinases were Glide, DOCK, ICM, LigandFit, FlexX, GOLD, respectively, and for matalloenzymes, the order was FlexX, LigandFit, Glide, FRED, and ICM. Additionally, there are other programs developed by pharmaceutical teams with comparable performance.

Other validation studies suggest that combining several scores into a consensus score may be better than any individual scoring function alone (82–86). Another popular strategy to make the scoring better is to reexamine the top scoring compounds. Because of the known shortcomings of the rapid high-throughput virtual screening scoring functions, one strategy is to use more complex, accurate methods to rescore these compounds. Additionally scoring functions can be improved by including solvation effects via an implicit solvent treatment such as Poisson Boltzmann, or its approximation: generalized Born. Scoring via a molecular mechanics Poisson-Boltzman (MMPBSA) or MMGBSA can help improve binding affinity predictions (38, 87–90). The variation in performance of docking programs and scoring functions suggests that molecular docking is an active area of research with room for improvements (91).

Flexible Receptor Docking

To quickly screen large virtual libraries effectively, the protein target is typically fixed. However, cases exist where the protein needs to respond to the binding of substrates to achieve efficient binding. In the case of kinases (92, 93), DHFR (94, 95), aldose reductase (96, 97), and members of the enolase superfamily (98), the protein backbone moves substantially to adapt to small molecule binding. The conformational changes may introduce new van der Waals, electrostatic, and hydrogen bond contacts or clear space for the ligand to fully enter the binding site. Addressing these changes can considerably affect docking pose generation and scoring. By simultaneously sampling the degrees of freedom in both the protein and the ligand, large changes in conformation can be explored and modeled more accurately. Certainly, a better representation of the conformational space should result in a better prediction, but it has an increased computational cost. The total conformational space available for exploration increases exponentially with each independent degree of freedom, regardless of whether it belongs to the protein or ligand. If no experimental data exists for a ligand–protein complex structure where the protein has conformationally responded to ligand binding, then this flexible protein docking method becomes more difficult to justify.

Some conformational changes can be relatively small, where only a few side-chains rearrange to accommodate ligand binding. While small, these changes may have a large impact on binding. Consider, for example, orienting polar groups to optimize a hydrogen bond network. If only a few residues move, sampling their degrees of freedom along with the ligand conformational space is the most practical approach. Some programs, such as AutoDock (47) and Glide (35) include options to specify flexible residues in their induced fit calculations. AutoDock performs the search of all degrees of freedom in one run while Glide undergoes a cycle of optimization of docking the ligand and optimizing flexible side-chains (99).

Another approach is to run molecular dynamics (MD) simulations on the free protein target to explore the conformational space available to the protein (relaxed complex method). The key advantage is that the receptor degrees of freedom are calculated once instead of for each ligand individually. While relatively new, some work has shown that this approach is viable. Wong et al. (100) performed MD simulations of the receptor and docked into trajectory snapshots. They found improved rankings of known ligands. Combining molecular dynamics and docking has also been proposed by other authors (101). Note, however, that large conformational changes, particularly those involving an entropic barrier, will be difficult to sample using molecular dynamics. Therefore, this approach is really best suited for relatively small changes at the side-chain level.

Larger conformational changes, whether local (loop rearrangements) or global (domain hinge motions), are more challenging to capture computationally. However, some groups have shown progress in this area. Wong and Jacobson (102) were able to predict loop rearrangements involved in ligand binding of up to 15 residues in length. They showed that the enrichment of known ligands improved when a database was docked against the predicted structures. In special cases, experimental evidence greatly simplifies the problem. For example, if a domain–domain rearrangement involves the two domains acting as rigid bodies around a small, flexible hinge region, then one can include only the relevant degrees of freedom in the docking calculation (103). Reducing the effective number of degrees of freedom is a viable approach, if one can identify them a priori.

Post-processing of molecular dynamics or more advanced sampling techniques can guide the selection of key receptor degrees of freedom. For example, normal mode or principal component analysis of a molecular dynamics simulation is one approach to identify large-scale concerted motions (104). The result of this type of analysis is a list of concerted motions that individually account for a fraction of the motions in the molecular dynamics simulation. The top 2 or 3 motions (normal modes or principal components) likely include the majority of variation in the conformational space.

An alternative to predicting large induced fit changes is to use a holo crystal structure where the changes are evident. While smaller changes may remain, even in domain–domain hinge regions, holo structures will likely capture the large conformational change while small perturbations may be adjusted at the side-chain level as discussed in this section above.

2.4.2. Identification of New Lead Compounds by Virtual Screening

The aim of virtual screening is to identify active compounds. In lead identification, accurate classification of hits as active compounds needs to be confirmed experimentally. Several groups have used virtual screening to find new lead compounds that bind to GyrB. Boehm et al. at Roche (26) used the binding interactions of the novobiocin-GyrB complex structure, two key hydrogen bond interactions of the ligand with an aspartate side chain and with a conserved crystallographic water molecule, as the required binding motif (see Fig. 7) to virtually screen a commercially available database and an in-house collection. They used LUDI and CATALYST (51) software to identify 3,000 compounds, from which 150 hits were confirmed and clustered into 14 classes. Seven classes (see Fig. 8) were validated as novel GyrB inhibitors: phenols, 2-amino-triazines, 4-amino-pyrimidines, 2-amino-pyrimidines, pyrrolopyrimidines, indazoles, and 2-hydroxymethyl-indoles. The molecular weights of these "lead-like" hits were lower than that of novobiocin, with activities 2–3 orders of magnitude lower than novobiocin. Of these lead-like hits, the Indazole A complexed with the 24 kDa N-terminal fragment of GyrB structure (see Fig. 9) was used to explore the potential van der Waals interactions of the ligand with the lipophilic residues Ile78, Pro79, and Ile94 to

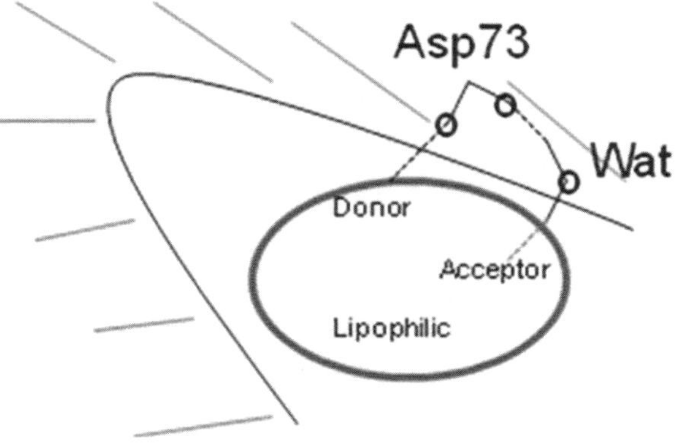

Fig. 7. An illustration of the binding hypothesis used for in silico screening.

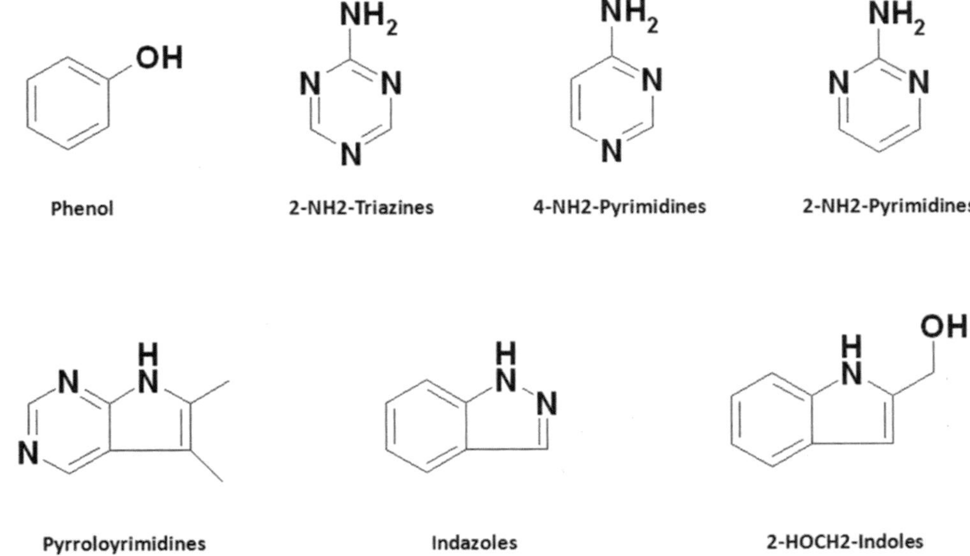

Fig. 8. Seven classes of lead compounds as novel inhibitors of DNA GyrB (26).

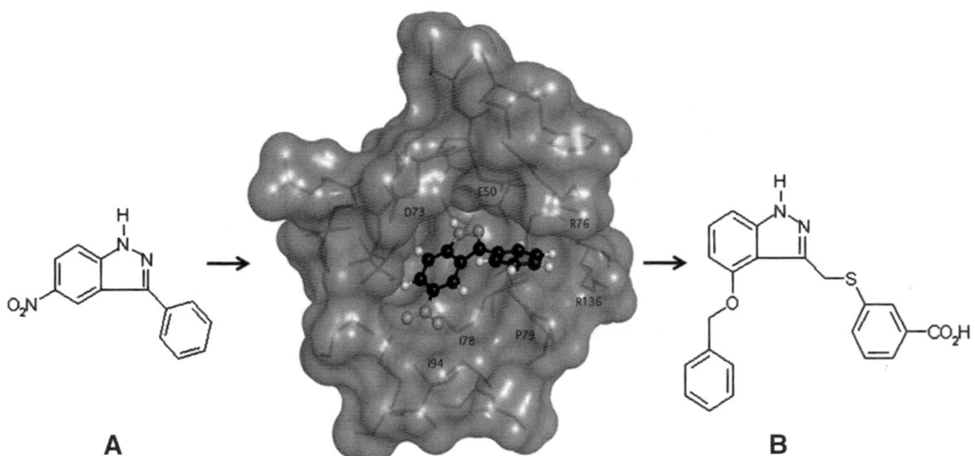

Fig. 9. Structure-guided design of a potent indazole B from the lead indazole A by maximizing inhibitor-receptor interactions (26).

produce a more potent Indazole B, which was 10 times more potent than Novobiocin. The binding hypothesis was confirmed crystallographically, where the indazole moiety hydrogen bonds with Asp73 and the conserved water molecule. The mercaptobenzoic acid side chain π-stacks with the Glu50-Arg76 salt bridge and hydrogen-bonds with Arg136. The benzyloxy side chain interacts with the lipophilic area encompassing Ile78-Pro79-Ile94.

Oblak et al. (105) also used LUDI to identify low-molecular-weight fragments of 2-aminobenzimidazole and indolin-2-one as potential GyrB inhibitors from the MDL-databases (see Fig. 10).

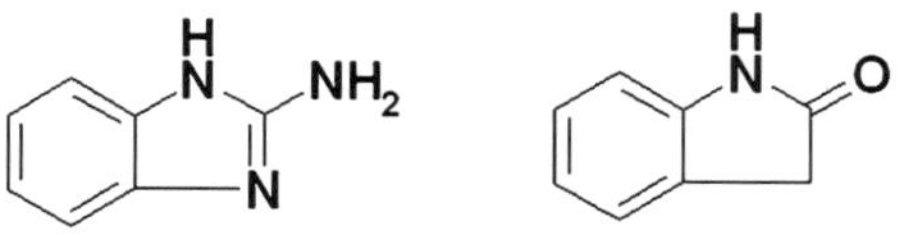

Fig. 10. Chemical structures of DNA GyrB lead compounds; aminobenzimidazole and indolin-2-one, respectively (105).

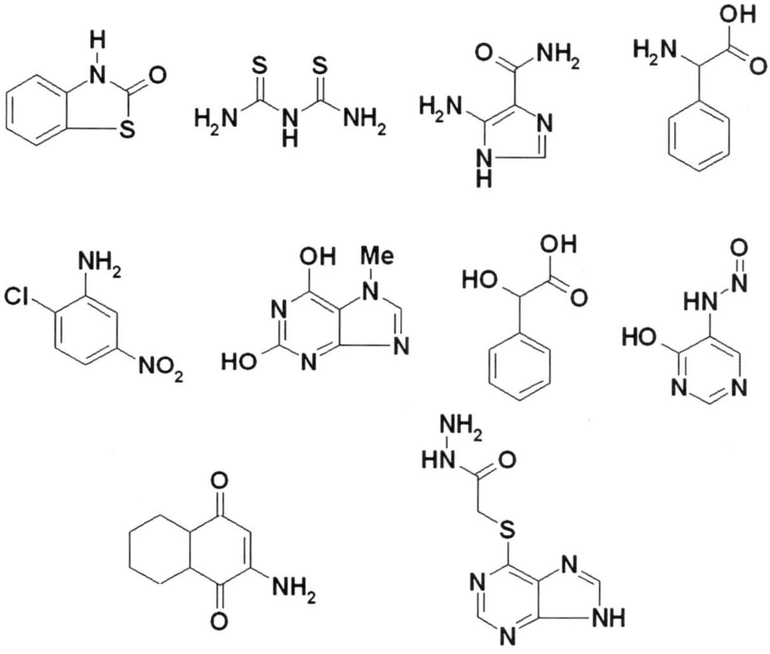

Fig. 11. Chemical structures identified by virtual screening using DOCK (107).

Because DNA gyrase is a member of the GHKL ATPase superfamily (106), it is structurally distinct from the larger family of eukaryotic kinases, where the indolinone hit is known as a potent inhibitor of many kinases (92). This fact improves the prospects for designing selectivity into subsequent generations of the indolinone scaffold.

Focusing on the previously unexplored structural pocket at the dimer interface of subunit A and a small area of the ATP binding pocket on subunit B targeted by coumarin and cyclothialidine, Ostrov et al. (107) used DOCK (v5.1.0) to screen 140,000 small molecules and identified the following chemotypes (see Fig. 11), such as methylxanthines, imidazole carboxamides, pyrimidine nitrous amides, benzothiazolinones, aminonapthoquinones, mandelic acids, aminobenzoacetic acids, and imidocarbonodithioic diamides as potential scaffolds for novel GyrB antibacterial leads targeting fluoroquinolone-resistant strains.

2.5. Structure-Based Lead Optimization

The main goal of lead optimization is to start with initial leads, selected from the computational and experimental screening hits, and optimize them to yield strongly binding, highly specific compounds. Once a lead is identified, more accurate computational methods can be employed to optimize the ligand. A structure-based approach focuses on improving binding affinity by suggesting perturbations to the lead compound. One approach is to use some of the rescoring tools such as MMGBSA or LIE methods. Another approach is to use free energy methods that are more general and potentially more accurate, but substantially more computationally expensive.

MMGBSA and LIE are both post processing methods that can be useful in lead optimization. Early in its development MMGBSA involved producing a thermodynamic ensemble of the protein–ligand complex and averaging the interaction energy of all the structures. An alternate approach is to calculate the binding energy for a single protein–ligand structure using the MM-GBSA energy model. Typically, the protein–ligand input to the single-point MMGBSA rescoring is the product of a docking calculation. The distinction between the two approaches is important because producing a thermodynamic ensemble of the protein–ligand complex is computationally intensive. There are two types of studies common in the literature: (1) those that focus on reproducing published data and (2) those that use these methods to discover new inhibitors. While these two methods are well known in academia, their use in pharmaceutical development is relatively new. Publications using MMGBSA or LIE for drug discovery have appeared in the literature only in the last 5–10 years.

De Amorim et al. (108) present an in depth discussion of the methods and applications of the LIE method. A retrospective study, using data for inhibitors of CDK2, Lck and p38, was used to fit parameters for a LIE model with continuum electrostatics. Using this method, Kolb et al. (109) found novel, low micromolar inhibitors of EphB4 and CDK2.

Rastelli et al. (110) performed a retrospective study using known DHFR inhibitors to test the usefulness of MMGBSA and MMPBSA in a virtual screening tool. They found that the methods were able to discriminate between binders and decoys in this diverse set. They found similar results when they performed ensemble averaging over snapshots of an MD trajectory vs. single point calculations on a single protein–ligand complex structure. Bag et al. (111) used MMGBSA estimates of binding free energy to discover novel inhibitors of DHFR. The correlation between the IC_{50}s of the inhibitors and the MMGBSA predictions was ~0.79. Henriksen et al. (112) identified novel bacterial histidine biosynthesis inhibitors using a combined MMGBSA/MMPBSA approach. Most importantly, the authors considered side-chain flexibility by running an MD simulation and averaging MMGBSA interaction energies from trajectory snapshots. They tested the compounds on a whole-cell assay directly without any initial enzymatic assay because of their confidence in the improved predictive power of these approaches. A combination of docking and MMPBSA ranking yielded new inhibitors of dengue virus methyltransferase (113).

Structure based drug design is particularly well suited for optimizing binding affinity, which frequently translates to higher potency. However, the effectiveness of a drug not only depends on potency but on other properties and biological activities, including but not limited to solubility, octanol–water partition coefficient, affinity for molecular pumps, ability to penetrate relevant tissues, lifetime in the blood stream, toxicology, and pharmacokinetic properties. Thus, many have combined multiple methods to maximize effectiveness. To this end, the understanding of the structure–activity relationships (SAR) is very important when applying SBDD methods for drug discovery. The rational design of experiments to generate biological data is a crucial step, providing possible comparisons of predictions and experimental results. The quantitative structure–activity relationships approach (QSAR) is a valuable drug design tool that can offer solutions based on the statistical analysis of relationships between chemical structure and biological activity in a quantitative and mechanism-based manner. Herein, the single-targeting GyrB approach is reported for the novobiocin and cyclothialidine lead compound generation and optimization by using structure-based and SAR-based techniques. Subsequently, the dual-targeting GyrB/ParE approach aimed at minimizing target-based resistance development is exemplified with the work from other groups that identified newer chemotypes beyond coumarins (e.g., novobiocin).

Musicki et al. (114) utilized the X-ray structures of a 24-kDa N-terminal fragment of GyrB to identify a larger hydrophobic region consisting of Val94 and Phe95, where the methyl groups of novobiocin were located. These two amino acids are conserved across 12 Gram-positive strains and in *E. coli*. Using the combined information, they modified the methyl groups of novobiocin to spirocyclopentyl moiety derivatives with lower molecular weights and improved physicochemical properties and pharmacokinetic profiles. The spirocyclopentylnoviose (RU79115) is shown in Fig. 12 with a MIC_{50} (against *S. aureus*) of 0.08 μg/mL, while the MIC_{50} of novobiocin is in the range of 0.3–40 μg/mL.

Angehrn et al. (115) from Hoffman-La Roche used SAR derived from various substitution patterns and X-ray structural analyses to identify the simple hydroxylated benzyl sulfide moiety as the active structural feature in cyclothialidine, a potent gyrase inhibitor. Inhibitors with this moiety demonstrated in vitro activity against Gram-positive bacteria. The best activities were shown by the 14-membered lactone C, which incorporated the benzyl sulfide as well as other features that improved pharmacokinetic properties

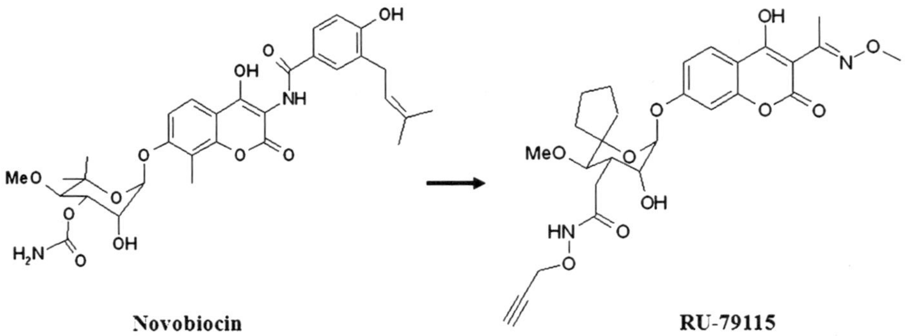

Novobiocin **RU-79115**

Fig. 12. Potent inhibitor (RU-79115) optimized from the natural product novobiocin (114).

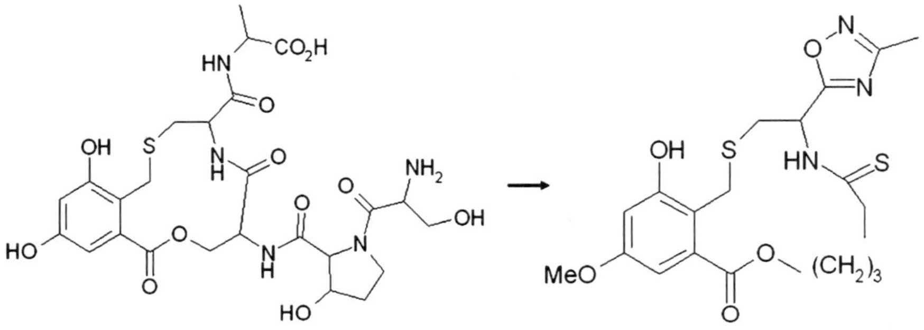

Cyclothialidine **C**

Fig. 13. A design of compound C from structural guidance and SAR (115).

and lipophilicity (see Fig. 13). Lactone C exhibits in vivo efficacy with an ED_{50} of 25 mg/kg against a *S. aureus* septicemia model and overcomes resistance against other marketed antibiotics.

Scientists at Vertex used the information from the structures of ParE complexed with ADPNP and novobiocin (described in Subheading 2.2) to design a series of aminobenzimidazoles (28) such as VRT-125853 and 752586 that demonstrate dual inhibition of GyrB and ParE (see Fig. 14). The K_i (μM) values of VRT-125853 against *E. coli* were 0.015 and 0.68 for GyrB and ParE, respectively. Similarly, the K_i (μM) values for VRT-752586 were <0.004 (GyrB) and 0.023 (ParE). Resistance incidence studies showed very low rates of resistance, consistent with the dual targeting profile of the inhibitors. As with novobiocin and ATP, the aminobenzimidazole core exploits the anchoring H-bonding interactions with the conserved Asp residue (D73) and structural water molecule that is coordinated by D73 and T165.

In a similar vein, Pfizer (116) and Evotec (117) teams reported a series of imidiazolo- and triazolo-pyridines as dual inhibitors of

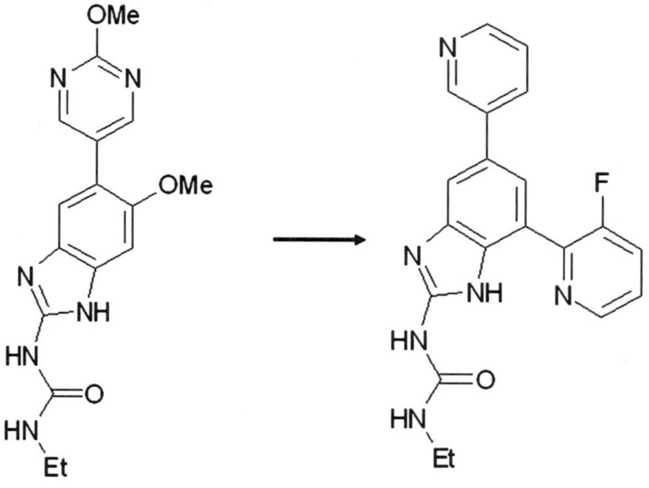

VRT-125853 **VRT-752586**

Fig. 14. Chemical structures of aminobenzimidazoles with dual *E. coli* GyrB/ParE activity.

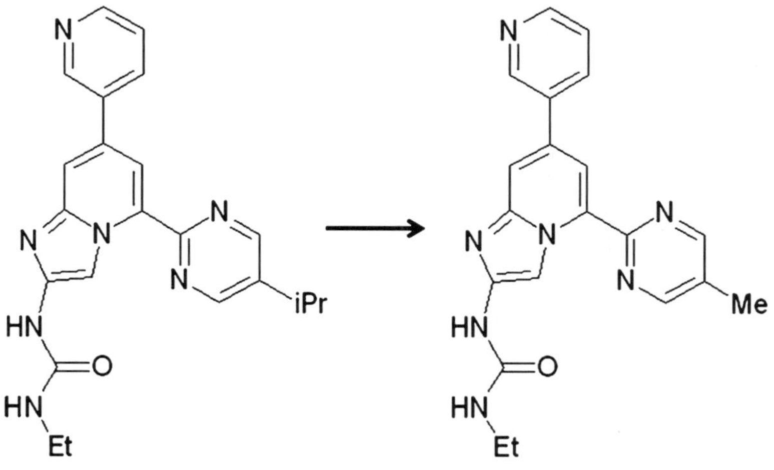

D **G**

Fig. 15. Chemical structures of imidazolo-pyridines reported by the Pfizer team.

bacterial GyrB and ParE (see Figs. 15 and 16). These synthetically challenging pyridines exhibit the same shape and pharmacophore as the aminobenzimidazoles developed by the Vertex group. The two groups reasoned that their 5,6-fused heterocyclic scaffolds could provide different SAR and optimization opportunities with alternative ADME, pharmacology, toxicology, and efficacy profiles. The authors pointed out that achievement of dual-targeting across

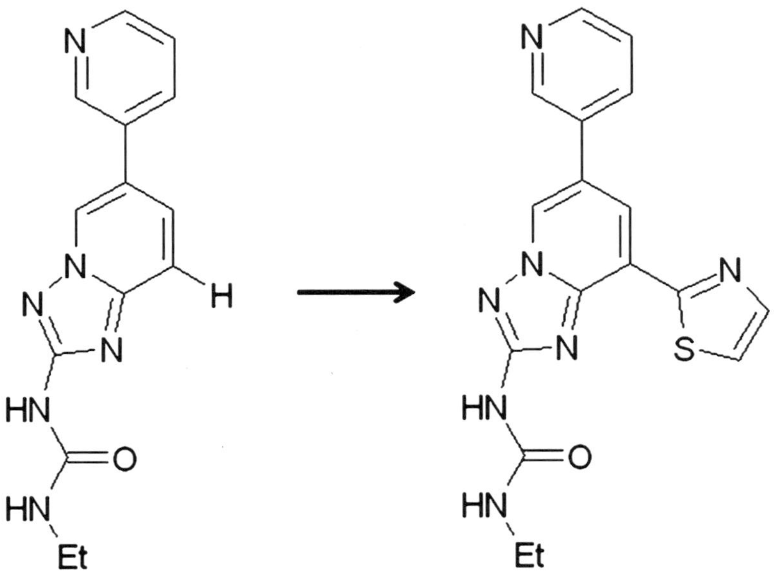

Fig. 16. Chemical structures of triazolo-pyridines reported by the Evotec team.

the series was challenging and only a few isolated molecules were observed with balanced, dual enzyme activity. The IC_{50} (μM) values of imidazolo-pyridines against *Spn* reported by the Pfizer team for compound D were 1.91 (GyrB) and 1.38 (ParE), and those of compound G were 0.117 (GyrB) and 0.147 (ParE). The IC_{50} (μM) values of triazolo-pyridines against *E. coli* reported by the Evotec team were 9.64 (GyrB, no activity measured for ParE) for compound M and 0.042 (GyrB) and 11 (ParE) for compound Z.

In recent years there has been a paradigm shift in antibiotic discovery from the development of single-target single-pharmacophore agents to multitarget single-pharmacophore agents. This trend is exemplified by the efforts directed toward the discovery of novel agents that inhibit the function of bacterial topoisomerases as described in this survey. Undoubtedly, structural biology and computational techniques have helped to attain the promising results.

3. Concluding Remarks

The availability of high-resolution crystal structures of GyrB and ParE and the opportunity for dual target inhibition to minimize the potential for resistance development have catalyzed numerous efforts to discover novel antibacterial agents against these targets. In conjunction with crystal structures, docking-based virtual screening applications have demonstrated effectiveness as a quick and inexpensive alternative to high throughput screening. Of the available methods, some have demonstrated reasonable accuracy and continue to be improved. At the time this chapter was being written, more than 30-million unique compounds were available in commercial libraries in which to search for leads. Invariably, in silico screening will identify numerous potential inhibitor scaffolds beyond the molecules surveyed here. The combination of computational SBDD methods and multiobjective ADMET SAR are valuable and effective drug design tools that identify and optimize lead compounds for tight binding, high specificity, and effectiveness. The chemical space diversity of dual-target topoisomerase inhibitors continues to expand beyond the coumarin and quinolone classes by utilizing these drug development tools.

In summary, the drug discovery process has steadily become more information driven. Thus, using the knowledge of drug resistance gained from the structure–function–activity relationships combined with continuing improvements in computational structure-based methods should help feed drug discovery pipelines with novel inhibitors with the desired properties for development into the next generation of antibiotics.

Acknowledgments

This work was performed under the auspices of the US Department of Energy by Lawrence Livermore National Laboratory under contract DE-AC52-07NA27344. Release number LLNL-JRNL-476952.

References

1. Lax E (2004) *The Mold On Dr. Florey's Coat: The Story of The Penicillin Miracle* (Henry Holt & Co., New York) p 307 pp.
2. Talbot GH, Bradley J, Edwards JJE, Bilbert D, Scheld M, & Barlett JG (2006) Bad Bugs Need Drugs: An Update on the Development Pipeline from the Antimicrobial Availability Task Force of the Infectious Diseases Society of America. *Clin. Infect. Dis.* **42**, 657–668.
3. Aspa J, Rajas O, Rodriguez de Castro F, Blanquer J, Zalacain R, Fenoll A, de Celes R, Vargas A, Salvanes FR, Espana PP, Rello J, & Torres A (2004) Drug-resistant pneumococcal pneumonia: clinical relevance and related factors. *Clin. Infect. Dis.* **38**, 787–798.
4. Barrett CT & Barrett JF (2003) Antibacterials: are the new entries enough to deal with the emerging resistance problems? *Curr. Opin. Biotech.* **14**, 621–626.
5. Jacobs MR (2003) Worldwide trends in antimicrobial resistance among common respiratory tract pathogens. *Pediatr. Infect. Dis. J.* **22**, S109–S119.
6. Livermore DM (2004) The need for new antibiotics. *Clin. Microbiol. Infect.* **10**, (Suppl. 4) 1–9.
7. Pottumarthy S, Fritsche TR, & Jones RN (2005) Comparative activity of oral and parenteral cephalosporins tested against multidrug-resistant Streptococcus pneumoniae:

report from the SENTRY Antimicrobial Surveillance Program (1997–2003). *Diagn. Microbiol. Infect. Dis.* **51**, 147–150.
8. Mutnick AH, Enne V, & Jones RN (2003) Linezolid resistance since 2001: SENTRY Antimicrobial Surveillance Program. *Ann. Pharmacother.* **37**, 769–774.
9. Schwartz BS, Ngo PD, & Guglielmo BJ (2008) Daptomycin treatment failure for vancomycin-resistant Enterococcus faecium infective endocarditis: impact of protein binding? *Ann. Pharmacother.* **42**, 289–290.
10. Cassell GH & Mekalanos J (2001) Development of antimicrobial agents in the era of new and reemerging infectious diseases and increasing antibiotic resistance. *J. Am. Med. Assoc.* **285**, 601–605.
11. Spratt BG (1994) Resistance to antibiotics mediated by target alterations. *Science* **264**, 388–393.
12. Ng EY, Trucksis M, & Hooper DC (1996) Quinolone resistance mutations in topoisomerase IV: relationship to the flqA locus and genetic evidence that topoisomerase IV is the primary target and DNA gyrase is the secondary target of fluoroquinolones in Staphylococcus aureus. *Antimicrob. Agents Chemother.* **40**, 1881–1888.
13. Chopra I (1998) Protein synthesis as a target for antibacterial drugs: current status and future opportunities. *Expert Opin. Investig. Drugs* 7, 1237–1244.
14. Silver L (2007) Multi-targeting by monotherapeutic antibacterials. *Nat. Rev. Drug Dis.* **6**, 41–55 and references cited therein.
15. Strahilevitz J & Hooper DC (2005) Dual targeting of topoisomerase IV and gyrase to reduce mutant selection: direct testing of the paradigm by using WCK-1734, a new fluoroquinolone, and ciprofloxacin. *Antimicrob. Agents Chemother.* **49**, 1949–1956.
16. Mills SD (2003) The role of genomics in antimicrobial discovery. *J. Antimicrob. Chemother.* **51**, 749–752.
17. Schneider G & Bohm HJ (2002) Virtual screening and fast automated docking methods. *Drug Discov. Today* 7, 64–70.
18. Bailey D & Brown D (2001) High-throughput chemistry and structure-based design: survival of the smartest. *Drug Discov. Today* **6**, 57–59.
19. Maxwell A (1997) DNA gyrase as s drug target. *Trends Microbiol.* **5**, 102–109.
20. Drlica K & Zhao XL (1997) DNA gyrase, topoisomerase IV, and the 4-quinolones. *Microbiol. Mol. Biol. Rev.* **61**, 377–392.
21. Wang JC (1996) DNA topoisomerases. *Annu. Rev. Biochem.* **65**, 635–692.
22. Hiasa H, Yousef DO, & Marians KJ (1996) DNA strand cleavage is required for replication fork arrest by a frozen topoisomerase-quinolone-DNA ternary complex. *J. Biol. Chem.* **271**, 26424–26429.
23. Brino L, Urzhumtsev A, Mousli M, Bronner C, Mitschler A, Oudet P, & Moras D (2000) Dimerization of Escherichia coli DNA-gyrase B provides a structural mechanism for activating the ATPase catalytic center. *J. Biol. Chem.* **275**, 9468–9475.
24. Hooper DC (1999) Mechanisms of quinolone resistance. *Drug Resistance Updates* **2**, 38–55.
25. Bradbury BJ & Pucci MJ (2008) Recent advances in bacterial topoisomerase inhibitors. *Current Opinion in Pharmacology* **8**, 574–581.
26. Boehm H-J, Boehringer M, Bur D, Gmuender H, Huber W, Klaus W, Kostrewa D, Kuehne H, Luebbers T, Meunier-Keller N, & Mueller F (2000) Novel inhibitors of DNA Gyrase: 3D structure based biased needle screening, hit validation by biophysical methods, and 3D guided optimization. A promising alternative to random screening. *J. Med. Chem.* **43**, 2664–2674.
27. Bellon S, D. PJ, Wei Y, Hayakawa K, Swenson LL, Charifson PS, Lippke JA, Aldape R, & Gross CH (2004) Crystal Structures of Escherichia coli Topoisomerase IV ParE Subunit (24 and 43 Kilodaltons): a Single Residue Dictates Differences in Novobiocin Potency against Topoisomerase IV DNA Gyrase. *Antimicrobial Agents and Chemotherapy* **48**, 1856–1864 and the references cited therein.
28. Charifson PS, Grillot A, Grossman TH, Parsons JD, Badia M, Bellon S, Deininger DD, Drumm JE, Gross CH, Letiran A, Liao Y, Mani N, Nicolau DP, Perola E, Ronkin S, Shannon D, Swenson LL, Tang Q, Tessier PR, Tian S, Trudeau M, Wang T, Wei Y, Zhang H, & Stamos D (2008) Novel Dual-targeting Benzimidazole Urea Inhibitors of DNA Gyrase and Topoisomerase IV Possessing Potent Antibacterial Activity: Intelligent Design and Evolution through the Judicious use of Structure-guided Design and Structure-activity Relationships. *J. Med. Chem.* **51**, 5243–5263.
29. Schechner M, Sirockin F, Stote RH, & Dejaegere AP (2004) Functionality Maps of the ATP Binding Site of DNA Gyrase B: Generation of a Consensus Model of Ligand Binding. *J. Med. Chem.* **47**, 4373–4390.
30. Miranker A & Karplus M (1991) Functionality maps of binding sites: a multiple copy simultaneous search method. *Prot.: Struct. Funct. and Genet.* **11**, 29–34.

31. McRobb FM, Capuano B, Crosby IT, Chalmers DK, & Yuriev E (2010) Homology Modeling and Docking Evaluation of Aminergic G Protein-Coupled Receptors. *Journal of Chemical Information and Modeling* **50**(4), 626–637.
32. Oshiro C, Bradley EK, Eksterowicz J, Evensen E, Lamb ML, Lanctot JK, Putta S, Stanton R, & Grootenhuis PDJ (2004) Performance of 3D-Database Molecular Docking Studies into Homology Models. *Journal of Medicinal Chemistry* **47**(3), 764–767.
33. Ferrara P & Jacoby E (2007) Evaluation of the utility of homology models in high throughput docking. *Journal of Molecular Modeling* **13**(8), 897–905.
34. Zhou Z, Felts A, Friesner R, & Levy R (2007) Comparative Performance of Several Flexible Docking Programs and Scoring Functions: Enrichment Studies for a Diverse Set of Pharmaceutically Relevant Targets. *Journal of Chemical Information and Modeling* **47**(4), 1599–1608.
35. Friesner RA, Banks JL, Murphy RB, Halgren TA, Klicic JJ, Mainz DT, Repasky MP, Knoll EH, Shelley M, Perry JK, Shaw DE, Francis P, & Shenkin PS (2004) Glide: A New Approach for Rapid, Accurate Docking and Scoring. 1. Method and Assessment of Docking Accuracy. *Journal of Medicinal Chemistry* **47**(7), 1739–1749.
36. Moustakas D, Lang P, Pegg S, Pettersen E, Kuntz I, Brooijmans N, & Rizzo R (2006) Development and validation of a modular, extensible docking program: DOCK 5. *Journal of Computer-Aided Molecular Design* **20**(10), 601–619.
37. Jones G, Willett P, Glen RC, Leach AR, & Taylor R (1997) Development and Validation of a Generic Algorithm for Flexible Docking. *J. Mol. Biol.* **267**, 727–748.
38. Guimarães CRW & Cardozo M (2008) MM-GB/SA Rescoring of Docking Poses in Structure-Based Lead Optimization. *Journal of Chemical Information and Modeling* **48**(5), 958–970.
39. Hansson T, Marelius J, & Åqvist J (1998) Ligand binding affinity prediction by linear interaction energy methods. *Journal of Computer-Aided Molecular Design* **12**(1), 27–35.
40. Osborne MJ, Schnell J, Benkovic SJ, Dyson HJ, & Wright PE (2001) Backbone Dynamics in Dihydrofolate Reductase Complexes: Role of Loop Flexibility in the Catalytic Mechanism†. *Biochemistry* **40**(33), 9846–9859.
41. Mancera R (2007) Molecular modeling of hydration in drug design. *Current Opinion in Drug Discovery and Development* **10**, 275–280.
42. Wong SE & Lightstone FC (2011) Accounting for water molecules in drug design. *Expert Opinion on Drug Discovery* **6**(1), 65–74.
43. Huang N & Shoichet BK (2008) Exploiting Ordered Waters in Molecular Docking. *Journal of Medicinal Chemistry* **51**(16), 4862–4865.
44. Minke WE, Diller DJ, Hol WGJ, & Verlinde CLMJ (1999) The Role of Waters in Docking Strategies with Incremental Flexibility for Carbohydrate Derivatives: Heat-Labile Enterotoxin, a Multivalent Test Case. *Journal of Medicinal Chemistry* **42**(10), 1778–1788.
45. Michel J, Tirado-Rives J, & Jorgensen WL (2009) Energetics of Displacing Water Molecules from Protein Binding Sites: Consequences for Ligand Optimization. *Journal of the American Chemical Society* **131**(42), 15403–15411.
46. Rarey M, Kramer B, Lengauer T, & Klebe G (1996) A Fast Flexible docking method using an incremental construction algorithm. *J. Mol. Biol.* **261**, 470–489.
47. Morris GM, Goodsell DS, Halliday RS, Huey R, Hart WE, Belew RK, & Olson AJ (1998) Automated docking using a Lamarckian genetic algorithm and an empirical binding free energy function. *Journal of Computational Chemistry* **19**(14), 1639–1662.
48. Jones G, Willett P, & Glen RC (1995) Molecular recognition of receptor sites using a genetic algorithm with a description of desolvation. *Journal of Molecular Biology* **245**(1), 43–53.
49. Corbeil CR & Moitessier N (2009) Docking Ligands into Flexible and Solvated Macromolecules. 3. Impact of Input Ligand Conformation, Protein Flexibility, and Water molecules on the Accuracy of Docking Programs. *J. Chem. Inf. Model.* **49**, 997–1009.
50. Liu M & Wang S (1999) MCDOCK: A Monte Carlo simulation approach to the molecular docking problem. *Journal of Computer-Aided Molecular Design* **13**(5), 435–451.
51. Accelrys Discovery Studio Suites, Accelrys, Inc, 10188 Telesis Court, Suite 100 San Diego, CA 92121, USA.
52. McMartin C & Bohacek RS (1997) QXP, Powerful, rapid computer algorithms for structure-based drug design. *J. Comput. Aided. Mol. Design* **11**, 333–344.
53. Miller MD, Kearsley SK, Underwood DJ, & Sheridan RP (1994) FLOG: A system to select 'quasi-flexible' ligands complementary to a receptor of known three-dimensional structure.

Journal of Computer-Aided Molecular Design **8**(2), 153–174.

54. FRED: Fast Rigid Exhaustive Docking, Version 2.2.5, OpenEye Scientific Software, Inc., Santa Fe, NM, USA.
55. Abagyan RT, & Kuznetsov, M. R., (1994) A new method for protein modeling and design: Applications to docking and structure prediction from the distorted native conformation. *J. Comput. Chem.* **15**, 488–506.
56. Blaney JM & Dixon JS, DockIt, Metaphorics, LLC, Mission Viejo, CA.
57. Hart TN & Read RJ (1992) A multiple-start Monte Carlo docking method. *Proteins* **13**, 206–222.
58. Hart TN, Ness, R. S., & Read, R. J., (1997) Critical evaluation of the research docking program for the CASP2 challenge. *Proteins,* Suppl **1**, 205–209.
59. Thomsen R & Christensen MH (2006) MolDock: A New Technique for High-Accuracy Molecular Docking. *J. Med. Chem.* **49**(11), 3315–3321.
60. Jain AN (2003) Surflex: Fully automatic flexible molecular docking using a molecular similarity-based search engine. *J. Med. Chem.* **46**, 499–511.
61. MOE: Molecular Operating Environment, Chemical Computing Group Inc., 1010 Sherbrooke, Street West, Suite 910, Montreal, Quebec, Canada.
62. Brooijmans N & Kuntz ID (2003) Molecular Recognition and Docking Algorithms. *Annual Review of Biophysics and Biomolecular Structure* **32**(1), 335–373.
63. Gilson M, GIven J, Bush B, & McCammon JA (1997) The statistical-thermodynamic basis for computation of binding affinities: a critical review. *Biophysical Journal* 72(3),1047–1069.
64. Wang J, Deng Y, & Roux B (2006) Absolute Binding Free Energy Calculations Using Molecular Dynamics Simulations with Restraining Potentials. **91**(8), 2798–2814.
65. Mobley DL, Graves AP, Chodera JD, McReynolds AC, Shoichet BK, & Dill KA (2007) Predicting Absolute Ligand Binding Free Energies to a Simple Model Site. *Journal of Molecular Biology* **371**(4), 1118–1134.
66. Morris GM, Goodsell DS, Huey R, & Olson AJ (1996) Distributed automated docking of flexible ligands to proteins: Parallel applications of AutoDock 2.4. *Journal of Computer-Aided Molecular Design* **10**(4), 293–304.
67. Muegge I (2000) A knowledge-based scoring function for protein-ligand interactions: Probing the reference state. *Perspectives in Drug Discovery and Design*, **20**, 99–114.
68. Gohlke H, Hendlich M, & Klebe G (2000) Knowledge-based Scoring Function to Predict Protein-Ligand Interactions. *Journal of Molecular Biology* **295**, 337–356.
69. Eldridge M, Murray C, Auton T, Paolini G, & Mee R (1997) Empirical scoring functions: I. The development of a fast empirical scoring function to estimate the binding affinity of ligands in receptor complexes. *Journal of Computer-Aided Molecular Design* **11**(5), 425–445.
70. Korb O, Stützle T, & Exner TE (2009) Empirical Scoring Functions for Advanced Protein–Ligand Docking with PLANTS. *Journal of Chemical Information and Modeling* **49**(1), 84–96.
71. Tirado-Rives J & Jorgensen WL (2006) Contribution of conformer focusing to the uncertainty in predicting free energies for protein-ligand binding. *J. Med. Chem.* **49**, 580–5884.
72. Bursulaya BD, Totrov M, Abagyan R, & Brooks III CL (2003) Comparative study of several algorithms for flexible ligand docking. *J. Comput. Aided. Mol. Design* **17**, 755–763.
73. Kellenberger E, Rodrigo J, Muller P, & Rognan D (2004) Comparative evaluation of eight docking tools for docking and virtual screening accuracy. *Proteins* **57**, 225–242.
74. Klon AE, Glick M, Thoma M, Acklin P, & Davies JW (2004) Finding more needles in the haystack: a simple and efficient method for improving high-throughput docking results. *J. Med. Chem.* **47**, 2743–2749.
75. Perola E, Walters WP, & Charifson PS (2004) A detailed comparison of current docking and scoring methods on systems of pharmaceutical relevance. *Proteins* **56**, 235–249.
76. Muege I & Enyedy I (2004) Virtual screening for kinase targets. *J. Curr. Med. Chem.* **11**, 693–707.
77. Cummings MD, DesJarlais RL, Gibbs AC, Mohan V, & Jaeger EP (2005) Comparison of automated docking programs as virtual screening tools. *J. Med. Chem.* **48**, 962–976.
78. Kontoyianni M, Sokol GS, & McClellan LM (2005) Evaluation of library ranking efficacy in virtual screening. *J. Comput. Chem.* **26**, 11–22.
79. Chen H, Lyne PD, Giordanetto F, Lovell T, & Li J (2006) On evaluating molecular docking methods for pose prediction and enrichment factors. *J. Chem. Inf. Model.* **46**, 401–415.
80. Warren GL, Andrews CW, Capelli A-M, Clarke B, LaLonde J, Lambert MH, Lindvall M, Nevins N, Semus SF, Senger S, Tedesco G,

Wall I, D., Woolven JM, Peishoff CE, & Head MS (2006) A critical assessment of docking programs and scoring functions. *J. Med. Chem.* **49**, 5912–5931.

81. Schulz-Gasch T & Stahl M (2003) Binding site characteristics in structure-based virtual screening: Evaluation of current docking tools. *J. Mol. Model* **9**, 47–57.
82. Wang R, Lu Y, Fang X, & Wang S (2004) An Extensive Test of 14 Scoring Functions Using the PDBbind Refined Set of 800 Protein–Ligand Complexes. *Journal of Chemical Information and Computer Sciences* **44**(6), 2114–2125.
83. Wang R, Lu Y, & Wang S (2003) Comparative Evaluation of 11 Scoring Functions for Molecular Docking. *Journal of Medicinal Chemistry* **46**(12), 2287–2303.
84. Clark RD, Strizhev A, Leonard JM, Blake JF, & Matthew JB (2002) Consensus scoring for ligand/protein interactions. *Journal of Molecular Graphics and Modelling* **20**(4), 281–295.
85. Gohlke H & Klebe G (2001) Statistical potentials and scoring functions applied to protein-ligand binding. *Current Opinion in Structural Biology* **11**(2), 231–235.
86. Charifson PS, Corkery JJ, Murcko MA, & Walters WP (1999) Consensus Scoring: A Method for Obtaining Improved Hit Rates from Docking Databases of Three-Dimensional Structures into Proteins. *Journal of Medicinal Chemistry* **42**(25), 5100–5109.
87. Graves AP, Shivakumar DM, Boyce SE, Jacobson MP, Case DA, & Shoichet BK (2008) Rescoring Docking Hit Lists for Model Cavity Sites: Predictions and Experimental Testing. *Journal of Molecular Biology* **377**(3), 914–934.
88. Zhong S, Zhang Y, & Xiu Z (2010) Rescoring ligand docking poses. *Current Opinion in Drug Discovery and Development* **13**, 326–334.
89. Lyne PD, Lamb ML, & Saeh JC (2006) Accurate Prediction of the Relative Potencies of Members of a Series of Kinase Inhibitors Using Molecular Docking and MM-GBSA Scoring. *Journal of Medicinal Chemistry* **49**(16), 4805–4808.
90. Srivastava M, Singh H, & Naik PK (2010) Molecular Modeling Evaluation of the Antimalarial Activity of Artemisinin Analogues: Molecular Docking and Rescoring using Prime/MM-GBSA Approach. *Current Research Journal of Biological Sciences* **2**(2), 83–102.
91. Cole JC, Murray CW, Nissink JW, Taylor RD, & Taylor R (2005) Comparing protein-ligand docking programs is difficult. *Proteins* **60**, 325–332.
92. Noble M, Barrett P, Endicott J, Johnson L, McDonnell J, Robertson G, & Zawaira A (2005) Exploiting structural principles to design cyclin-dependent kinase inhibitors. *Biochimica et Biophysica Acta (BBA) - Proteins & Proteomics* **1754**(1–2), 58–64.
93. Cavasotto CN & Abagyan RA (2004) Protein Flexibility in Ligand Docking and Virtual Screening to Protein Kinases. *Journal of Molecular Biology* **337**(1), 209–225.
94. Bowman AL, Lerner MG, & Carlson HA (2007) Protein Flexibility and Species Specificity in Structure-Based Drug Discovery: Dihydrofolate Reductase as a Test System. *Journal of the American Chemical Society* **129**(12), 3634–3640.
95. Then RL (2004) Antimicrobial Dihydrofolate Reductase Inhibitors - Achievements and Future Options: Review. *Journal of Chemotherapy* **16**, 3–12.
96. Klebe G, Krämer O, & Sotriffer C (2004) Strategies for the design of inhibitors of aldose reductase, an enzyme showing pronounced induced-fit adaptations. *Cellular and Molecular Life Sciences* **61**(7), 783–793.
97. Sotriffer CA, Krämer O, & Klebe G (2004) Probing flexibility and "induced-fit" phenomena in aldose reductase by comparative crystal structure analysis and molecular dynamics simulations. *Proteins: Structure, Function, and Bioinformatics* **56**(1), 52–66.
98. Babbitt PC, Hasson MS, Wedekind JE, Palmer DRJ, Barrett WC, Reed GH, Rayment I, Ringe D, Kenyon GL, & Gerlt JA (1996) The Enolase Superfamily: A General Strategy for Enzyme-Catalyzed Abstraction of the Œ-Protons of Carboxylic Acids, *Biochemistry* **35**(51), 16489–16501.
99. Fernández-Recio J, Totrov M, & Abagyan R (2002) Soft protein–protein docking in internal coordinates. *Protein Science* **11**(2), 280–291.
100. Wong CF, Kua J, Zhang Y, Straatsma TP, & McCammon JA (2005) Molecular docking of balanol to dynamics snapshots of protein kinase A. *Proteins: Structure, Function, and Bioinformatics* **61**(4), 850–858.
101. Alonso H, Bliznyuk AA, & Gready JE (2006) Combining Docking and Molecular Dynamics Simulations in Drug Design. *Medicinal Research Reviews* **26**(5), 531–568.
102. Wong S & Jacobson MP (2008) Conformational selection in silico: Loop latching motions and ligand binding in enzymes. *Proteins: Structure, Function, and Bioinformatics* **71**(1), 153–164.

103. Sandak B, Wolfson HJ, & Nussinov R (1998) Flexible Docking Allowing Induced Fit in Proteins: Insights From an Open to Closed Conformational Isomers. *Proteins: Structure, Function, and Bioinformatics* **32**, 159–174.
104. Cavasotto CN, Kovacs JA, & Abagyan RA (2005) Representing Receptor Flexibility in Ligand Docking through Relevant Normal Modes. *Journal of the American Chemical Society* **127**(26), 9632–9640.
105. Oblak M, Gdadolnik SG, Kotnik M, Jerala R, Filipic M, & Solmajer T (2005) In silico fragment-based discovery of indolin-2-one analogues as potent DNA gyrase inhibitors. *Bioorganic & Medicinal Chemistry Letters* **15**, 5207–5210.
106. Dutta R & Inouye M (2000) GHKL, an emergent ATPase/kinase superfamily. *Trends Biochem. Sci.* **25**, 24–28.
107. Ostrov DA, Prada JAH, Corsino PE, Finton KA, Le N, & Rowe TC (2007) Discovery of novel DNA Gyrase inhibitors by high-throughput virtual screening. *Antimicrobial Agents and Chemotherapy* **51**, 3688–3698.
108. De Amorim HLN, Caceres RA, & Netz PA (2008) Linear Interaction Energy (LIE) Method in Lead Discovery and Optimization. *Current Drug Targets* **9**, 1100–1105.
109. Kolb P, Huang D, Dey F, & Caflisch A (2008) Discovery of Kinase Inhibitors by High-Throughput Docking and Scoring Based on a Transferable Linear Interaction Energy Model. *Journal of Medicinal Chemistry* **51**(5), 1179–1188.
110. Rastelli G, Rio AD, Degliesposti G, & Sgobba M (2010) Fast and accurate predictions of binding free energies using MM-PBSA and MM-GBSA. *Journal of Computational Chemistry* **31**(4), 797–810.
111. Bag S, Tawari NR, Degani MS, & Queener SF (2010) Design, synthesis, biological evaluation and computational investigation of novel inhibitors of dihydrofolate reductase of opportunistic pathogens. *Bioorganic & Medicinal Chemistry* **18**(9), 3187–3197.
112. Henriksen ST, Liu J, Estiu G, Oltvai ZN, & Wiest O (2010) Identification of novel bacterial histidine biosynthesis inhibitors using docking, ensemble rescoring, and whole-cell assays. *Bioorganic & Medicinal Chemistry* **18**(14), 5148–5156.
113. Podvinec M, Lim SP, Schmidt T, Scarsi M, Wen D, Sonntag L-S, Sanschagrin P, Shenkin PS, & Schwede T (2010) Novel Inhibitors of Dengue Virus Methyltransferase: Discovery by in Vitro-Driven Virtual Screening on a Desktop Computer Grid. *Journal of Medicinal Chemistry* **53**(4), 1483–1495.
114. Musicki B, Periers A, Laurin P, Ferroud D, Benedetti Y, Lachaud S, Chatreaux F, Haesslein J, Iltis A, Pierre C, Khider J, Tessot N, Airault M, Demassey J, Dupuis-Hamelin C, Lassaigne P, Bonnefoy A, Vicat P, & Klich M (2000) Improved antibacterial activities of coumarin antibiotics bearing 5', 5'-dialkylnoviose: Biological activity of RU79115. *Biorg. Med. Chem. Lett.* **10**, 1695–1699.
115. Angehrn P, Buchmann S, Funk C, Goetschi E, Gmuender H, Hebeisen P, Kostrewa D, Link H, Luebbers T, Masciadri R, Nielsen J, Reindl P, Ricklin F, Schmitt-Hoffmann A, & Theil FP (2004) New antibacterial agents derived from the DNA gyrase inhibitor cyclothialidine. *J. Med. Chem.* **47**, 1487–1513.
116. Starr JT, Sciotti RJ, Hanna DL, Huband M, D., Mullins LM, Cai H, Gage JW, Lockard M, Rauckhorst MR, Owen RM, Lall MS, Tomilo M, Chen H, McCurdy SP, & Barbachyn MR (2009) 5-(2-Pyrimidinyl)-imidazol[1,2-a] pyridines are antibacterial agents targeting the ATPase domains of DNA gyrase and topoisomerase IV. *Biorg. Med. Chem. Lett.* **19**, 5302–5306.
117. East SP, White CB, Barker O, Barker S, Bennett J, Brown D, Boyd EA, Brennan C, Chowdhury C, Collins I, Convers-Reignier E, Dymock BW, Fletcher R, Haydon DJ, Gardiner M, Hatcher S, Ingram P, Lancett P, Mortenson P, Papadopuolos K, Smee C, Thomaides-Brears HB, Tye H, Workman, J., & Czaplewski LG (2009) DNA gyrase (GyrB)/topoisomerase IV (pare) inhibitors: Synthesis and antibacterial activity. *Biorg. Med. Chem. Lett.* **19**, 894–899.

Chapter 10

Chemical Informatics: Using Molecular Shape Descriptors in Structure-Based Drug Design

Andy Jennings

Abstract

The concept of molecular shape has been considered in various forms in the context of drug design. The following chapter details the application of molecular shape to the design of compound libraries for assessment of potential biological activity. Whilst the utility of shape descriptors is well documented in the area of ligand similarity, the use of shape descriptors is equally applicable to protein structures. Indeed, work has been published using various descriptors to compare proteins but little published where protein shape descriptors have been used to investigate ligand selectivity.

Key words: Molecular Shape, Structure-based drug design, Virtual Screening, Combinatorial design, Computational Chemistry, Bioisostere.

1. Introduction

1.1. Shape as a Concept

Shape is an important attribute of biologically active small molecules. When referring to shape in the design of biologically active small molecules it is customary to consider the "best fit" between such a molecule and its biological target. Emil Fischer was the most notable, if not the first, to suggest that both the biological target (enzyme) and the natural substrate possess specific and complementary geometric shapes that fit exactly into one another (1). This is often referred to as "the lock and key" model. This concept of shape complementarity underlies the vast majority of drug design, whether it is to improve potency against the target of interest or to reduce the biological activity toward undesirable countertargets. Here, we shall consider the design of chemical arrays or libraries with an emphasis on shape, although electrostatics are often also key to a successful design.

In the field of medicinal chemistry, "bioisosteres" are substituents or groups with similar physicochemical properties and that

Leslie W. Tari (ed.), *Structure-Based Drug Discovery*, Methods in Molecular Biology, vol. 841,
DOI 10.1007/978-1-61779-520-6_10, © Springer Science+Business Media, LLC 2012

may have broadly similar binding to a biological target (2–6). Bioisosteres are frequently sought in order to provide different *A*bsorption, *D*istribution, *M*etabolism, or *E*xcretion (ADME) properties (7–14). If the most important attribute of the group to be replaced is electrostatic and not steric, the bioisostere only need be similar electrostatically rather than having to mimic all physicochemical properties simultaneously. This is an important distinction, as it is often mistakenly believed that a bioisostere is required to be similar across all chemical properties. A classic example is the substitution of a fluorine atom for a metabolically labile/sensitive hydrogen atom to potentially prevent metabolism at a metabolically labile site. The fluorine and hydrogen atoms are similar in size but often possess different metabolic properties, so a longer biological half-life may to be expected.

1.2. The Rationale for Using Shape as a Design Tool

As previously noted, shape complementarity is a prerequisite for binding between a ligand and protein. Although electrostatic attraction can be significant, and operate over large distances, steric repulsion can overwhelm any electrostatic attraction. Whereas electrostatic attraction reaches a maximum when the two point charges approach, the competing effect is steric repulsion with particles of radius R. The net effect is that any two charged particles of radius R will experience an increased attraction as they approach one another until their surfaces meet, at which point steric repulsion rapidly equals the electrostatic attraction.

1.3. Computationally Designed Compound Libraries

The term "compound library" has been used to describe a set of compounds designed around a common scaffold or core. Historically, a compound library was synthesized and tested as a combinatorial library mixture (so-called mixed-pool synthesis). These combinatorial library mixtures required either subsequent deconvolution based upon synthesis of sub-libraries, or decoding of active component by incorporation of chemical and/or photocleavable tags. Alternatively, the same compounds could be synthesized as discretes, and this approach has often been referred to as high throughput organic synthesis (HTOS) or parallel synthesis, to distinguish from the historical concept of combinatorial library.

At present, the term "compound library" encompasses both approaches, but is generally understood to mean those libraries derived from parallel synthesis, particularly as the interest in the mixed-pool concept has waned. Another use of the term is to describe a set of compounds that span multiple chemotypes (so-called targeted- or gene-family libraries). These libraries are often targeted toward a specific biological protein class, such as kinases, and represent specific pharmacophore features known to be required for binding (e.g., the hydrogen bond donor/acceptor pattern present in the hinge binding regions present in the protein kinase family). Finally, the term compound library can also be used to describe a general set of compounds (so-called "universal libraries") for use in high-throughput screening and can number many hundreds of thousands or even millions of compounds, and may span a range of drug-like (15) or lead-like (16) properties, independent of target class.

There are three main purposes for carrying out computational (or virtual) compound library-design:

1. To expand/investigate the SAR (Structure–Activity Relationship) of a single chemotype. For example, one might begin with the core structure of a bioactive molecule and design a library to probe the physicochemical requirements at each position.
2. To target a specific protein class. With the increase in availability of protein 3D structure information, it is often of value to incorporate this information when designing a library. The protein structure provides guidance as to which positions of a chemotype are likely to yield the most informative molecules. Alternatively, the protein can guide the selection of a set of likely bioactive molecules from across a number of chemotypes. This latter definition is often the driver behind commercial or in-house libraries to probe a specific target class (e.g., a kinase set, a GPCR set, etc.).
3. A diverse or so-called universal compound library. By comparing the shapes of molecules, a library of compounds can be constructed that is maximally dissimilar to others in the set. This is useful for a general probe library (e.g., an X-ray co-complex fragment library or a hit-generation library).

1.4. The Concept of Similarity

There are many molecular representations and similarity metrics used to assist in the medicinal chemist in drug discovery, and the concept of shape is one of the most important. Specifically, it is not simply considering the volume of a molecule but how the molecule displays its atoms in a complex 3D representation. The concept of shape similarity is the key driver used to compare the shape of molecules, yet the quality of the results is as subjective as the similarity measure chosen.

Shape itself has been described in various ways in the field of computational chemistry. For example, pharmacophores, may embrace the notion of shape if the pharmacophore places geometric constraints on how molecular features must be displayed.

Presented within are strategies for how shape may be used in the library-design process, both in general terms and for the purpose of probing a specific protein target.

Similarity is a subjective assessment, and one can only say with certainty that two compounds are identical (100% similar). For example, pyridine and benzene are relatively similar in shape but much less similar in terms of electrostatics. Therefore, care must be taken when assessing similarity; a specific group can be replaced by

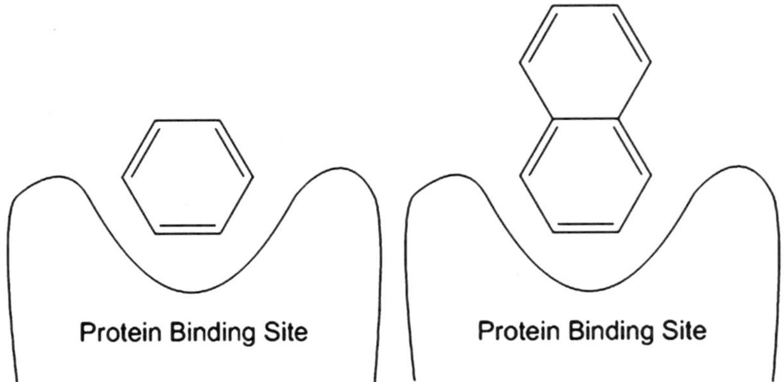

Fig. 1. Schematic showing how benzene and naphthalene may be considered identical in the environment of the binding site of a protein.

one that is similar in one property, while another property may greatly vary. In contrast, naphthalene and benzene would not be considered as similar in a global sense, but may be viewed as very similar when comparing atoms of each molecule that are involved in ligand binding to the protein of interest (see Fig. 1).

A key consideration is what context and scope is being considered when shape features are being used as a design tool or filter, and that our similarity assessment should be appropriate to the task. In the case of benzene vs. naphthalene, the whole molecule shape similarity between the two molecules is of less utility than the shape similarity of the portion of each molecule that is interacting with the ligand-binding site.

2. Materials

2.1. OpenEye Software

2.1.1. Shape Toolkit (TK)

The Shape TK (17) facilitates the calculation of molecular descriptors for shape (steric multipoles), volume overlap between molecules, and spatial similarity of chemical groups (color force field), as well as the optimization of the latter two quantities.

2.1.2. Rapid Overlay of Chemical Structures

The Rapid Overlay of Chemical Structures (ROCS) (18) application provides a fast and accurate (19) method for superimposing molecules, and is built on top of the Shape toolkit. It has proven useful in drug discovery campaigns (20).

2.1.3. OEChem Toolkit (TK)

The OEChem TK is a fast and flexible programming library for chemistry and cheminformatics. OEChem TK is available in C++ and is wrapped for Python and Java. OEChem TK is required to access all other OpenEye toolkits and provides molecule handling in all the OpenEye applications.

2.2. R Software

In order to carry out subsequent analysis, clustering, statistics, and results visualization, one may make use of the R package. R is a programming language and software environment for statistical computing and graphics. It is an implementation of the S programming language with lexical scoping semantics inspired by Scheme. The R language has become the standard (21, 22) among statisticians for the development of statistical software, and is widely used for statistical software development and data analysis (22). R is part of the GNU project (23).

3. Methods

3.1. Overview of Shape-Based Methods

A review of all shape-based methods is beyond the scope of this chapter, but three of the more common methods are worthy of brief discussion.

3.1.1. Hard-Sphere Models

Traditionally, molecules are constructed using space-filling models, the atoms described as fused spheres. This simplistic view provides a powerful visual aid but presents significant mathematical problems when calculating the volume of a molecule. The difficulties arise from calculating the volume of the fused spheres without counting any overlap more than once. The complexity of the calculations as one gets increasing number of atoms can make the volume calculation prohibitively expensive, from a computational point of view.

Pharmacophore models can also be considered a hard-sphere model, as they usually employ spheres to describe molecular features. Pharmacophore models do suffer from the same issues of calculating overlaps as the space-filling models, although don't become so prohibitive due to the fewer number of spheres to overlap. However, the reduction in complexity is also what limits the value and utility of pharmacophore methods when comparing molecular shapes. That is, to describe any molecule as accurately as possible, and hence to obtain an accurate shape similarity, one would require one sphere per atom. Consequently, this would result in an identical problem with space-filling methods, the computationally intensive and prohibitive cost of calculating numerous intersecting spheres.

There have been numerous publications that employ so-called "Alpha shapes" (24) to describe the shape of molecules (25). Alpha shapes are generalizations of the convex hull, where a parameter "alpha" allows description of shape to varying levels of detail, for a given set of points "*S*." For sufficiently large alpha, the alpha shape is identical to the convex hull of the shape one is trying to model. When *S* are points located at the centers of the atoms of a molecule, and alpha is equal to the atomic radii, the alpha shape is the geometric dual of the space-filling model of that molecule.

3.1.2. Grid-Based Models

Another approach is to describe each shape/volume by a grid and then compare these molecular grids. Typically, a volumetric grid enclosing the atoms is defined, and the interior grid points are determined (26). The geometric center of the interior points is found and the principal axes computed. The center and the maximum and minimum extent lengths in three dimensions are used as a gross way to describe the overall shape and size of a molecule. The query and reference structures are aligned based upon their principal axis, and the grid volumes of both are compared to determine shape similarity.

This approach can also lead to computationally expensive calculations unless coarse grids are employed. However, these coarse grids produce very rough assessments of shape similarity and will often miss subtle features.

3.1.3. Continuous Function Models

The concept of using a sum of continuous functions (i.e., a Gaussian) is another alternative to describe molecular shape. In this method, overlapping atom-centered Gaussians take the place of intersecting solid spheres (27). By defining the parameters of a Gaussian appropriately, the atomic volume for an atom can be accurately reproduced. Also, the overlap terms between any two atoms, and hence any higher-order overlaps, are all Gaussian functions themselves. As a result, expressions for the molecular volume and surface area of this Gaussian model can be obtained as the gradient and Hessian matrix of the nuclear coordinate derivatives. With this methodology it has been shown that the solid sphere volume can be obtained with a high level of accuracy. In addition, the overlap of volumes generated using this methodology are also Gaussians, hence can be computed efficiently. OpenEye have implemented this into their software (Shape Toolkit and ROCS) and hence rapid assessments of both volume/shape similarity and the matrices to produce such an optimal overlap can be obtained.

Recently, OpenEye software has been used to determine if molecules can be represented and compared effectively using shape and electrostatic distribution (28) (see Subheading 3.3).

3.2. Library-Design Procedure

3.2.1. Choice of Library Type

As previously mentioned, there are two main types of libraries: those selected from preexisting molecules, and those designed and synthesized specifically for a given biological target. Often the former will be carried out and those molecules identified as hits will be the starting point for the latter.

3.2.2. Core and Reagent Selection

If the library in question is to be designed, rather than simply selected from preexisting molecules, the first step is to identify the core upon which the library will be based. Cores may be identified from various sources but are often derived from qualified hits from an in vitro screening campaign. The compounds identified as hits are studied to determine the minimal structure of interest (the core).

The core may be elaborated at sites amenable to modification to probe specific physicochemical attributes of the final molecules, including shape.

3.2.3. Construction of Virtual Molecules

Numerous tools and methods exist to support *in silico* enumeration of a library based upon a structure of the core, structures of the reagents, and some method to define the reaction. For instance, one can use an enumeration engine written in OEChem and use SMILES (29) for the structures of cores and reagents. SMARTS is used to encode the reaction information. The choice of enumeration method, and the format of the final molecules, is an entirely subjective choice.

3.2.4. Subsequent Filtering of Products

When an enumerated library contains many more compounds than can possibly be synthesized and/or screened, one must be judicious in the choice of compounds targeted for synthesis. As previously discussed, shape enumeration and comparison is computationally expensive and should therefore represent one of the last steps to be carried out in the process of library design. Where appropriate, the faster—or less subjective—methods for library filtering should be employed. Depending upon the purpose of the library, one can make use of drug-like or lead-like criteria. These include molecular weight and oxygen/nitrogen atom count. Both are very rapid filters. In this way, a relatively large set of compounds may be significantly reduced in number with only a moderate amount of computing time and power. Simply knowing the order to apply filters can make the entire library-design workflow much more effective, and may allow much higher quality (usually more computationally expensive) methods to be employed as the number of molecules under consideration is reduced to a manageable number.

3.2.5. Use of Query Shape as a Positive/Negative Filter

In drug discovery, it is not uncommon that the 3D structure of the protein's target-binding site is unknown. In this case, the structures of known binders can be used as a guide to identify new molecules. Computing the shape similarity between compounds in the library and a known binder allows one to select particular ranges of similarity. For instance, one might choose the majority of library compounds possessing similar shapes to one another but also include a smaller number which possess very different shapes. The inclusion of compounds less similar in shape can be a good way to extrapolate from the known binder and prevent the library being too limited in its scope.

3.2.6. Use of Protein Shape as a Positive/Negative Filter

Increasingly, the 3D structures of target proteins are becoming available (65,967 structures; 18 February 2011) (30). In the same way that the shape of a reference ligand may be used as a positive filter, the shape of the ligand-binding site may be used as a negative filter.

That is to say, the best compounds are those whose overlap with the protein atoms is minimal. It should be clear that if the library is targeted toward a particular protein target that the shape should be complementary, but another protein binding site can also be used as a negative filter if the protein is an unwanted counter-target. By judicious use of the target of interest combined with the shape of any counter-target proteins, compounds may be designed that possess not only affinity but also selectivity.

It is possible to employ shape descriptors of a protein to determine how well a ligand "fits" to a protein site, and hence how well it is expected to bind that protein. Consequently, if this holds true we can use shape descriptors of proteins to also determine the selectivity profile of a ligand.

Shape, in the context of a protein, can refer to a wider range of attributes than for comparing ligands. This is due to the increased size, increased complexity, increased flexibility, differing function, availability and quality of structural data, and the intent behind using shape. This last point is important as we are now not just comparing two or more systems but we are also attempting to determine the significance of any differences/similarities observed.

A review of the literature reveals the varied use of "shape descriptors" applied to proteins, and the majority applied to the comparison of protein structures to determine nonobvious functional relationships or protein–protein interactions. While these areas are of great interest, the methods employed are often not appropriate and/or not fine-grain enough for our purposes where the addition of a single heavy atom to a molecule can significantly impact its binding affinity and hence selectivity profile.

The most frequent use of shape descriptors in ligand selection is the field of docking. In this technique, the atom coordinates of the protein target are used to predict the likely pose of a bound ligand and provide some measure of "fit" (which should ideally correlate with binding affinity). A review of this field is beyond the scope of this discussion, and many excellent reviews have been published.

The use of Gaussians can be employed to describe the shape of a protein binding site—either by using the Gaussian to describe the protein residues or to describe the empty void described by the protein residues. Problems exist for both approaches: the former requires the ligand to be oriented into the appropriate frame of reference to the protein, and the latter relies on the appropriate choice of how to define the limits of any pocket.

3.2.7. Diversity Libraries

Another approach is the subsequent use of an all-*vs.*-all comparison of shapes for those virtual compounds deemed to be suitable for the target protein. In determining these relationships, clustering may be performed in order to reduce the number of molecules that require physical synthesis, thereby saving time, money, and assay materials. Clustering is commonly employed in general library-design and will not be covered in detail here. If one is interested in designing compounds suitable for a target class (e.g., kinases) one can also use shape as a filter to improve the profile of the final set. For instance, by the sequential use of multiple binding-site shapes from the target class, a set of molecules may be selected that will be compatible with most or all of the class.

Suffice to say that the choice of clustering level, and the number of compounds to synthesize from each cluster is highly subjective and the capacity of the assay or the economics of synthesis are often a driver for the decision of library composition.

3.2.8. Assessment of the Proposed Library

Once the use of shape similarity for library selection is complete, additional filters may be applied. For instance, the final set may be subject to predictions or measurements of ADME properties, in order to further refine the set. These tools are improving in predictive power and represent a significant opportunity to improve the quality of the final compounds. Another common tool employed is protein–ligand docking (31, 32). Here, the library compounds undergo conformational sampling in the environment of the protein and intramolecular attraction computed. Comparative studies have shown that these techniques can provide useful guidance, but their effectiveness varies by protein family and method (33).

3.3. Shape and Electrostatics, in the Context of Bioisosterism

Recently, shape and electrostatics have been used in conjunction to determine molecular similarity for bioisosteres (28). The aim of this work was to assess how well shape, and electrostatic fields, could recapitulate observations from medicinal chemistry literature regarding these small biostere systems.

Two simple data sets were explored in this work in order to discern the feasibility of using such methods. The first set of molecules included a small number of commonly used linker groups (see Table 1). The second set was composed of either one or two fused ring systems containing no rotatable bonds (see Table 2), and were taken from the Maybridge heterocyclic ring-numbering chart (34). Conformational freedom was associated only with ring flips and rotational and translational degrees of freedom. These molecules were chosen to validate the methodology as they provide a simple and intuitive test set.

In order to prepare the molecules for comparison, formal charges were manually assigned and conformations of the molecules generated in OMEGA using an energy window of 5 kcal/mol and duplicate removal RMSD of 0.6 Å. The program Quacpac (17) was used to assign single-point AM1-BCC atomic partial charges for all conformations, and these conformers superimposed in an all-vs.-all manner using ROCS (17). Each overlay was scored with EON (17) using a terminal rotor spin value of 30°. The best pairwise scores were written to a square matrix, and the similarity

Table 1
Common linker groups examined in this work

Molecule structures

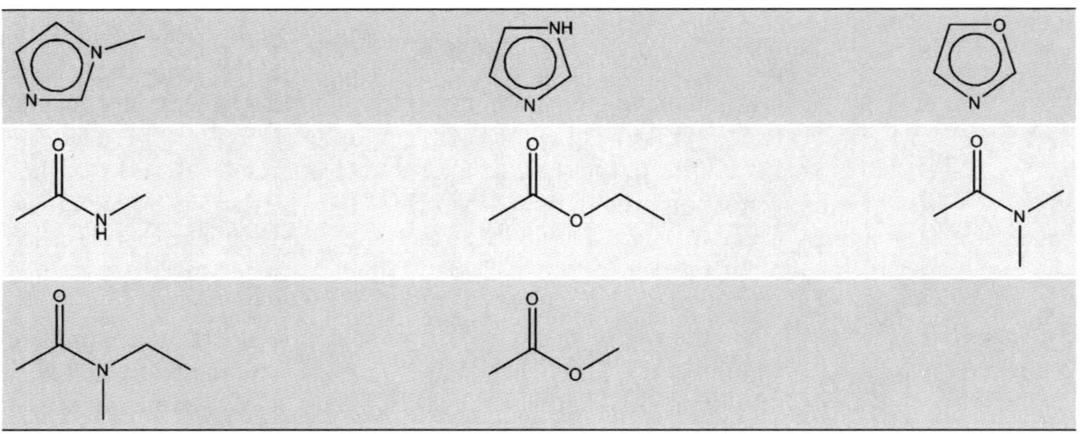

Table 2
The compounds represented by the Maybridge identifier

Cycles	Ring structures
1	

(continued)

Table 2 (continued)

Cycles	Ring structures
2	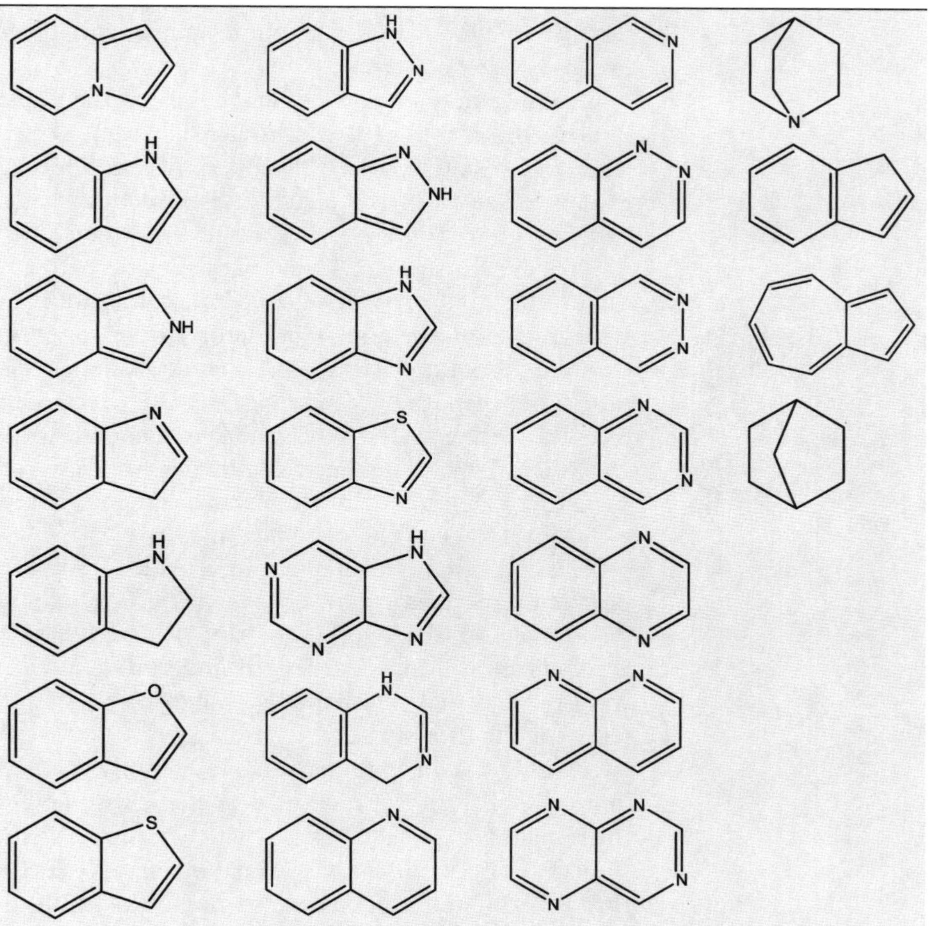

score between conformations *i* and *j* was set to the maximum value of the score between either *i,j* or *j,i*.

Five similarity matrices were ultimately generated. These represented, for the best overlay of a conformation from molecules A and B, the shape similarity, the electrostatic similarity using a Poisson–Boltzmann (PB) electrostatic model, the electrostatic similarity using a Coulombic electrostatic model, and the products of shape and PB and shape and Coulombic.

Similarities were also calculated using Daylight fingerprints (29, 35) and the corresponding Tanimoto metric, in order to provide a comparison with the shape/electrostatic method. Daylight fingerprints are a much-used 2D method of profiling molecules based upon bond paths and are also extensively used in computational chemistry to identify (dis)similar molecules. Clustering of the dissimilarity matrices was performed using the package R (21, 22), with the agnes method in the "cluster" package (36). Agglomerative clusters were generated using the single-linkage and Ward's methods. Visualization of the cluster trees was performed directly in R. Maximum similarity matrices were viewed using in-house code.

A commonly used set of eight linkers (see Table 1) was constructed in order to explore whether shape alone or in combination with electrostatics could classify simple commonly used linker groups. Analysis of the results indicated that shape alone is not sufficient to distinguish between these molecules.

The electrostatic (ET) matrices provide better resolution but the product of shape*ET (shape Tanimoto multiplied by the electrostatic Tanimoto, either Coulombic or Poisson–Boltzmann) gave the best resolution (data not shown). Minimal differences were noted between the Coulombic and PB electrostatic Tanimoto matrices. Further studies were carried out using the shape*PB matrices. It was felt that the PB electrostatic measure was more appropriate than the Coulombic measure since aqueous-based phenomena were of relevance. These studies show that shape is of key importance but is often not sufficient to provide an accurate comparison between molecules.

In the published work, the shape*PB method was compared to a classification using path-length fingerprints as the descriptor. It was expected that the shape*ET method should perform comparably for the intended use. The maximum similarity matrix calculated using Daylight fingerprints, effectively classified the dataset into two subclusters: rings and non-rings, with no linkage between the subclusters. This contrasts with the shape*PB matrix, where similarities between the sets were apparent. It is these similarities that are exploited in SAR development. The results from this small data set showed that the methods can be applied to classifying similar small linkers, and using a combination of shape and electrostatics, it is possible to pick up useful similarities that would have been otherwise missed using a standard path-descriptor method. The same methodology was subsequently applied to the larger Maybridge set (data not shown).

We concluded from this work (the linkers in Table 1 and the ring systems in Table 2) that employing physically realistic shape and electrostatic field similarity metrics provide a way to assign similarities between molecules that would not necessarily be found using classical path-length fingerprint descriptors. It was also shown that both charge and cluster methods affect the results of the similarity analyses and clusters, sometimes drastically. The various clustering schemes can be used to explore different aspects of SAR design, from series explosion to hole-filling exercises, however caution should be used when looking at potential false-positives and -negatives.

The shape*ET method found molecules that one would deem to be similar based upon both intuition and empirical data. Appropriate ring and non-ring systems similarities were captured using this method and one would therefore expect to be more broadly applicable to the problem of bioisostere identification and design. Finally, this work explored and highlighted the failure of 2D fingerprint methods to meet the level of performance possible with shape*ET, and highlights the utility of employing methods based upon real-world physics rather than somewhat artificial concepts like bond-path information.

4. Notes

4.1. Issues

Although the use of shape is exceptionally powerful, the measure of similarity remains subjective. In addition, electrostatics, ligand flexibility, and protein flexibility contribute to binding. Highlighted below are some of the deficiencies of using shape as a design criterion.

4.1.1. Use of Shape Alone

While shape is a key contributor to binding, it is not the only factor in ligand binding. Additionally, the electrostatic profile of the ligand must be complementary with the protein target. If the ligand is too polar, desolvation (37) will also negatively affect the observed binding affinity. The sterics of a molecule operate over short distances whereas electrostatics can operate over very long distances. Therefore, it is appropriate to consider the electrostatic profile in addition to shape for evaluation of any compound.

4.1.2. Use of a Static Protein Model

Proteins are not static systems, yet the majority of docking algorithms consider them to be. Unfortunately, knowing this limitation it is not clear how best to take this into consideration for a shape-based filter. Common options are to relax the rejection criteria for a shape filter, and/or to consider multiple conformations of the protein if that information exists.

4.1.3. Conformational Flexibility of the Library Compounds

The possibility of conformers of each library compound must also be considered. Compounds with a moderate degree of flexibility will make the assessment of shape similarity/dissimilarity challenging. The design of a diverse library will be made even that much more challenging when the components have significant flexibility. Consider where molecule A has 5 distinct conformers and molecule B has 10 distinct conformers: the total number of comparisons just for these two molecules will be 50 (5×10).

4.1.4. Shape Is Not Equally Important at All Positions of a Molecule

The benzene-naphthalene example (see Fig. 1) highlights the problem with assuming that all parts of a molecule's shape are equally important. By the use of focused shape searching, where only the interesting region of the query molecule are weighted for consideration, may help to prevent overlooking potentially relevant compounds.

4.1.5. Interpolation with Use of a Known Hit Compound

By using a known hit as the comparator for shape comparisons, there is the risk that all subsequent molecules of interest identified may be the same size (or smaller). When starting with a larger drug-like molecule this may not represent a significant shortcoming, however if one were following up on something more lead-like, or worse yet fragment-like, the quality of the library may be severely degraded. For instance, new and interesting regions of the target protein may never be probed by the computational hit molecules. Even if the technique used for shape matching were to allow larger-than-query hits to be ranked highly, there is no way to assess if these are suitable for the target protein unless coupled with some method incorporating this information.

4.1.6. Drawback of Using Multiple Sites from a Target Class

Any molecules that fit all sites in a range of proteins from a target class in a comparable way are likely to be small and noncomplex, therefore most likely to be of low affinity. However, this may not represent a significant limitation when the compounds are designed for fragment-based X-ray screening given the high concentrations involved and the information that can be derived from the resulting structures.

5. Summary

The perspective provided here, with an emphasis on the effect of shape on library design and bioisosterism, is somewhat subjective. The philosophy of library design very much depends upon the application of the library, and the resources of the organization itself, and will therefore differ between scientific groups or indeed individuals within that group.

What is clear, both from theory and from practice, is that the shape is important in constructing a library. Whether one's goal is to produce a maximally (dis)similar library, or to produce one that is specific for a protein or protein family, shape must be carefully considered and incorporated into the design workflow. The approaches available to do this, and the software that implements these approaches, are now both numerous and mature and the researcher can choose these tools to aid their work.

References

1. Fischer, E. (1894) Einfluss der Configuration auf die Wirkung der Enzyme. *Ber. Dt. Chem. Ges.* **27**, 2985–93.
2. Friedman, H.L. (1951) Influence of Isosteric Replacements Upon Biological Activity **206**. *National Academy of Sciences-USA: Washington, DC.* 295–300.
3. Thornber, C. W. (1979) Isosterism and Molecular Modification in Drug Design. *Quart. Rev. Chem.* **8**, 563–579.
4. Olesen P. H. (2001) The use of bioisosteric groups in lead optimization. *Curr. Opin. Drug Discov. Devel.* **4** (4), 471–8.
5. Wagener M., Lommerse J. P. (2006) The quest for bioisosteric replacements. *J. Chem. Inf. Model.* **46** (2), 677–85.
6. Schuffenhauer A., Gillet V. J., Willett P. (2000). Similarity searching in files of three-dimensional chemical structures: analysis of the BIOSTER database using two-dimensional fingerprints and molecular field descriptors. *J. Chem. Inf. Comput. Sci.*, **40** (2), 295–307.
7. Burger, A. (1991) Isosterism and Bioisosterism in Drug Design. *Prog. Drug Res.* **37**, 287–367.
8. Patani, G. A., LaVoie, E. J. (1996) Bioisosterism: A Rational Approach in Drug Design. *Chem. Rev.* **96**, 3147–3176.
9. Kubinyi, H. (1998) Similarity and Dissimilarity: A Medicinal Chemist's View. *Perspect. Drug Discovery Des.* **9**-**11**, 225–232.
10. Olesen, P. H. (2001) The Use of Bioisosteric Groups in Lead Optimisation. *Curr. Opin. Drug Discovery Dev.* **4**, 471–478.
11. Boehm, H.-J., Flohr, A., Stahl, M. (2004) Scaffold Hopping. *Drug Discovery Today: Technol.* **1**, 217–224.
12. Lima, L. M., Barreiro, E. J. (2005) Bioisosterism: A Useful Strategy for Molecular Modification and Drug Design. *Curr. Med. Chem.* **12**, 23–49.
13. Schneider, G., Schneider, P., Renner, S. (2006) Scaffold-Hopping: How Far Can You Jump. *QSAR Comb. Sci.* **25**, 1162–171.
14. Brown, N., Jacoby, E. (2006) On Scaffolds and Hopping in Medicinal Chemistry. *Mini-Rev. Med. Chem.* **6**, 1217–1229.
15. C.A. Lipinski; F. Lombardo; B.W. Dominy and P.J. Feeney (1997). "Experimental and computational approaches to estimate solubility and permeability in drug discovery and development settings". *Adv. Drug Del. Rev.* **23,** 3–25.
16. Teague, S.J., Davis, A.M., Leeson, P.D., Oprea, T. I. (1999) The design of leadlike combinatorial libraries. *Angewandte Chemie, International Edition.* **38**, 3743–3748.
17. http://www.eyesopen.com.
18. Grant, J. A., Gallardo, M. A., Pickup, B. T. (1996) A fast method of molecular shape comparison. A simple application of a Gaussian description of molecular shape. *J. Comp. Chem.* **17**, 1653–1666.
19. Boström J., Greenwood J. R., Gottfries J. (2003) Assessing the performance of OMEGA with respect to retrieving bioactive conformations. *J. Mol. Graph. Mod.* **21**, 449–462.
20. Rush, T.S., Grant, J.A., Mosyak, L., Nicholls, A. (2005) A Shape-Based 3-D Scaffold Hopping Method and its Application to a Bacterial Protein-Protein Interaction. *J. Med. Chem.* **48**, 1489–95.
21. Fox, J., Andersen, R. (January 2005) (PDF) Using the R Statistical Computing Environment to Teach Social Statistics Courses. Department of Sociology, McMaster University. http://www.unt.edu/rss/Teaching-with-R.pdf. Retrieved 2006-08-03.
22. Vance, A. (2009-01-06). "Data Analysts Captivated by R's Power". New York Times. http://www.nytimes.com/2009/01/07/technology/business-computing/07program.html. Retrieved 2009-08-25. "R is also the name of a popular programming language used by a growing number of data analysts inside corporations and academia. It is becoming their lingua franca..."
23. "What is R?". http://www.r-project.org/about.html. Retrieved 2009-08-25.
24. Edelsbrunner, H., Kirkpatrick, D.G., Seidel, R. (1983) On the shape of a set of points in a plane. *IEEE Trans. Inf. Theory.* **IT-29** (4), 551–559.
25. Wilson, J.A., Bender, A., Kaya, T., Clemons, P.A. (Sept 23, 2009) Alpha Shapes Applied to Molecular Shape Characterization Exhibit Novel Properties Compared to Established Shape Descriptors. *J. Chem. Inf. Model.* web published.
26. Hahn, M. (1997) Three-dimensional shape-based searching of conformationally flexible compounds. *J. Chem. Inf. Comput. Sci.* **37**, 80–86.
27. Grant, J.A., Pickup. B.T. (1995) A gaussian description of molecular shape. *J. Phys. Chem.* **99**, 3503–3510.
28. Jennings, A., Tennant. M. (2007) Selection of molecules based on shape and electrostatic similarity: proof of concept of "electroforms". J. Chem. Inf. Model. **47** (5), 1829–38.

29. Anderson, E., Veith, G.D., Weininger, D. (1987) SMILES: A line notation and computerized interpreter for chemical structures. *Report No. EPA/600/M-87/021. U.S. EPA*, Environmental Research Laboratory-Duluth, Duluth, MN 55804.
30. Berman, H. M., Westbrook, J., Feng, Z., Gilliland, G., Bhat, T. N., Weissig, H.,Shindyalov, I. N., Bourne, P. E. (2000). The Protein Data Bank. *Nucleic Acids Res.* **28** (1), 235–242.
31. Lengauer, T., Rarey, M. (1996). Computational methods for biomolecular docking. *Curr. Opin. Struct. Biol.* **6** (3), 402–6.
32. Kitchen, D. B., Decornez, H., Furr, J. R., Bajorath, J. (2004). Docking and scoring in virtual screening for drug discovery: methods and applications. *Nat. Rev. Drug Disc.* **3** (11), 935–49.
33. Warren, G. L., Andrews, C. W., Capelli, A. M., Clarke, B., LaMonde, J., Lambert M. H., Lindvall, M., Nevins, N., Semus, S. F., Senger, S., Tedesco, G., Wall, I. D., Woolven, J. M., Peishoff, C. E., Head, M. S. (2006) A critical assessment of docking programs and scoring functions, *J. Med. Chem.* **49**, 5912–5931.
34. http://www.maybridge.com/Images/pdfs/ring_numbering.pdf.
35. Daylight Chemical Information Systems, Inc., Mission Viejo, CA. http://www.daylight.com.
36. Maechler, M., Rousseeuw, P., Struyf, A., Hubert, M. (2005). Cluster Analysis Basics and Extensions; unpublished.
37. Cabani, S., Gianni, P., Mollica, V, Lepori, L., (1981) Group contributions to the thermodynamic properties of non-ionic organic solutes in dilute aqueous solution, *J. Solut. Chem.* **10**, 563–595.

Chapter 11

Accounting for Solvent in Structure-Based Drug Design

Leslie W. Tari

Abstract

Water plays a crucial role in the mediation of protein–ligand interactions, as underscored by the fact that most X-ray crystal structures (of sufficient resolution) of protein–ligand complexes possess water molecules at the protein–ligand interface. In this chapter, the accuracy and reliability of ordered waters observed in crystal structures is discussed. Additionally, the thermodynamic aspects of the inclusion of water in ligand binding to proteins is described, with the goal of providing practical guidelines for dealing with ordered water molecules during structure-guided lead optimization.

Key words: X-ray crystallography, Solvent and Structure-based drug design, Water-binding thermodynamics, water in protein-ligand interfaces

1. Introduction

Water plays a crucial and often enigmatic role in protein structure and function. It is the solvent of most biological macromolecules, acts as a mediator that dictates the specificity of biomolecular interactions, and a substrate (or product) in the reactions catalyzed by many enzymes. Nature despises a vacuum, and water molecules seek to occupy all spaces not occupied by proteins. Water is important in defining the shapes of proteins and stabilities of folded protein structures by promoting the coalescence of hydrophobic surfaces and via the mediation of hydrogen-bonding interactions between polar side chains and backbone atoms in a polypeptide chain. Water comprises a single oxygen atom and a pair of protons, allowing it to accommodate a tetrahedral constellation of two hydrogen-bond donors and two hydrogen-bond acceptors (see Fig. 1). Its small size affords it mobility, and a large, multifunctional interaction surface relative to its molecular volume. These properties

Leslie W. Tari (ed.), *Structure-Based Drug Discovery*, Methods in Molecular Biology, vol. 841,
DOI 10.1007/978-1-61779-520-6_11, © Springer Science+Business Media, LLC 2012

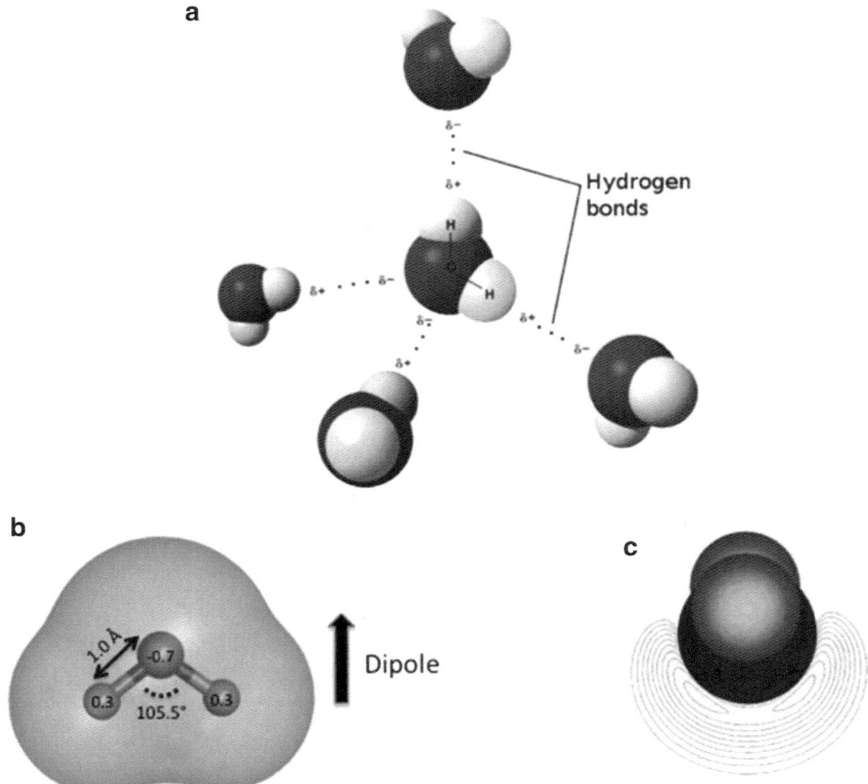

Fig. 1. (**a**) The oxygen on a water molecule is *sp*^{3} hybridized, creating a tetrahedral coordination geometry that allows it to coordinate four atoms with hydrogen-bonds: two H-bond donors and two H-bond acceptors (adapted from the image on http://en.wikipedia.org/wiki/Properties_of_water). (**b**) Geometry and charge distribution for a liquid water molecule from ab initio calculations (1). (**c**) The lone pairs of electrons on a water molecule do not form "rabbit ears." Instead, the electron density is broadly distributed on the face of the oxygen atom opposite to the two protons (2). The electrostatic potential around the oxygen atom is shown in the figure. While it is broad, two minima exist in the expected positions.

make it a versatile utility molecule in proteins that helps to stabilize their active conformations, supplement and/or diversify the physicochemical landscape of active-site or receptor-binding pockets, and modify substrate specificity in enzymes. Water also acts as a "lubricant" that facilitates protein folding as well as conformational adjustments important for protein function by forming networks that link distant protein residues.

Despite the importance of water in protein structure and function, its effects are often largely ignored in structure-based drug design efforts, due to the difficulty in accounting for the structural and thermodynamic effects of interaction with bound water or displacement of bound waters at protein–ligand interfaces. However, as several well-characterized systems demonstrate (described later), careful consideration of solvent effects markedly improves the probability of success in a rational drug design program. In this chapter, some of the thermodynamic and structural implications of solvent inclusion in structure-based drug design and ligand optimization are highlighted, with an emphasis on strategies for engaging or displacing ordered solvent in drug-binding pockets.

2. Identification of Water Molecules in Protein–Ligand Complexes

To characterize the effects of water molecules on the binding of a small molecule ligand to a protein, the positions and orientations of bound waters at the protein–ligand interface must be accurately determined. This is a daunting task: The interference from bulk solvent and the experimental time scale of structural studies on proteins in solution generally preclude detection of bound waters. However, structures of protein–ligand complexes determined using X-ray crystallographic techniques are very effective for revealing the positions of bound water molecules. How is this possible, given the fact that average residence times for a bound waters are in the millisecond to picosecond range (3, 4) and a crystallographic experiment occurs over a period of hours? A crystal structure represents a time- and space-averaged (over trillions of molecular copies) picture of the three-dimensional electron density distribution of the protein–ligand complex in the crystal. Peaks describing discrete water density arise because the potential mean force for water molecules at those positions are at local minima (5). Restated, the free energy of a water molecule at nearby positions would be unfavorably high, creating a steep energy well. This increases the average occupancy compared with bulk solvent at the energy minima by water molecules over the course of the experiment, resulting in an electron density peak above background. When using crystallographic methods to determine hydration structure, one must do so while being mindful of the following caveats: to discriminate waters with certainty, accurate crystallographic data to a resolution of better than 2.5 Å is necessary. Typically, about one ordered water molecule is observed per protein residue at 2.0 Å, with the number rising to 1.6–1.7 per water at 1.0 Å (6). There is no dependence on the observed number of water molecules and data collection temperature, and surprisingly, very little dependence between the fraction of protein polar/apolar surface area and the number of ordered waters observed (6). The positions of water protons must be inferred rather than observed, unless neutron diffraction studies can be performed. However, the inability to experimentally observe protons is generally not too detrimental, since a careful analysis of the hydrogen-bond donors and acceptors surrounding the water molecule in question can be used to model the proton positions with a high degree of certainty in most cases. Analysis of atomic thermal-factors (*B*-factors, see Chapter 1) and examination of the components of the first coordination shell around a putative water will readily distinguish waters from heavier elements, such as metal cations. Heavier elements placed in water peaks will refine with anomalously high *B*-factors (to attenuate their scattering contributions) compared with surrounding atoms, and will usually be associated with negative density peaks in difference electron density maps. In the opposite scenario, where a water's oxygen is placed in a density peak that should be occupied by a heavier element, the water will refine with an anomalously low (or even negative!) *B*-factor and will usually be associated with positive difference density peaks, even after refinement. Care must be taken to distinguish waters from isoelectronic species, such as ordered sodium and ammonium ions, which are common constituents in protein buffers and crystallization media. In these cases, examination of the coordination geometry and composition in the shell surrounding the electron density peak in question usually provides the circumstantial evidence necessary to distinguish waters from their isoelectronic counterparts. When adding water molecules to electron density, the crystallographer must always exercise care; addition of solvent introduces four adjustable parameters (*x, y, z* coordinates and an isotropic *B*-factor) into the model that can "soak-up" errors in the experimental data and/or the atomic model (7, 8). To avoid fitting noise, waters should only be placed into electron density peaks that are within 3.4 Å of at least one chemical group (including other bound waters) that is capable of donating or accepting a hydrogen-bond.

When multiple structures of a target protein are available, particularly in different crystal forms, it is a useful exercise to compare the ordered hydration shells between the different structures. Waters with lower *B*-factors (relative to the other waters in the same structure) that appear in the same positions across the different structures can be deemed reliable, while waters with larger *B*-factors that are not observed in the different structures must be scrutinized with caution. By necessity, there is a fair degree of subjectivity exercised by crystallographers when adding waters to electron density, which can lead to discrepancies in hydration structure between crystal structures of the same protein characterized in different laboratories. For example, structures of interleukin 1β determined independently in four laboratories at similar resolution contained between 83 and 168 ordered waters, but only 29 of those were common between the four structures (9). Surprisingly, while all the common waters were observed in the first hydration shell, some were relatively solvent exposed and not buried in the protein interior. The lesson from examples like this and many others is that numerous factors have to be considered when analyzing water molecules from crystal structures that are to be exploited in structure-based-drug design: does the water make hydrogen-bonds with at least one neighboring atom? Is the water far from any crystal-packing interface? Is the *B*-factor of the water reasonably similar to its immediate neighbors? Does the water not interact with too many hydrogen-bond acceptors (i.e., as a metal cation would)? If the answer to all the questions posed above is negative, then it is probably safe to assume that the water molecule is "real" and not a refinement artifact or incorrectly assigned ion.

3. Water and the Thermodynamics of Ligand Binding

In the context of rational drug design, protein receptor sites are generally viewed as cavities filled with water molecules. To bind, groups on a small molecule ligand must displace water, either through apposition of structurally matched hydrophobic surfaces or through replacement of ordered waters by suitable hydrogen-bonding moieties. It is assumed that ligand affinity is dictated by the favorable entropy associated with liberating the trapped waters in the receptor pocket, coupled with the enthalpic gains arising from various electrostatic-based interactions between ligand and protein. While this model is compelling in its simplicity, and in a gross sense accurate, it glosses over a number of important subtleties, including the fact that water molecules are not always displaced when small molecules bind; in some cases waters from bulk solvent are recruited to the binding interface. Water molecules are highly versatile molecular entities that can play distinct roles in the mediation of ligand binding. By forming ordered networks, waters can behave as removable or extensible modules in a receptor-binding pocket that can modify the specificity of a protein to make it highly adaptable and promiscuous or, alternatively, highly selective. The mechanism employed by the oligopeptide binding protein OppA to bind small peptides of two to five residues in a sequence independent manner is a classic example highlighting the role water molecules can play in expanding the specificity of a protein receptor. As shown in Fig. 2, the promiscuity of OppA stems from the fact that it possesses a capacious binding-pocket where the interactions between peptide side chains and the protein can either be direct or water-mediated: variable constellations of water molecules act as extensions of the receptor-binding surface, forming well-defined hydrogen-bonding networks between peptide and binding-pocket (10, 11). On the opposite end of the spectrum, water molecules are used in many systems to impart exquisite selectivity to biomolecular interactions. Structural studies on the complex of hen egg-white lysozyme and a heterologously expressed Fv fragment from the monoclonal antibody that binds it show over 20 water molecules in and around the antibody–antigen interface, mediating antibody–antigen interactions (12, 13). Surprisingly, a comparison of the free and antigen-bound structures reveals that there is no net loss of ordered solvent at the antibody-combining

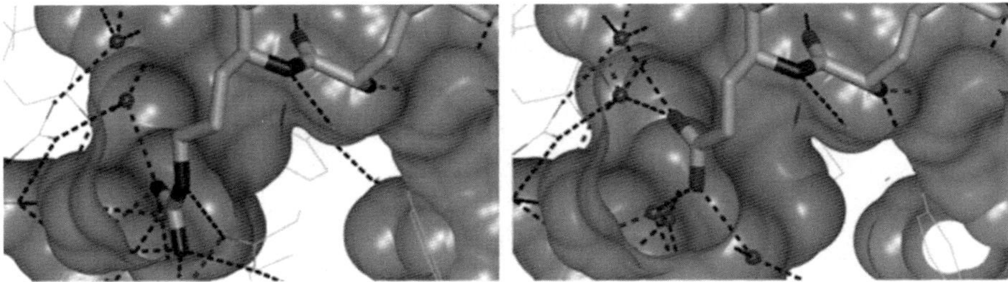

Fig. 2. Identical views of the receptor-binding pockets from complexes of OppA from *Salmonella typhimurium* with KRK (*left*, PDB code 1qka) and KEK (*right*, PDB code 1jeu). Potential H-bonds are displayed as *dotted lines*. Both peptides bind to the receptor with similar affinities, despite presenting central amino acid residues that differ in size, shape, and charge. The Glu side chain in the KEK complex engages the receptor pocket exclusively through water-mediated interactions. The larger Arg side chain in the KRK complex displaces some of the waters observed in the KEK complex and interacts with receptor residues directly, while still employing some of the waters observed in the KEK complex and recruiting new water molecules not observed in the KEK complex for water-mediated interactions with the receptor. A Glu residue in the receptor pocket (Glu32) also adopts different conformations between the two complexes to accommodate the different ligands. Water molecules act as a plastic receptor matrix that extend and modify the information content of the receptor pocket, engaging different ligands through specific hydrogen-bonding interactions.

site upon complex formation (13). To the contrary, ordered water molecules are recruited to the complex in and around the antibody–antigen interface, which both mediate contacts and stabilize conformational changes in the Fv fragment that improve the specificity and integrity of antigen binding and recognition. Other examples of ordered water enhancing the selectivity of protein–ligand interactions are in the binding of some protease inhibitors to subtilisin (14), as well as in the peptide-binding site of the SH2 domain of Src kinase (15). As the examples above illustrate, the role of water in the mediation of protein–ligand interactions is remarkably diverse, and accounting for the effects of water on the thermodynamics of small molecule binding to proteins is not simple.

The free energy associated with the formation of a protein–ligand complex is described in Eq. 11.1 and its relation to the binding constant of the ligand (at equilibrium) is described in Eq. 11.2:

$$\Delta G_{\text{binding}} = \Delta H - T\Delta S \tag{1}$$

$$\Delta G^{\circ} = \Delta H^{\circ} - T\Delta S^{\circ} = -\,\mathrm{RT}\log(K_{\text{eq}}) \tag{2}$$

A hallmark of complexes whose formation is mediated by weak intermolecular interactions like those observed between proteins, ligands and solvent, is that the enthalpic and entropic terms in the binding free energy equation act in opposition to one another to create what is called enthalpy–entropy compensation (16). Stronger ligand-binding, which is accompanied by a favorable (more negative) enthalpy, is correlated with a larger decrease in entropy due to the formation of a complex that tightly constrains the ligand. Weaker ligand-binding constrains the ligand to a lesser extent, and is thus associated with a smaller entropy decrease than the strong-binding case described above, but at the expense of a less favorable (less negative) enthalpy of interaction. So, despite the fact that the changes in enthalpy and entropy accompanying a ligand-binding event to a protein can individually be quite large, the opposing nature of the two terms generally results in a net free energy change that is relatively small.

Explicit incorporation of the contribution made by water molecules to Eq. 11.1 requires the inclusion of additional terms (Eq. 11.3).

$$\Delta G_{\text{binding}} = \Delta G_{\text{complex}} - \Delta G_{\text{desolvation,ligand}} - \Delta G_{\text{desolvation,receptor}} \qquad (3)$$

$\Delta G_{\text{complex}}$ is the interaction free energy (including trapped solvent) of the protein–ligand complex, $\Delta G_{\text{desolvation,ligand}}$ is the free energy of desolvation of the ligand upon binding to the protein, and $\Delta G_{\text{desolvation,receptor}}$ is the free energy of shielding the protein active-site from water molecules (17). The magnitudes of the free energy terms in Eq. 11.3 are very difficult to estimate. The relative contributions of the discrete interacting forces are strongly dependent on many difficult to model factors, including active-site and overall protein plasticity, ligand strain, and protein and ligand hydration before and after ligand binding. Also, due to enthalpy–entropy compensation, the individual terms can be quite large, while the net change in free energy is usually small, making the process prone to errors. Access to high-resolution structures of the protein of interest in ligand-free and multiple ligand-bound states coupled with accurate thermodynamic measurements can improve prospects of addressing the key questions surrounding the incorporation of solvent in structure-based drug design. The key questions that need to be addressed are multifold: (a) Will desolvated (upon receptor binding) polar residues on the ligand be located in a protein environment that sufficiently replace the polar interactions lost with bulk solvent? (b) Will bound water molecules present in the ligand-free protein be accessible by groups on the ligand? If they are, can groups be designed on the ligand that are positioned to hydrogen-bond with the water in such a way that they either do not disturb or strengthen the hydrogen-bonding network between the water in question and the surrounding elements to obtain a favorable change in free energy? Alternatively, would it be advantageous to supplant the bound water with a chemical entity that engages surrounding elements in an energetically equivalent or superior fashion to the previously bound water molecule? (c) In general, but particularly in cases where a bound water molecule is observed at a protein–ligand interface that was not observed in the ligand-free protein structure, what is the entropic cost associated with trapping a water molecule? Some examples highlighting the importance of each of the concepts described above follow.

The importance of desolvation in ligand binding is well established. From a thermodynamic standpoint, desolvation of a small molecule ligand upon binding to a protein is accompanied by a favorable entropy change, since it is associated with the release of conformationally restrained water molecules to bulk solvent. However, the entropic benefits of desolvation can be negated if they are associated with an enthalpic cost that is prohibitive: if the interactions between desolvated polar regions of the small molecule and the protein are inferior in quantity and/or quality compared with the interactions between those same regions of the small molecule and water in a solvated system, the benefits of releasing trapped solvent can be reduced or lost. Studies on the endopeptidase thermolysin (summarized in Fig. 3) with a series of similar inhibitors revealed that one of the inhibitors engages a

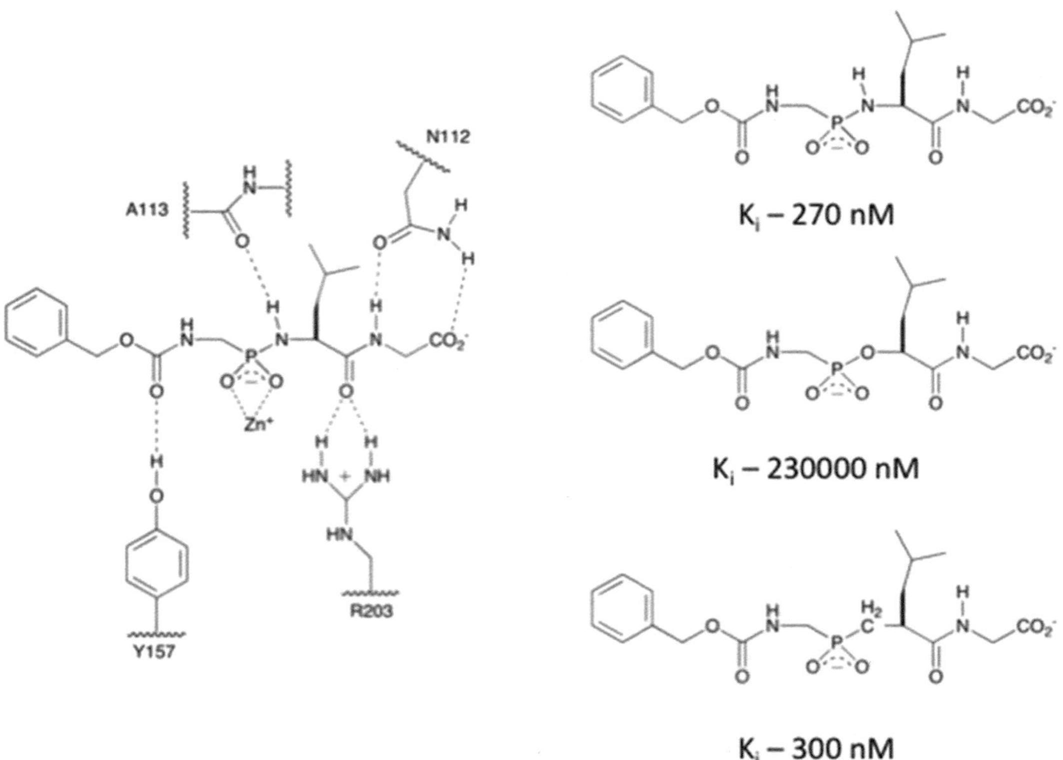

Fig. 3. *Left*: Schematic showing the key polar interactions between thermolysin and a phosphonamidate transition-state analog inhibitor. Potential H-bonds are depicted as *dotted lines*. A series of analogs of the inhibitor shown on the *left* were made replacing the –NH– that interacts with A113 with an –O– and $-CH_2-$, respectively (shown on the *right*). While the affinities of the –NH– and –O– containing inhibitors differ by ~1,000-fold, the affinities of the –NH– and $-CH_2-$ based inhibitors are virtually identical (21). An explanation of for the SAR is provided in the text.

carbonyl oxygen from the peptide backbone of the protein with an –NH– group (18–21). As would be expected, swapping the –NH– for an –O– reduced the potency of the inhibitor by ~1,000-fold, since the –O– replaces a hydrogen-bonding interaction with an electrostatically repulsive one between two oxygen atoms. However, replacement of the –NH– by a –CH_2– generates inhibitors that are equipotent, despite the fact that the –CH_2– is unable to form a hydrogen-bond with the backbone carbonyl (21). Using free energy perturbation calculations and molecular modeling the parity between the –NH– and –CH_2– inhibitors was predicted (22), and traced to two factors: (1) The neutral –CH_2– moiety, while unable to hydrogen-bond, does not have a repulsive interaction with the receptor carbonyl. (2) The solvation free energy for the –CH_2– is estimated to be much lower than it is for the –NH– (or –O–). The dual hydrogen-bond donor/acceptor character of the –NH– allows it to interact with multiple waters in solution. However, in the thermolysin active-site, the –NH– is only observed making a single hydrogen-bond with surrounding residues in the receptor pocket, resulting in an unfavorable free energy of desolvation. As examples like this demonstrate, ligand desolvation must be carefully accounted for in structure-based drug design to avoid the pitfall of overestimating of the affinity gained by adding polar character to compounds during ligand optimization.

In addition to factoring desolvation into the ligand optimization equation, ordered solvent molecules in the protein target must be dealt with carefully. Many proteins employ ordered solvent molecules to stabilize their folded conformations, to bind substrates, or both. Small molecule ligands must be designed that either engage ordered solvent with hydrogen-bonding interactions, or displace it altogether. If ordered solvent molecules can be displaced by groups on the small molecule (favorable entropy) that capture the same surrounding elements in the receptor with hydrogen-bonds of superior quality (favorable enthalpy), the thermodynamic benefits can be substantial. However, solvent displacement is not always possible, due to steric and/or electronic constraints in the vicinity of the bound solvent molecule that preclude its replacement by different groups. In those cases, molecules must be designed that engage unutilized hydrogen-bonding potential on the bound water in an optimal fashion. Inhibitors designed against two important anti-infective drug targets, DNA gyrase and HIV protease, highlight the different scenarios described above.

DNA gyrase B (GyrB) and topoisomerase IV (ParE) (described in Chapter 1), are highly conserved ATPases that play key roles in bacterial DNA replication (23). Since they are essential enzymes in bacteria, and possess no close eukaryotic counterparts, they are prime candidates for the development of new classes of antibiotics (24). In the ATP-binding pockets of GyrB and ParE, the "anchoring" interaction between ATP and the enzyme utilizes a conserved

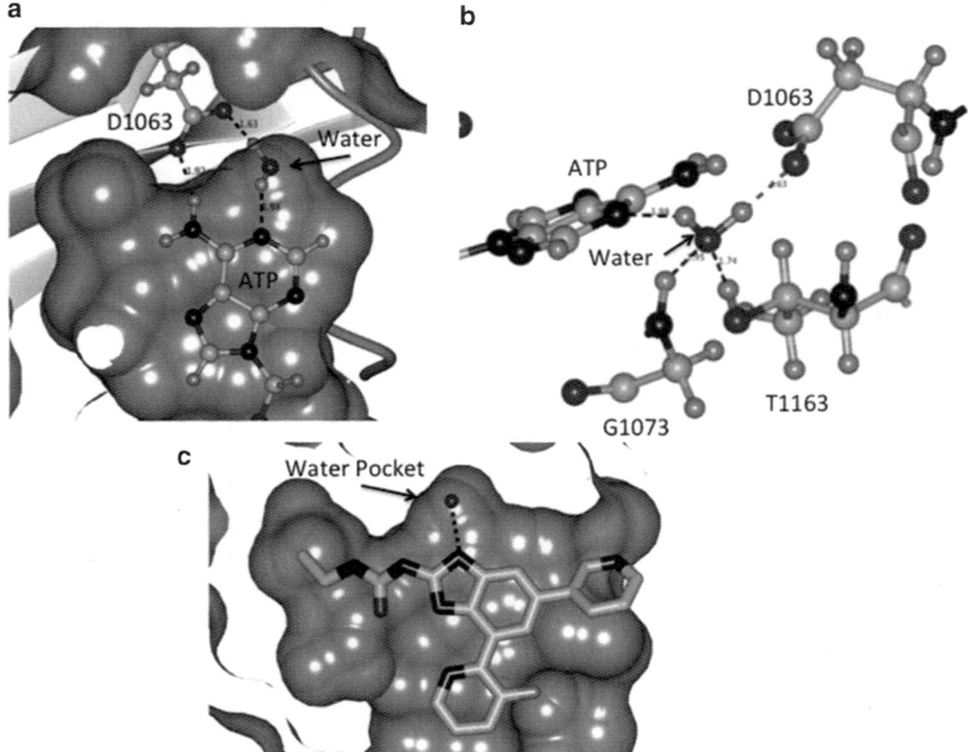

Fig. 4. (**a**) Cutaway of the solvent accessible surface of the ATP-binding domain of *Escherichia coli* topoisomerase IV complexed with AMPPNP from a crystal structure (25). The adenine base donates an H-bond to D1063, and accepts an H-bond from a tightly bound water molecule. (**b**) A close-up view of the coordination sphere around the substrate-binding water molecule. Almost all available H-bond acceptor potential of the water is exploited by the enzyme, via H-bonds donated by the main chain amide of G1073 and the T1163 side chain hydroxyl. One of the water protons engages D1063 with a charge-assisted H-bond, leaving the second water proton available to interact with N1 of the adenine base. The H-bond donors and acceptors surrounding the water form an almost perfect tetrahedral constellation. In addition to improving the fidelity of the interaction of ATP with the active-site pocket, the water molecule appears crucial in stabilizing the folded conformation of the protein by bridging several key residues. (**c**) Cutaway of the solvent accessible surface of the ATP-binding domain of *Enterococcus faecalis* GyrB complexed with a benzimidazole inhibitor from a crystal structure (D. Bensen and L. Tari, unpublished results). The counterpart to the conserved water described in (**b**) is shown for context. The inhibitor engages the water with an H-bonding interaction (shown as a *dotted line*). The water is coordinated in a deep, narrow cavity. The steric and electronic constraints the water pocket places on ligand binding make the water very difficult to displace, so that all reported inhibitor design efforts directed toward GyrB/ParE described to date treat the water molecule as part of the enzyme.

aspartic acid and a tightly bound water molecule to coordinate the adenine base (see Fig. 4). The substrate-binding water is coordinated by an almost perfect tetrahedral arrangement of two hydrogen-bond donors and an anionic hydrogen-bond acceptor (from the conserved Asp residue). A single proton faces solvent, and interacts with N1 of the adenine base. To evaluate the prospects for displacing the water with an inhibitor, the thermodynamic consequences of trapping the water, steric accessibility of the water-binding site and constellation of hydrogen-bonding residues surrounding the water need to be evaluated. Dunitz has determined the theoretical upper limit for entropy loss incurred when trapping a water molecule using data on the transfer of a water molecule from bulk solvent to a site where it is positionally frozen (26). The standard entropy of liquid water is 16.7 cal/mol at 298 K. Since a water bound to a protein is not likely to be constrained more than it is in a crystalline hydrate salt, where its entropy is approximately 10 cal/mol/K, the maximum entropic cost associated with transferring a water molecule from bulk solvent to a protein is equal to the difference, or about 7 cal/mol/K. This corresponds to a maximal free energy cost of about 2 kcal/mol at 300 K for a tightly bound water (26). Returning to the GyrB example, the free energy stored in the trapped substrate-binding water can only be harvested if the interactions between the water molecule and receptor pocket are captured by group(s) on the inhibitor to a similar extent. As shown in Fig. 4, GyrB water pocket is deep and narrow, and can only accommodate a group with a specific constellation of hydrogen-bond donors and acceptors, i.e., a hydroxyl group. Thus, all synthetic inhibitors targeting GyrB/ParE that have been developed to date (or natural product inhibitors that have been characterized) do not displace the water, but engage it with an hydrogen-bond acceptor in a similar fashion to ATP (24). Numerous molecular scaffolds that inhibit GyrB have been developed that treat the bound water

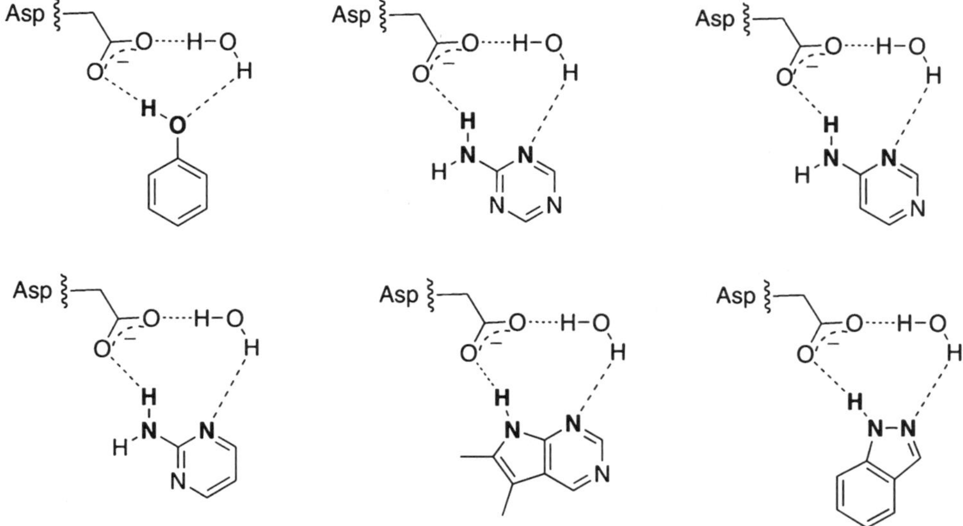

Fig. 5. Examples of potential GyrB/ParE inhibitor scaffolds. Potential H-bonds are depicted as *dotted lines*. The scaffolds were generated via *in silico* screening using a pharmacophore motif with two key interactions: an H-bond donor to interact with the carboxylate of the conserved aspartate residue, and an H-bond acceptor of the H-bond supplied by the conserved water molecule (27). The H-bond donor and acceptor moieties on the inhibitors are highlighted in *bold-type*. The scaffolds shown in the figure were selected from a pool of 3,000 *in silico* hits that were validated by several biophysical methods, including surface plasmon resonance, ultracentrifugation, NMR spectroscopy and X-ray crystallography. Several of the scaffolds shown were optimized to create potent inhibitor leads against the protein target.

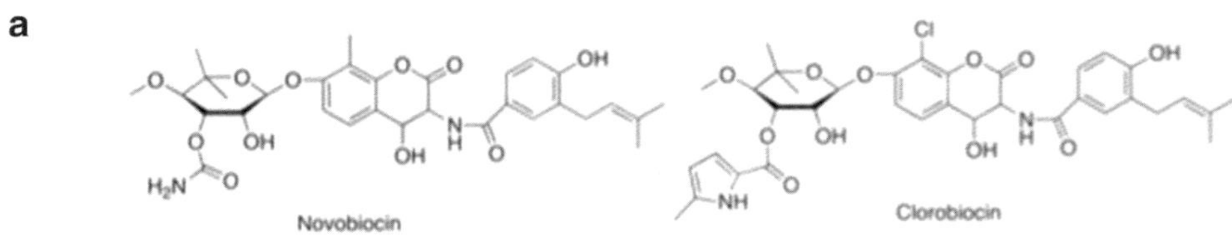

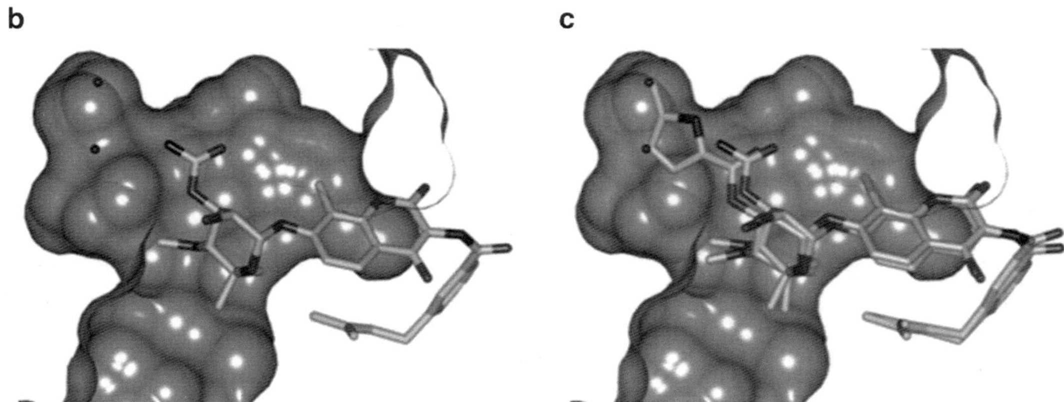

Fig. 6. (**a**) Chemical structures of novobiocin (*left*) and clorobiocin (*right*). The clorobiocin replaces the amine substituent on the carbamate moiety with a methyl pyrrole. (**b**) Cutaway view of the solvent accessible surface of *E. coli* GyrB complexed with novobiocin from the crystal structure of the 24 kDa ATP-binding domain of *E. coli* (R136H) GyrB complexed with novobiocin (28). (**c**) Same as (**b**), but with the clorobiocin ligand from the structure of the 24 kDa ATP-binding domain of *E. coli* GyrB complexed with clorobiocin (29) superimposed. The ordered waters from the novobiocin complex that are displaced by clorobiocin are shown in (**b**) and (**c**) for context. The methyl pyrrole moiety displaces the ordered waters from the pocket interior while still capturing the key H-bonding interactions observed in the novobiocin complex. As a result, clorobiocin binds to GyrB with ~40× greater potency than novobiocin.

molecule as a part of the protein architecture. Some examples are shown in Fig. 5.

In addition to the "structural" water in the GyrB active-site described above, several other ordered water sites are present in GyrB that need to be considered to design potent GyrB inhibitors. Structural and thermodynamic data on complexes of the natural product coumarin antibiotics novobiocin and clorobiocin with *Escherichia coli* GyrB are illustrative of the importance of bound water to binding thermodynamics (28–30). As shown in Fig. 6, the crystal structure of *E. coli* GyrB complexed with novobiocin reveals two ordered water molecules that bridge the carbamate of novobiocin to the active-site pocket interior. The waters bridge the carbamate to the conserved substrate-binding Asp with a strong hydrogen-bond, and interact with the peptide backbone via geometrically less optimal hydrogen-bonds. Most of the pocket surface surrounding the waters is lipophilic. The nonideal coordination environment surrounding these waters (when compared to the environment surrounding the "structural" water in GyrB described above) makes these waters prime candidates for displacement by a ligand. Nature has employed this strategy with the inhibitor clorobiocin, substituting the NH_2 group in novobiocin with a methyl pyrrole (see Fig. 6). The bulkier group occludes the space occupied by the two water molecules in the novobiocin complex (29). Clorobiocin enjoys a 37-fold increase in affinity for GyrB when compared with novobiocin, with the improved binding due not to enthalpy, which favors novobiocin (by approximately 2.7 kcal/mol) but instead due to entropy. The entropic contribution, $-T\Delta S$, is 4.8 kcal/mol more favorable for the clorobiocin complex at 300 K (30). So, the large change in affinity observed between clorobiocin and novobiocin appears to be derived largely from the entropic gains correlated with the displacement of the ordered waters from the GyrB pocket interior (28, 29).

A detailed analysis of the ordered water molecules in DNA gyrase using thermodynamic integration simulations (and corroborated by thermodynamic data) suggest that the entropy ($T\Delta S$) loss associated with the binding of water to a protein interior increases by linear increments for each of the first three hydrogen-bonds formed, as a result of the loss in degrees of freedom for each hydrogen-bond made (30). Adding a fourth hydrogen-bond is calculated to have a minimal effect on entropy, suggesting that there is little additional entropic penalty for forming the fourth hydrogen-bond. Assuming these results are representative for waters commonly found at protein/inhibitor interfaces, when coupled with detailed structural data, they provide a useful guide for evaluating the entropic benefit of displacing coordinated water molecules during ligand optimization. As would be expected, there is a greater potential thermodynamic benefit in displacing waters that engage their surrounding with a greater number of hydrogen-bonds.

A well-known example highlighting the utility in exploiting advantageously positioned structural water molecules to design potent inhibitors is the structure-guided design of cyclic urea inhibitors of human immunodeficiency virus-1 protease (HIV-PR), a validated therapeutic target in the treatment of acquired immunodeficiency syndrome (AIDS) (31). HIV-PR is a homodimeric aspartic protease featuring a twofold symmetric active-site (32). A distinguishing feature of the retroviral protease, which has no counterpart in mammalian aspartic proteases, is the presence of a tetrahedrally coordinated solvent molecule which mediates interactions between the peptide substrate and flaps from symmetry related monomeric units in the active homodimer (see Fig. 7). The water accepts two hydrogen-bonds from the main chain nitrogen atoms of Ile50 residues from symmetry related monomers, and donates two hydrogen-bonds to carbonyl oxygens of peptide substrates or peptidomimetic inhibitors (32). While it had long been recognized that improvements in inhibitor potency and selectivity would be possible if inhibitors could be designed that structurally mimic the water molecule (33), the first inhibitors to achieve this goal were designed by scientists at Dupont Merck (34). As shown in Fig. 7, the urea carbonyl from inhibitors based on a cyclic urea

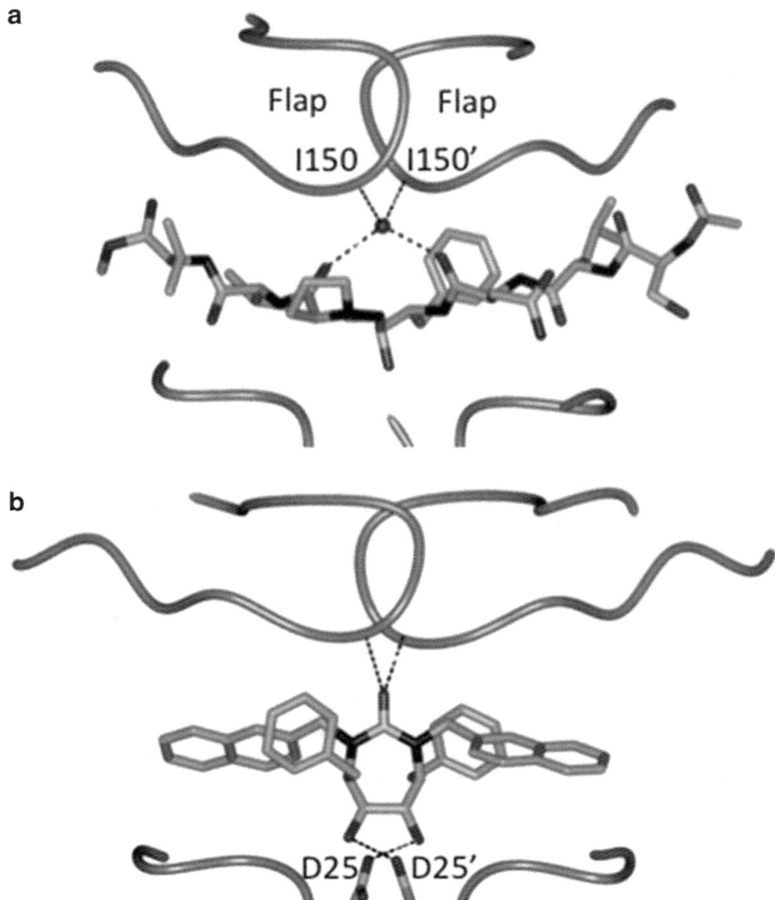

Fig. 7. Two inhibitor bound structures of human immunodeficiency virus-1 protease (HIV-PR), showing some key features of the *C2*-symmetric dimer complexes. (**a**) Structure of the complex of HIV-PR with a peptidomimetic inhibitor (35). The inhibitor coordinates an immobilized water molecule that bridges flexible enzyme flaps via hydrogen-bonding interactions with main chain N–H groups from symmetry related Ile150 residues. The water donates two hydrogen-bonds to the inhibitor. The highly potent cyclic urea inhibitor shown in (**b**) derives its potency by displacing the immobilized water with the urea carbonyl and retaining strong hydrogen-bonds with the Ile150 flap residues. In addition, the inhibitor engages the catalytic aspartates (25 and 25′ in the figure) with hydrogen-bonding interactions (36).

core is able to displace the flap water molecule from the HIV protease dimer and successfully mimic the hydrogen bonding features of the displaced water molecule. In addition to reaping the entropic benefits of displacing a tightly bound water molecule, inhibitors in the class were designed with a diol functionality opposite the urea carbonyl to act as a transition state mimic that engages the catalytic aspartates. Ultimately, cyclic urea inhibitors were developed with exquisite enzyme potency, in vitro antiviral activity and significant oral bioavailability, which were selected for further development and advanced into phase I clinical studies (34).

4. Summary

Water acts as a moderator and mediator in protein–ligand interactions, and plays a key role in dictating ligand-binding affinity and selectivity. The importance of water in mediating ligand-binding is underscored by a recent analysis of deposited crystal structures of protein–ligand complexes, where over 85% of the structures have at least one ordered water molecule at the protein ligand interface (36). From the standpoint of SBDD, the problem exists that in most cases, the thermodynamic implications of water inclusion in ligand-binding are still very difficult to predict. However, the situation continues to improve. As examples like the DNA gyrase studies described above demonstrate, detailed structural and thermodynamic analysis of water-binding sites on systems of interest for drug discovery can dramatically improve our understanding of how to deal with ordered waters during ligand optimization.

References

1. Silvestrelli, P. L. and Parrinello, M. (1999) Structural, electronic, and bonding properties of liquid water from first principles. *J. Chem. Phys.* **111**, 3572–3580.
2. Ball, P. (2008) Water as an active constituent in cell biology. *Chem. Rev.* **108**, 74–108.
3. Likic, V. A. and Prendergast, F. G. (2001) Dynamics of internal water in fatty acid binding protein: computer simulations and comparison with experiments. *Proteins* **43**, 65–72.
4. Denisov, V. P. and Halle, B. (1995) Hydrogen exchange and protein hydration: the deuteron spin relaxation dispersions of bovine pancreatic trypsin inhibitor and ubiquitin. *J. Mol. Biol.* **245**, 698–709.
5. Levitt, M. and Park, B. H. (1993) Water: now you see it, now you don't. *Structure* **1**, 223–226.
6. Carugo, O. and Bordo, D. (1990) How many water molecules can be detected by protein crystallography? *Acta Crystallogr. Sect. D: Biol. Crystallogr.* **D55**, 479–483.
7. Kleywegt, G. J. (2000) Validation of Protein Crystal Structures. *Acta Crystallogr. Sect. D: Biol. Crystallogr.* **D56**, 249–265.
8. Kleywegt, G. J. and Jones, T. A. (1995) Where freedom is given, liberties are taken. *Structure* **3**, 535–540.
9. Ohlendorf, D. H. (1994) Accuracy of refined protein structures. II. Comparison of four independently refined models of human interleukin 1β. *Acta Crystallogr. Sect. D: Biol. Crystallogr.* **D50**, 808–812.
10. Tame, J. R., Murshudov, G. N., Dodson, E. J. Neil, T. K., Dodson, G. G., Higgins, C. F. and Wilkinson, A. J. (1994) The structural basis of sequence-independent peptide binding by OppA protein. *Science* **264**, 1578–1581.
11. Tame, J. R., Sleigh, S. H., Wilkinson, A. J. and Ladbury, J. E. (1996) The role of water in sequence-independent ligand binding by an oligopeptide transporter protein. *Nat. Struct. Biol.* **12**, 998–1001.
12. Bhat, T. N., Bentley, G. A., Fischmann, T. O., Boulot, G. and Poljak, R. J. (1990) Small rearrangements in structure of Fv and Fab fragments of antibody D1.3 on antigen binding. *Nature* **347**, 483–485.
13. Bhat, T. N., Bentley, G. A., Boulot, G., Greene, M. I., Tello, D., Dall'Acqua, W., Souchon, H., Schwarz, F. P., Mariuzza, R. A. and Poljak, R. J. (1994) Bound water molecules and conformational stabilization help mediate an antigen-antibody association. *Proc. Natl. Acad. Sci. USA.* **91**, 1089–1093.
14. McPhalen, C. A., and James, M. N. G. (1988) Structural comparison of two serine proteinase-protein inhibitor complexes: Eglin-C-subtilisin Carlsberg and Cl-2-subtilisin Novo. *Biochemistry.* **27**, 6582–6598.
15. Chung, E., Henriques, D., Renzoni, D., Zvelebil, M., Bradshaw, J. M., Waksman, G., Robinson, C. V. and Ladbury, J. E. (1998) Mass spectrometric and thermodynamic studies reveal the role of water molecules in complexes formed between SH2 domains and tyrosyl phosphopeptides. *Structure.* **6**, 1141–1151.

16. Dunitz, J. D. (1995) Win some, lose some: enthalpy-entropy compensation in weak intermolecular interactions. *Chem. Biol.* **2**, 709–712.
17. Shoichet, B. K., Leach, A. R. and Kuntz, I. D. (1999) Ligand solvation in molecular docking. *Proteins.* **34**, 4–16.
18. Bartlett, P. A. and Marlowe, C. K. (1987) Evaluation of intrinsic binding energy from a hydrogen bonding group in an enzyme inhibitor. *Science.* **235**, 569–571.
19. Tronrud, D. E., Holden, H. M. and Matthews, B. W. (1987) Structures of two thermolysin-inhibitor complexes that differ by a single hydrogen bond. *Science.* **235**, 571–574.
20. Bash, P. A., Singh, U. C., Brown, F. K., Langridge, R. and Kollman, P. A. (1987) Calculation of the relative change in binding free energy of a protein-inhibitor complex. *Science.* **235**, 574–576.
21. Morgan, B. P., Scholtz, J. M., Ballinger, M. D., Zipkin, I., D. and Bartlett, P. A. (1991) Differential binding energy: A detailed evaluation of the influence of hydrogen-bonding and hydrophobic groups on the inhibition of thermolysin by phosphorous-containing inhibitors. *J. Am. Chem. Soc.* **113**, 297–307.
22. Merz, K. M. and Kollman, P. A. (1989) Free energy perturbation simulations of the inhibition of thermolysin: Prediction of the free energy of binding of a new inhibitor. *J. Am. Chem. Soc.* **111**, 5649–5658.
23. Champoux, J. J. (2001) DNA topoisomerases: structure, function and mechanism. *Annu. Rev. Biochem.* **70**, 369–413.
24. Oblak, M., Kotnik, M. and Solmajer, T. (2007) Discovery and Development of ATPase Inhibitors of DNA Gyrase as Antibacterial Agents. *Curr. Med. Chem.* **14**, 2033–2047.
25. Bellon, S., Parsons, J. D., Wei, Y., Hayakawa, K., Swenson, L. L., Charifson, P. S., Lipple, J. A., Aldape, R. and Gross, C. H. (2004) Crystal structures of *Escherichia coli* topoisomerase IV ParE subunit (24 and 43 kilodaltons): A single residue dictates differences in novobiocin potency against topoisomerase IV and DNA gyrase. *Antimicrob. Agents Chemother.* **48**, 1856–1864.
26. Dunitz, J. D. (1994) The entropic cost of bound water in crystals and biomolecules. *Science.* **264**, 670.
27. Boehm, H. J., Boehringer, M., Bur, D., Gmeunder, H., Huber, W., Klaus, W., Kostrewa, D., Kuehne, H., Luebbers, T., Meunier-Keller, N. and Mueller, F. (2000) Novel inhibitors of DNA gyrase: 3D structure based biased needle screening, hit validation by biophysical methods, and 3D guided optimization. A promising alternative to random screening. *J. Med. Chem.* **43**, 2664–2674.
28. Holdgate, G. A., Tunnicliffe, A., Ward, W. H. J., Weston, S. A., Rosenbrock, G., Barth, P. T., Taylor, I. W. F., Pauptit, R. A. and Timms, D. (1997) The entropic penalty of ordered water accounts for weaker binding of the antibiotic novobiocin to a resistant mutant of DNA gyrase: A thermodynamic and Crystallographic Study. *Biochemistry.* **36**, 9663–9673.
29. Lafitte, D., Lamour, V., Tsvetkov, P. O., Makarov, A. A., Klich, M., Deprez, P., Moras, D., Briand, C. and Gilli, R. (2002) DNA gyrase interaction with coumarin-based inhibitors: The role of the hydroxybenzoate isopentyl moiety and the 5'-methyl group of the noviose. *Biochemistry.* **41**, 7217–7223.
30. Yu, H., and Rick, S. W. (2009) Free energies and entropies of water molecules at the inhibitor-protein interface of DNA gyrase. *J. Am. Chem. Soc.* **131**, 6608–6613.
31. Brik, A. and Wong, C.-H. (2003) HIV-1 protease: mechanism and drug discovery. *Org. Biomol. Chem.* **1**, 5–14.
32. Wlodawer, A. and Vondrasek, (1998) J. Inhibitors of HIV-1 protease: a major success of structure-assisted drug design. *Annu. Rev. Biophys. Biom.* **27**, 249–284.
33. Erickson, J., Neidhart, D. J., VanDrie, J., Kempf, D. J., Wang, X. C., Norbeck, D. W., Plattner, J. J., Rittenhouse, J. W., Turon, M. and Wideburg, N. (1990) Design, activity and 2.8 Å crystal structure of a C2 symmetric inhibitor complexed to HIV-1 protease. *Science.* **249**, 527–533.
34. Lam, P. Y. S., Jahdav, P. K., Eyermann, C. J., Hodge, C. N., Ru, Y., Bacheler, L. T., Meek, J. L., Otto, M. J., Rayner, M. M., Wong, Y. N., Chang, C-H., Weber, P. C., Jackson, D. A., Sharpe, T. R. and Erickson-Viitanen, S. (1994) Rational design of potent, bioavailable, nonpeptide cyclic ureas as HIV protease inhibitors. *Science.* **263**, 380–384.
35. Swain, A. L., Miller, M. M., Green, J., Rich, D. H., Scneider, J., Kent, S. B. and Wlodawer, A. (1990) X-ray crystallographic structure of a complex between a synthetic protease of human immunodeficiency virus 1 and a substrate-based hydroxyethylamine inhibitor. *Proc. Natl. Acad. Sci. USA.* **87**, 8805–8809.
36. Lu, Y., Wang, R., Yang, C. Y. and Wang, S. (2007) Analysis of ligand-bound water molecules in high-resolution crystal structures of protein-ligand complexes. *J. Chem. Inf. Model.* **47**, 668–675.

Chapter 12

Structure-Based Drug Design on Membrane Protein Targets: Human Integral Membrane Protein 5-Lipoxygenase-Activating Protein

Andrew D. Ferguson

Abstract

Leukotrienes are biologically active lipid metabolites of arachidonic acid that are involved in inflammation and play a significant role in respiratory and cardiovascular disease. The integral nuclear membrane protein 5-lipoxygenase-activating protein (FLAP) is essential for leukotriene biosynthesis in response to cellular activation. The crystal structures of human FLAP with two inhibitors were recently determined. Inhibitors are bound within the lipid-exposed portion of FLAP, and the unexpected location of the inhibitor-binding site suggests a transport mechanism for arachidonic acid and provides functional insights into leukotriene biosynthesis. This chapter describes how this human integral membrane crystal structure was solved by pushing the limits of low-resolution structure determination and refinement, demonstrating how a low-resolution structure can impact biology and chemistry, and discusses future opportunities for structure-based drug design for this therapeutic target.

Key words: X-ray crystallography, Structure-based drug design, FLAP, Leukotrienes, Arachidonic acid, MAPEG, Membrane protein

1. Introduction

Leukotrienes are lipid mediators of inflammation that play a significant role in respiratory and cardiovascular disease (for excellent reviews see refs. 1–6). Cellular activation by immune complexes and other stimuli initiates leukotriene biosynthesis with the release of arachidonic acid from the nuclear membrane by cytosolic phospholipase A_2 (cPLA_2) (2, 5–8). Arachidonic acid is first oxygenated to the hydroperoxide 5-(S)-hydroperoxy-6,8,11,14-eicosatetraenoic acid (5-HpETE) and then dehydrated to the unstable epoxide LTA_4 in a reaction that is catalyzed by 5-lipoxygenase (5-LO) (1, 3). Conversion of arachidonic acid to LTA_4 is driven by the

Leslie W. Tari (ed.), *Structure-Based Drug Discovery*, Methods in Molecular Biology, vol. 841, DOI 10.1007/978-1-61779-520-6_12, calcium-dependent translocation of 5-LO to the nuclear membrane (9–11). Conjugation of glutathione to LTA_4 by leukotriene C_4 synthase generates LTC_4 (the precursor of LTD_4 and LTE_4), whereas leukotriene A_4 hydrolase produces LTB_4. LTA_4 can be exported from the cell for transcellular biosynthesis of other leukotrienes (12).

In the late 1980s and early 1990s, Merck identified an indole-class compound, designated MK-886 (see Fig. 1a), that inhibited in vitro and in vivo leukotriene biosynthesis without affecting 5-LO, $cPLA_2$ or other nonselective mechanisms of inhibiting leukotriene biosynthesis (13). Bayer had also identified a quinoline-class inhibitor, named BAY X1005 (now known as DG-031) (see Fig. 1b) with similar properties (14). Second-generation quinoline–indole hybrid compounds exemplified by MK-591 were also designed (see Fig. 1c) (15, 16). To determine the protein target of these compounds, a photoaffinity analogue of MK-886 and Sephadex beads coupled to MK-886 were generated (17). These reagents were used to identify an integral nuclear membrane protein that specifically bound to MK-886. This protein was purified, its cDNA cloned, and antisera raised against peptide epitopes derived from this protein was used to immunoprecipitate photoaffinity-labeled protein from human leukocyte membranes. As this 18-kDa integral nuclear membrane protein was required for 5-LO activity, it was named the 5-lipoxygenase-activating protein or

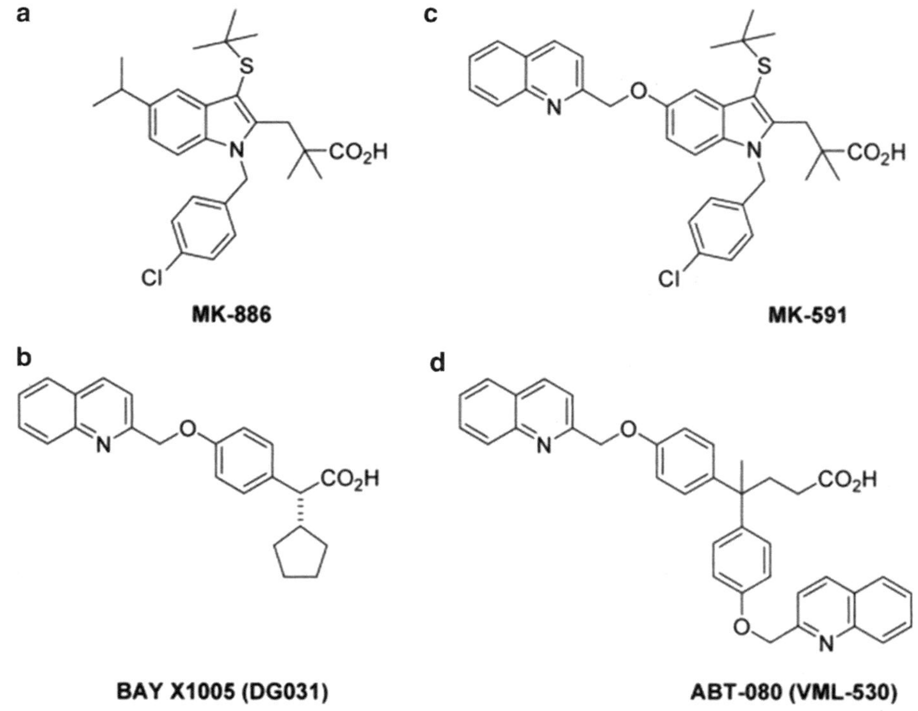

Fig. 1. Chemical structures of (**a**) MK-886, (**b**) BAY X1005 (DG-031), (**c**) MK-591, and (**d**) ABT-080 (VML-530).

FLAP (9, 10, 17). The catalytic activity of 5-LO is significantly enhanced by its interaction with FLAP (5, 9, 10, 18). MK-886 also prevents the concentration-dependent binding of arachidonic acid to FLAP (19, 20), and the association of FLAP with 5-LO at higher concentrations (21).

Following the identification of FLAP, five additional human proteins and a further 130 prokaryotic proteins with similar protein sequences and gene structures were identified (22, 23). These six human integral membrane proteins compose the MAPEG (*M*embrane-*A*ssociated *P*roteins in *E*icosanoid and *G*lutathione metabolism) superfamily, which includes three microsomal glutathione transferases (mGST-1, mGST-2, and mGST-3), leukotriene C_4 synthase (LTC_4 synthase), and microsomal prostaglandin E synthase (mPGES-1)—all of which utilize glutathione for their enzymatic activity. In contrast to other MAPEG proteins, FLAP has not been shown to have enzymatic activity and it is unclear if its function is modulated by glutathione.

It is not understood how FLAP activates 5-LO, but it appears to function as a nuclear membrane anchor scaffold for 5-LO (18) and as an arachidonic acid binding protein (19). Activation of 5-LO involves a physical interaction with FLAP (24), implying that leukotriene biosynthesis inhibitors compete with arachidonic acid for binding to FLAP or cause an undefined conformational change in FLAP that decreases its affinity for arachidonic acid and/or 5-LO (20). Pharmacological inhibition of either 5-LO or FLAP results in the complete inhibition of leukotriene biosynthesis (3, 5, 6). To understand the molecular basis of FLAP's role in leukotriene biosynthesis and how inhibitors regulate its function, the three-dimensional structure of FLAP with bound inhibitors was determined (25).

2. Results

2.1. Molecular Biology

The gene encoding human FLAP (ALOX5AP) was PCR amplified using the following primers: 5′-ATA<u>CCATGG</u>GCATGGATCAAGAAACTGTAGGC-3′ and 5′-CG<u>GGATCC</u>TTAATGATGATGATGATGATGGGGAATGAGAAGTAGAGGG-3′, introducing a noncleavable C-terminal 6×His-tag following the full-length FLAP protein sequence. An alternative construct with a thrombin-cleavage site before the C-terminal 6×His-tag was amplified using the following primers: 5′-ATA<u>CCATGG</u>GCATGGATCAAGAAACTGTAGGC-3′ and 5′-CG<u>GGATCC</u>TTAATGATGATGATGATGATGGCTGCCGCGCGGCACCAGGGGAATGAGAAGTAGAGGGG-3′. The resulting PCR products were digested with *Nco*I and *Bam*HI, and ligated into the bacterial expression vector pET28a, producing $FLAP_{1-161}$6×His/pET28a and $FLAP_{1-161}$ thrombin/6×His/pET28a.

A baculovirus expression construct for full-length human FLAP was generated by introducing a linker into the *Bam*HI site of expression vector pBAC.1, using the following primers: 5′-GATCGCGCCACCATGCACCACCACCACCACCACC TGGTGCCACGCGGTTCTCATATGGAAATTCCG-3′ and 5′-GATCCGGAATTTCCATATGAGAACCGCGTGGCA CCAGGTGGTGGTGGTGGTGGTGCATGGTGGCGC-3′. This linker removes the 5′ *Bam*HI site of vector pBAC.1, creates a Kozak sequence, and introduces a starting methionine codon before the N-terminal 6×His-tag followed by a thrombin-cleavage site and an *Nde*I restriction site. The FLAP gene sequence was PCR amplified using the following primers: 5′-CATATGGATC AAGAAACTGTAGGC-3′ and 5′- GAATTCTTAGGGAATGAG AAGTAGAG-3′. The resulting product was digested with *Nde*I and *Eco*RI, and ligated into the modified pBAC.1 vector, producing $FLAP_{1-161}$thrombin/6×His/pBAC.1.

2.2. Insect Cell Expression of FLAP

Sf9 cells were transfected with $FLAP_{1-161}$thrombin/6×His/pBAC.1. 2 μg of recombinant pBAC.1 plasmid and 0.5 μg of linearized BaculoGold virus DNA were added to a T25 flask containing ≈3×10^6/mL of *Sf9* cells in 5 mL of serum-supplemented media for 1 week at 27°C. A low multiplicity of infection viral stock (≈1×10^8 pfu/mL) was prepared by infecting 400 mL of *Sf9* cells at ≈1.2×10^6/mL with 0.5 mL of primary virus, and growth was continued until the majority of cells were no longer viable (7–10 days). Baculovirus-infected *Sf9* cells were used for expression and were grown at 27°C. Samples were generated by infecting 400 mL of *Sf9* cells at ≈1×10^6/mL with 8 mL of primary virus. Cells were harvested by 4–5 days after infection by centrifugation at 2,500 × *g* for 10 min, and frozen at −20°C.

2.3. Bacterial Expression of FLAP

Bacterial expression vectors were transformed into *E. coli* BL21(DE3). The cells were grown in LB media supplemented with 1× PBS pH 7, 10 g/L glycine, and 2% glycerol supplemented with 50 mg/mL kanamycin at 37°C until the OD_{600} reached ≈1. Protein expression was induced by the addition of 1.5 mM IPTG for 2 h at 30°C. The cells were harvested by centrifugation, washed once with PBS pH 7, and stored at −80°C. Selenomethionyl-labeled protein was obtained by expressing FLAP in *E. coli* B834(DE3) cells grown in M9 minimal media supplemented with 40 mg/L selenomethionine (26) and the protein was prepared in the same way as the native protein.

2.4. Purification of FLAP

Bacterial or insect cell pellets were thawed, resuspended in lysis buffer [20 mM Tris pH 7.4, 50 mM NaCl, 10% glycerol, 1 mM TCEP, and EDTA-free protease inhibitor tablets], and lysed using an EmulsiFlex-C5 system. Crude cell lysate was cleared by centrifugation at 7,000 rpm for 30 min at 4°C. Membranes were purified by centrifugation at 35,000 ×*g* for 2.5 h at 4°C, and resuspended in lysis buffer. Recombinant FLAP was solubilized with 2% dodecyl-β-D-maltopyranoside (DDM) for 45 min at room temperature with gentle agitation and centrifuged at 40,000 ×*g* for 30 min at 4°C. Supernatant containing His-tagged FLAP was mixed with 2 mL of NiNTA agarose for 1 h at 4°C. The slurry was loaded by gravity into an empty column and washed with 20 mM Tris pH 7.4, 150 mM NaCl, 20 mM imidazole, 1 mM TCEP, 10% glycerol, and 0.1% DDM. Detergent-exchange was performed by extensively washing the NiNTA agarose with 20 mM Tris pH 7.4, 150 mM NaCl, 1 mM TCEP, 10% glycerol, and 0.25% octaethylene glycol monododecyl ether ($C_{12}E_8$) prior to elution with 20 mM Tris pH 7.4, 150 mM NaCl, 250 mM imidazole, 1 mM TCEP, 10% glycerol, and 0.25% $C_{12}E_8$. FLAP was further purified by size exclusion chromatography using a Superdex 200 column equilibrated with 5 mM Tris pH 7.4, 20 mM NaCl, 1 mM TCEP, 2% glycerol, and 0.1% $C_{12}E_8$. Fractions were pooled and concentrated using a 10-kDa molecular weight cutoff centrifugal filter device. Protein homogeneity and identity were confirmed by SDS-PAGE, Western blotting, mass spectrometry, and N-terminal sequencing. A competitive radioligand-binding assay was used to confirm the inhibitor binding profile of detergent-solubilized FLAP, and purified membranes containing FLAP (27, 28).

2.5. Crystallization of FLAP

Concentrated protein solution was equilibrated with MK-591 or the iodinated structural analogue of MK-591 using a 1:1.1 protein-to-inhibitor ratio for 4 h at 4°C. Extensive crystallization screening failed to identify any promising crystal growth conditions in the absence of inhibitors for FLAP. Attempts to remove the C-terminal 6×His-tag using thrombin unexpectedly produced several lower molecular weight cleavage products. Sparse matrix screening was initially carried out with $FLAP_{1-161}$thrombin/6×His protein in a variety of primary detergents. Preliminary crystals appeared under several conditions using hanging drop vapor diffusion at 20°C. However, all initial crystals diffracted poorly. Additional sparse matrix screening was subsequently performed with $FLAP_{1-161}$6×His protein. Although both recombinant proteins crystallize under similar conditions, the diffraction properties of $FLAP_{1-161}$6×His crystals were superior to those of $FLAP_{1-161}$thrombin/6×His crystals.

The source of recombinant protein (insect vs. bacterial cells) and expected differences in bound endogenous lipids that might affect the internal crystalline order of $FLAP_{1-161}$6×His crystals were also evaluated. The absence or presence of endogenous lipids in purified integral membrane protein samples has a well documented effect on membrane protein crystallization (29–31). Mass spectrometry analysis of purified $FLAP_{1-161}$6×His protein expressed in insect cells revealed the presence of bound phosphoinositides.

In contrast, no bound lipids were observed with $FLAP_{1-161}$6×His protein expressed in *E. coli*. PIP strips (nitrocellulose membranes to which specific lipids have been noncovalently immobilized) were used to define which phosphoinositides bound to FLAP. Although individual and combinations of these phospholipids were added to $FLAP_{1-161}$6×His protein expressed in *E. coli* prior to crystallization, these experiments failed to improve the diffraction properties of these crystals. Moreover, $FLAP_{1-161}$6×His purified from insect cell membranes failed to crystallize under similar conditions.

Surface entropy reduction engineering was utilized in an attempt to obtain crystals with improved diffraction properties (32, 33). Two lysine residues (K116 and K148) and one aspartate residue (D62) were mutated to alanine, and both cysteine residues (C60 and C78) found in FLAP were jointly mutated to either alanine or serine, generating [K116A] $FLAP_{1-161}$6×His, [K148A] $FLAP_{1-161}$6×His, [D62A] $FLAP_{1-161}$6×His, [C60A/C78A] $FLAP_{1-161}$6×His, and [C60S/C78S] $FLAP_{1-161}$6×His. Given the apparent importance of the uncleaved C-terminal His-tag for the crystallization of $FLAP_{1-161}$6×His, two C-terminal deletion FLAP mutants (Δ148–161 and Δ154–161) in the [C60A/C78A] $FLAP_{1-161}$6×His background, producing [C60A/C78A] $FLAP_{1-147}$6×His and [C60A/C78A] $FLAP_{1-153}$6×His were also designed. Crystallization trials with this entire series of recombinant proteins revealed substantial differences in crystal morphology, size, reproducibility, and diffraction properties. No crystals were obtained with either [C60A/C78A] $FLAP_{1-147}$6×His or [C60A/C78A] $FLAP_{1-153}$6×His.

The highest quality crystals were obtained by mixing equal volumes (1 + 1 μL) of [K148A] $FLAP_{1-161}$6×His protein at 8 mg/L with a reservoir solution containing 0.1 M sodium citrate pH 5.6, 0.32 M $LiCl_2$, 6% PEG 6000, 1 mM TCEP, 0.25% C_8E_4, and 0.1% $C_{12}E_8$ at 20°C (34). It should be noted that [K148A] $FLAP_{1-161}$6×His protein purified in one-step (IMAC only) generated crystals with superior diffraction properties as compared to protein that had been purified in two-steps (IMAC followed by SEC). Size exclusion chromatography likely removes endogenous lipids that are needed to form superior quality protein crystals. Optimization of crystal size, appearance, mechanical stability, and reproducibility were conducted using sitting-drop vapor diffusion crystallization plates. Small crystals appeared overnight and grew to maximum dimensions of 400 × 200 × 200 μm within 2 weeks. Selenomethionyl-labeled [K148A] $FLAP_{1-161}$6×His crystals behaved similarly to the native protein.

2.6. Data Collection and Processing

Prior to data collection, the glycerol concentration in the reservoir solution was slowly raised to 20% over 2 h at 4°C (5% incremental additions), after which crystals were mounted in cryoloops. There was considerable variation in crystal quality, and extensive in-house screening of crystals was necessary to find suitable candidates for data collection. All X-ray diffraction data were collected at cryogenic temperature using synchrotron radiation. X-ray diffraction data was collected from native [K148A] $FLAP_{1-161}$6×His crystals in complex with MK-591. SAD diffraction data was collected from a selenomethionyl-labeled [K148A] $FLAP_{1-161}$6×His crystal in complex with the iodinated structural analogue of MK-591 at the peak of the selenium X-ray fluorescence spectrum (35). All diffraction data were processed with XDS (36). These crystals belong to the primitive tetragonal space group $P42_12$, with six molecules per asymmetric unit, and have a Matthews' coefficient of 4.97 Å^3/Da (corresponding to an estimated solvent content of 75%).

2.7. Low-Resolution Structure Determination and Refinement

Given the lack of a suitable structural homolog for FLAP, experimental phasing was required to solve the structure (25). Phase information was derived from a SAD diffraction experiment using selenomethionyl-substituted FLAP in complex with the iodinated structural analogue of MK-591. This iodinated structural analogue was specifically designed to provide an additional source of experimental phase information (see Fig. 2a, b). Eighteen selenium sites

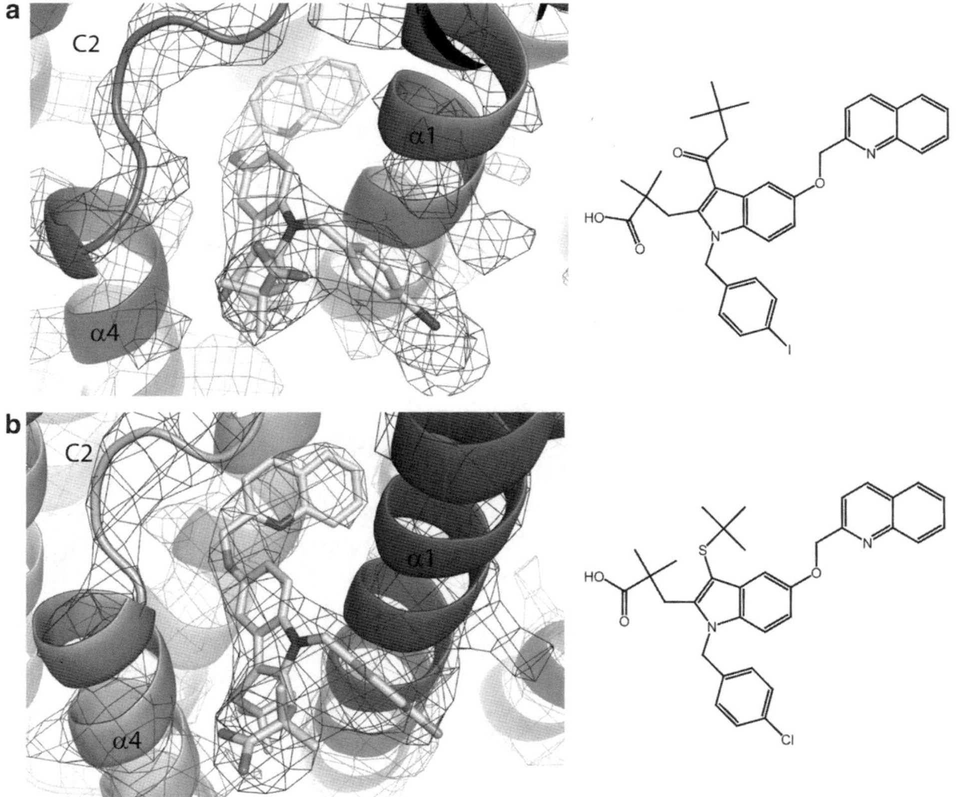

Fig. 2. The inhibitor binding site of FLAP. (**a**) The anomalous difference electron density map (contoured at 9σ) for FLAP in complex with the iodinated structural analogue of MK-591 is shown as mesh. The final sixfold NCS averaged $2F_o-F_c$ electron density map (contoured at 2σ) is shown as mesh. (**b**) The final sixfold NCS averaged $2F_o-F_c$ electron density map (contoured at 2σ) with MK-591 is shown as mesh.

were located and refined with SHARP (37, 38). The hand of the molecule was established by unambiguous differences observed in experimental electron density maps calculated from both phases of the heavy atom enantiomorphs. The selenium sites were used to define the threefold rotational symmetry of the trimer as well as the relationship between two trimers in the asymmetric unit. Six additional peaks in the anomalous difference map could not be associated with protein-related features. It was concluded that these peaks arose from the iodine atoms in the iodinated MK-591 structural analogue based on the following observations. Given that this diffraction data was collected at the peak of the selenium X-ray fluorescence spectrum, iodine has a limited anomalous signal at this wavelength (theoretical anomalous scattering values of $f' \approx 0$ and $f'' \approx 3$). The relative peak heights of these six anomalous peaks when compared to the 18 selenium sites are consistent with them corresponding to iodine atoms. The experimental electron density surrounding these six anomalous peaks also showed the expected features for the inhibitor (see Fig. 2a). The addition of the positions of the iodine atoms to the experimental phase refinement had a minimal effect on the refinement statistics and map quality, presumably due to the weak phasing power of iodine anomalous scattering at this wavelength.

The experimental electron density map was improved with density modification using a solvent fraction of 75%. The helix building module of ARP/wARP (37–41) was then used to automatically build two distinct helical bundles into the experimental electron density map, representing approximately 40% of the expected contents of the asymmetric unit. Those regions that could be automatically built with helices composed of alanine residues form the lipid-embedded regions of the transmembrane helices. However, no continuous segments of the FLAP protein sequence could be automatically fit into the experimental electron density map using experimental phases. The clarity of the experimental electron density map was further improved by iterative sixfold NCS averaging and temperature factor sharpening (42). The two bundles of polyalanine helices automatically built by ARP/wARP and the positions of the 18 selenomethionine residues and six bound inhibitor molecules were used as a guide to iteratively fit the protein sequence into the experimental electron density map. Cocrystallization of selenomethionyl-labeled [K148A] $FLAP_{1-161}$6×His with the iodinated structural analogue of MK-591 was essential to correctly interpret the experimental electron density map, assign the protein sequence, build the model, and to define the location of the inhibitor binding sites.

Macromolecular refinement utilized all reflection data and Hendrickson-Lattman coefficients, and was carried out using the BUSTER software package from Global Phasing (43). Rigid body refinement was used to (1) optimize the relative positions of the transmembrane helices that compose each FLAP monomer; (2) position the six copies of the FLAP monomer that are found in the asymmetric unit; and (3) establish the orientations of the two trimers in the asymmetric unit. It is not possible to interpret this packing arrangement as a biologically relevant dimer-of-trimers as both trimers are positioned essentially perpendicular to each other. Sixfold NCS restraints (main chain and side chain atoms) were applied to all residues during the early stages of refinement. NCS constraints were not used at any stage of the refinement. Once the protein sequence had been assigned, NCS restraints were released from the lumenal helix, the cytosolic loops, the lumenal loop, and all residues beyond residue 140. Although this partial release of NCS restraints results in distinct conformations for both cytosolic loops, the conformations and relative positions of the lumenal helices and loops remain unchanged. A drop in R_{free} was observed at this stage of the refinement. During the later stages of refinement, only partial NCS restraints (main chain atoms only) were maintained on the transmembrane helices (residues 11–37, 44–77, 80–101, and 116–140), and application of this restraint scheme produced an additional drop in R_{free}. Given the relatively modest resolution of this data additional rounds of refinement without NCS restraints were not used. Split main chain and side chain temperature factor refinement was used at the final stage of refinement, resulting in a marginal drop in R_{free} (42).

The structure of FLAP in complex with MK-591 was solved by molecular replacement using the coordinates of FLAP in complex with the iodinated structural analogue of MK-591 as the search model. Examination of the difference density map for FLAP in complex with MK-591 showed connected difference density with the expected molecular features of this inhibitor. The final refined models of both complexes were confirmed by inspection of omit maps, and have reasonable geometry. Bond lengths and angles are not distorted and the protein main chain has reasonable angles, with ≈80% of residues in the accepted regions of the Ramachandran plot, and most of the remaining approximately 20% of residues falling within the generously allowed regions. Sequential rounds of model building and refinement produced models for FLAP in complex with both inhibitors that contain all residues except for approximately 20 C-terminal residues as the number of visible residues in the electron density maps differs for each monomer.

3. Discussion

3.1. Description of the FLAP Structure

Each FLAP monomer is composed of four transmembrane α-helices (α1–α4) that are connected by two elongated cytosolic loops (C1 and C2) and one short lumenal loop (L1) (see Fig. 3a).

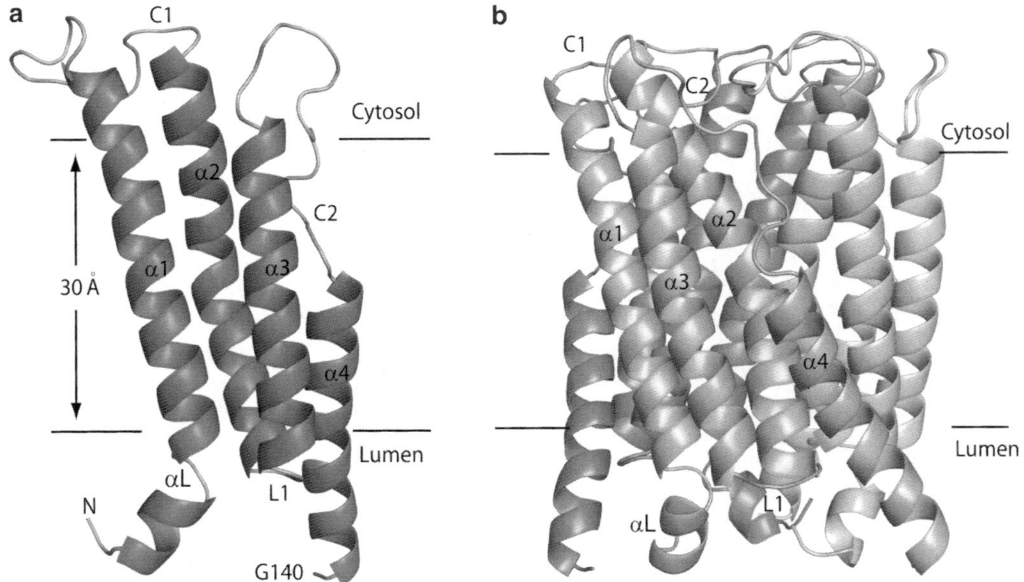

Fig. 3. Fold of the monomer. (**a**) Ribbon representation of the FLAP monomer as viewed parallel to the nuclear membrane. Structure of the trimer. (**b**) Ribbon representation of the FLAP trimer as viewed parallel to the nuclear membrane.

The orientations of the cytosolic and lumenal loops of FLAP are analogous to those of mGST-1 (44, 45) and LTC_4 synthase (46–48), and follow the "positive-inside" rule (49), in which positively charged residues are found more frequently on the cytosolic side of membranes. Lumenal helix αL (residues 3–8) is located in the lumen and precedes transmembrane helix α1 (residues 10–37). The lumenal helix is linked to transmembrane helix α1 by a single residue linker just outside the lower leaflet of the membrane. Cytosolic loop C1 (residues 38–47) connects the first to the second transmembrane helix α2 (residues 48–77), which both extend above the upper leaflet of the membrane. Transmembrane helix α2 also has a distinctive bend in it that is centered at residue P65 at the midpoint of the membrane. Lumenal loop L1 (residues 78–80) connects the second to the third transmembrane helix α3 (residues 81–101), which is perpendicular to the plane of the membrane. Cytosolic loop C2 (residues 102–115) connects the third to the fourth transmembrane helix α4 (residues 116–138). In contrast to the other helices, transmembrane helix α4 begins in the middle of the membrane and continues into the lumen. The C-terminus of FLAP is disordered beyond residue G140.

FLAP forms a homotrimer (see Fig. 3b), which is consistent with previous data indicating the trimeric arrangement of FLAP and other MAPEG proteins (25, 44–48). Moreover, the fold of FLAP is in agreement with the electron crystallographic structure of rat mGST-1 (50), the projection maps for human LTC_4 synthase

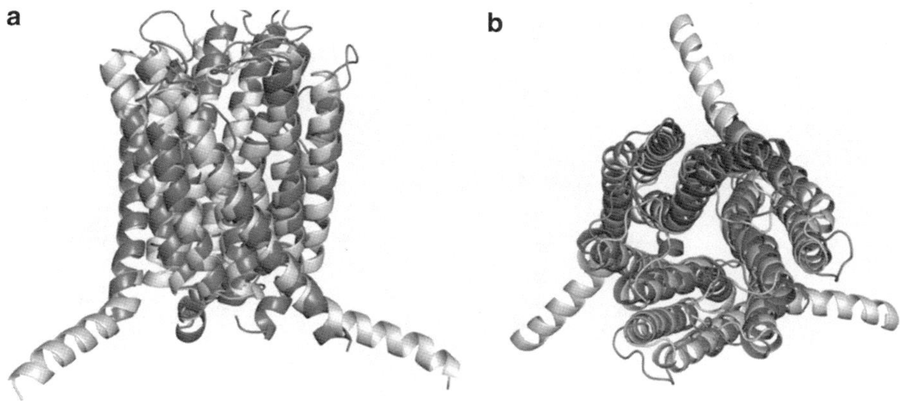

Fig. 4. Structural comparison of the FLAP (*dark*) and LTC_4 synthase (*light*) trimer. (**a**) The view is given parallel to the nuclear membrane; and (**b**) The view is given from the cytosol.

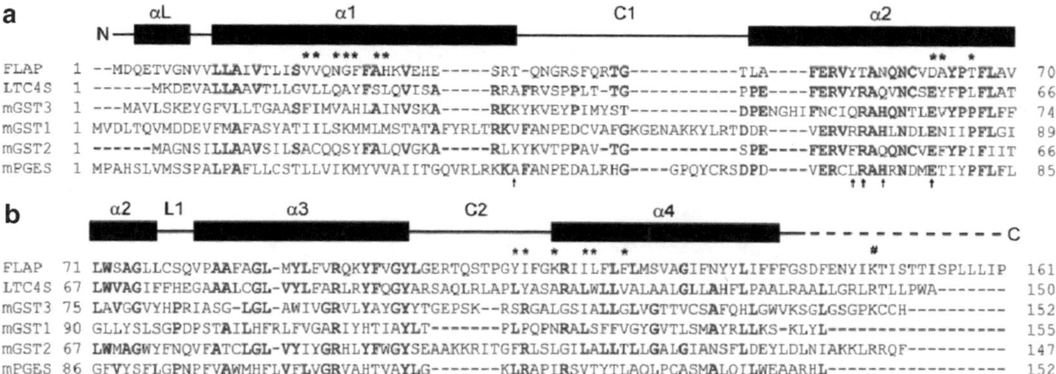

Fig. 5. Sequence alignment of the MAPEG superfamily. *Black boxes* above the sequence alignment correspond to α-helices. *Bold sequence* coloring indicates those regions of substantial primary sequence homology. Those residues shown to be involved in glutathione binding in other MAPEGs are indicated with an *arrow*. An *asterisk* has been placed above those residues that are involved in inhibitor binding in FLAP. The Derewenda mutation at residue K148A is indicated. The *dashed line* represents those residues (141–161) that are disordered in the FLAP structure.

(51) and mPGES-1 (52), and the crystal structure of human LTC_4 synthase (47, 48). Following the determination of the human FLAP structure (25), two independently determined crystal structures of human LTC_4 synthase with bound glutathione were solved (47, 48). Comparing these MAPEG structures shows that they are structural homologues, even though they share minimal primary sequence identity (a testament to the painstaking structural validation that was used to confirm the FLAP structure) (see Figs. 4 and 5). It should be noted that the residues reported to form the glutathione binding site in the structure of rat mGST-1 (50) and human LTC_4 synthase (47, 48) are not conserved in FLAP (see Fig. 5). Structural comparison of the FLAP and LTC_4 synthase structures demonstrates that the glutathione and MK-591 binding sites are spatial distinct, and those residues that form interactions

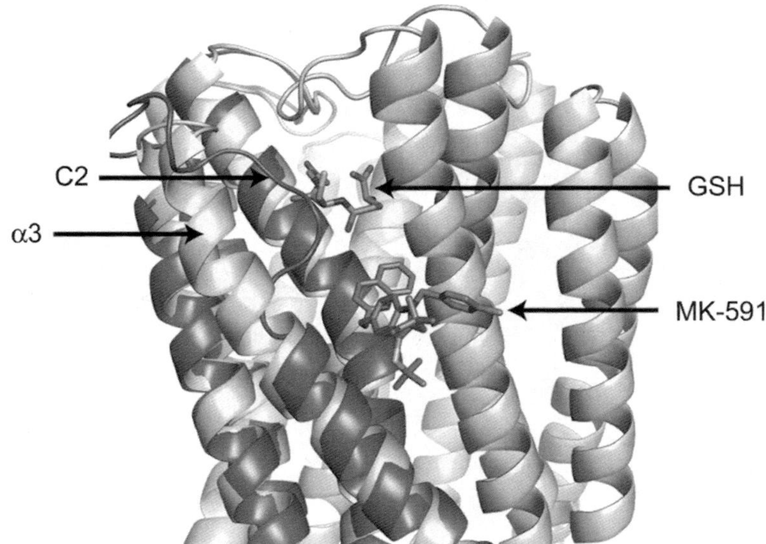

Fig. 6. Comparison of the MK-591 and glutathione binding sites. FLAP (*dark*) and LTC_4 synthase (*light*).

with glutathione in LTC_4 synthase are not conserved in FLAP (see Fig. 6). Cytosolic loop C2 of FLAP physically overlaps with the glutathione molecule found in LTC_4 synthase. The relative lengths of transmembrane helix α3 and cytosolic loop C2 represent the principle structural differences between FLAP and LTC_4 synthase (see Fig. 6).

The FLAP trimer has a flattened cytosolic top and a slightly pointed lumenal base (see Fig. 3b). The cytosolic loops close the top of the trimer, and the lumenal helices and loops form the bottom of the molecule. The electron density and temperature factors associated with the cytosolic loops indicate that they are dynamic and inherently flexible. The threefold symmetry axis of the trimer is perpendicular to the plane of the membrane, and when viewed along that plane, the FLAP trimer resembles a cylinder. When viewed in this orientation, the molecule is 60°Å high and 36°Å wide and is positioned in the center of the lipid bilayer, protruding approximately 15°Å from either side of the membrane.

The trimer forms extensive intersubunit contacts with each monomer burying approximately 4,900°Å^2. On both sides of the membrane and within the bilayer, the cytosolic loops, the lumenal helix, and transmembrane helices α1, α2, and α4 form an elaborate network of intermolecular contacts. Unexpectedly, transmembrane helix α3 and the lumenal loop do not form any intermolecular contacts within the membrane. There are also three polar interactions (positioned near the rotational symmetry axis of the trimer) that are formed between adjacent monomers (N23-T66, Q58-Q58,

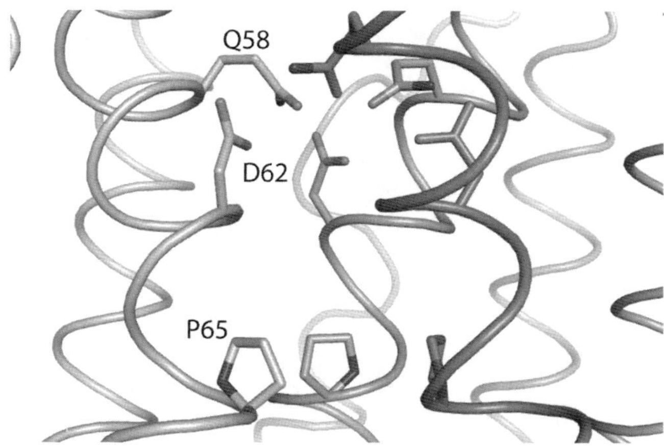

Fig. 7. Close-up of the membrane-embedded constriction rings of FLAP. A symmetrical ring of P65 residues is placed below the negatively charged symmetrical ring of Q58 and D62 residues.

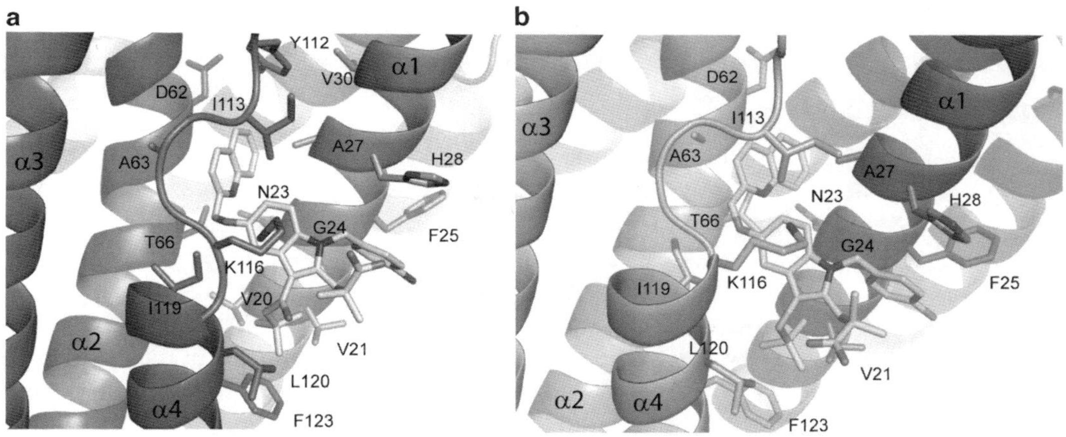

Fig. 8. The inhibitor binding site of FLAP in complex with (**a**) the iodinated structural analogue of MK-591, and (**b**) MK-591. All side chains within 4 Å of these inhibitors are shown.

and Q58-D62) (see Fig. 7). Those hydrophobic residues line the membrane-embedded portion of the intersubunit interfaces and form extensive van der Waals contacts.

3.2. Inhibitor Binding Site

Three 750 Å^3 grooves are located between adjacent FLAP monomers on the lipid-exposed surface of the trimer (see Fig. 8a, b). Within these hydrophobic surface grooves, clear connected non-protein electron density was observed—strongly suggesting that these were the inhibitor binding sites. This assignment was based upon three observations: (1) anomalous difference maps define the positions of six noncovalently bound iodinated MK-591 structural analogue molecules in the asymmetric unit (one inhibitor molecule per FLAP monomer); (2) similar nonprotein features are observed at the equivalent positions in the electron density maps of FLAP in complex with both inhibitors; and (3) the electron density in this region is the correct size and expected shape for these inhibitors.

Inhibitors intercalate between the monomers in the lipid-exposed hydrophobic grooves of the trimer, and form interactions with a common set of residues (see Fig. 8a, b). These inhibitors form primarily nonpolar van der Waals interactions with residues V20, V21, F25, and A27 from transmembrane helix α1, residues Y112 and I113 from cytosolic loop C2, residue A63 from transmembrane helix α2, and residues I119, L120, and F123 from transmembrane helix α4 of the adjacent monomer. In addition, a limited number of weak polar interactions are also formed with residue N23 from transmembrane helix α1, residues D62 and T66 from transmembrane helix α2, and residue K116 from transmembrane helix α4 of the neighboring subunit.

Given the modest resolution of the electron density maps and the unexpected location of the inhibitor binding site, extensive directed mutagenesis was necessary to validate these findings by measuring the binding of radiolabeled MK-591 to mutant forms of FLAP (Table 1). These data demonstrated that mutating those residues in close proximity (≤4 Å) to the inhibitor causes a dramatic decrease in binding affinity. Unexpectedly, several directed mutations increased inhibitor binding affinity (V20A, I113A, K116A, and Δ1–10). The insensitivity of mutations distant from the bound inhibitor (>4 Å) suggests that the observed loss in inhibitor binding affinity was a direct result of mutating proximal residues and not due to aberrant folding of these mutants. Collectively, these data provided strong support for the biological relevance of the inhibitor binding site identified in the structure.

Previous studies defined regions of FLAP that were important for inhibitor binding (20, 53). Inhibitor binding is dramatically reduced or abolished by removing residues 37–53, 42–61, 48–53, 48–61, 52–58, 53–54, 55–56, 57–58, 59–61, and by generating the directed mutants D62N from the membrane-embedded portion of transmembrane helix α2 and Y101A from the cytosolic end of transmembrane helix α3. In contrast, removing residues 33–36, 33–41, 37–41, 42–47, 106–108, 148–161, and directed mutants C60A, D62Q, and W72F from transmembrane helix α2, and D142A and E144A following transmembrane helix α4 have no effect. When viewed in the context of the FLAP structure, most of these loss-of-function deletions are at the cytosolic surface of the trimer. Jointly removing the cytosolic portions of transmembrane helices α1 and α2 and cytosolic loop C1 would substantially alter the FLAP structure, providing a likely explanation for the observed loss in inhibitor binding. In contrast, removing a short cytosolic piece of transmembrane helix α1 and/or cytosolic loop C1 or three residues from cytosolic loop C2 would cause a minimal structural

Table 1
Effect of directed mutagenesis on inhibitor binding to FLAP

	Mutation	Position	IC_{50} (nM)
Group 1	WT	–	4.2
	V20A	α1	0.2
	V21A	α1	35.2
	G24A	α1	7.9
	F25A	α1	8.2
	A27V	α1	198.2
	H28A	α1	3.7
	D62A	α2	39.7
	A63V	α2	14.4
	T66A	α2	330.6
	Y112A	C2	366.8
	I113A	C2	1.8
	K116A	C2	0.05
	I119A	α4	5.8
	L120A	α4	7.5
	F123A	α4	22.3
Group 2	F26A	α1	4.7
	K29A	α1	15.0
	V30A	α1	270.1
	E31A	α1	9.6
	H32A	α1	6.2
	Δ1-10	αL	0.8

Western blotting was used to demonstrate that equal amounts of protein are expressed for these directed mutants. All residues found within 4 Å of the bound iodinated structural analogue of MK-591 belong to Group 1. Those residues positioned greater than 4 Å from the inhibitor molecule belong to Group 2

perturbation and would not affect inhibitor binding. Of the previously described directed mutants, only residue D62 is positioned within 4 Å of these inhibitors.

Inhibitors must access their binding sites by initially entering the cytosolic end of FLAP. Once sequestered from the solvent, they would move down the inward-facing cytosolic portions of transmembrane helices α1, α2, and α4 of the trimer toward their binding sites. The flexibility of the cytosolic loops of FLAP suggests that they may facilitate this process. The dramatic loss in binding affinity for directed mutants V30A from transmembrane helix α1 and Y112A from cytosolic loop C2, provide support for this proposal (Table 1). When these inhibitors reach their binding sites, FLAP assumes a conformation which no longer binds arachidonic acid.

The hypothesis for partially overlapping binding sites for inhibitors and arachidonic acid is further supported by an earlier study that defined regions of FLAP that are essential for leukotriene biosynthesis catalyzed by 5-LO (54). FLAP deletion mutants

Δ106–108 and Δ148–161 are unable to stimulate arachidonic acid utilization by 5-LO, although they remain fully capable of binding inhibitors—an observation which is in agreement with the composition of the inhibitor binding site. Removing residues 106–108 from cytosolic loop C2 may prevent the formation of the 5-LO/FLAP complex, which in turn would stop the transfer of arachidonic acid to 5-LO.

3.3. Modelling Compounds into the Inhibitor-Binding Site

Using the FLAP structure in complex with MK-591 as the template, MK-886 (13), DG-031 (14) and VML-530 (55) were modeled into the inhibitor binding site of FLAP (see Fig. 9a–d) (3).

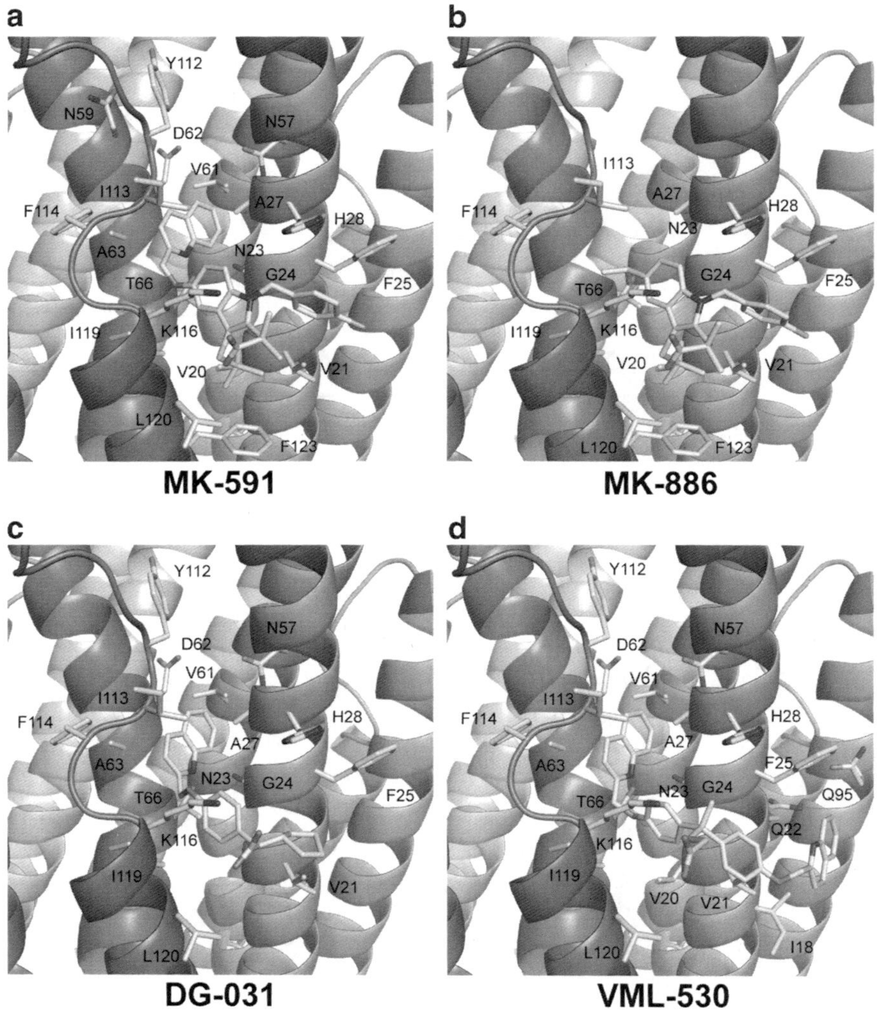

Fig. 9. Inhibitor binding site of FLAP. (**a**) MK-591, (**b**) MK-886, (**c**) BAY X1005 (DG-031), and (**d**) ABT-080 (VML-530). All residues within 4 Å of the bound (or modeled) inhibitor are shown as sticks.

Expectedly, a common set of residues forms interactions with these inhibitors. DG-031 contains a chiral center and the (*R*)-enantiomer is considerably more potent compared to the (*S*)-enantiomer. These results are consistent with this observation as the cyclopentyl ring in the (*R*)-enantiomer occupies a similar position as the chlorobenzyl group of MK-591 and the acid groups have similar orientations. In contrast, the cyclopentyl ring in the (*S*)-enantiomer would occupy a less favorable position. VML-530 is predicted to bind so that its acid group is bound in an equivalent orientation as the acid group of MK-591. The second quinoline group of VML-530 would be directed toward the lipid-exposed region of the inhibitor binding site.

Figure 10 shows the key binding elements for MK-591 binding to FLAP, and explains why bulky replacements for the quinoline group of MK-591 are not tolerated as the binding pocket is spatially restricted. The structure also clarifies why FLAP could be purified with a MK-886 analogue coupled to Sephadex beads when attached to the left hand acid group, which is positioned toward the membrane (17). It is clear from the structure that substituents appended to the chlorobenzyl group of MK-886 or MK-591 will be directed toward the membrane. A new series of indole-class FLAP inhibitors with substituents at this position maintain potency for both FLAP binding and leukotriene biosynthesis inhibition (56). One of these third-generation indole-class compounds AM-103 has completed a Phase I clinical study in healthy volunteers and shown effective inhibition of leukotriene biosynthesis (57). FLAP inhibitor AM-103 (GSK-2190914) and a related inhibitor GSK-2190915 are being

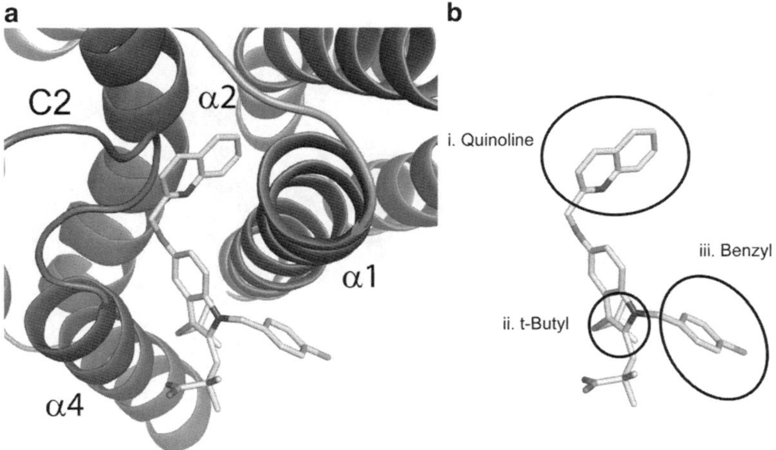

Fig. 10. Inhibitor binding site of FLAP. (**a**) Overview of the spatial restrictions for the design of novel FLAP inhibitors as viewed from the cytosol. (**b**) The key inhibitor regions are circled. (i) The quinoline binding pocket is positioned deep within the FLAP trimer, and there is a limited amount of space available for chemical modification. (ii) The tert-butyl binding pocket is formed by a series of lipophilic residues. (iii) The benzyl group is placed in close proximity to transmembrane helix α1 and is directly exposed to the membrane.

progressed to Phase II clinical trials in patients with asthma by GlaxoSmithKline under license from Amira Pharmaceuticals (www.amirapharm.com) (58, 59). Recent work has lead to the development of another potent FLAP inhibitor with excellent physical properties, designated AM-679 (60).

3.4. Central Pocket of FLAP

Perhaps the most remarkable feature of the FLAP structure is a conspicuous pocket with an internal volume of approximately 3,200 Å^3 found at the bottom of the trimer that is open to the lumen (see Fig. 11a–d). The entrance of this cone-shaped pocket has an internal diameter of approximately 6 Å and is formed by the lumenal helices and loops that define three distinctive entry crevices on the lumenal surface of the trimer. The central pocket extends to the middle of the membrane where it is physically

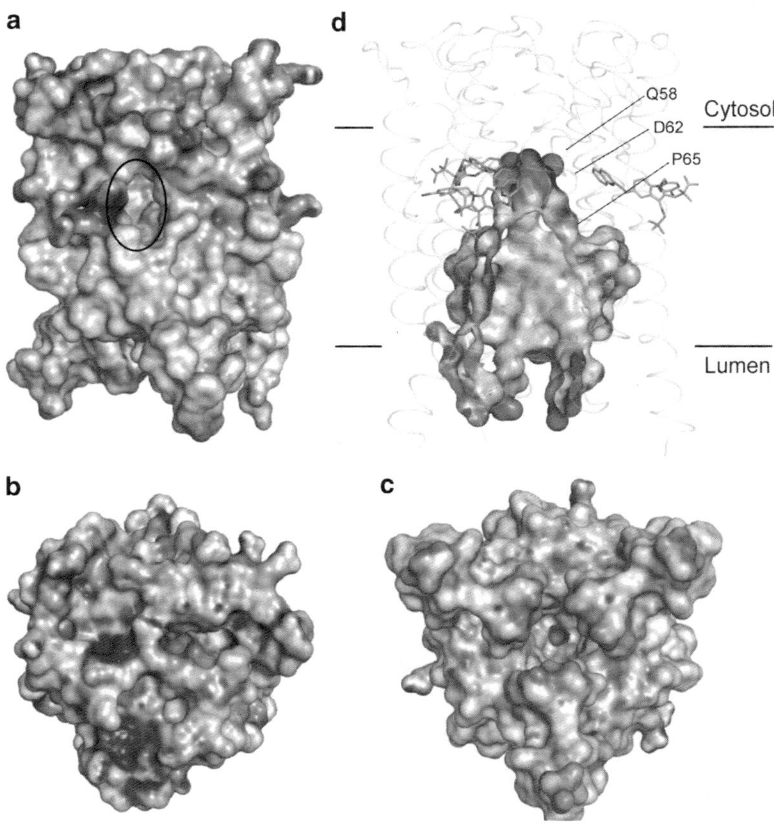

Fig. 11. Molecular surface of the FLAP trimer. The view is given (**a**) parallel to the nuclear membrane, (**b**) from the cytosol, and (**c**) from the lumen. One of the three ~750 Å^3 hydrophobic grooves found on the lipid-exposed surface of the trimer has been *circled*. The asymmetric distribution of electrostatic potential demonstrates that the cytosolic and lumenal ends of the trimer are positively charged and negatively charged, respectively. (**d**) Central pocket of FLAP. The front of this pocket has been removed for clarity. The negatively charged lumenal entrance to this pocket is formed by the lumenal helices. The negatively charged constriction within the membrane is formed by residues Q58 and D62.

constricted by a symmetric ring of P65 residues (minimum radius of approximately 0.5 Å); coinciding with the bend in transmembrane helix α2. Proline-induced distortions in transmembrane helices are known to be involved in signal transduction and solute transport across membranes by acting as molecular hinges (61). An additional symmetric ring, formed by residues Q58 and D62 from transmembrane helix α2, sits atop this point (see Figs. 7 and 11c). The asymmetric distribution of the electrostatic potential of this pocket is striking—its lumenal entrance and membrane-embedded constriction point are negatively charged, whereas the interior lining is largely hydrophobic (see Fig. 11d).

3.5. Proposed Mechanism of Arachidonic Acid Transfer

The structures presented here establish the inhibited conformation of FLAP, which is incapable of binding arachidonic acid. Inhibitors compete with arachidonic acid for binding to FLAP and block the transfer of arachidonic acid to 5-LO. Despite years of intensive study, the molecular basis of how FLAP binds and transfers arachidonic acid to 5-LO remains undefined. By combining this structural analysis with previous data, a model can be proposed to explain how FLAP transfers arachidonic acid to 5-LO and how inhibitors prevent this process (see Fig. 12).

The transfer begins with the calcium-dependent release of arachidonic acid by $cPLA_2$ from the nuclear membrane (7, 8). Arachidonic acid molecules then laterally diffuse within the plane of the membrane and bind in the three hydrophobic surface grooves that contain bound inhibitors. 5-LO is now capable of binding FLAP, and the positively charged cytosolic loops of the trimer likely mediate this interaction (62). Formation of the 5-LO/FLAP complex then drives the translocation of arachidonic acid. Although the precise trigger for these events remains to be established,

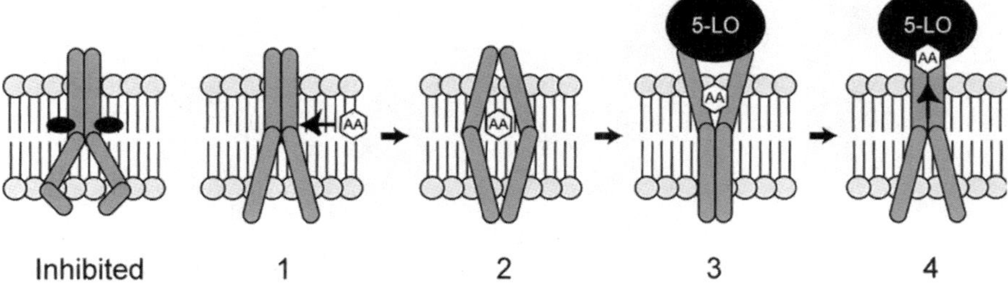

Fig. 12. Proposed arachidonic acid transfer mechanism. In this cartoon, the structure of inhibited FLAP is represented on the *left* and *black ovals* represent bound inhibitors. In this conformation, FLAP in incapable of binding arachidonic acid. *Step 1* FLAP-mediated transport of arachidonic acid begins with the release of arachidonic acid by $cPLA_2$ from the nuclear membrane. *Step 2* Arachidonic acid then enters the inhibitor binding pocket of FLAP. Following arachidonic acid binding, FLAP presumably undergoes a conformational change. *Step 3* 5-LO binds to the cytosolic loops of FLAP. Formation of the 5-LO/FLAP complex triggers the translocation of arachidonic acid to 5-LO. *Step 4* Once loaded with arachidonic acid, 5-LO converts arachidonic acid to LTA_4. LTA_4 may then be exported from the cell for transcellular biosynthesis of leukotrienes or converted to LTB_4 by LTA_4 hydrolase. Alternatively, LTA_4 may be transferred back into the cytosolic pocket of FLAP and then diffuse toward the lipid-exposed inhibitor binding site, and then transferred laterally to LTC_4 synthase.

5-LO binding to the cytosolic surface of FLAP provides a logical candidate.

Fatty acid binding is not strictly specific for arachidonic acid as FLAP can bind other unsaturated fatty acids with similar affinity, and these fatty acids can also compete with inhibitors for specific binding to FLAP in a concentration-dependent manner (54, 63, 64). Although the presence of a free carboxyl group on the fatty acid is not required for binding, the degree of saturation does significantly affect binding (63). Previous mutagenesis studies suggest that there is partial overlap between the arachidonic acid and inhibitor binding sites of FLAP (20, 28, 53). Access to these lipid-exposed binding grooves may be facilitated by the inherent flexibility of cytosolic loop C2 (see Fig. 6), which extends to just above the midpoint of the membrane and is followed by transmembrane helix α4. These structural features suggest that there is a point within the plane of the membrane, the inhibitor-binding site, for arachidonic acid to bind FLAP.

Can the structure of LTC_4 synthase provide any insights into the location of the arachidonic acid binding site of FLAP? Interestingly, there is a detergent molecule bound in cytosolic vestibule of LTC_4 synthase. Comparing the chemical structures of arachidonic acid and the substrate for LTC_4 synthase, LTA_4, illustrates clear structural similarities. Moreover, the acyl chain of the detergent physically overlaps with the MK-591 molecule in FLAP (see Fig. 13). The hypothesis for partially overlapping binding sites in FLAP for inhibitors and arachidonic acid is further supported by these structural observations, and suggests that arachidonic acid and LTA_4 likely bind in this region of FLAP.

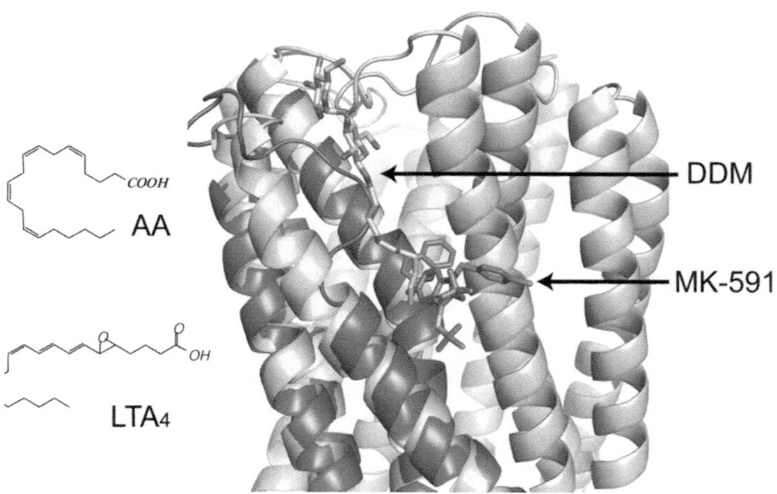

Fig. 13. Potential arachidonic acid binding site of FLAP. Comparison of the chemical structures of arachidonic acid and LTA_4; and the relative positions of bound DDM molecule in LTC_4 synthase (*light*) and MK-591 in FLAP (*dark*).

Once loaded with arachidonic acid, 5-LO converts arachidonic acid to 5-HpETE and then LTA_4. LTA_4 may be exported from the cell for transcellular biosynthesis of other leukotrienes or be released into the cytosol for conversion to LTB_4 by leukotriene A_4 hydrolase (12). It seems plausible that LTA_4 may be released back into the cytosolic pocket of FLAP, from where it diffuses toward the lipid-embedded inhibitor binding site. The lateral transfer of LTA_4 from FLAP to LTC_4 synthase within the nuclear membrane is fully consistent with recent observations showing that these proteins colocalize in the outer nuclear membrane, the relative alignment of the glutathione binding site of LTC_4 synthase and the inhibitor binding site of FLAP, and that these MAPEGs form functional macromolecular complexes in vivo and in vitro (65–67). The physical interaction between FLAP and LTC_4 synthase appears to be mediated by the membrane spanning portions of the respective homotrimers (67, 68).

4. Summary

This chapter describes the structure of human FLAP in complex with two inhibitors. This structure also rationally explains how FLAP can function as both a nuclear membrane scaffold protein for 5-LO and an arachidonic acid transport protein. However, many important biological questions remain unanswered, including the structural determinants of 5-LO binding, the precise details of the arachidonic acid translocation pathway, and most importantly, the conformational changes that accompany each step in the transport cycle. Leukotrienes have an established role in respiratory and cardiovascular disease. This structure provides a novel and valuable tool for the development of novel therapeutics and for the understanding of leukotriene biosynthesis.

References

1. Funk, C.D. (2001) Prostaglandins and leukotrienes: advances in eicosanoid biology. *Science* **294**, 1871–1875.
2. Funk, C.D. (2005) Leukotriene modifiers as potential therapeutics for cardiovascular disease. *Nat Drug Disc* **4**, 664–672.
3. Evans, J.F., Ferguson, A.D., Mosley, R.T., and Hutchinson, J.H. (2008) What's all the FLAP about?: 5-lipoxygenase-activating protein inhibitors for inflammatory diseases. *Trends Pharmacol Sci* **292**, 72–78.
4. Riccioni, G., Capra, V., D'Orazio, N., Bucciarelli, T., and Bazzano, L.A. (2008) Leukotriene modifiers in the treatment of cardiovascular diseases. *J Leukoc Biol.* **8**, 1374–1378.
5. Rådmark, O., and Samuelsson, B. (2009). 5-Lipoxygenase: mechanisms of regulation. *J Lipid Res* **50**, 40–45.
6. Sampson, A.P. (2009) FLAP inhibitors for the treatment of inflammatory diseases. *Curr Opin Invest Drugs* **10**, 1163–1172.
7. Evans, J., Spencer, D., Zweifach, A., and Leslie, C. (2001) Intracellular calcium signals regulating cytosolic phospholipase A2 translocation to internal membranes. *J Biol Chem* **276**, 30150–30160.

8. Luo, M., Flamand, N., and Brock, T. (2006) Metabolism of arachidonic acid to eicosanoids within the nucleus. *Biochim Biophys Acta* **1761**, 618–625.
9. Dixon, R.A.F. *et al.* (1990) Requirement of a 5-lipoxygenase-activating protein for leukotriene synthesis. *Nature* **343**, 282–284.
10. Miller, D.K. *et al.* (1990) Identification and isolation of a membrane protein necessary for leukotriene production. *Nature* **343**, 278–281.
11. Brock, T.G., McNish, R.W., and Peters-Golden, M. (1995) Translocation and leukotriene synthetic capacity of nuclear 5-lipoxygenase in rat basophilic leukemia cells and alveolar macrophages. *J Biol Chem* **270**, 21652–21658.
12. Folco, G., and Murphy, R.C. (2006) Eicosanoid transcellular biosynthesis: from cell-cell interactions to in vivo tissue responses. *Pharmacol Rev* **58**, 375–388.
13. Gillard, J. *et al.* (1989) L-663,536 (MK-886) (3-[1-(4-chlorobenzyl)-3-t-butyl-thio-5-isopropylindol-2-yl]-2,2-dimethylpropanoic acid), a novel, orally active leukotriene biosynthesis inhibitor. *Can J Physiol Pharmacol* **67**, 456–464.
14. Müller-Peddinghaus, R. *et al.* (1993) BAY X1005, a new inhibitor of leukotriene synthesis: *In vivo* inflammation pharmacology and pharmacokinetics. *J Pharm Exp Ther* **267**, 51–57.
15. Brideau, C. *et al.* (1992) Pharmacology of MK-0591 (3-[1-(4-chlorobenzyl)-3-(t-butylthio)-5-(quinolin-2-ylmethoxy)-indol-2-yl]-2,2-dimethylpropanoic acid), a potent, orally active leukotriene biosynthesis inhibitor. *Can J Physiol Pharmacol* **70**, 799–807.
16. Prasit, P. *et al.* (1993) A new class of leukotriene biosynthesis inhibitor: the development of MK-0591. *J Lipid Mediat* **6**, 239–244.
17. Evans, J.F. *et al.* (1991) 5-Lipoxygenase-activating protein is the target of a quinoline class of leukotriene synthesis inhibitors. *Mol Pharmacol* **40**, 22–27.
18. Woods, J. *et al.* (1993) 5-lipoxygenase and 5-lipoxygenase-activating protein are localized in the nuclear envelope of activated human leukocytes. *J Exp Med* **178**, 1935–1946.
19. Mancini, J. *et al.* (1993) 5-lipoxygenase-activating protein is an arachidonate binding protein. *FEBS Letters* **318**, 277–281.
20. Mancini, J., Coppolino, M., Klassen, J., Charleson, S., and Vickers, P. (1994) The binding of leukotriene biosynthesis inhibitors to site-directed mutants of human 5-lipoxygenase-activating protein. *Life Sci* **54**, 137–142.
21. Rouzer, C., Ford-Hutchinson, A., Morton, H., and Gillard, J. (1990) MK886, a potent and specific leukotriene biosynthesis inhibitor blocks and reverses the membrane association of 5-lipoxygenase in ionophore-challenged leukocytes. *J Biol Chem* **265**, 1436–1442.
22. Jakobsson, P.J., Morgenstern, R., Mancini, J., Ford-Hutchinson, A., and Persson, B. (1999) Common structural features of MAPEG - a widespread superfamily of membrane associated proteins with highly divergent functions in eicosanoid and glutathione metabolism. *Protein Sci* **8**, 689–692.
23. Bresell, A. *et al.* (2005). Bioinformatic and enzymatic characterization of the MAPEG superfamily. *FEBS J* **272**, 1688–1703.
24. Plante, H., Picard, S., Mancini, J., and Borgeat, P. (2006) 5-Lipoxygenase-activating protein homodimer in human neutrophils: evidence for a role in leukotriene biosynthesis. *Biochem J* **393**, 211–218.
25. Ferguson, A.D. *et al.* (2007) Crystal structure of inhibitor bound 5-lipoxygenase-activating protein. *Science* **317**, 510–512.
26. Doublié, S. (1997) Preparation of selenomethionyl proteins for phase determination. *Meth Enzymol* **276**, 523–530.
27. Charleson, S. *et al.* (1992) Characterization of a 5-lipoxygenase-activating protein binding assay: Correlation of affinity for 5-lipoxygenase-activating protein with leukotriene synthesis inhibition. *Mol Pharmacol* **41**, 873–879.
28. Vickers, P.J. *et al.* (1992) Identification of amino acid residues of 5-lipoxygenase-activating protein essential for the binding of leukotriene biosynthesis inhibitors. *Mol Pharmacol* **42**, 94–102.
29. Seddon, A.M., Curnow, P., and Booth, P.J. (2004). Membrane proteins, lipids and detergents: not just a soap opera. *Biochimica et Biophysica Acta* **1666**, 105–117.
30. Palsdottir, H., and Hunte, C. (2004). Lipids in membrane protein structures *Biochimica et Biophysica Acta* **1666**, 2–18.
31. Lee, A.G. (2004). How lipids affect the activities of integral membrane proteins. *Biochimica et Biophysica Acta* **1666**, 62–87.
32. Derewenda, Z.S. (2004). Rational protein crystallization by mutational surface engineering. *Structure* **12**, 529–5535.
33. Derewenda, Z.S., and Vekilov, P.G. (2006). Entropy and surface engineering in protein crystallization. *Acta Cryst D* **62**, 116–124.
34. Xu, S. *et al.* (2007) Expression, purification and crystallization of human 5-lipoxygenase-activating protein with leukotriene-biosynthesis inhibitors. *Acta Cryst F* **63**, 1054–1057.
35. Hendrickson, W.A. (1991) Determination of macromolecular structures from anomalous

diffraction of synchrotron radiation. *Science* **254**, 51–58.

36. Kabsch, W. (2010) *XDS. Acta Cryst D* **66**, 125–132.
37. Bricogne, G., Vonrhein, C., Flensburg, C., Schiltz, M., and Paciorek, W. (2003) Generation, representation and flow of phase information in structure determination: recent developments in and around SHARP 2.0. *Acta Cryst D* **59**, 2023–2030.
38. Vonrhein, C., Blanc, E., Roversi, P., and Bricogne, G. (2006). Automated structure solution with autoSHARP. *Meth Mol Biol* **364**, 215–230.
39. Morris, R.J., Perrakis, A., and Lamzin, V.S. (2003) ARP/wARP and automatic interpretation of protein electron density maps. *Meth Enzymol* **374**, 229–244.
40. Morris, R.J. *et al.* (2004) Breaking good resolutions with ARP/wARP. *J Synchrotron Radiat* **11**, 56–59.
41. Cohen, S.X. *et al.* (2004) Towards complete validated models in the next generation of ARP/wARP. *Acta Cryst D* **60**, 2222–2229.
42. Delabarre, B., and Brunger, A.T. (2006) Considerations for the refinement of low-resolution crystal structures. *Acta Cryst D* **62**, 923–932.
43. Blanc, E. *et al.* (2004) Refinement of severely incomplete structures with maximum likelihood in BUSTER-TNT. *Acta Cryst D* **60**, 2210–2221.
44. Andersson, C., Weinander, R., Lundqvist, G., DePierre, J.W., and Morgenstern, R. (1994) Functional and structural membrane topology of rat liver microsomal glutathione transferase. *Biochim Biophys Acta* **1204**, 298–304.
45. Weinander, R. *et al.* (1997) Structural and functional aspects of rat microsomal glutathione transferase. The roles of cysteine 49, arginine 107, lysine 67, histidine, and tyrosine residues. *J Biol Chem* **272**, 8871–8877.
46. Lam, B.K., Penrose, J.F., Xu, K., Baldasaro, M.H., and Austen, K.F. (1997) Site-directed mutagenesis of human leukotriene C4 synthase. *J Biol Chem* **272**, 13923–13928.
47. Ago, H. *et al.* (2007) Crystal structure of a human membrane protein involved in cysteinyl leukotriene biosynthesis. *Nature* **448**, 609–612.
48. Molina, D.M. *et al.* (2007) Structural basis for synthesis of inflammatory mediators by human leukotriene C_4 synthase. *Nature* **448**, 613–616.
49. von Heijne, G. (2006) Membrane-protein topology. *Nat Rev Mol Cell Biol* 7, 909–918.
50. Holm, P. *et al.* (2006) Structural basis for detoxification and oxidative stress protection in membranes. *J Mol Biol* **360**, 934–945.
51. Schmidt-Krey, I. *et al.* (2004) Human leukotriene C(4) synthase at 4.5 A resolution in projection. *Structure* **12**, 2009–2014.
52. Thoren, S. *et al.* (2003) Human microsomal prostaglandin E synthase-1: purification, functional characterization, and projection structure determination. *J Biol Chem* **278**, 22199–22209.
53. Vickers, P. *et al.* (1993) Amino acid residues of 5-lipoxygenase-activating protein critical for the binding of leukotriene biosynthesis inhibitors. *J Lipid Mediat* **6**, 31–42.
54. Mancini, J.A., Waterman, H., and Riendeau, D. (1998) Cellular oxygenation of 12-hydroxyeicosatetraenoic acid and 15-hydroxyeicosatetraenoic acid by 5-lipoxygenase is stimulated by 5-lipoxygenase-activating protein. *J Biol Chem* **273**, 32842–32847.
55. Kolasa, T. *et al.* (2000) Symmetrical Bis (heteroarylmethoxyphenyl)alkylcarboxylic acids as inhibitors of leukotriene biosynthesis. *J Med Chem* **43**, 3322–3334.
56. Macdonald, D. *et al.* (2008) Substituted 2,2-bisaryl-bicycloheptanes as novel and potent inhibitors of 5-lipoxygenase activating protein. *Bioorg Med Chem Lett* **18**, 2023–2027.
57. Hutchinson, J.H. *et al.* (2009) 5-lipoxygenase-activating protein inhibitors: development of 3-[3-tert-butylsulfanyl-1-[4-(6-methoxy-pyridin-3-yl)-benzyl]-5-(pyridin-2-yl-methoxy)-1 H-indol-2-yl]-2,2-dimethyl-propionic acid (AM103). *J Med Chem* **52**, 5803–5815.
58. Lorrain, D.S. *et al.* (2009) Pharmacological characterization of 3-[3-tert-butylsulfanyl-1-[4-(6-methoxy-pyridin-3-yl)-benzyl]-5-(pyridin-2-ylmethoxy)-1 H-indol-2-yl]-2,2-dimethyl-propionic acid (AM103), a novel selective 5-lipoxygenase-activating protein inhibitor that reduces acute and chronic inflammation. *J Pharmacol Exp Ther* **331**, 1042–1050.
59. Lorrain, D.S. *et al.* (2010) Pharmacology of AM803, a novel selective five-lipoxygenase-activating protein (FLAP) inhibitor in rodent models of acute inflammation. *Eur J Pharmacol* **640**, 211–218.
60. Stock, N. *et al.* (2010) 5-Lipoxygenase-activating protein inhibitors. Part 2: 3-{5-((S)-1-Acetyl-2,3-dihydro-1 H-indol-2-ylmethoxy)-3-tert-butylsulfanyl-1-[4-(5-methoxy-pyrimidin-2-yl)-benzyl]-1 H-indol-2-yl}-2,2-dimethyl-propionic acid (AM679)-a potent FLAP inhibitor. *Bioorg Med Chem Lett* **20**, 213–217.
61. Cordes, F., Bright, J., and Sansom, M. (2002) Proline-induced distortions of transmembrane helices. *J Mol Biol* **323**, 951–960.
62. Charleson, S. *et al.* (1994) Structural requirements for the binding of fatty acids to

5-lipoxygenase-activating protein. *Eur J Pharmacol* **267**, 275–280.

63. Flamand, N., Lefebvre, J., Surette, M., Picard, S., and Borgeat, P. (2006) Arachidonic acid regulates the translocation of 5-lipoxygenase to the nuclear membranes in human neutrophils. *J Biol Chem* **281**, 129–136.
64. Kulkarni, S., Das, S., Funk, C., Murray, D., and Cho, W. (2002) Molecular basis of the specific subcellular localization of the C2-like domain of 5-lipoxygenase. *J Biol Chem* **277**, 13167–13174.
65. Soberman, R., and Christmas, P. (2003) The organization and consequences of eicosanoid signaling. *J Clin Invest* **111**, 1107–1113.
66. Mandal, A.K. *et al.* (2004) The membrane organization of leukotriene synthesis. *Proc Nat Acad Sci USA* **101**, 6587–6592.
67. Strid, T., Svartz, J., Franck, N., Hallin, E., Ingelsson, B., Söderström, and M., Hammarström, S. (2009) Distinct parts of leukotriene C(4) synthase interact with 5-lipoxygenase and 5-lipoxygenase activating protein. *Biochem Biophys Res Commun* **381**, 518–522.
68. Newcomer, M.E., and Gilbert, N.C. (2010) Location, location, location: compartmentalization of early events in leukotriene biosynthesis. *J Biol Chem* **285**, 25109–25114.

Chapter 13

Application of SBDD to the Discovery of New Antibacterial Drugs

John Finn

Abstract

The emergence of bacteria that are multiply resistant to commonly used antibiotics has created the medical need for novel classes of antibacterial agents. The unique challenges to the discovery of new antibacterial drugs include the following: spectrum, selectivity, low emergence of new resistance, and high potency. With the emergence of genomic information, dozens of antibacterial targets have been pursued over the last 2 decades often using SBDD. This chapter reviews the application of structure-based drug design approaches on a selected group of antibacterial targets (DHFR, DHNA, PDF, and FabI) where significant progress has been made. We compare and contrast the different approaches and evaluate the results in terms of the biological profiles of the leads produced. Several common themes have emerged from this survey, resulting in a set of recommendations.

Key words: Structure-Based Drug Design, Antibiotics, Ligand Efficiency, Drug-Properties

1. Introduction

Bacterial resistance to existing antibacterial drugs has increased dramatically over the last 2 decades (1–3) (see www.idsociety.org). This causes problems in treating bacterial infections because empirical treatment with the wrong agent has a significant adverse impact on therapeutic results, including a doubling of mortality rates in life-threatening hospital infections. This problem is particularly severe for infections caused by Gram-positive bacteria. For example, over 60% of the hospital-acquired *Staphylococcus* infections (the most common hospital-acquired infection) are now caused by methicillin-resistant *Staphylococcus aureus* (MRSA). Likewise, multidrug resistance is now emerging as a problem in Gram-negative bacteria, especially resistance to carbapenems and fluoroquinolones. Many resistant strains, including MRSA, are multiply resistant to all of the major classes (beta-lactams, quinolones, and

Leslie W. Tari (ed.), *Structure-Based Drug Discovery*, Methods in Molecular Biology, vol. 841,
DOI 10.1007/978-1-61779-520-6_13, © Springer Science+Business Media, LLC 2012

macrolides) of antibacterial drugs and represent a serious and growing threat to public health. Infections caused by resistant bacteria are more lethal (over 70% of deaths due to hospital acquired infections are caused by resistant bacteria) and more costly to treat. Consequently, there is an acute medical need for new therapeutic agents that can be used as first and second line treatments in infectious disease.

Unfortunately, the pharmaceutical industry has struggled to introduce new antibiotic drugs (4, 5). Despite significant efforts, only two new drug classes have been approved by the FDA in the last 4 decades. These drugs (the oxazolidinone linezolid and the lipopeptide daptomycin) were discovered using the traditional method of screening compounds for antibacterial activity followed by optimization guided mainly by antibacterial activity. The increasing medical need for new antibacterial drugs combined with the low productivity of traditional cell-based screening methods, fueled interest in alternative approaches for discovering antibacterial drugs. Starting in the mid-1990s, many companies pursued target-based drug discovery approaches to antibacterial discovery by applying the tools of bacterial genomics, high-throughput screening, and structure-based drug design (SBDD). While many programs generated screening hits, hit optimization proved to be very difficult due to issues specific to this therapeutic area, such as bacterial cell penetration, the need for wide antibacterial spectrum, and the need for low emergence of new resistant strains (6, 7).

From a technology evaluation standpoint, antibacterial discovery, due to its many challenges and proven difficulties, provides a rigorous test of the ability of SBDD to solve multiple drug discovery problems. Many different antibacterial discovery groups have used SBDD, taking advantage of the ability to readily crystallize many important bacterial targets. In this chapter, we explore the degree of success that SBDD campaigns have had in solving the issues posed in advancing useful antibacterial agents into the clinic. Selected antibacterial targets that were the objects of SBDD campaigns are described and the role of SBDD in the ligand optimization process is examined. Rather than trying to provide a comprehensive listing of all SBDD programs in the area, the targets selected in this chapter were selected to exemplify specific challenges. In particular, the chapter focuses on the ability of structural information to solve the issues of spectrum, selectivity, potency, and resistance. While many of these SBDD guided programs generated interesting antibacterial agents, significant issues remain. The goal of this chapter is to highlight the strengths and weaknesses of the technology and provide suggestions for future antibacterial SBDD efforts.

2. Dihydrofolate Reductase (DHFR): Addressing the Issue of Resistance to the First-Generation Drug

Bacterial dihydrofolate reductase (DHFR) (EC 1.5.1.3) has been exploited as an antibacterial target for over 5 decades and has led to the discovery of the drug trimethoprim (8) (TMP, see Fig. 1). This drug is a potent and selective inhibitor of bacterial DHFR (K_i of 0.9 nM vs. *S. aureus* DHFR and K_i of 19 μM vs. human DHFR). At the cellular level, TMP has good activity against a variety of Gram-positive and Gram-negative strains including *S. aureus* (minimum inhibitory concentration, MIC = 1 μg/mL). A very interesting aspect of DHFR antibacterial agents is that they can be combined in a synergistic fashion with a sulfonamide drug, most typically sulfamethoxazole (9) (SMX, see Fig. 1). SMX is an inhibitor of dihydropteroate synthase (DHPS), an enzyme in the same pathway as DHFR, and the TMP–SMX combination reduces by about 1/10th the amounts of both drugs needed to achieve the same MIC as either of the agents alone. Both TMP and the synergistic combination of TMP and SMX (sold as Bactrim™) are used widely to treat infections. Not surprisingly, resistance to both TMP and SMX has emerged, limiting the effectiveness of these drugs. The extent of resistance varies widely depending on bacterial species and location. For example, a 2002 study found that in the USA approximately 20% of the *Staphylococcus pneumoniae* strains were resistant to Bactrim, while in Korea 70% of the *S. pneumoniae* strains were found to be Bactrim resistant (10). Thus, there is an absolute requirement for potency against these TMP-resistant strains for any new DHFR-targeting antibacterial drug.

From an SBDD target selection standpoint, DHFR represents an attractive choice: an easily crystallized target, strong clinical validation of the utility of DHFR-based drugs, and clarity of the problem that needs to be addressed (coverage of TMP-resistant strains). Consequently, there have been multiple efforts directed towards inventing antibacterial DHFR inhibitors that have potency on resistant strains. Further, diverse strategies were employed in these efforts and the resulting lead molecules that emerged have different biological profiles.

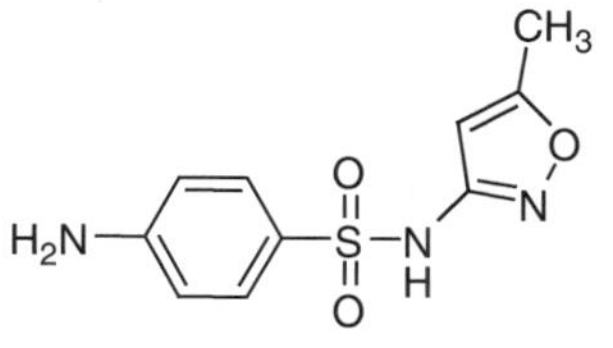

sulfamethoxazole (SMX)
1

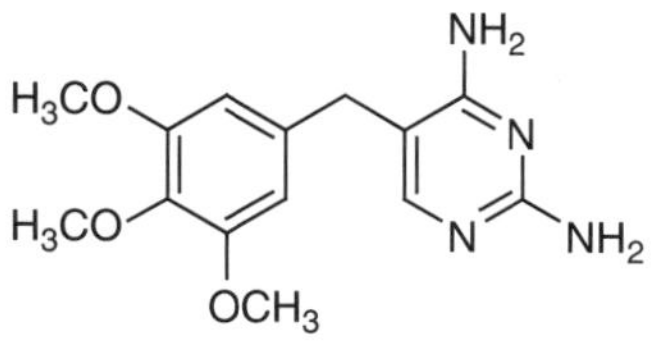

Trimethoprim (TMP)
2

Fig. 1. Chemical structures of sulfamethoxazole and trimethoprim.

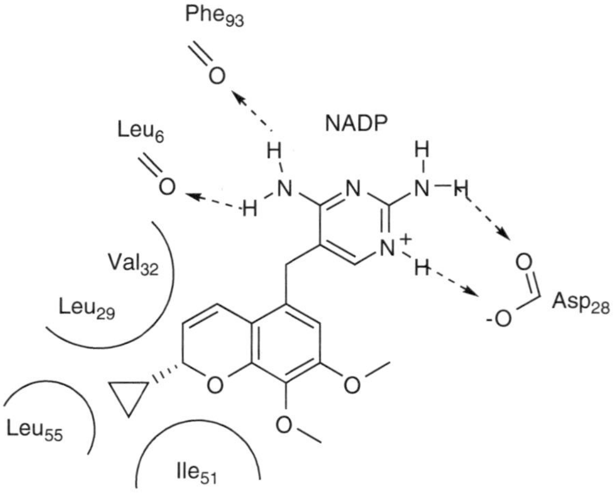

Fig. 2. Schematic outlining the key interactions between Iclaprim and *Staphylococcus aureus* DHFR.

Almost all recent DHFR programs are structure-based. Researchers at Roche were the first to report the structures of resistant DHFR enzymes starting with the *Escherichia coli* DHFR mutant F99Y (11). This mutation is common to many bacterial strains and is responsible for the majority of the TMP-resistant *S. aureus* strains (12). At the structural level, this mutation disrupts the hydrogen-bonding network that is formed between a diamino-pyrimidine and DHFR (see Fig. 2). In a wild-type DHFR-diaminopyrimidine structure, the hydrogens of the diaminopyrimidine bind to a conserved Asp and to two backbone carbonyls (Leu_6 and Phe_{93} using *S. aureus* DHFR numbering). In enzymes containing the F99Y mutation, the hydroxyl of the tyrosine binds to the backbone carbonyl of Leu_6, disrupting the H-bonding interaction to the diaminopyrimidine core. To overcome this resistance, attempts have been made to remove the 4-amino group and find replacements for the 2,4-diaminopyrimidine core (13, 14). In general, the compounds identified by these replacement efforts suffered reduced potency. A more popular approach to addressing resistance relies on designing ligands that add favorable interactions outside the diaminopyrimidine pocket to compensate for the energy required to disrupt the Tyr_{99}-Leu_6 hydrogen bond and reestablish the hydrogen-bonding network seen between the diaminopyrimidine and DHFR observed in the wild type strains (the “ram your way in” strategy). By definition, the ramming strategy should still result in compounds with reduced potency on the F99Y-resistant strains compared to wild-type strains, so this strategy relies heavily on designing high-potency DHFR ligands.

The approach of designing highly potent diaminopyrimidine ligands (the ramming strategy) has been heavily utilized, leading to multiple unique series of DHFR inhibitors. Figure 3 shows selected

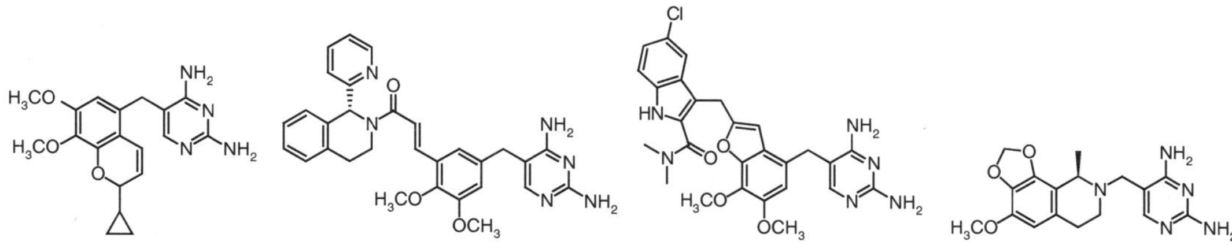

3 Iclaprim (Icl) **4** Ro-62-6091 **5** AR-709 **6**

Fig. 3. Structures of several key DHFR ligands resulting from SBDD programs.

Table 1
Enzymatic and antibacterial potencies for selected DHFR antibacterial agents

	DHFR K_i or IC_{50}[a] (nM)					MIC (μg/mL) or MIC_{90}[b]			
Compound	**Sa WT**	**Sa F99Y**	**Spn WT**	**Spn TMPr**	**Human**	**Sa WT**	**Sa F99Y**	**Spn WT**	**Spn TMPr**
1 Tmp	1.2	90	6	900	19,000	1	64	4	>64
3	0.08	0.9	8	43	775	0.06	2	0.06	2
4	8[a]		22		>10,000	0.03[b]	0.5[b]	0.125[b]	16[b]
5	8[a]	90[a]	27[a]	9[a]		0.125	2	0.06	0.06
6	1,300[a]		9.8[a]	2.8[a]	1,200[a]			0.125	0.125

Sa *Staphylococcus aureus*; Spn *Staphylococcus pneumonia*; *WT* wild-type strain; *F99Y* F99Y mutant strain; *TMPr* trimethoprim resistant strain

[a] IC_{50} is used versus k_i

[b] MIC_{90} is used versus MIC

key compounds with activity against TMP-resistant strains that emerged from these efforts including Iclaprim, Ro-62-6091, AR-709, and **6**. Table 1 shows data for enzyme inhibition and antibacterial activity for these four compounds and TMP. The TMP and Iclaprim data were measured at Trius; the data for Ro-62-6091, AR-709, and **6** is taken from published work described below. Table 2 shows calculated properties for the compounds described in Table 1.

Iclaprim, (compound **3** in Fig. 3), is the most studied of these four compounds (15, 16). This compound represents a relatively conservative change from TMP where the added hydrophobicity of the dihydropyran ring and the cyclopropyl substituent occupy a hydrophobic pocket. The structure of the Iclaprim *S. aureus* DHFR F99Y complex shows both the hydrophobic interactions between the dihydropyrano ring and residues Leu_{29}, Ile_{54}, Leu_{55} (*S. aureus* DHFR numbering) and the retention of the hydrogen-bonding network between diaminopyrimidine core and the enzyme (17, 18). Because this structure demonstrates that Iclaprim disrupts the Tyr_{99}–Leu_6 hydrogen bond seen in the apo-F99Y DHFR structure, it validates the SBDD strategy where the design of additional

Table 2
Calculated properties for DHFR antibacterial agents described in Table 1 (J. Finn, unpublished results)

Compound	Molecular weight (g/mol)	*A*log *P*	Predicted oral absorption	Solubility	LE Sa (Spn)[a]
1 Tmp	290	1.44	Good	−3.14	2.20
3	354	2.51	Good	−4.60	2.04
4	522	3.78	Very poor	−5.53	1.51
5	535	4.27	Very poor	−7.40	1.55
6	345	−0.19	Good	−2.54	1.14 (1.64)

[a] *LE* ligand Efficiency is calculated by the equation LE = Δ*G*/#HA (the number of heavy atoms)

favorable interactions in one part of the binding pocket can compensate for the loss of activity in another part of the binding pocket. As a result, Iclaprim is significantly more potent against both the wild-type (15X) and the F99Y mutant (100X) strains when compared to TMP. This potency translates into highly improved antibacterial activity vs. both wild-type and TMP-resistant *S. aureus* strains. There are, however, differences (around 10X) between the potency of Iclaprim against wild-type vs. resistant strains as expected due to the design strategy. Significantly, Iclaprim still maintains excellent selectivity, with relatively weak activity against human DHFR and a selectivity ratio of 10,000. Iclaprim also has a very favorable profile in the generation of resistant strains under the selection pressure of the drug (19). In spontaneous resistance studies with *S. aureus*, no mutants were detected in the presence of Iclaprim at 4X the MIC, consistent with a resistance incidence level of $<10^{-10}$ cfu/mL. By contrast, TMP generated resistant mutants at the rate of 10^{-9} cfu/mL at the same concentration of drug.

From a properties standpoint, Iclaprim has favorable features. The molecular weight is increased 20% relative to TMP but is still relatively low. Consequently, the score for ligand efficiency of 2.04 is still very high indicating that the new hydrophobic interactions are well designed. The compound is more hydrophobic and less soluble than TMP (the calculated solubility is in log scale; thus, Iclaprim is calculated to be around 1/20th as water soluble as TMP). Iclaprim has calculated log *P* and polar surface area values consistent with those expected for compounds with oral bioavailability. This is confirmed in human studies of pharmacokinetics where the measured human oral bioavailability is approximately 40% (20). By comparison, TMP (a smaller, more soluble molecule) has human oral bioavailability of >90% (20).

Based on the enhanced activity against *S. aureus* and other Gram-positive pathogens, Iclaprim was advanced into clinical trials and an NDA was submitted to the FDA. It was reviewed at an advisory committee meeting in November 2008. The clinical data from the two pivotal Phase 3 trials for treatment of patients with complicated skin and soft structure infections (cSSSI) demonstrated that Iclaprim is efficacious and well tolerated (21). The degree of efficacy, however, failed by a small margin to achieve the FDA mandated level of noninferiority to the comparator antibiotic linezolid. While there might be many reasons for the lower efficacy rates in the Iclaprim arms of the trials, a likely explanation is the dose was too low. Dose selection is a balance between potency and safety and posed several hard choices:

- *Potency*: The 0.8 mg/kg dose selected for Iclaprim in these trials was aimed to treat the majority of *S. aureus* strains including MRSA that are TMP sensitive *but not the trimethoprim-resistant strains*. Thus, the observed C_{max} for this dose was 0.83 μg/mL, well below the MIC for the *S. aureus* F99Y-resistant strains. While only a handful of the clinical isolates were identified as TMP resistant and most of these were isolated from patients with successful outcomes, the difference between the observed 12% and the required 10% noninferiority limits statistically is only 4–6 patients. Consequently, if only a handful of Iclaprim failures were due to infections caused by TMP-resistant strains, this could statistically lower the success rate below the 10% delta.
- *Safety*: Iclaprim has an observed potency against hERG (in vitro patch clamp assay) of 0.86 μM (IC_{50}) (vs. an IC_{50} >20 μM for TMP) and there was significant concern about Qtc prolongation. Fear of this side effect probably influenced the decision to adopt a relatively low dose in human trials.

As illustrated above, it is always desirable to aim for compounds with high potency against resistant strains. In retrospect, the significant difference between the MICs against TMP-sensitive and TMP-resistant strains created a dilemma for clinical trial design: Do you overdose the majority of the patients who have an infection caused by a TMP-sensitive strain to ensure coverage of the patients with an infection caused by a resistant strain and in doing so risk safety issues? Or do you risk efficacy by dosing too low to cure patients with TMP-resistant infections? On the safety front, Iclaprim's potency against hERG is likely to be a result of the increased lipophilicity of the molecule (when compared to TMP). Thus, the strategy of designing highly potent but more lipophilic compounds probably resulted in multiple issues including reduced solubility, reduced oral bioavailability and a diminished safety margin. Consequently, the Iclaprim SBDD strategy resulted in a compound with significant differences in potency between TMP-sensitive and TMP-resistant strains and increased potency against hERG. This led to downstream issues that halted the development of this drug.

Ro-62-6091, (compound **4** in Fig. 3), represents the first super-potent diaminopyrimidine inhibitor and was reported by the Roche team at the 1999 ICAAC meeting (22, 23). The design principle relies on building out from TMP with hydrophobic fragments that bind to residues in the more solvent exposed regions of the binding site. Although this results in a very potent ligand, it also results in significantly lower ligand efficiency (LE = 1.51), indicating that interactions with these more remote regions are not as favorable energetically as those hydrophobic interactions involved in the binding of Iclaprim and TMP. The high enzymatic potency for Ro-62-6091 results in a strong anti-*Staphylococcal* profile with MIC_{90} values of 0.03 and 0.5 μg/mL vs. TMP-sensitive and TMP-resistant *S. aureus*, respectively. As with Iclaprim, we see that the ramming strategy reduces the overall MIC value on TMP-resistant strains, but still suffers from a 16X potency differential with the TMP-sensitive strains. So, as a consequence of building ligands with additional hydrophobic interactions, Ro-62-6091 suffers from the standpoint of physicochemical properties, with higher $\log P$ and reduced aqueous solubility. This compound was not advanced to the clinic, presumably due to the inability to develop a suitable IV or oral formulation. As before, a focus on adding mainly hydrophobic interactions carries a significant price in reduced drug properties.

TMP resistance in *S. pneumoniae* is particularly difficult for new DHFR agents to overcome. For example, even though Iclaprim is enzymatically more potent on *S. pneumoniae* DHFR in both sensitive and resistant forms, the MIC_{90} values for Iclaprim and Ro-62-6091 on TMP-resistant *S. pneumoniae* strains are 16 μg/mL, considerably higher than what would be considered clinically useful. To specifically meet the challenge of *S. pneumoniae*, researchers at Arpida developed AR-709, compound **5** in Fig. 3 (24, 25). Similar to Ro-62-6091, the SBDD strategy centered on building additional, mainly hydrophobic, interactions directed towards the more exterior regions of the binding pocket. To specifically address the resistance issue, the optimization strategy prioritized compounds with potency against the TMP-resistant *S. pneumoniae* DHFR enzyme. The resulting lead compound AR-709 is more potent against the TMP-resistant than the TMP-sensitive *S. pneumoniae* DHFR enzyme (K_i 9 vs. 27 nM). Given the similar enzymatic potencies, it is not surprising that AR-709 has similar antibacterial potencies against both TMP-resistant and TMP-sensitive strains. The initial antibacterial results were encouraging. MIC tests against large panels of *S. pneumoniae* isolates with AR-709 demonstrated MIC_{90} values of 0.5 μg/mL or less;

confirming the utility of this DHFR agent as an antistreptococcal drug. While the additional binding interactions outside of the diaminopyrimidine binding site might have been expected to lead to higher frequencies of resistance emergence, this was not the case. In contrast to TMP where resistance incidence in *S. pneumoniae* is very high ($RI = 10^{-6}$ cfu/mL), spontaneous resistance to AR-709 is very low ($RI = <10^{-9}$ cfu/mL) (26).

While AR-709 is a clearly effective antibacterial agent, issues remain to be addressed to prove that it will be clinically useful. The compound design strategy lead to a drug with relatively high MW, low predicted solubility, poor predicted oral bioavailability and high $\log P$. The prediction for low oral bioavailability was confirmed in human microdosing studies where the measured oral bioavailability was 2.5% (27). These studies also showed significant distribution (well above plasma levels) into key compartments of lung tissue, a very desirable feature for a respiratory tract drug. Given the hydrophobic nature of this compound, it remains to be seen if a reasonable IV formulation can be developed. As an alternative, an inhalation formulation may be explored. The safety profile of this potential drug has not been extensively reported.

Compared to the strategy relying on structure-based optimization of the trimethoprim core that gave rise to compounds **3–5**, compound **6** (see Fig. 3) was the result of a structure-guided combinatorial chemistry approach (28). In this approach, a synthetic route (see Fig. 4) was developed that allowed for the synthesis of a wide variety of substituted 5-aminomethyl-2,4-diaminopyrimidines **8** from the pyridinium salt 7. This route is fairly robust, and incorporation of 9,000 secondary amines using this reaction was investigated. The first method to select amines was docking of a virtual library of the proposed analogs into *S. aureus* DHFR and subsequent scoring and ranking of the analogs. Alternative approaches included selecting a set of low scoring analogs (as a control) and using diversity methods. Although the authors were looking for compounds with good potency against trimethoprim-resistant *S. pneumoniae*, the use of the wild-type *S. aureus* DHFR for docking studies was rationalized based on the high sequence conservation between the two enzymes and the availability of the *S. aureus* DHFR structure.

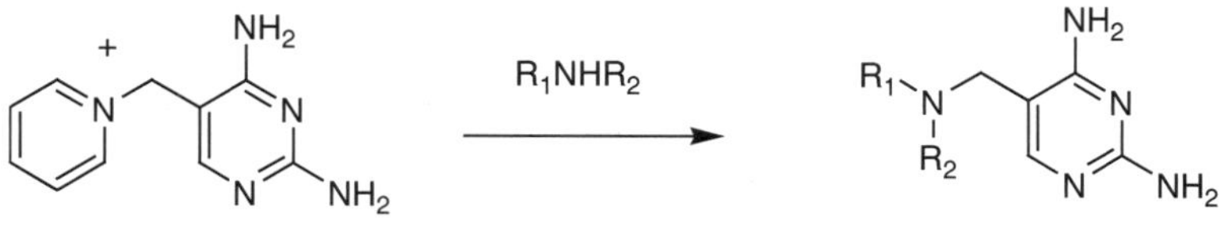

Fig. 4. High-throughput synthesis approach for synthesis of DHFR antibacterial agents.

Table 3
Hit rates from high-throughput synthesis

Library	Size	# of inhibitors	Hit rate %
Structure-based high scoring	252	54	21
Structure-based low scoring (control)	269	4	1
Diversity	501	17	3

The compounds from high-throughput synthesis were evaluated for both enzyme inhibition and antibacterial activity. The criteria for a hit was enzyme inhibition of >50% at 10 μM and antibacterial activity at 25 μM. To remove compounds with off-target antibacterial activity, the antibacterial activity needed to be antagonized by thymidine (a classic proof of DHFR-mediated antibacterial activity). To remove highly hydrophobic compounds, antibacterial activity was measured in the presence of 10% human serum. One feature that expedited this work was the ability to screen the crude reaction mixtures in both enzymatic and antibacterial assays. Table 3 provides the results of this study, and as one would expect, the highest hit rates came from the structure-based virtual screen. Of the several compounds profiled in this paper, compound **6** (see Fig. 3) was highlighted by the authors. The enzymatic profile of this compound is fairly distinct from any of the 5-benzyl-diaminopyrimidines (such as TMP, and compounds **3–5** from Fig. 3). Compound **6** (R) is more potent against TMP-resistant *S. pneumoniae* DHFR (IC_{50} 3 nM) than against wild type *S. pneumoniae* DHFR (IC_{50} 10 nM). Consequently, the compound has equal antibacterial potency against TMP-resistant and TMP-sensitive *S. pneumoniae* strains. Another interesting feature of this compound is the loss of potency against *S. aureus* DHFR (IC_{50} = 1.3 μM) relative to *S. pneumoniae* DHFR. Given that the compound was selected by screening on *S. aureus* DHFR this result is unexpected. Further, the vast differences between the enzymatic potencies of the compound against these two DHFR orthologs demonstrates that despite the high sequence conservation observed between the two proteins there are significant subtle differences between the shapes of their respective binding pockets.

Examination of the properties for compound **6** in Table 3 demonstrates additional advantages of the structure guided high-throughput synthesis approach. Compound **6** has several desirable calculated features including lower log *P*, high solubility, low MW, and predicted oral bioavailability. These enhanced drug-like properties combined with the comparable antibacterial potency against both wild type and TMP-resistant strains suggest that this series has significant potential. Overall, the approach demonstrates a nice synergy between structural, synthetic and assay methodologies. The resulting compounds such as **6** demonstrate the power of serendipity to find compounds with a molecular shape that fits better with the resistant form of the enzyme thereby affording equal potency. As seen in the Iclaprim example, the ability to avoid significant potency disparities between bacterial strains is very important. While considerable work remains (e.g., spectrum, selectivity) to convert interesting leads such as **6** into drugs, there is no apparent reason why an SBDD approach would not yield compounds with improved profiles.

In summary, the experience with SBDD generated DHFR antibacterial agents has demonstrated that potency against resistant strains is a very achievable goal using several different approaches. At first glance, many were skeptical that the SBDD "ramming strategy" would ever yield agents with clinically useful antibacterial potencies against both TMP-sensitive and TMP-resistant strains; this has now been clearly demonstrated with agents such as AR-709. What appears to be more problematic is to solve the resistance issues while building drug-like properties. In retrospect, the strategy of building bigger, more hydrophobic compounds to add enzymatic and antibacterial potency succeeded in most cases at the expense of drug properties. In the future, consideration of the drug properties of the lead series needs to rise in importance even at the early stages of selecting a lead compound. The high-throughput synthesis approach provided some of the more interesting work in this area but exploitation of these initial leads has not been reported.

3. Dihydroneopterin Aldolase (DHNA): Bad Target or Bad Drug?

Dihydroneopterin aldolase (DHNA: EC 4.1.2.25) catalyzes a retroaldol reaction that generates 6-hydroxymethyl-7,8-dihydropterin **10** from 7,8-dihydroneopterin **9** (see Fig. 5). This reaction is upstream of DHFR in the folate biosynthetic pathway and is an essential bacterial enzyme that is not present in mammals. Based on genomic information suggesting high sequence conservation across bacterial species, this target was investigated by a group at Abbott for the possibility of generating a novel class of antibiotics with a novel mechanism (29). High-throughput X-ray crystallographic screening (CrystalLEAD) provided several small fragment inhibitors including **11–14** (see Fig. 5). Each of these compounds mimics key hydrogen-bonding interactions that were identified from the cocrystal structure of *S. aureus* DHNA with compound **9**, and have good ligand efficiency. Compound **14** was selected for optimization and analogs with increased enzymatic potency vs. *S. aureus* DHNA were identified including compounds **15–16**.

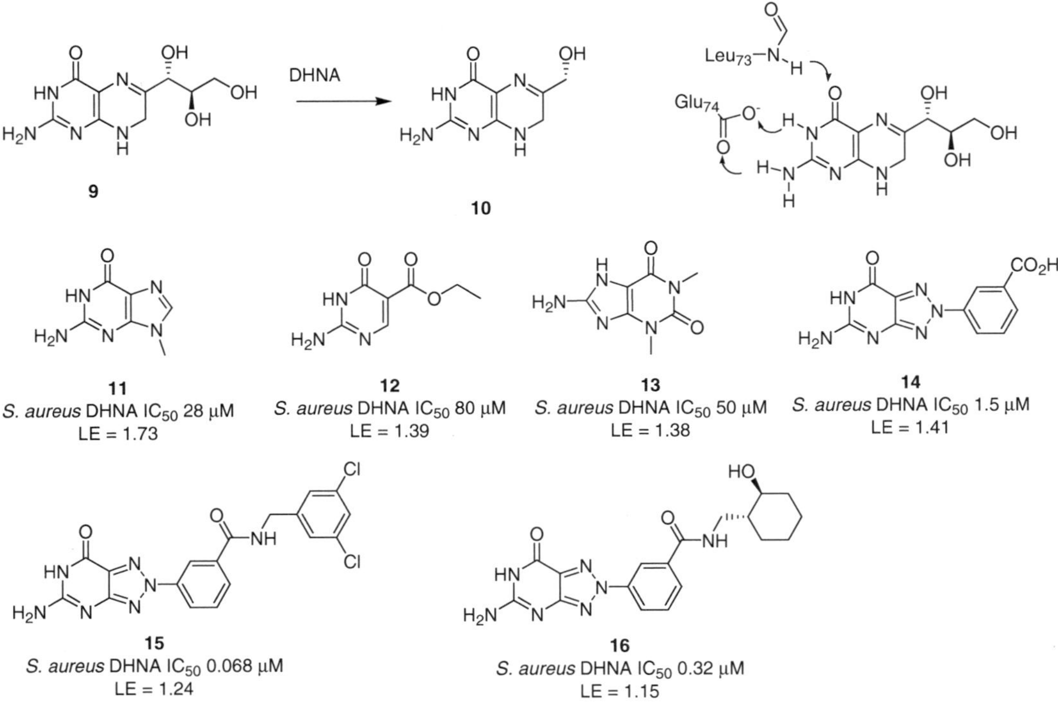

Fig. 5. The reaction catalyzed by DHNA, with key binding interactions and SBDD-generated inhibitors highlighted.

The analog set was tested for antibacterial activity against a wide panel of Gram-positive and Gram-negative strains with none of the compounds showing antibacterial activity. Then, antibacterial activity was investigated against permeabilized and pump-knock out strains of *E. coli* and against Gram-positive strains in the presence of permeability enhancing agents such as poylmyxin nonapeptide. Again, no antibacterial effects were seen, leading the investigators to speculate on the validity of DHNA as a bacterial target. One explanation for the lack of antibacterial activity was that the intracellular concentration of the substrate for DHNA could be quite high relative to the K_m of the enzyme. In this case, compounds would need to be much more potent to show bacterial growth inhibition effects.

This example was selected to demonstrate the issues encountered with developing antibacterial agents against novel targets. In the case of DHNA, the target is in a biochemical pathway validated at two other enzymes (DHFR and DHPS) targeted by clinically used drugs. Thus, inhibition of the pathway at DHNA should be correlated with antibacterial activity. However, all targets are not equal in value in target based antibacterial drug discovery. Considerations of flux through a pathway and substrate concentrations can significantly affect the level of potency required to support antibacterial activity. Additionally, some targets may be present at much higher levels than required to support bacterial growth. Thus, there may be targets where the inhibition of the first 95% of the enzyme's capacity does not impact bacterial growth, with other targets which are working at capacity, so that any level of inhibition impacts bacterial growth.

Given these concerns, there are significant questions as to the ability of this one study to rule out DHNA as an antibacterial target. The most potent compound described has an IC_{50} of 68 nM, and based on the literature a typical DHFR inhibitor of this potency would have fairly weak antibacterial potency (MIC > 8 μg/mL). Further, different chemical classes behave differently in their ability to generate antibacterial activity. A good example of this is the work from Welcome where a series of DHPS inhibitors were discovered for a target that is validated by the sulfonamide class of antibacterial agents (30, 31). The lead compounds in this series had equal enzyme potency to the sulfonamides but no antibacterial activity was observed, highlighting the fact that issues beyond enzymatic potency, such as the ability to penetrate the bacterial cell wall, are needed for antibacterial activity. Although the Abbott team tested their compounds on different bacterial species, they did not develop enzyme assays for these different orthologs. As was documented in multiple DHFR series, very subtle structural changes in the active site can dramatically impact enzyme potency and reliance on sequence data is not an assurance of a broad-spectrum enzyme profile. Consequently, the ability to rule out a target based on a negative result in obtaining antibacterial activity is limited. The best case scenario would be to have multiple diverse drug-like chemotypes with potent enzymatic activity on different bacterial orthologs but this is a significant hurdle that very few companies are willing to invest in.

4. Peptide Deformylase (PDF): Different Strategies for Peptidomimics

Bacterial protein synthesis is initiated with formyl-methionine-tRNA and all polypeptides synthesized by the bacterial ribososme initially contain a *N*-formyl-methionine terminus. Bacteria then process these proteins by the removal of the formyl group by the enzyme peptide deformylase, PDF (EC 3.4.11.18). Because mammalian systems do not utilize the formyl-methionine route, PDF falls into the attractive class of antibacterial targets that are essential for bacterial growth and lack a mammalian counterpart. For this reason, PDF has been the subject of numerous investigations that utilized a SBDD approach in optimization. PDF programs are of particular interest because they provide an example of target with numerous issues including spectrum, selectivity, and resistance. Additionally, since the substrates and initial natural product leads were peptides, it provides a case study of different strategies to modify a peptide lead into a small molecule drug. For the purpose of this review, this section limits the discussion to those few programs that have progressed to advanced candidates. Other reviews describe the efforts in this area with a wider scope (32, 33).

From a structural design standpoint, spectrum is a significant issue because there are two distinct forms of PDF proteins with *E. coli* and *Haemophilus influenzae* being Type 1 and with *S. aureus* and *S. pneumoniae* being Type 2 (34). The sequence homology between the two ortholog sets is fairly low with a homology between *S. aureus* and *E. coli* PDFs of 23%. There is an abundance structural data on PDF available, with PDF structures from nine different orthologs in the Protein Databank. Consequently, the design of wide-spectrum inhibitors requires careful attention to structural diversity observed between target binding-pockets and consideration of conformational flexibility in both the target binding-pockets and designed ligands to enable binding to both Type 1 and Type 2 enzymes. The differences between the type 1 and type 2 enzymes can be readily seen in the superimposed complexes of PDF with a common ligand. Figure 6 highlights these differences by comparing an early inhibitor, the natural product actinonin, bound to both *S. aureus* and *E. coli* PDF. There are significant differences between the two PDF active sites and this is reflected in the differences in the binding conformation of actinonin. There is no overlap in the hydroxyproline residue due to a significant differences in the 3D shape of the binding pockets. Consequently, PDF inhibitors need to have conformational flexibility to achieve spectrum. While flexibility is important to achieve spectrum, specificity is a major issue given the need for a scaffold based on a metalloprotease inhibition. Thus, the challenge to researchers in this area is to design compounds that are flexible enough for spectrum, but not at the expense of selectivity.

To a large extent, researchers have been successful in designing broad-spectrum inhibitors. At least three programs, (British Biotech-Genesoft, Vicuron-Novartis and GSK) have advanced compounds into the clinic. The key compounds from these three programs are shown in Fig. 7 along with the original natural product lead actinonin. The experimental data for these compounds is compiled in Table 4 and the calculated properties are compiled in Table 5.

Most PDF programs have initiated optimization efforts starting from the natural product actinonin. Actinonin was originally isolated by following whole cell antibacterial activity. The mechanism of this compound was not known. Probably due to a short half life *in vivo*, this compound failed to show efficacy in animal infection models. Upon initiation of work in the PDF area, the researchers at Vicuron (then Versicor) determined the mechanism of action of this compound was via inhibition of PDF. The cocrystal

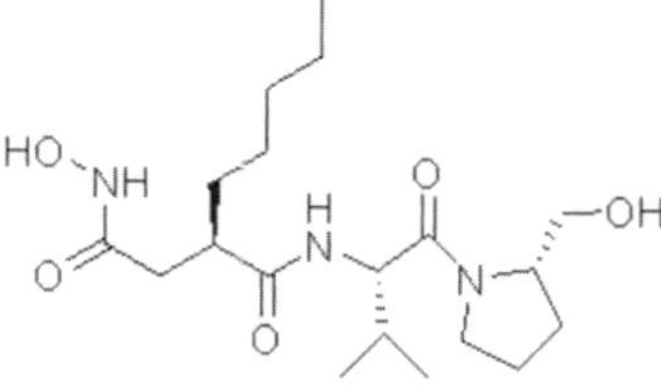

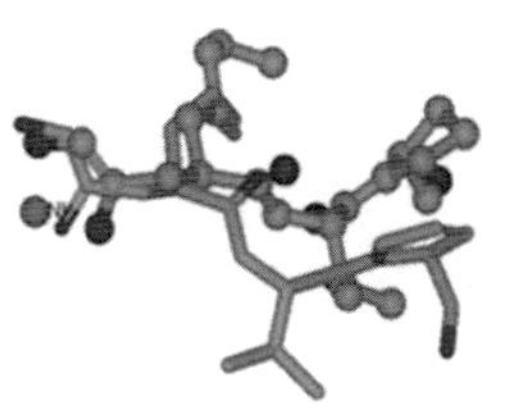

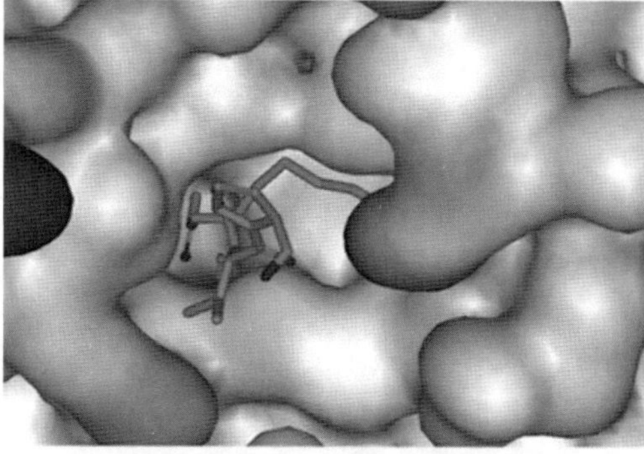

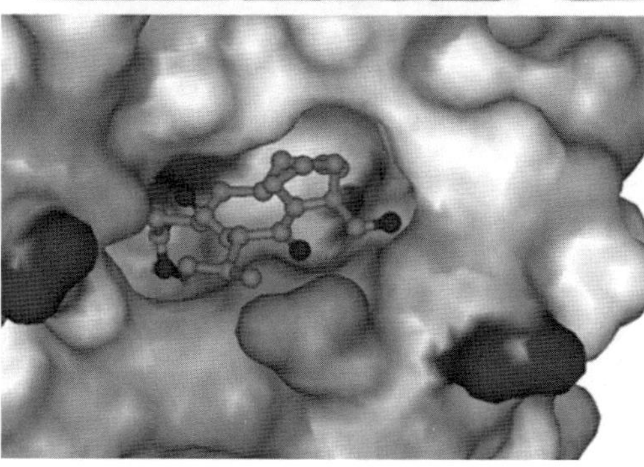

Fig. 6. ***Top left***: the structure of actinonin; ***Top right***: Overlay of the bound structures of actinonin from the ***S. aureus*** PDF-actinonin structure (***sticks***) and the ***E. coli*** PDF-actinonin structure (***ball*** and ***stick***); ***Middle***: ***S. aureus*** PDF-actinonin structure (1Q1Y); ***Bottom***: ***E. coli*** PDF-actinonin structure (1G2A).

structure of actinonin bound to PDF revealed the key interactions required for binding. As shown in Fig. 8, PDF is a metalloenzyme and the hydroxamic acid of the inhibitor forms a tight complex to metal (Zn^{2+} or Ni^{2+}). Given that the enzyme needs to bind selectively to methionine, the S1′ pocket is designed to fit a small hydrophobic residue (i.e., methylthioethyl, *n*-pentyl). The other two side chains fit S2′ and S3′ pockets that are not as deeply buried and not as well defined. Further, the structural conservation among the different PDF orthologs is quite high at the metal binding and the S1 pockets, but differs significantly at S2′ and S3′, particularly between Type 1 and Type 2 PDF enzymes. This sets up a particularly tricky set of problems where there is limited room for modification of the anchoring hydroxamic acid P1 regions of the ligand and

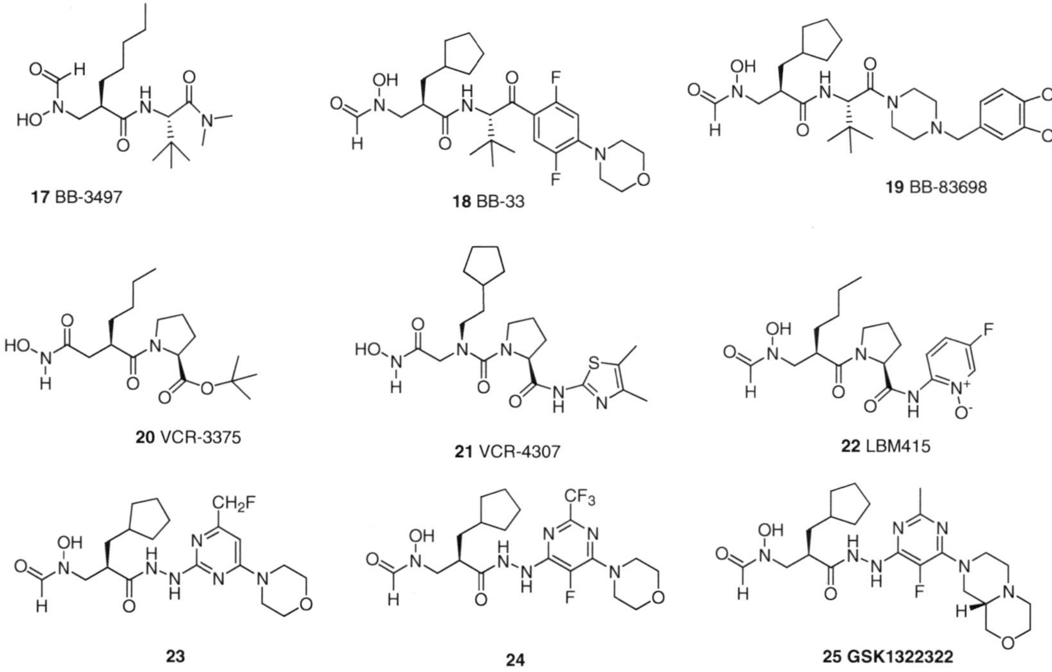

Fig. 7. Key peptide deformylase antibacterial agents.

Table 4
Enzymatic and antibacterial potencies for PDF antibacterial agents

	PDF K_i or IC_{50}[a] (nM)				MIC (μg/mL) or MIC_{90}[b]				
Compound	**Sa**	**Spn**	**Hi**	**Ec**	**Sa**	**Ef**	**Spn**	**Hi**	**Ec**
Actinonin				10	8–16		8	1–2	64
17				7	4–16	8–32	8	0.25	128
18				0.5			0.25–0.5	0.25–2	
19				10			0.125	2–16	
20				4	1		32	0.5	
21		8		2	0.5–1	2–4	0.5–4	2–4	64
22					2[b]	4[b]	1[b]	4[b]	
23					8		1	4	
24	24	<2	12		0.25–1		0.06–0.25	2–4	
25	11	3	3		0.5–1	48	0.25–1	0.5–2	16

References provided in text
Sa *S. aureus*; Spn *S. pneumoniae*; Hi *Haemophilus influenzae*; Ec *Escherichia coli*
[a] IC_{50} is used versus k_i
[b] MIC_{90} is used versus MIC

Table 5
Calculated properties for PDF antibacterial agents (J. Finn, unpublished results)

Compound	Molecular weight (g/mol)	*A* log *P*	Predicted oral absorption	Solubility	LE (*E. coli* or *S. aureus*)[a]
Actinonin	385	1.27	Good	–1.30	1.51
17	343	1.60	Good	–1.52	1.74
18	530	2.79	Good	–3.19	1.35
19	509	3.73	Good	–4.33	1.08
20	342	1.71	Good	–2.14	1.88
21	436	2.26	Good	–3.31	1.44
22	396	0.47	Good	–1.31	
23	424	1.90	Good	–3.10	
24	460	2.72	Good	–4.52	1.29[a]
25	480	–0.05	Good	–1.97	1.20[a]

[a]*LE* ligand efficiency

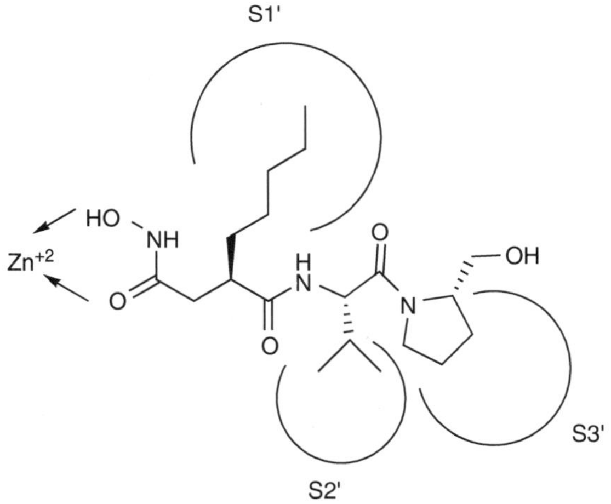

Fig. 8. Schematic representation of the PDF binding site from the complex of PDF with actinonin.

limited opportunities to increase binding energies dramatically by groups occupying the S2′ and S3′ pockets.

A group at British Biotech in collaboration with Genesoft pursued a fairly classical SAR-driven optimization approach (35–38). In place of the hydroxamic acid of actinonin, an *N*-formyl-hydoxylamine was used to bind to the metal. The *N*-formyl-hydroxylamine interacts with the PDF enzyme via a hydrogen-bonding network that is nearly identical to the network observed for hydroxamic acid.

A methylene spacer between the *N*-formyl-hydroxyamine and the side chain that fits the S1′ pocket is needed for optimal potency. A small set of hydrophobic side chains (*n*-butyl and methylene-cyclopentyl) were found to be optimal for fitting the hydrophobic S1′ pocket. Several cocrystal structures demonstrated that these regions were well optimized and that further work in this region was not warranted. The bulk of the optimization program focused on S2′ and S3′ regions. These pockets are more solvent exposed and as a result, a very flat SAR was observed for many analogs. For the P2′ fragment, the *t*-butyl side chain of *tert*-leucine provided the best antibacterial activity from over 20 different amino acids examined. Demonstrating the open nature of this pocket, the enzymatic data for *E. coli* PDF were fairly similar, ranging between 20 and 100 nM in IC_{50} values for most analogs. The program then concentrated in the P3′ region, where it was found that a simple dimethyl amide, **17** (BB-3497 in Fig. 7), had interesting antibacterial activity. This group could be replaced by ketones, thereby reducing the peptidic nature of the compounds. Compound **18** is typical of a potent aryl ketone. Eventually, a piperdine amide, compound **19** (BB-83698), was selected as a clinical candidate as this compound had an attractive antibacterial profile and is effective *in vivo* for treating bacterial pneumonia. In humans, this compound displayed reasonable pharmacokinetics (half-life of 8.5 h at a 475 mg dose) for once a day treatment of infections without serious side effects (39). Little has been published on this compound since 2004; thus, it is likely that the development of this compound was discontinued.

A group at Vicuron in collaboration with Novartis, utilized a combinatorial chemistry approach to identify lead compounds for PDF (40–41). They prepared libraries using three different metalloprotease warheads and measured the resulting compounds for both enzymatic and antibacterial activity. From the 528 membered hydroxamate library, compound **21** was identified as a lead. This compound contains a proline at the P2’ position. This provides a nice contrast to the British Biotech efforts that focused on t-leucine at this position. Interestingly, the t-leucine group was not included in the Vicuron library; likewise proline was not used in the British Biotech SAR. The nice feature of **21** as a lead is its low molecular weight. Further investigations of proline amides, lead to the urea, compound **22**, and the first clinical candidate **22** (LMB-415). Compounds in these series had good antibacterial potency against important respiratory pathogens such as *S. pneumoniae* and *H. influenzae*.

GSK has had a long standing program on the development of PDF inhibitors. In contrast to the British Biotech and Vicuron efforts there has been much less published to document the results of these efforts. At the 2010 ICAAC meeting, GSK researchers described their hydrazine-pyrimidine series (42, 43). In these posters, they clearly demonstrated the value of using enzymatic assays from multiple PDF orthologs to aid in the optimization of spectrum. Compounds such as **23** (see Fig. 7) have good antibacterial potency against *S. pneumoniae* and *H. influenzae* but weaker activity on *S. aureus*. Subtle changes (such as changing the CH_2F group of **23** to the CF_3 group of **24**) provided a better balance in MIC potencies. Further optimization led to compound **25** (GSK1322322) with an attractive antibacterial profile. From a structure standpoint, the 2-hydrazino-pyrimidine scaffold fits nicely into the S2′ and S3′ regions, providing ligands with good enzyme properties and favorable drug properties including oral bioavailability. At the 2010 ICAAC meeting, GSK announced that compound **25** was advanced into clinical trials.

These three programs offer a striking contrast in approaches and results. While all three essentially start with the same lead, actinonin, they ended up with three different series. Owing to the nature of the target, all three series utilize an *N*-formyl-hydroxylamine with a methylene spacer to a methionine-like amino acid residue. The S2′–S3′ fitting portion of the molecules, however, differ completely with British Biotech and Vicuron using a second amino acid fragment and GSK using a heterocyclic fragment. Even between British Biotech and Vicuron there are significant differences in the portion of the molecule binding to the S2′ and S3′ regions with the traditional medicinal chemistry analog approach taken by British Biotech yielding larger less ligand-efficient molecules than those discovered at Vicuron where a combinatorial approach was used at an early stage. The GSK group produced the most non-peptide-like lead molecules, but the drug properties of these compounds are comparable to the other two groups.

Interestingly, the reported potencies at the enzyme level for molecules from these three different programs level out in the 1–10 nM range. This probably indicates that the majority of the binding energy is derived from the metal binding and S1′ site and that the series are fairly similar in this region. Although initial sequence conservation studies and structural analysis suggest that broad spectrum agents against this target would be difficult to design, these series are successful at adding the Gram-negative pathogen *H. influenzae* to a broad Gram-positive spectrum. From a properties standpoint, the polar peptide starting point has favorable drug properties as do the resulting lead compounds derived from it. As shown in Table 5, these leads have favorable calculated properties including oral bioavailability and aqueous solubility. On the negative side, as a class this series of compounds seems limited in potency and spectrum. Further, there appears to limited room to change the *N*-formyl-hydroxylamine fragment so crucial to the key anchoring interaction or the peptide-like nature of the molecule that imparts flexibility to achieve the desired spectrum. These constraints may lead to nonspecific inhibition of other metalloenzymes and possible side effects. In addition to this, there are concerns about mutation/deletion of the entire *N*-formylmethionine pathway as a mechanism of bacterial resistance (44).

From a standpoint of advancing from a novel target to compounds advanced into the clinic, the three PDF programs demonstrate several important advantages gained by using SBDD. First, there is clear proof that structural information accelerated progress and was particularly useful in removing much of the peptidic nature of the initial lead. From a potency and spectrum standpoint, all three programs were successful in generating lead molecules with antibacterial potency and a Gram-positive spectrum with antibacterial profiles that were deemed sufficiently attractive to advance into clinical trials. This is no small accomplishment. Careful design considerations were needed to achieve enzymatic spectrum across orthologs that have significant differences, especially in the S2′–S3′ regions. SBDD led to compounds with reasonable properties and in general, resulted in smaller more ligand efficient molecules than the natural product lead. Issues such as the emergence of PDF resistance and selectivity vs. mammalian metalloproteases are areas of concern. It remains to be seen how the first-generation compounds fare in the clinic and if this will be an approach to bring new antibacterial agents into the market to combat resistance. Unfortunately, little data have been published in the last couple years documenting the fate of the PDF clinical candidates.

5. Fatty Acid Synthesis (FabI): Narrow Spectrum by Design (or Accident)

For a long time, fatty acid biosynthesis has been an attractive target for the discovery of new antibacterial agents (45). One attractive feature of this pathway is the significant differences between the mammalian and bacterial enzymes providing the potential for discovering highly selective inhibitors. In mammals, the fatty acid synthase (FAS) type I system consists of a large protein that is composed of several distinct domains that carry out the multiple enzymatic steps. By contrast, bacteria have a type II FAS system comprising discrete proteins that have little homology with the components of the mammalian system. This pathway has been validated as a viable source of antibacterial targets by a number of compounds including cerulenin, thiolactomycin, and Triclosan (45). Thus, targets in the fatty acid biosynthesis pathway have been the subject of multiple antibacterial discovery efforts by different groups.

The enzyme FabI (EC 1.3.1.9; enoyl-acyl carrier protein [ACP] reductase) converts enoyl-ACP to acyl-ACP, which is then used in a condensation reaction to form β-ketoesters. The FabI enzyme is considered to be a key regulatory step in fatty acid synthesis and is the target of Triclosan. Triclosan is a broad-spectrum antibiotic with reasonable potencies across Gram-positive and Gram-negative bacteria. There are toxicity issues when Triclosan is used systemically, thus it is limited to topical use in consumer products including antibacterial soaps and various disinfectants. Nevertheless, the interesting antibacterial activity of Triclosan highlights the potential of FabI as an antibacterial target.

The first significant effort in finding new FabI-based inhibitors was conducted by the GSK group (46). Their initial genomic analysis indicated that FabI was fairly well conserved across Gram-positive and Gram-negative bacteria. HTS-compatible assays were developed and a major screening effort for new inhibitors was conducted about 10 years ago (47, 48). This effort led to the discovery of attractive small molecule hits and structure-guided optimization yielded FabI targeting antibacterial agents. During the optimization process, GSK conducted additional genomic sequencing and analysis and data emerged that demonstrated that several species of bacteria utilize a different enzyme (FabK) to carry out the enoyl-ACP reduction. Not surprisingly, given the significant sequence differences between FabI and FabK, the newly discovered FabI antibacterial agents lacked spectrum in those bacteria with the FabK enzyme. Consequently, the overall interest in this target switched to the development of narrow-spectrum agents and the research efforts moved to smaller companies including Affinium and Crystal Genomics who focused on the development of narrow spectrum antibiotics. In particular, FabI antibacterial agents are being pursued as *S. aureus*-specific antibacterial agents. The lack of activity against other bacteria may be advantageous, as it avoids disruption of bacterial flora in the GI tract.

In a classical example of how SBDD can be used to rapidly improve potency, the GSK group systematically optimized the benzodiazepine derivative **26** (see Fig. 9), which was found in an HTS screen (49). The initial truncation studies showed that the key part of the benzodiazepine ring is the *para*-amino group and

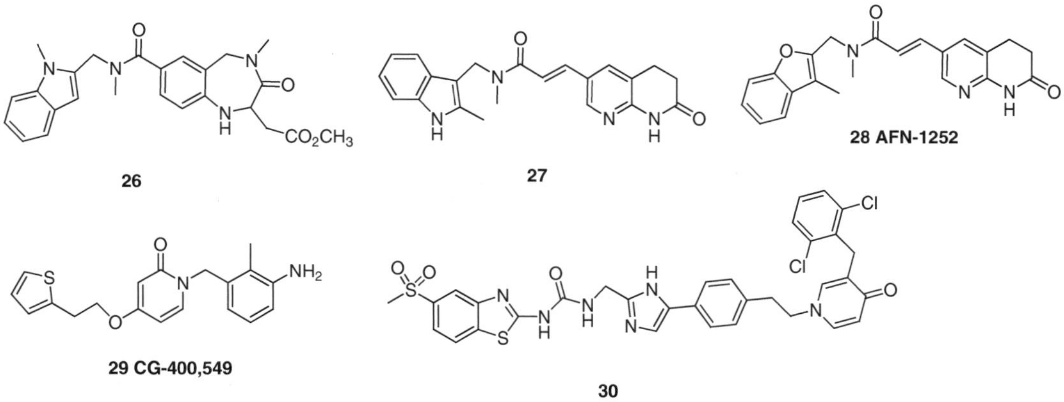

26

27

28 AFN-1252

29 CG-400,549

30

Fig. 9. Key FabI antibacterial agents.

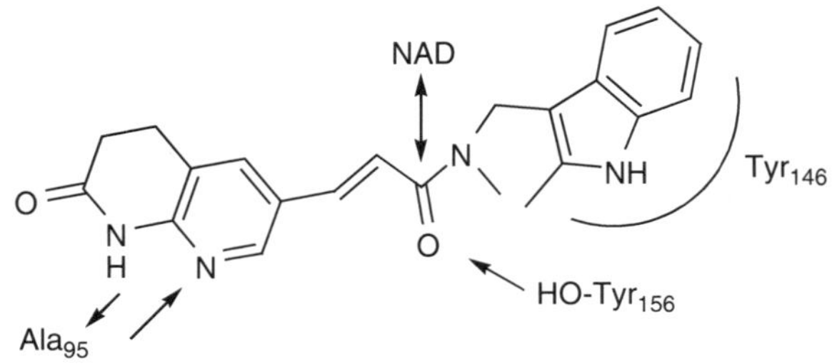

Fig. 10. Schematic showing the binding of **27** to *E. coli* FabI.

Table 6
Enzymatic and antibacterial potencies for FabI/FabK antibacterial agents

	IC_{50} FabI (μM)			IC_{50} FabK (μM)	MIC (μg/mL) or MIC_{90}[a]				
Compound	**Sa**	**Hi**	**Ec**	**Spn**	**Sa**	**Ef**	**Spn**	**Hi**	**Mc**
26	17.1	6.9		>30	>64	>64	>64	>64	32
27	0.047	0.13	<0.06	2	0.016	16	16	1	<0.06
28	0.014				0.016[a]				
29	0.064				0.25[a]	>128	>128		
30	0.38			0.0045	>32		0.5		

References provided in text
Sa *S. aureus*; Spn *S. pneumoniae*; Ec *E. coli*; Ef *Enterococuss faecalis*; Hi *H. influenzae*; Mc *Moraxella cattarhalis*

that the rest of the ring could be truncated without significant lost of potency. They were able to cocrystallize an early (*S. aureus* IC_{50} 6.7 μM) analog of **26** with *E. coli* FabI, and this structure showed that the central amide π-stacks with the NAD^+ cofactor in a fashion that resembles the reaction's transition state. This structure also revealed hydrophobic pockets on each side of the amide providing insights for the design of improved amides including Compound **27**, one of the highlighted leads resulting from this work (50, 51). Figure 10 provides a schematic highlighting the key interactions of amide **27** with *E. coli* FabI. It is also noteworthy that the GSK group measured FabI potency against both Gram-positive and Gram-negative orthologs, providing confidence that analoging strategies based on analysis of *E. coli* cocrystal structures would also lead to potency gains in *S. aureus*. Comparing the potency of hit **26** to lead **27** (see Table 6) punctuates the effectiveness of this SBDD effort: enzyme potency was increased by 300X and antibacterial activity was observed where none was observed for the original hit. Comparing the properties of these two compounds in Table 7, this effort was accomplished with a net reduction in MW,

Table 7
Calculated properties for FabI/FabK antibacterial agents (J. Finn, unpublished results)

Compound	Molecular weight (g/mol)	*A*log *P*	Predicted oral absorption	Solubility	LE Sa[a]
26	448	2.40	Good	–.3.88	1.20
27	374	2.57	Good	–4.24	1.95
28	375	2.99	Good	–4.61	2.06
29	326	1.63	Good	–2.67	2.34
30	706	6.13	None	–6.02	1.05

[a]*LE* ligand efficiency

significant increases in ligand efficiency and similar properties (predicted oral absorption, *A*log*P*, and solubility). Although compound **27** is a potent FabI inhibitor, it has only modest potency against FabK. This potency difference is expected due to structural changes between these enzymes. As a consequence of the lack of FabK potency, amide **27** has little antibacterial activity against *S. pneumoniae*.

Affinium licenced the FabI program from GSK and advanced amide **28** (AFN-1252) into human microdosing studies (52, 53). AFN-1252 is a very close analog of **27** containing a benzofuran in place of an indole. AFN-1252 is slightly more potent (3X) and has very similar properties to **26**. This compound is being examined as an anti-*S. aureus* drug, in particular as an anti-MRSA drug that can be formulated both orally and intravenously. Large MIC panels demonstrate the high level of potency against *S. aureus* for AFN-1252, and microdosing studies have indicated that the compound has reasonable human oral bioavailability.

A Korean company, Crystal Genomics, has also worked on the FabI target and has advanced pyrimidone **29** (CG-400,549) into clinical studies (54, 55). Comparison of **29** to **28** shows that while CG-400,549 is weaker in both enzymatic and antibacterial potency, it compares favorably in ligand efficiency and drug-like properties. This focus on properties differs from most SBDD approaches where the focus is mainly on potency. In the case of FabI antibacterial agents, such as AFN-1252, antibacterial potency is already very good and shifting focus to other areas such as properties could provide formulation advantages. Additional clinical studies are needed to determine the relative strengths and weaknesses of these two compounds.

In an attempt to build broad-spectrum back into this area, researchers at Meiji Seika developed dual FabI–FabK inhibitors including compound **30** (56). This work demonstrated that development of a balanced potency inhibitor is very difficult; the lead compound is a much better inhibitor of FabK than FabI. As a consequence, there is little antibacterial activity against *S. aureus*. As shown in Table 7, the strategy of fusing two inhibitor pharmacophores can lead to a high molecular weight compound with poor drug properties.

From the standpoint of evaluating the value of SBDD in antibacterial discovery, FabI provides a very strong example of the efficiency of the approach. In this target area, gains were quickly made that resulted in compounds interesting enough to advance into the clinic. While the discovery of the FabK enzyme in some bacterial species has diminished some of the enthusiasm for this area, the overall antibacterial potency against *S. aureus* is impressive. Further, the design strategies were consistent with producing reasonable drug-properties and this has resulted in compounds with dual IV/oral activity. The commercial and therapeutic value of advancing a narrow spectrum into the clinic is debatable and it will be interesting to see if these compounds are pursued into Phase 2 and 3 trials.

6. Lesson Learned

SBDD is a promising approach in antibacterial discovery: While this survey is limited to a narrow set of targets, it is abundantly clear that multiple new classes of antibacterial agents are being generated by SBDD programs. Beyond the set of targets in this review, impressive progress has been made using SBDD approaches in several other targets, including GyrB, LpxC, MetRS, and even the bacterial ribosome. This is in sharp contrast to approaches that rely on screening for antibacterial activity using natural products or synthetic collections or optimization of enzymatic hits without access to structural information. SBDD technology has proven its value to the field and will continue to be central to the discovery of new classes of antibacterial agents.

Target selection plays a major role in the profile of the leads generated in the program: While SBDD can often improve potency, spectrum, and drug properties (see below), the potency and spectrum of the leads is closely tied to the selection of the target. Thus, DHFR lead compounds have mainly Gram-positive spectrum and can reach MIC potencies between 0.03 and 0.1 μg/mL against *S. aureus*, while PDF compounds typically have a narrower spectrum and typically reach MIC potencies between 0.5 and 4 μg/mL. For FabI compounds, potencies of <0.01 μg/mL against *S. aureus* are reached and are considered a *S. aureus* only drug class. The profiles of these leads seem independent of the SBDD approach used.

If a quality target is selected, SBDD programs typically deliver potent antibacterial compounds: This survey was biased towards targets (DHFR, PDF, FabI) that have yielded advanced preclinical or clinical compounds. Therefore, by definition, target selection was not a problem with the exception of DHNA, which was selected to illustrate the problems working with a nonvalidated novel target. What is striking in this survey, is how often different groups working on the same target (DHFR, PDF, FabI) all were successful in using SBDD to discover novel antibacterial agents despite using different approaches. In most cases, compounds with potent enzymatic and antibacterial activity were discovered fairly rapidly. By contrast, the non-SBDD approach of optimizing hits from HTS has been relatively ineffective.

SBDD can be used to tailor spectrum (within limits); however, key tools such as enzymatic assays on more than one enzyme ortholog and crystallography using multiple orthologs are often ignored: While potency is typically not an issue in antibacterial programs, achieving the proper (commercially viable) spectrum is a major issue. The issue of enzymatic spectrum is often not reported (and probably not measured). Thus, one is left guessing if the lack of spectrum is due to the lack of enzyme potency or other issues such as bacterial cell wall penetration. The DHFR area provides the most information on enzymatic and antibacterial spectrum and there is a very clear connection between antibacterial and enzymatic potency. Further, enzymatic spectrum often has a very subtle SAR where significant differences in potency are found even though at the sequence level the orthologs are quite similar. Compound **6** (see Fig. 3) is an example of this: It was found using a virtual screen against the *S. aureus* DHFR enzyme but has 100X more potency against *S. pneumoniae* DHFR than *S. aureus* DHFR! PDF presents and interesting case where there was known structural diversity between Type 1 and Type 2 PDF enzymes and where careful ligand design, incorporating ligand flexibility, was used to improve spectrum. It is hard to judge the degree of success of these approaches because the enzymatic activity of PDF inhibitors is typically published using only one protein ortholog. Judging by antibacterial potency, spectrum is a major issue in the PDF area with most compounds limited to a Gram-positive plus an *H. influenzae* profile. Even among the Gram-positives, there is spotty spectrum with many compounds lacking good potency against *S. aureus* and/or *Enterococcus faecalis*. In the case of FabI inhibitors, compounds with high potency against *S. aureus* FabI lack potency against *S. pneumoniae* FabK due to significant differences in the binding sites. There is a clear need for multiple enzyme assays using different orthologs to understand spectrum. Ideally, these assays are in place at the beginning of the program. In cases where enzymatic spectrum is an issue, it is highly recommended to cocrystallize compounds with multiple orthologs and use the structural information to design spectrum.

The SBDD approach used influences drug properties: For DHFR, a focus on obtaining potency by filling hydrophobic pockets led to less soluble compounds more hydrophobic compounds. Not surprisingly, these compounds encountered significant downstream problems such as toxicity (increased hERG activity) and drug properties (poorer oral absorption, decreased aqueous solubility). By contrast, the use of structure-guided combinatorial chemistry provided more diverse leads with improved properties. Likewise, in PDF area, the use of combinatorial chemistry early in the program resulted in ligands with lower molecular weight. Consequently, one needs to look at the overall profile of the desired candidate in selecting an SBDD strategy. Often times, this will mean avoiding strategies that add too much hydrophobicity to the lead and adapting a strategy that selects other (more polar) sites that will lead to the design of more drug-like molecules. While there is no one approach that will suit all circumstances, the overall trend is that *an aggressive exploration of diversity early in the program can provide better more drug-like leads*. Also, a focus on optimizing ligand efficiency early can avoid the pitfall of adding potency by filling less important enzyme pockets, yielding oversized molecules with poor properties.

7. Summary

- Pick the target carefully, as the choice of target has a large influence on the spectrum and potency of the resulting antibacterial agents.
- Measure the enzymatic activity on a panel of orthologs to know the enzymatic spectrum.
- Cocrystallize multiple ligands with the target and incorporate multiple ortholog structures to understand the key conserved molecular interactions and the need for ligand flexibility.
- Measure antibacterial activity on a panel of both sensitive and resistant pathogens.
- Incorporate design concepts that improve drug-like properties of new ligands.
- Explore diversity early in a SBDD program to identify leads with better drug properties.
- Design highly ligand-efficient analogs.

References

1. Arias, C. A. and Murray B. (2009) Antibiotic-Resistant Bugs in the 21st Century: A Clinical Super-Challenge. *N. Engl. J. Med.* **360**, 439–443.
2. Boucher, H.W., Talbot, G. H., Bradley, J.S., Edwards, J. E., Gilbert, D., Rice, L. B., Scheld, M., Spellberg, B. and Bartlett, J. (2009) Bad bugs, no drugs: no ESKAPE! An update from the Infectious Diseases Society of America. *Clin. Infect. Diseases* **48**, 1–12.
3. Giske, C., Monnet, D., Cars, O. and Carmeli, Y. (2008) Clinical and Economic Impact of Common Multidrug-Resistant Gram-Negative Bacilli. *Antimicrob. Agents Chemother.* **52**, 813–821.
4. Overbye, K. and Barrett, J. (2005) Antibiotics: where did we go wrong? *Drug Discovery Today* **10**, 45–52.
5. Bush, K. (2004) Antibacterial drug discovery in the 21st century. *Clin. Microbiol. Infect. Suppl. 4*, 10–17.
6. Payne, D., Gwynn, M., Holmes, D. and Pompliano, D. (2007) Drugs for bad bugs: confronting the challenges of antibacterial discovery. *Nat. Rev. Drug Disc.* **6**, 29–40.
7. Stein, J. (2005) Innovative antibacterial drugs: nothing ventured, nothing gained. *Expert Opin. Investig. Drugs* **14**, 107.
8. Kompis, I., Islam, K. and Then, R. (2005) DNA and RNA Synthesis: Antifolates *Chem. Rev.* **105**, 593–620.
9. Bushby, S. R. (1973) Trimethoprim-sulfamethoxazole. *In vitro* microbiological aspects. *J. Infect. Diseases* **128**(Suppl.) S442–S462.
10. Canton, R., Loza, E., Morosini, M. and Baquero, F. (2002) Antimicrobial resistance amongst isolates of *Streptococcus pyogenes* and *Staphylococcus aureus* in the PROTEKT antimicrobial surveillance program during 1999-2000. *J. Antimicrob. Chemother.* **50(Suppl. S1)**, 9–24.
11. Dale, G., Broger, C., D'Arcy, A., Hartman, P., DeHoogt, R., Jolidon, S., Kompis, I., Labhardt, A., Langen, H., Locher, H., Page, M., Stueber, D., Then, R., Wipf, B. and Oefner, C. (1997) A single amino acid substitution in *Staphylococcus aureus* dihydrofolate reductase determines trimethoprim resistance. *J. Mol. Bio.* **266**, 23–30.
12. Vickers, A., Potter, N., Fishwick, C., Chopra, I. and O'Neill, A. (2009) Analysis of mutational resistance to trimethoprim in *Staphylococcus aureus* by genetic and structural modelling techniques. *J. Antimicrob. Chemother.* **63**, 1112–1117.
13. Kujundzic, N., Kovacevic, K., Jakovina M. and Gluncic, B. (1988) Synthesis and antibacterial effect of derivatives of 5-(3,4,5-trimethoxybenzyl)pyrimidine, -tetrahydropyrimidine, -hexahydropyrimidine and -hydantoin. *Croatica Chemica Acta* **61**, 121–35.
14. Summerfield, R., Daigle, D., Mayer, S., Mallik, D., Hughes, D., Jackson, S., Sulek, M., Organ, M., Brown, E. and Junop, M. (2006) Pharmacophoric Features of Biguanide Derivatives: An Electronic and Structural Analysis. *J. Med. Chem.* **49**, 6977–6986
15. Kohlhoff, S. and Sharma, R. (2007) Iclaprim. *Expert Opin. Invest. Drugs* **16**, 1441–1448.
16. Peppard, W. and Schuenke, C. (2008) Iclaprim, a diaminopyrimidine dihydrofolate reductase inhibitor for the potential treatment of antibiotic-resistant staphylococcal infections. *Curr. Opin. Invest. Drugs* **9**, 210–225.
17. Schneider, P., Hawser, S. and Islam, K. (2003) Iclaprim, a novel diaminopyrimidine with potent activity on trimethoprim sensitive and resistant bacteria. *Bioorg. Med. Chem. Lett.* **13**, 4217–4221.
18. Oefner, C., Bandera, M., Haldimann, A., Laue, H., Schulz, H., Mukhija, S., Parisi, S., Weiss, L., Lociuro, S. and Dale, G. (2009) Increased hydrophobic interactions of Iclaprim with *Staphylococcus aureus* dihydrofolate reductase are responsible for the increase in affinity and antibacterial activity. *J. Antimicrob. Chemother.* **63**, 687–698.
19. Hawser, S., Haldimann, A., Parisi, S., Gillessen, D. and Islam, K. (2002) "AR-100, A Novel Diaminopyrimidine Compound: resistance studies in Trimethoprim-Sensitive and –Resistant Staphylococcus *aureus*" *42nd ICAAC Meeting*; San Diego, CA (poster F-2028).
20. Brandt, R., Neuenhofer, D., Thomsen, T., Hadvary, P. and Islam, K. (2007) "Pharmacokinetics and bioavailability of Iclaprim oral and intravenous formulations in humans" *47th ICAAC Meeting*; Chicago, IL (poster A-806).
21. Krievins, D., Brandt, R., Hawser, S., Hadvary, P. and Islam, K. (2009) Multicenter, randomized study of the efficacy and safety of intravenous iclaprim in complicated skin and skin structure infections. *Antimicrob. Agents Chemother.* **53**, 2834–2840.
22. Wyss, P., Guerry, P., Hartman, P., Hubschwerlen, C., Jolidon, S., Locher, H., Specklin, J. and Stalder, H. (1999) "Anti-MRSA Dihydrofolate Reductase Inhibitors: Synthesis and SAR" *39th ICAAC Meeting*; San Francisco, CA (poster F-1800).
23. Locher, H., Wyss, P., Then, R. and Hartman, P. (1999) "Anti-MRSA Dihydrofolate Reductase Inhibitors: Biological Characterization" *39th ICAAC Meeting*; San Francisco, CA (poster F-1801).

24. Mukhija, S., Bandera, M., Parisi, S., Rigo, S., Lieb, S., Lociuro, S., Gillessen, D. and Islam, K. (2005) "AR-709 – An Investigational diaminopyrimidine: Inhibition, Binding and Mode of Action" *47th ICAAC Meeting*; Chicago, IL (poster F1-1955).

25. Jansen, W., Verel, A., Verhoef, J. and Milatovic, D. (2008) *In vitro* activity of AR - 709 against *Streptococcus pneumoniae. Antimicrobial Agents and Chemotherapy* **52**, 1182–1183.

26. Hawser, S., Bihr, M., Weiss, L., Islam, K. and Lociuro, S. (2007) "AR-709, A Novel Diaminopyrimidine Compound: Resistance Studies in Trimethoprim-sensitive and -resistant Bacteria" *47th ICAAC Meeting*; Chicago, IL (poster F1-1960).

27. Lappin, G., Warrington, S., Sanghera, D., Dowen, S., Lister, N., Islam, K. and Lociuro, S. (2007) "Plasma Pharmacokinetics Of AR-709 Administered to Male Healthy Volunteers as Microdoses by the Intravenous and Oral Route" *47th ICAAC Meeting*; Chicago, IL (poster F1-939).

28. Wyss, P., Gerber, P., Hartman, P., Hubschwerlen, C., Locher, H., Marty, H. and Stahl, M. (2003) Novel Dihydrofolate Reductase Inhibitors. Structure-Based versus Diversity-Based Library Design and High-Throughput Synthesis and Screening. *J. Med. Chem.* **46**, 2304–2312.

29. Sanders, W., Nienaber, V., Lerner, C., McCall, J., Merrick, S., Swanson, S., Harlan, J., Stoll, V., Stamper, G., Betz, S., Condroski, K., Meadows, R., Severin, J., Walter, K., Magdalinos, P., Jakob, C., Wagner, R. and Beutel, B. (2004) Discovery of Potent Inhibitors of Dihydroneopterin Aldolase Using CrystaLEAD High-Throughput X-ray Crystallographic Screening and Structure-Directed Lead Optimization. *J. Med. Chem.* **47**, 1709–1718.

30. Lever, O. W., Bell, L. N., Hyman, C., McGuire, H. M. and Ferone, R. (1986) Inhibitors of dihydropteroate synthase: substituent effects in the side-chain aromatic ring of 6-[[3-(aryloxy) propyl]amino]-5-nitrosoisocytosines and synthesis and inhibitory potency of bridged 5-nitrosoisocytosine-p-aminobenzoic acid analogs. *J. Med. Chem.* **29**, 665–670.

31. Lever, O. W., Bell, L. N., McGuire, H. M. and Ferone, R. (1985) Monocyclic pteridine analogs. Inhibition of *Escherichia coli* dihydropteroate synthase by 6-amino-5-nitrosoisocytosines. *J. Med. Chem.* **28**, 1870–1874.

32. Aubart, K. and Zalacain, M. (2006) Peptide Deformylase Inhibitors *Prog. Med. Chem.* **44** 109–143.

33. Jain, R., Chen, D., White, R.J., Patel, D.V. and Yuan, Z. (2005) Bacterial Peptide Deformylase Inhibitors: A New Class of Antibacterial Agents *Curr. Med. Chem.* **12** 1607–1621.

34. Smith, K., Petit, C., Aubart, K., Smyth, M., McManus, E., Jones, J., Fosberry, A., Lewis, C., Lonetto, M. and Christensen, S. (2003) Structural variation and inhibitor binding in polypeptide deformylase from four different bacterial species. *Prot. Sci.* **12,** 349–360.

35. East, S., Beckett, R., Brookings, D., Clements, J., Doel, S., Keavey, K., Pain, G., Smith, H., Thomas, W., Thompson, A., Todd, R. and Whittaker, M. (2004) Peptide deformylase inhibitors with activity against respiratory tract pathogens. *Bioorg. Med. Chem. Lett.* **14,** 59–62.

36. Azoulay-Dupuis, E., Mohler, J. and Bedos, J. (2004) Efficacy of BB-83698, a Novel Peptide Deformylase Inhibitor, in a Mouse Model of Pneumococcal Pneumonia. *Antimicrob. Agents Chemother.* **48,** 80–85.

37. Clements, J., Beckett, R., Brown, A., Catlin, G., Lobell, M., Palan, S., Thomas, W., Whittaker, M., Wood, St., Salama, S., Baker, P., Rodgers, H., Barynin, V., Rice, D. and Hunter, M. (2001) Antibiotic Activity and Characterization of BB-3497, a Novel Peptide Deformylase Inhibitor. *Antimicrob. Agents Chemother.* **45,** 563–570.

38. Davies, S., Ayscough, A., Beckett, R., Clements, J., Doel, S., Pratt, L., Spavold, Z., Thomas, S. and Whittaker, M. (2003) Structure-Activity Relationships of the Peptide Deformylase Inhibitor BB-3497: Modification of the P2′ and P3′ Side Chains. *Bioorg. Med. Chem. Lett.* **13**, 2715–2718.

39. Ramanathan-Girish, S., McColm, J., Clements, J., Taupin, P., Barrowcliffe, S., Hevizi, J., Safrin, S., Moore, C., Patou, G., Moser, H., Gadd, A., Hoch, Ute, Jiang, V., Lofland, D. and Johnson, K. (2004) Pharmacokinetics in Animals and Humans of a First-in-Class Peptide Deformylase Inhibitor. *Antimicrob. Agents Chemother.* **48,** 4835–4842.

40. Chen, D., Hackbarth, C., Ni, Z., Wu, C., Wang, W., Jain, R., He, Y., Bracken, K., Weidmann, B., Patel, D., Trias, J., White, R. and Yuan, Z. (2004) Peptide Deformylase Inhibitors as Antibacterial agents: Identification of BRC3375, a Proline-3-Alkylsuccinyl Hydroxamate Derivative, by Using an Integrated Combinatorial and Medicinal Chemistry Approach. *Antimicrob. Agents Chemother.* **48,** 250–261.

41. Hackbarth, C., Chen, D., Lewis, J., Clark, K., Mangold, J., Cramer, J., Margolis, P., Wang, W., Koehn, J., Wu, C., Lopez, S., Withers, G., Gu, H., Dunn, E., Kulathila, R., Pan, S., Porter, W., Jacobs, J., Trias, J., Patel, D., Weidmann, B., White, R. and Yuan, Z. (2002) *N*-Alkyl Urea Hydroxamic Acids as a New Class of Peptide Deformylase Inhibitors with Antibacterial Activity. *Antimicrob. Agents Chemother.* **46,** 2752–2764.

42. Aubart, K., Benowitz, A., Campobasso, N., Dreabit, J., Fang, Y., Karpinski, J., Kelly, S., Liao, X., Lee, J., Mercer, D., Lewandowski, T., VanAller, G., Zonis, R., Christensen, S. and Zalacain, M. (2010) "Hydrazinopyrimidines as a New Class of Peptide Deformylase Inhibitors" *50th ICAAC Meeting:* Boston, MA (poster F1-2110).
43. Qin, D., Fang, Y., Benowitz, A., Liao, X., Karpinski, J., Dreabit, J., Knox, A., Kelly, S., Axten, J., Mercer, D., Kulkarni, S., Campobasso, N., Zonis, R., VanAller, G., Christensen, S., Zalacain, M. and Aubart, K. (2010) "Peptide Deformylase Inhibitors: Discovery of a Clinical Candidate from a Novel Chemical Class" *50th ICAAC Meeting:* Boston, MA (poster F1-2111).
44. Margolis, P., Hackbarth, C., Young, D., Wang, W., Chen, D., Yuan, Z., White, R. and Trias, J. (2000) Peptide Deformylase in *Staphylococcus aureus*: Resistance to Inhibition Is Mediated by Mutations in the Formyltransferase Gene. *Antimicrob. Agents Chemother.* **44,** 1825–1831.
45. Lu, H. and Tonge, P. (2008) Inhibitors of FabI, an Enzyme Drug Target in the Bacterial Fatty Acid Biosynthesis Pathway. *Accounts Chem. Res.* **41**, 11–20.
46. Payne, D. J., Warren, P. V., Holmes, D. J., Ji, Y. and Lonsdale, J. T. (2001) Bacterial fatty-acid biosynthesis: a genomics-driven target for antibacterial drug discovery. *Drug Disc. Today* **6**, 537–544.
47. Seefeld, M. A., Miller, W. H., Newlander, K. A., Burgess, W. J., Payne, D. J., Rittenhouse, S. F., Moore, T. D., DeWolf, W. E., Keller, P. M., Qiu, X., Janson, C. A., Vaidya, K., Fosberry, A. P., Smyth, M. G., Jaworski, D. D., Slater-Radosti, C. and Huffman, W. F. (2001) Inhibitors of bacterial enoyl acyl carrier protein reductase (FabI): 2,9-disubstituted 1,2,3,4-tetrahydropyrido[3,4-b]indoles as potential antibacterial agents. *Bioorg. Med. Chem. Lett.* **11**, 2241–2244.
48. Heerding, D. A., Chan, G., DeWolf, W. E., Fosberry, A. P., Janson, C. A., Jaworski, D. D., McManus, E., Miller, W. H., Moore, T. D., Payne, D. J., Qiu, X., Rittenhouse, S. F., Slater-Radosti, C., Smith, W., Takata, D. T., Vaidya, K. S., Yuan, C. C. K. and Huffman, W. F. (2001) 1,4-Disubstituted imidazoles are potential antibacterial agents functioning as inhibitors of enoyl acyl carrier protein reductase (FabI). *Bioorg. Med. Chem. Lett.* **11**, 2061–2065.
49. Miller, W. H., Seefeld, M. A., Newlander, K. A., Uzinskas, I. N., Burgess, W. J., Heerding, D. A., Yuan, C. C. K., Head, M. S., Payne, D. J., Rittenhouse, S. F., Moore, T. D., Pearson, S. C., Berry, V., DeWolf, W. E., Jr., Keller, P. M., Polizzi, B. J., Qiu, X., Janson, C. A. and Huffman, W. F. (2002) Discovery of Aminopyridine-Based Inhibitors of Bacterial Enoyl-ACP Reductase (FabI). *J. Med. Chem.* **45**, 3246–3256.
50. Seefeld, M. A., Miller, W. H., Newlander, K. A., Burgess, W. J., DeWolf, W. E. Jr., Elkins, P. A., Head, M. S., Jakas, D. R., Janson, C. A., Keller, P. M., Manley, P. J., Moore, T. D., Payne, D. J., Pearson, S., Polizzi, B. J., Qiu, X., Rittenhouse, S. F., Uzinskas, I. N., Wallis, N. G. and Huffman, W. F. (2003) Indole Naphthyridinones as Inhibitors of Bacterial Enoyl-ACP Reductases FabI and FabK. *J. Med. Chem.* **46**, 1627–1635.
51. Payne, D. J., Miller, W. H., Berry, V., Brosky, J., Burgess, W. J., Chen, E., DeWolf, W. E., Jr., Fosberry, A. P., Greenwood, R., Head, M. S., Heerding, D. A., Janson, C. A., Jaworski, D. D., Keller, P. M., Manley, P. J. Moore, T. D., Newlander, K. A., Pearson, S., Polizzi, B. J., Qiu, X., Rittenhouse, S. F., Slater-Radosti, C., Salyers, K. L., Seefeld, M. A., Smyth, M. G., Takata, D. T., Uzinskas, I. N., Vaidya, K., Wallis, N. G., Winram, S. B., Yuan, C. C. K. and Huffman, W. F. (2002) Discovery of a novel and potent class of FabI-directed antibacterial agents. *Antimicrob. Agents Chemother.* **46**, 3118–3124.
52. Karlowsky, J. A., Laing, N. M., Baudry, T., Kaplan, N., Vaughan, D., Hoban, D. J. and Zhanel, G. G. (2007) *In vitro* activity of API - 1252, a novel FabI inhibitor, against clinical isolates of Staphylococcus aureus and Staphylococcus epidermidis. *Antimicrob. Agents Chemother.* **51**, 1580–1581.
53. Karlowsky J. A., Kaplan N., Hafkin B., Hoban D. J. and Zhanel G. G. (2009) AFN- 1252, a FabI inhibitor, demonstrates a Staphylococcus-specific spectrum of activity. *Antimicrob. Agents Chemother.* **53**, 3544–3548.
54. Yum, J. H., Kim, C. K., Yong, D., Lee, K., Chong, Y., Kim, C. M., Kim, J. M. Ro, S. and Cho, J. M. (2007) *In vitro* activities of CG400549, a novel FabI inhibitor, against recently isolated clinical staphylococcal strains in Korea. *Antimicrob. Agents Chemother.* **51**, 2591–2593.
55. Bogdanovich, T., Clark, C., Kosowska-Shick, K., Dewasse, B., McGhee, P. and Appelbaum, P. C. (2007) Antistaphylococcal activity of CG400549, a new experimental FabI inhibitor, compared with that of other agents. *Antimicrob. Agents Chemother.* **51**, 4191–4195.
56. Kitagawa, H., Ozawa, T., Takahata, S., Iida, M., Saito, J. and Yamada, M. (2007) Phenylimidazole Derivatives of 4-Pyridone as Dual Inhibitors of Bacterial Enoyl-Acyl Carrier Protein Reductases FabI and FabK. *J. Med. Chem.* **50**, 4710–4720.

Chapter 14

Leveraging SBDD in Protein Therapeutic Development: Antibody Engineering

Gary L. Gilliland, Jinquan Luo, Omid Vafa, and Juan Carlos Almagro

Abstract

Antibodies make up the largest, growing segment of protein therapeutics in the pharmaceutical and biotechnology industries. The development or engineering of therapeutic antibodies is based to a large extent on our knowledge of antibody structure and requires sophisticated methods that continue to evolve. In this chapter, after a review of what is known about the structure and functional properties of antibodies, the current, state-of-the-art antibody engineering methods are described. These methods include antibody humanization, antigen-affinity optimization, Fc engineering for modulated effector function and extended half-life, and engineering for improved stability and biophysical properties. X-ray crystallographic structures of antibody fragments and their complexes can play a critical role in guiding and, in some cases, accelerating these processes. These approaches represent guidelines for developing antibody therapeutics with the desired affinity, effector function, and biophysical properties.

Key words: Antibody Engineering, Fc Engineering, Antibody Humanization, Affinity Maturation, Developability, Fc Effector Function., Antibody Structure

1. Introduction

Protein therapeutics or biologics comprise an ever-increasing segment of almost all major pharmaceutical and biotechnology company portfolios. A wide range of protein products are being developed that include antibodies, growth factors, enzymes, toxins, etc. Of these families of proteins, antibodies make up the largest, growing component. The reasons for this are many. Antibodies have excellent properties that include extremely tight binding, high specificities and efficacies, long half-lives, good biophysical properties, and minimal immunogenicity (1). Currently, monoclonal antibodies (mAbs) comprise more than 30% of the therapeutic proteins in clinical trials and are the second largest biologics class after vaccines (2). The innate properties and further engineering of

Leslie W. Tari (ed.), *Structure-Based Drug Discovery*, Methods in Molecular Biology, vol. 841,
DOI 10.1007/978-1-61779-520-6_14, © Springer Science+Business Media, LLC 2012

this unique class of proteins have provided the biopharmaceutical sector with 30 marketed mAbs and nearly 300 more in clinical development across all major therapeutic indications (3).

Antibodies have revolutionized traditional medicine, after technological advances in hybridoma biology (4), in phage/cellular display technologies (5–7), in protein engineering (described in this chapter), and in manufacturing, involving large-scale production, purification, and formulation (8). Fc structure and function analyses have enabled the identification of key amino acid residues and glycoforms that can be engineered to further impact therapeutic antibody efficacy and safety (9–11). While IgG1 has been the standard isotype of choice for the majority of the therapeutics (due to manufacturability and stability), engineered IgG2 and IgG4 subtypes and chimeras have also been developed, offering distinct advantages. Of course, antibodies do have limitations that impact their use with certain indications. For example in oncology, antibodies may be reduced in their efficacy because of restricted tumor penetration because of their size.

Interestingly, because of the success of antibodies and because of their limitations, new classes of biologics are being developed with similar binding properties, but with quite different structures, and hence quite different functional properties. These can be roughly divided into two different general categories of scaffolds; those that are based on immunoglobulin-like domains (12) and those that are nonantibody, novel binding proteins (13). Despite their promise, members of these potential classes of therapeutics have been slow in reaching the market.

Antibodies will not be replaced any time soon, despite the emergence of a possible new generation of biologics based on alternative scaffolds. Notwithstanding the maturity of the field, the methodologies are continuing to be developed for engineering antibodies with the appropriate functional and biophysical properties. With this in mind, this chapter focuses on antibody engineering. Specifically, after a short review of what we know about antibody structure and functional properties, the current state-of-the-art in engineering is described, ending with a brief discussion on where the field is heading. It should be noted that the approaches used for engineering antibodies are very similar to many of those being used in the alternative scaffold arena, so they have general applicability.

2. Antibody Structure

The field of antibody engineering is predicated on our knowledge of the structure of antibodies. Our detailed knowledge of this family of proteins is based in large part on 3D crystal structure determinations that began in the 1970s and continue even today (for reviews see refs. (14–18)). Antibodies are "Y"-shaped molecules comprised of two heavy chains (HC's) and two light chains (LC's) held together covalently through disulfide bridges. An example of an intact antibody structure is shown in Fig. 1. The two antigen-binding Fab's are formed by the pairing of the two LC domains with the two N-terminal HC domains. The pairing of the remaining HC domains forms the Fc region found at the base of the "Y." The structures of intact antibodies have been difficult to obtain, and when they are, the structures are determined at relatively low resolution. The reason for this is due to the flexibility of the molecule resulting from the Fabs' ability to change their orientation with respect to the Fc. However, it was discovered early on that antibodies could be selectively proteolyzed into Fc and Fab fragments (19). This approach led to the first medium-resolution crystal structures of Fab's and Fc's. Today, the use of molecular biology methods allows laboratories to routinely produce Fab's, Fc's, and other antibody fragment constructs. This has led to the determination and deposition in the Protein Data Bank (PDB) (20) of >1,100 crystal structures of the individual fragments but also complexes of these macromolecules with proteins,

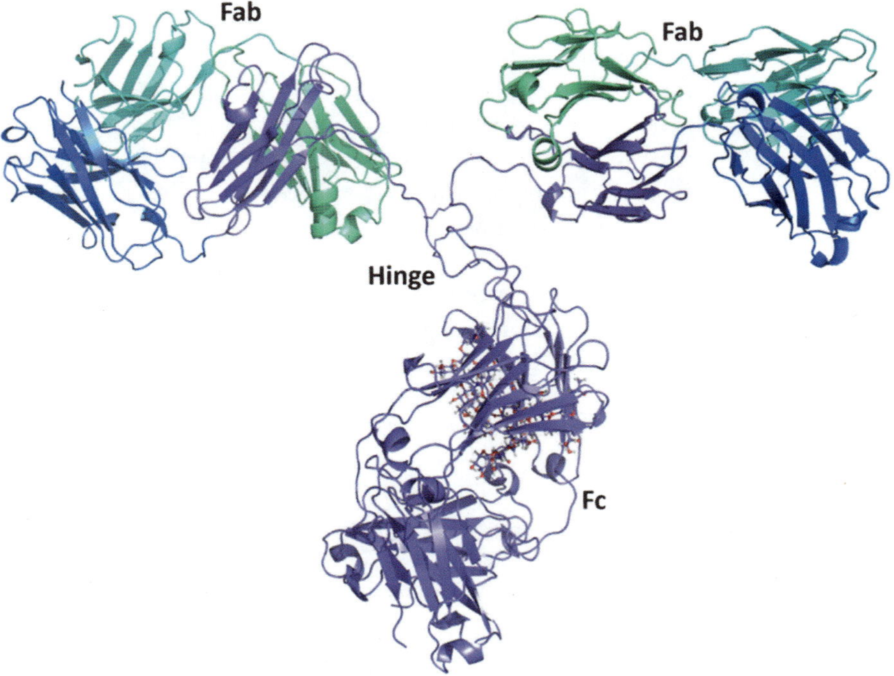

Fig. 1. An intact murine IgG2a antibody (PDBid 1igt, see ref. (143)) shown as a ribbon representation with the carbohydrate in the Fc shown with balls and sticks. The heavy chain V_H is shown in *marine* and constant regions are shown in *slate*. The light chain V_L domains are shown in *cyan* and the constant domain is shown in *green-cyan* (all figures of macromolecule structures were produced using PyMol (DeLano WL. 2002. *The PyMOL molecular graphics system*. Delano Scientific, San Carlos, CA)).

peptides, haptens, and other ligands. This wealth of data continues to provide, through data mining, new insights into how antibody structure relates to function and provides the foundation for antibody engineering. Before describing the current state-of-the-art of antibody engineering, a description of the antibody structure is provided.

2.1. Antibody Structural Features

Immunoglobulins are divided into five classes, IgA, IgD, IgE, IgG, and IgM, based on their HC's. The HC's can pair with two classes of light chains, κ or λ. The HC's have an N-terminal variable domain (V_H) followed by a number of constant domains depending upon the class. IgA's, IgD's, and IgG's have three constant domains (C_{H1}, C_{H2}, C_{H3}), and IgE's and IgM's have four constant domains (C_{H1}, C_{H2}, C_{H3}, C_{H4}) and an additional J chain. The κ and λ LC's have an N-terminal variable domain (V_L) and a single constant domain (C_L). Further description of the antibody structure focuses on the IgG class, since this is the primary form that is engineered to produce therapeutic antibodies and it is the class that we have the most complete structural picture.

2.2. The Immunoglobulin Fold

The domains of the heavy and light chains have approximately 110 amino acid residues and have a folding pattern that is referred to as the "immunoglobulin fold." Each domain has a structural motif called a Greek key barrel formed from two tightly packed β-sheets. The amino acid residues in the β-strands are part of what are called the framework residues. These residues, often invariant, are involved in the tight packing of the two β-sheets and are responsible for domain–domain interactions in the IgG structure. In the constant domains, the four-stranded and three-stranded β-sheets are formed from β-strands ↓A ↑B ↓E ↑D, and from β-strands, ↓C ↑F ↓G (the arrows represent the direction of the polypeptide chain from N- to C-terminus), respectively. These domains typically have short loops connecting the β-strands making them quite compact. The N-terminal variable domains in contrast to the constant domains have two β-sheets, one four-stranded, ↓A ↑B ↓E ↑D, and the other five-stranded, ↓C" ↑C' ↓C ↑F ↓B. The variable domains, in general, have longer loops connecting the β-strands. Both domains have cysteine (Cys) residues in ↑B and ↑F β-strands that form an intradomain disulfide bridge.

2.3. The Fab

Immunoglobulin Fab's are formed by the interaction of the V_H and C_{H1} domains with the V_L and C_L domains, respectively. The Fab interacts with its target through its antigen combining site formed by the packing of the V_H and V_L through the interaction of the two ↓C" ↑C' ↓C ↑F ↓B β-sheets forming a nearly twofold symmetry. This folding pattern is illustrated in Fig. 2. This barrel-like structure interacts with antigen through its loops (on the N-terminal side) or complementarity-determining regions

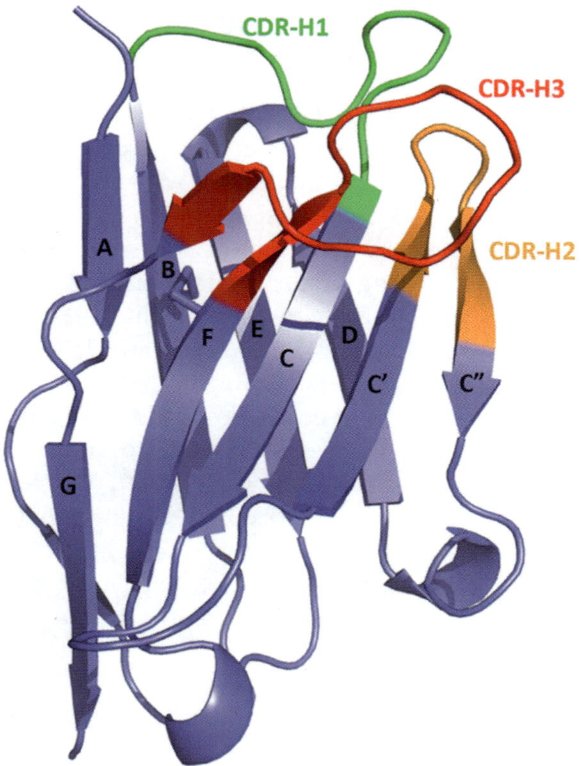

Fig. 2. A ribbon representation of the V_H of an Fab (PDBid, 3L7E) (66) with the β-strands A–G labeled. The three complementarity-determining regions (CDR) are shown in *green* (H1), *orange* (H2), and *red* (H3). The disulfide bond in between is shown in stick representation between β-strands B and F.

(CDR's) (discussed below). Opposite from the variable domains interactions, the constant domains, C_{H1} and C_L, pack together in a nearly perpendicular fashion via the exposed faces of the ↓A ↑B ↓E ↑D β-sheets.

An extended region of the polypeptide, referred to as the switch region, connects the Fab constant and variable domains of each chain. The switch region links the V ↑F and C ↓A β-strands. The orientation of the Fab V domains with respect to the C domains can be defined by what is termed the elbow angle (21, 22). A recent survey of elbow angles for Fab's found a range of 116–226° for those with λ light chains and much smaller values for those with κ light chains (23). A structurally conserved feature discovered in the HC termed the ball-and-socket joint is thought to allow pivoting while restricting the elbow angle to a limited set of values (24). The ball is composed of two C_{H1} residues, Phe and Pro, and the socket is formed by Leu/Val, Thr, and Ser in V_L. A later analysis of elbow angle found values of >180°. In these structures, the ball and socket are separated, allowing the domains to assume larger angles (23).

2.4. The Antigen Combining Site

One of the focal points of antibody engineering is the antigen combining site. It is formed by the pairing of the variable domains, V_H and V_L. The two domains each contribute three CDR's, CDR-L1, CDR-L2, and CDR-L3 for V_L and CDR-H1, CDR-H2, and CDR-H3 for V_H. These hypervariable regions were originally identified by protein sequence analysis studies (25, 26) prior to the determination of the 3D structure of the antibody. The three CDR's from the two V domains are brought together by the packing of the β-sheet surfaces as described above (Fv region). The CDR's are found at the N-terminal end of the Fv. The two β-sheets and the nonhypervariable loops are often referred to as the Framework Region (FR) (see Figs. 2 and 3).

The CDR's in each domain approximately correspond to the residues in the loops connecting β-strands ↑B and ↓C for CDR-L1 and CDR-H1, ↑C' and ↓C" for CDR-L2 and CDR-H2, and ↑F and ↓G for CDR-L3 and CDR-H3 (see Subheading 4.2 below). The CDR's vary both in their amino acid sequence and number of amino acid residues. (The antibody sequence diversity comes from the genetic recombination of the IGHV, IGHD, and IGHJ gene

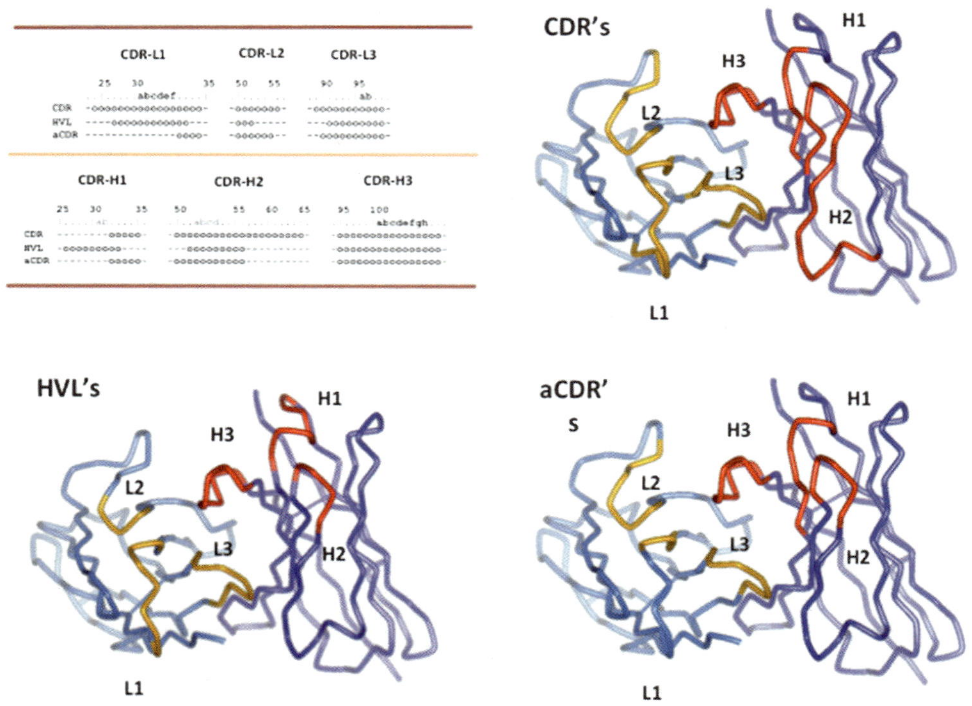

Fig. 3. Antigen-binding site definitions. *Left top*: linear representation of antigen-binding site definitions using Chothia's numbering (27). Residues defining the CDR's (*top*), Hypervariable loops (HVL, *center*), and abbreviated CDR's (aCDR's, *bottom*) are represented by "o." *Right top*: CDR's (in *orange* and *red* for V_L and V_H, respectively) mapped onto the trace of an Fv structure seen from antigen perspective. *Left bottom*: HVL's mapped onto the trace of the same structure. *Right bottom*: aCDR's. Note the differences in the antigen-binding site definitions between CDRs and HVL at H2 and H1, and between CDR's and aCDR's at L1 and H1 (the PDBid of the coordinates used to illustrate the definitions is 1FVA (144)).

segments for V_H and IG(K/L)V and IG(K/L)J gene segments for V_L with subsequent somatic hypermutation in mature B cells). As mentioned earlier, the antigen combining site is formed by the packing of the ↓C" ↑C' ↓C ↑F ↓B β-sheets in the variable region of the Fab bringing the three CDR's of each domain into close proximity.

2.5. Canonical CDR Structures

Early analyses of the antigen combining sites of the small set of structures of immunoglobulin fragments available at the time showed that five out of the six CDR's, with the exception of CDR-H3, had a limited set of main-chain conformations or *canonical structures* (27, 28). This finding represented a paradigm shift in the field, dispelling the belief that each antibody has unique hypervariable loop (HVL) conformations. Canonical structures are defined by the conserved residues located in the HVL, the framework residues and the loop length. Subsequently, antibody sequence analysis revealed that the number of possible combinations of canonical structures was limited (29–31) and that even though the CDR-H3 conformation is variable, the residues in the loop nearest the framework (torso) also have defined conformations (32, 33). These findings suggested that structural restrictions at the antigen-binding site may affect antigen recognition. Additional studies found that the lengths of the HVL's are the determinants of antigen-combining site topography (34), since they are the major factor influencing canonical structures (27, 35, 36).

The limited set of CDR canonical structures allowed their eventual inclusion in macromolecular modeling strategies for antibody structures (37). Approaches for antibody modeling continue to evolve along with the field of protein structure prediction (38–41). Antibody 3D modeling has been highly facilitated by the discovery of the canonical structures (27), as well as by the large number of antibody crystal structures available in the PDB (20). In fact, different antibody modeling methods have recently been compared in a blinded study (41), including two homology modeling strategies independently developed by CCG (Chemical Computer Group) and Accelrys Inc., and two automated servers, namely, Rosetta Antibody (39) and PIGS (Prediction of ImmunoGlobulin Structure) (40), a fully automated server that predicts Fv structures based on the canonical structure model. In the study, nine structures were used as benchmarks to compare the modeling methodologies. With the exception of CDR-H3, good agreement between the models and X-ray structures was found in most cases for all of the modeling approaches.

2.6. The Hinge

The region of the HC connecting the C_{H1} and C_{H2} domains is termed the *antibody hinge*, which separates the two Fab's from the Fc. The hinge allows the Fab's to independently change their orientations with respect to the Fc. The details of the hinge region for the human IgG are limited, since only three structures provide details of its conformations. These include the structures of two intact immunoglobulins, IgG1 Kol (42) and IgG1 b12 (43). The hinge has been divided into three regions: the upper hinge provides the Fab the freedom to rotate, the core hinge in which the disulfide bonds link the two HC's, and the lower hinge allows movement of the Fc. The observed structural features include a helical region in the upper hinge, a poly-L-proline helix for the core and extended structure for the lower hinge (42, 44).

2.7. The Fc

The Fc domain that mediates effector function is formed by the pairing of the C_{H2} and C_{H3} of one HC with the same domains in the other HC. The initial structure of the Fc at medium resolution (45) followed by higher resolution structures (e.g., see ref. (46)) have provided a consistent picture of the structural features of this region of the antibody. The C_{H2} domains do not interact directly with each other, and carbohydrate residues, covalently attached at N297, fill the gap between the two domains. In some of the Fc structures, hydrogen bonding is observed between sugar residues, or water and the sugar residues in the two carbohydrate chains.

3. Antibody Engineering

Antibody Engineering as mentioned above has its foundation in our knowledge of the 3D structure of members of this family of proteins. This knowledge is used to guide the conversion of antibodies from virtually any vertebrate including mouse, rabbit, camelids, or even sharks into human-like antibodies via humanization of antibodies. It has also provided the foundations for engineering antibodies with enhanced potency and efficacy via affinity optimization, as well as modulating half-life and effector functions via Fc engineering. Below, we review these processes.

4. Antibody Humanization

4.1. Humanization Approaches

Humanization of antibodies is designed to increase the human content of a V region obtained from sources other than human, for instance, mouse or rabbit hybridomas. The first humanized antibody with therapeutic value, Alemtuzumab, was obtained in the 1980s by CDR grafting (47). As of July 2010, the FDA has approved 28 antibodies for therapeutic applications in humans and as many as 13 are humanized molecules (48). The success of humanization has fueled the diversification of humanization methods. Some of the methods, often called rational methods (49), include

Deimmunization (50), Specificity-Determining Residues (SDR) grafting (51), Superhumanization (52), Human String Content Optimization (53), and germ-line humanization (54). These methods have in common the design, based on sequence and structural considerations, of a few humanized variants to be tested for binding or any other property of interest. If the designed variants prove to be unsatisfactory, a new design cycle and binding assessment is initiated.

Other humanization methods, sometimes called empirical methods (55), rely on selection rather than on the design cycle. These methods emerged with the invention of phage display and high-throughput screening (HTS) techniques during the 1990s. Phage display and HTS offered efficient tools to explore combinatorial libraries of billions of antibody variants and select those of interest with relative ease. A typical example of an empirical method is Guided Selection (56). This method enabled the discovery of the first phage display antibody approved by the FDA called Adalimumab (48) for treatment of rheumatoid arthritis and Crohn's disease. In contrast to the rational methods to humanize antibodies, Guided Selection makes no assumptions on the impact of mutations on the antibody structure and binding. Another example of an empirical method is Framework (FR) shuffling (57), which is based on the generation of a library of humanized antibody variants by combining nonhuman CDR's with a repertoire of mixed human germ-line FR's, followed by screening for binding to antigen.

The three main drivers of the expansion and development of humanization methods have been (1) to increase the human germ-line gene content of the therapeutic antibodies to minimize immunogenic reactions, (2) to preserve the binding profile of the parental nonhuman antibody molecule to retain potency, efficacy and minimize costs and time in further engineering processes, and (3) to avoid infringement of patents protecting CDR grafting and derived humanization methods. In the next section, we provide an overview of the CDR grafting method while discussing how the increase in the human content and preservation of the binding profile have been guiding the evolution of humanization methods. A discussion of the antibody humanization Intellectual Property (IP) landscape is beyond the scope of this chapter.

4.2. CDR Grafting

The first implementation of the CDR grafting method was based, as its name indicates, in identifying the CDR of a nonhuman antibody and grafting these regions into a human FR (47). As mentioned above, the CDR's were defined by Wu and Kabat (25), prior to the solution of the Fv domain's structure, as the regions with highest variability values in a multiple alignment of antibody sequences. Subsequent resolution of the Fv structures and antigen–antibody complexes (15) validated Wu and Kabat's predictions.

Analyses of the first atomic structures of Fab and V_L fragments revealed that the V domains are composed of well-conserved FR's and loops that vary in structure, called HVLs. The HVLs are stabilized by residues with low variability that, through their packing, hydrogen bonding, or the ability to assume unusual φ, ψ, or ω polypeptide backbone torsion angles, are primarily responsible for the main-chain conformations of the HVL's. The CDR's, whose definitions are based on sequence comparisons, contain all the HVL except for the first CDR of V_H (see Fig. 3).

Yet, another definition of binding site came from the analysis of antigen–antibody complexes by Padlan et al. (58). Their study indicated that only about one third of the CDR residues are typically involved in the interaction with antigen. Comparison of the residues in contact with antigen and sequence variability values pinpointed residues in contact with the antigen as the most variable ones. These residues were defined as SDR's and the regions containing the SDR's were termed abbreviated CDR's (aCDR's) (see Fig. 3). By targeting these residues for grafting instead of the CDR or HLV, a method, designed to increase the human content of the humanized antibodies, called SDR grafting (51) was developed.

4.3. Framework Selection

Several sources of human sequences have been utilized in CDR grafting including consensus sequences, sequences product of immune responses and germ-line gene sequences. The latter source has increasingly been utilized as a template to generate CDR grafted antibodies. The rationale is that consensus sequences are created by artificial combination of different genes, and thus, being nonnatural products could be immunogenic. Mature sequences are products of immune responses, and thus, more often than not, carry somatic mutations that are not under species selection, and hence, might be immunogenic.

The physical maps of the human IGH and IGL/IGK loci were elucidated during the 1990's and the functional germ-line gene repertoires they encode have been thoroughly characterized since then. This information has been compiled in The Immnogenetics Database (IMGT) at the URL: http://imgt.cines.fr/. The repertoire of human functional IGHV, IGHD, and IGHJ germ-line genes is composed of 39 functional IGHV genes assorted into 7 gene families, 27 IGDH genes also classified in 7 gene families and 6 IGHJ genes (30, 59, 60). The repertoire of human IGKV germline genes comprises 30 functional genes distributed in 6 IGKV families and 5 IGKJ genes (30, 61). Since the mouse antibodies are made of 95% κ-type light chains, IGLV and IGLJ are seldom used in humanization protocols.

One method to select human germ-line genes to be used as FR donors is by blasting the amino acid sequence of the parental nonhuman V region against the repertoire of human germ-line genes.

The human genes are then ranked based on their homology with respect to the nonhuman V region and genes with highest homology are selected as FR donors. This procedure of FR selection is known in the field as best fit or homology method (62). Of course, the higher the homology between the nonhuman V parental region amino acid sequence and the human V genes, the number of potential incompatibilities between the nonhuman CDR's and human FR's are fewer, hence, the higher the chances of retaining the affinity of the nonhuman antibody. Nevertheless, other criteria in addition to homology could be factored into the selection process. For instance, genes encoding the same canonical structures as the nonhuman antibody can be favored over those encoding different ones. Also, the sequences could be analyzed to remove genes encoding developability liabilities, e.g., potential glycosylation sites, unpaired Cys residues, deamidation sites, and exposed methionine (Met) residues or other residues potentially compromising further development of the antibody (see Subheading 6 below). Furthermore genes shown to be frequently used in vivo could be ranked with a higher score over genes with poor or marginal expression. For instance, analysis of human IGHV genes amplified from single B-cells of multiple human donors (63, 64) indicate that families IGHV1, IGHV3, IGHV4, and IGHV5 constitute the majority of the productively rearranged IGHV segments with a few genes, e.g., IGHV1-69, IGHV3-23, and IGHV5-51, comprising 30% or more of the total. For human IGKV genes, those derived primarily from families IGKV-1, IGKV-2, and IGKV-3 are overused, with a single gene, IGKV3-30 (also known as A27), accounting for more than 10% or more of the total. These genes have been used in several humanization protocols as well as scaffolds to build antibody phage-displayed libraries for antibody discovery (65). Thus, a wealth of information is available for these genes, which facilitate the sequence analyses of the engineered variants, 3D modeling and further engineering processes of antibodies bearing FR's derived from these genes.

4.4. FR Libraries

Originally, after grafting nonhuman CDR's onto the selected human FR's, several mutations were introduced in the human FR's (back mutations) to accommodate the CDR's and avoid losses in binding (62). With the characterization of the human germ-line gene repertoire, an alternative to avoid back mutations and thus, increase the human content of the final product has been implemented (66). In this strategy Human FR Libraries (HFL's), which combine the parental CDR's with different human FR's are designed and tested for binding and/or other variables of interest. By designing HRL's the chances to select the proper combination of residues that naturally accommodate the CDR's increases, and hence the likelihood of retaining binding after CDR grafting.

The size of the HFL can vary. In the example described by Fransson and coworkers, a mouse antibody specific for human IL13 was humanized by combining 6 V_H chains and 15 V_L chains. These variants were designed based on three human FR's from the IGVH2 germ-line gene family, two from IGVH4 family and one from IGVH3 family. For FR-4 the human IGJH-1 was chosen. The V_L library was built by combining 11 genes from the IGKV-1 gene family and two each from IGKV-2 and IGKV-3 families with IGHJ-1, 2, 4 and 5 genes.

All 19 variants were produced in IgG1 format with each V_L variant paired with each V_H variant. The resulting library was initially screened using direct binding to biotinylated antigen in solution and the ability to inhibit the target antigen from binding to its cognate receptor. Both assays showed similar results, although the receptor inhibition assay allowed better discrimination of differences among the V_L and V_H combinations. One V_H variant exhibited the best binding profile in both assays combined with most V_L chains. To further characterize the interaction with antigen, the best five V_L chains were evaluated by BIAcore showing similar K_D values (250–370 pM) with five-fold affinity loss with respect to the mouse parent antibody (50 pM). The X-ray structure of the complex between the parental antibody and IL13 (66) indicated that most of the interactions with the antigen were though residues in V_H, thus explaining the promiscuity of V_L usage.

4.5. Engineering CDR's for Enhanced Human Content

CDR grafting protocols typically do not modify the CDR's of the parental nonhuman antibody, which carry nonhuman amino acids and hence a risk of immunogenicity (67). In recent years methodologies have been developed to identify nonhuman residues and/or immunogenic spots in the CDRs and mutate them to human germ-line gene residues without compromising binding (68, 69). In one methodology (68) the closest matching human germ-line sequence to the humanized antibody was identified based on sequence comparison, and all possible single substitutions that increase the sequence identity of the engineered antibody sequence to the closest human germ-line sequence were screened for binding. As a result, the post-CDR grafting engineered antibodies were indistinguishable from human mAbs isolated from transgenic mice or human antibody libraries. Alternatively, overlapping peptides scanning the CDR's have been tested for their ability to induce proliferative responses with CD4+ T cells and dendritic cells from community donors (69). The peptides carrying T-cell epitopes were indentified and deimmunized by modifying the amino acid sequence of the CDR's based on 3D modeling, alanine scanning data or selection of variants that were shown a priori to not impact affinity based on an empirical antigen-binding assay.

4.6. Affinity Optimization

Affinity losses are a frequent side effect of humanization (70). In addition, most of the antibodies isolated from phage display libraries have affinities in the low nanomolar or high picomolar range (65, 71, 72). Since higher affinity can have an impact in efficacy in some indications and/or antibodies with higher affinity can be administered in lower or less frequent doses, increasing the affinity to the low picomolar or even femtomolar range may be desirable. Affinity optimization is typically accomplished by creating a library of antibody variants using the parent molecule as the V region template and display technologies, e.g., phage (73), ribosome (74), or yeast display (75) to select the variants of interest from the library. Diversity can either be introduced randomly across the parental V gene (76, 77) or targeted to specific regions (78). The inherent properties of ribosome display, i.e., PCR amplification between selection rounds, make it well suited for random approaches. Focused diversity is typically applied to phage and yeast display, and has the advantage over random mutagenesis in that one has more control over the consequences of changes introduced into the V gene. This is critical when optimizing a therapeutic antibody, since indiscriminate introduction of mutations could generate immunogenic spots, destabilizing mutations and developability liabilities (see below). An additional advantage of focused diversity is that variation can be concentrated in a given area of the V region—typically the binding site, and thus the sequence space of a set of predetermined positions can be exhaustively explored using saturation mutagenesis. Generation of a set of diverse variants retaining or improving affinity can also provide the raw material to help in selecting antibodies carrying germ-line gene amino acids and/or less liable amino acids in a position of interest.

Given the limitation of display technologies in sampling a library of $>10^{10}$ variants and since CDR's are on average nine residues long, the sequence space generated by saturation mutagenesis cannot be exhaustively explored if more than one CDR is targeted for diversification. Therefore, libraries of individual CDR's are cloned and selected in parallel against the target (79). The best variants of this first round of selection are then combined and screened for improved binding. An alternative to this parallel approach is to conduct sequential selections by choosing the best variant in one CDR library and use it as the starting point to optimize the next CDR. The latter strategy has consistently yielded variants with improved affinity (79). Alternatively, only a few residues within a given CDR can be diversified, which enables the exploration of more than one CDR at a time. One strategy, called Specificity-Determining Residues Usage Optimization (SDRU) targets for varying those residues observed to be in contact with the antigen (66). This strategy has the potential to explore synergistic combinations of residues coming from different CDR's.

5. Antibody Fc Engineering

5.1. Nature's Fcs: IgG Isotypes

Complimenting the natural diversity of target/antigen specificity conferred by the CDR's of the Fab domain, specific human IgG isotypes (e.g., IgG1, IgG2, IgG3, and IgG4) have evolved differential capacities to recruit immune effector functions, such as antibody-dependent cellular cytotoxicity (ADCC, e.g., IgG1 and IgG3), antibody-dependent cellular phagocytosis (ADCP, e.g., IgG1, IgG2, IgG3, and IgG4), and complement dependent cytotoxicity (CDC, e.g., IgG1, IgG3). Isotype-specific engagement of such immune functions is based on selective Fc receptor interactions on distinct immune cell populations (e.g., natural killer cells, neutrophils, and macrophages) and the ability to bind C1q, fix complement, and activate the assembly of a membrane attack complex (MAC). Among the various isotypes, the relative affinity for Fcγ receptors (e.g., FcγRI (CD64), FcγRIIa/b/c (CD32), FcγRIIIa/b (CD16)) is high for IgG1 and IgG3; however, there is minimal affinity for IgG2 (restricted to FcγRIIa 131H polymorphism), and the only receptor with distinct affinity for IgG4 is FcγRI (80). Such selective affinities for FcγRs have been mapped to IgG's lower hinge/C_{H2} region, and cocrystal structures have revealed key contact residues that vary among isotypes (81–83). Thus, antibodies function as "adaptor" molecules that can engage Fab/CDR-mediated opsonization of foreign antigens, while using isotype-specific Fc's to recruit and activate specific leukocytes of the immune system to eliminate foreign or infectious cells and particles.

Complement activation after opsonization with IgG can culminate in effector functions, including CDC, after the assembly of the MAC or complement dependent phagocytosis, through specific receptors binding complement components on effector macrophages. Consistent with their ability to bind the elemental C1q of complement, different isotypes have shown high to low levels of activity for complement fixation or target cell lysis (CDC) as IgG1 > IgG3 > IgG2 > IgG4. While the physiological relevance of IgG2 and IgG4 in CDC remains disputable, IgG2 capacity to activate complement on microbes is well-documented (84, 85). Moreover, therapeutic IgG1 and IgG2 capacities to bind C1q and activate complement may be the culprits of infusion reactions, with apparent lack of such activities associated with an IgG4 backbone (86). Thus, complement-mediated lysis contributing to the therapeutic efficacy of anti-CD20 (rituximab) or anti-CD52 (almetuzumab) may also cause unwanted toxicity, such as infusion reactions. Conceivably, biopharmaceutical companies may consider Fc engineering to eliminate complement activity while retaining other effector functions, as a strategy to minimize infusion reactions.

5.2. Optimized Fcs

The advent of molecular biology has enabled drug developers to tailor therapeutic mAb Fc's for optimal function (9–11). Depending on therapeutic need for effector function, isotype selection and/or chimerization and amino acid substitutions spanning the lower hinge/C_{H2} region have been used to enhance or reduce affinities for Fc receptors and the C1q component of complement (81–83, 87–89). Of note, the S239D/I332E combination of substitutions confers over 100-fold enhanced binding to FcγRIIIa and offers significant improvements in ADCC (89). The locations of these variants in the Fc structure are illustrated in Fig. 4. Additionally, since glycosylation of IgG at N297 (EU numbering system) in the C_{H2} domain is critical for Fc structure and effector function (90), numerous glycol-engineering efforts since have modulated affinities to FcγRs and C1q through cell-based enzymatic deletion or

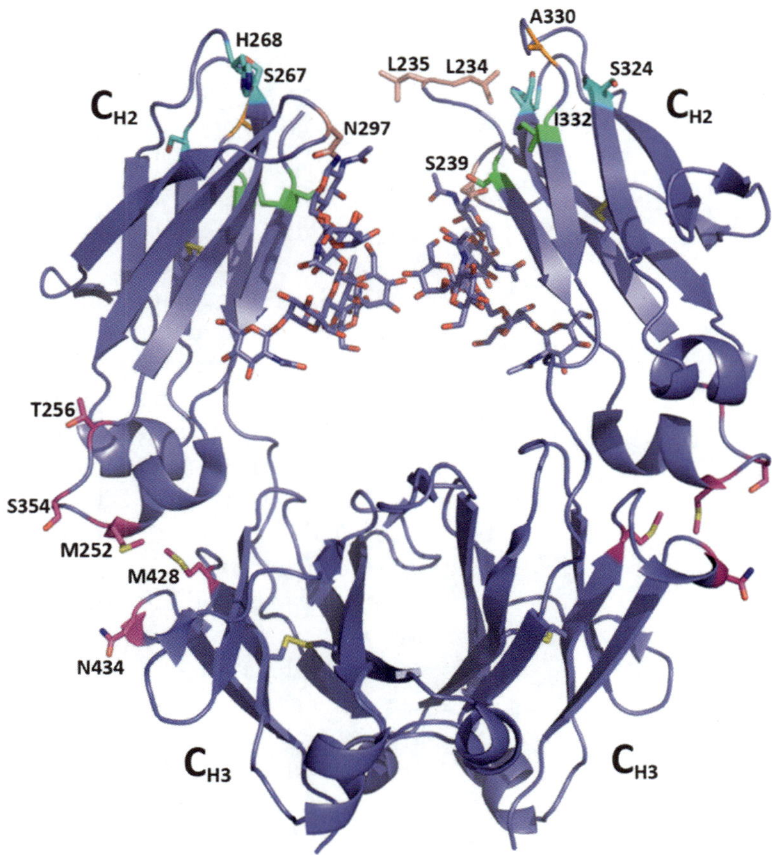

Fig. 4. The Fc of human IgG1 (PDBid 2DTQ, see ref. (46)) Residue side chains highlighted include I332E/S239D (*green*) that enhance FγcR binding leading to increases in antibody-dependent cellular cytotoxicity (ADCC) and antibody-dependent cellular phagocytosis (ADCP); N297A (aglycosylated form), L234A/L235A (*salmon*) FcγR reduced binding; A330L (*orange*) eliminate complement dependent cytotoxicity (CDC); S267E/H268/S324T (*cyan*) increased C1q binding and CDC; and M252Y, S254T, T256E, and M428L/N434L (*magenta*) neonatal Fc receptor (FcRn) mediated half-life extension.

modification of intramolecular N-linked oligosaccharide residues (i.e., lowering fucose content, introducing bisecting GlcNac, increasing sialic acid, increasing mannose) (9, 91). Importantly, the deletion of the N297 glycosylation site or its substitution (N297A) can significantly reduce or eliminate IgG affinity for FcγRs (90), further emphasizing the key role of glycosylation, and by extension oligosaccharide composition, in selectively enhancing or reducing FcγR interactions. Additional amino acid substitutions, such as the A330L, can abrogate CDC (89). Conversely, amino acid substitutions of S267E/H268/S324T (92) can provide additive enhancements of in vitro CDC for therapeutic antibodies.

Therapeutic Fc engineering requires understanding of the biology of the target (e.g., soluble ligand or cell surface), as well as an understanding of the target distribution, prevalence, and accessibility. Fc optimization by amino acid substitutions can increase Fc receptor affinity and enhance cytotoxicity against cancer cells that differentially express high-copy number antigens (e.g., anti-CD20 for lymphoma and anti-HER2/neu (trastuzumab) for breast cancer). However, such Fc platforms could potentially induce undesirable toxicity in alternative indications where blocking a cell surface receptor from interacting with its ligand or blocking receptor mediated signaling is sufficient for efficacy and where recruitment of immune cell functions is unwanted (93). Potential toxicity may arise from unnecessary activation of FcγRs by immune complexes (e.g., IgG opsonized multimeric targets, aggregates, or foreign particle) or cell-surface bound therapeutic mAbs that may trigger a cascade of cytokine release from activated immune cells (e.g., NKs, mast cells, neutrophils, monocytes, or macrophages). Such events may induce cytokine release syndrome (CRS) or platelet activation (e.g., through FcγRIIa) leading to thrombotic or adverse events that may impact drug safety and favorable pharmacodynamics, further limiting the maximal and safe therapeutic dose of treatment. To minimize such adverse events without compromising antibody efficacy, the combination of hinge-C_{H2} spanning amino acid substitutions can significantly reduce binding to FcγRs and immune cell activations.

Fc engineering has yielded various Fc backbones with minimal effector function and conserved half-life, some of which include using IgG2 and its chimeras, aglycosylated IgG1 (90), IgG4 S228P/F234A/L235A (82, 90), IgG2m4 (94). With regard to muted versions of IgG4, specific affinity for FcγRI has been eliminated using the L234A/L245A substitutions and S228P at the core hinge (also found in IgG1 and IgG2 in that position) further stabilizes the IgG4 molecule to prevent Fab arm exchange (95). Alternatively, isotype chimerization (e.g., IgG2/IgG4), as for eculizumab (anti-C5 complement protein), has also been used to minimize effector function. Nevertheless, bivalent cross-linking of target antigen may trigger cytotoxicity, so eliminating effector function is part of the equation. Univalent platforms lacking an Fc (e.g., Fab domain) should also be considered, when extended half-life is not critical to the pharmacodynamics of the antibody therapeutic. Chemical modification through PEGylation of Fab's has been used as a compensatory strategy to confer half-life, in such cases.

Thus, antibody design has resulted in the identification of selective and tunable Fc properties for the desired "drugable" therapeutic target, aimed at improving efficacy and safety profiles, depending on the need for complement, FcγR binding and immune cell engagement. Tailored Fc's remain alternatives to complete removal of the Fc domain which would drastically alter pharmacokinetics of the Fab or Fab'2, due to the central role of the C_{H2}–C_{H3} region of the Fc domain that is required for interactions with FcRn, the neonatal Fc receptor, conferring serum half-life in circulation. Not surprisingly, recent efforts have also been directed at and modulating FcRn binding through protein engineering, to further increase or reduce antibody half-life.

5.3. Fc Engineering for Extended Half-Life

Aside from effector functions, the Fc confers antibody half-life through interactions with the FcRn, also known as the Brambell receptor, in recognition of Francis Brambell, who predicted the first receptor system for IgG (96). Endothelial pinocytosis of circulating IgG engages a salvaging pathway that is mediated by pH sensitive binding to endosomal FcRn (pH 6.0) and subsequent exocytosis and release back into circulation at neutral pH (7.4) (97). The neonatal Fc receptor (FcRn), as the name implies, also enables transfer of maternal IgG to the fetus and has been implicated in intestinal sampling of immune complexes mediating oral tolerance. Engineering efforts have been directed at enhancing pH sensitive binding (tighter binding at pH 6.0 and release at pH 7.4) of key residue spanning the C_{H2}–C_{H3} domain of the Fc. One combination of substitutions commonly referred to as YTE (M252Y, S254T, T256E) increased half-life relative to IgG1 in cynomolgus monkeys (98). Importantly, an alternative combination, M248L/N434L, conferred a threefold extension in half-life in cynomolgous monkeys, relative to IgG1, and demonstrated improved efficacy using anti-EGFR in mouse transgenic models of human FcRn with human tumor xenografts (99). Noteworthy, recent studies have implicated FcγR's and FcRn in enhanced antigen presentation (100). The advance of optimized Fc's into the clinic will reveal the potential impact of such engineered Fc and half-life extensions on therapeutic immunogenicity, if any.

6. Engineering Antibody Stability and Biophysical Properties

Developability or manufacturability has been used to describe the overall property of an antibody to be manufactured, formulated and stabilized to achieve therapeutic effects. Therapeutic formulations of antibodies often require very high concentrations (50 mg/mL and above). Antibodies, like other proteins, tend to aggregate at these high concentrations. A variety of factors, including the intrinsic biophysical and biochemical properties of an antibody as well as salt, pH, and formulation additives, contribute to the successful formulation and long term stability (the ability of the antibody to retain therapeutic efficacy without aggregation and/or other chemical and physical changes). Posttranslational modifications and the biophysical and biochemical stability of an antibody are thus important considerations during the development of antibody therapeutics.

6.1. V-Region Glycosylation

Antibodies are glycoproteins with a well-characterized glycosylation at N297 in the C_{H2} domain. The removal of the carbohydrates from C_{H2} leads to a significant decrease of the overall thermal stability and half-life (90). In addition, the V region sequences can contain N-linked glycosylation sequence motifs that are often modified. Some of these motifs are germ-line encoded whereas others are a consequence of somatic hypermutations or random sequence combinations in synthetic phage antibody libraries. Approximately 25% of normal IgG is glycosylated in the V regions (101–103). Depending upon the location of the attachment, V-region glycans can impact antigen binding (103–108). Variable region glycosylation also plays a role in the serum half-life, and immunogenicity (108–110). For example, the therapeutic antibody, Erbitux, is glycosylated in the framework 3 of the V_H domain (111). Most of the various attached glycans are incompletely sialylated (111), which are susceptible to recognition by asialoglycoprotein receptors for degradation. The serum half-life is ~4.5 days compared to the average of 21 days for a normal antibody (112, 113). This suggests that the V region glycans play a role in its rapid clearance. Because of the heterogeneity that V-region glycosylation introduces into the antibody with undesirable or unpredictable pharmacokinetic properties, it has become a prudent practice to eliminate the V region glycosylation motifs if at all possible.

6.2. Chemical Modifications

A number of protein residues are chemically labile and susceptible to covalent modifications that lead to changes in the local or global protein structure and properties (114). These changes, primarily in the V regions, can have a significant impact on the bioactivity, serum half-life, immunogenicity, and aggregation propensity of a therapeutic antibody. In addition, these modifications may also affect antigen binding directly or indirectly depending upon the sites of modifications. The most common reactions are deamidation, isomerization, oxidation, disulfide breakage and formation, and peptide hydrolysis. The potential sites can be identified from a simple amino acid sequence analysis, but the 3D structure of the Fab provides a direct way to determine side chain accessibility as well as local structural restraints such as hydrogen bonding and hence the liability.

Deamidation eliminates the amide functional group from an aspargine (Asn) or glutamine (Gln) side chain (115). This reaction will result in a net negative charge increase. For asparagines, local structural changes can also occur when deamidation converts an Asn to an iso-Asp, in which the backbone is modified (116). The rate of deamidation is strongly influenced by buffer conditions. The tendency for Asn (and to a lesser extent Gln) to go through deamidation is determined by a number of factors. The relative position of Asn and/or Gln in a protein's structure and the sequence context at a deamidation site are the main determinants. For example, among the sequence combinations AsnGly, AsnSer, AsnAla, and AsnHis, commonly observed for Asn, AsnGly is the most reactive. The rate of reactivity drops more than 50-fold for AsnLeu in model peptides (117). For the Asn residue, there have been attempts to obtain empirical relationships to estimate its deamidation potential based upon the protein primary and 3D structure (115). Similar work is accumulating in the literature for Gln (118). Asp isomerization is similar to Asn deamidation in that the local backbone structure is modified although there is no change in net charge. The sequence and structure motifs are both important for the reaction.

Oxidation occurs most often in residues with reactive side chains: methionine (Met), cysteine (Cys), histidine (His), tryptophan (Trp), and tyrosine (Tyr). Exposed Met residues are easily oxidized to Met sulfoxide under normal storage conditions and to Met sulfone under extreme oxidative conditions (119). The resulting heterogeneity is often deleterious to binding and efficacy of the antibody. For example, oxidation of Met252 in the C_{H2} domains can reduce the FcRn binding, and thus, shorten the half-life (120). For Met residues in the V domains, oxidation can also impact antigen binding or surface properties of the antibody. Hence, the positions and solvent exposure of these residues needs to be carefully examined by structural studies or molecular modeling.

Free Cys sulfhydryl is also easily oxidized to form Cys disulfide bonds. During long-term storage, free sulfhydryl groups may catalyze intra- or interchain disulfide exchanges, which may cross-link multiple antibody molecules that eventually result in protein aggregation. The human IGHV7 family (121) and two genes from IGKV1 family (1-8 and 1D-8) (30) have unpaired Cys residues in FR-3. Occasionally, free Cys residues also occur in CDR-H3.

To avoid complications and potential aggregation mediated by disulfide cross-linking, it is prudent to avoid the use of these germline genes for humanization. It is usually relatively easy to identify and eliminate unpaired Cys residues from antibody CDR's.

Trp residues are notorious for being oxidized by reactive oxygen species and photooxidation, leading to the formation of 5-hydroxy-Trp, oxy-indole alanine, kynurenine, and *N*-formylkynurenine (119). Rate-limiting factors include the primary sequence of the protein, pH and the solvent accessibility of the Trp. The Trp residue has a large side-chain that provides a significant surface area for protein–protein interactions. It is often a major determinant for antigen binding affinity. Thus, its susceptibility to oxidation deserves careful evaluation during the course of an antibody development.

6.3. Physical Stability

The conformation (2°, 3°, and 4°) of a protein is dynamic in solution and is influenced by a number of factors such as temperature, pH, pressure, denaturing agents (e.g., guanidine hydrochloride, surfactants), and mechanical disruption (e.g., shaking and shearing) as well as its intrinsic stability. Large-scale fluctuations in the conformation during the manufacturing process may lead to denaturation, aggregation, and adsorption of the protein to surfaces and ultimately loss of product and/or protein function. Thermodynamic stability is an important intrinsic property for determining a protein's potential for denaturation and aggregation. Proteins with low stability tend to unfold and nucleate aggregation more easily. One measure of thermodynamic stability is the midpoint temperature of thermal unfolding (T_m), which is also known as thermal stability. It is often determined by differential scanning calorimetry (DSC), circular dichroism (CD) or other methods. It has been shown that the T_m of a protein is inversely correlated with its susceptibility to unfolding and denaturation in solution and to unfolding-dependent degradation processes (122).

The different domains of an IgG, Fv, C_{H1}/C_L, C_{H2}, and C_{H3} are relatively structurally independent. The thermal transitions of the constant domains are largely determined by the IgG isotype (123). Of the two common therapeutic isotypes (IgG1 and IgG4), the former has a more stable Fc based on the T_m of the C_{H2} and C_{H3} domains. The T_m of Fab (thermal transitions of Fv and C_{H1}/C_L are difficult to measure separately) spans a wide range from 57 to 82°C. Antibodies with low Fab T_m have been found to aggregate more easily and express poorly.

The first approach to enhance thermostability is class switching, i.e., combining the Fv with constant domains of a different class from the parent antibody (124). Because the antigen binding Fv domains are identical to the parental antibody, the overall stability of the chimeric antibody is largely determined by the intrinsic stabilities of the constant domains. Further enhancements are dependent upon the stabilization of the V domains (125–127). Several strategies to stabilize V domains have emerged: (1) grafting CDR's onto more stable FR's such as consensus FR's, or incorporating stabilizing mutations into FR's (128, 129) to increase the intrinsic stabilities of the V_H and V_L domains; (2) improving the V_H–V_L interface interactions by using disulfide bonds to cross-link V_H/V_L domains (130); and (3) random mutagenesis including gene shuffling followed by selection in harsh conditions (131). However, these methods have not been routinely applied to therapeutic mAbs because of technical or immunogenicity considerations.

Recently, a "germ-line design" approach has been proposed for improving the stability of the Fv (132). It was hypothesized that germ-line sequences evolved to encode highly stable antibodies, which can tolerate potentially destabilizing mutations during somatic hypermutation (133). The method consists of five steps: (1) structural studies of the Fab molecule, or accurate molecular modeling, (2) identification of structural imperfections such as internal voids in V_H or V_L or at the V_H–V_L interface, (3) comparison of the V_H and V_L sequences with their respective germ-line sequences to identify nonbinding residues that differ from the germ-line counterparts, (4) substitution of germ-line residues back into the lead antibody molecules individually or in combination, and (5) assay of stability and activity of resulting antibody variants. This approach for an antibody, mAb15, produced variants exhibiting T_m increases of more than 12°C (132). At the same time, the germ-line content of the resulting mAb variants increased, which may lead to lower overall immunogenicity.

6.4. Aggregation

Therapeutic antibodies are often stored over long periods of time at very high concentrations before administration. These conditions add to the tendency that proteins form aggregates which lead to loss of activity and increased immunogenicity risks. Many external factors such as buffers, salts and excipients and intrinsic properties of the antibody influence its aggregation potential (reviewed in ref. (134)). Nonspecific hydrophobic–hydrophobic interactions between protein molecules are often considered to be one of the dominant forces driving aggregation. Reducing antibody surface hydrophobicity has been successful in improving protein solubility (135). Recently, spatial-aggregation-propensity (SAP) has been proposed to measure quantitatively protein and surface hydrophobicity and its potential contribution to aggregation (136–138). Using this structure-based computational approach, the authors were able to identify surface patches with excessive hydrophobicity in therapeutic antibodies. They then made mutations in the constant domains that led to variants with improved solubility and long-term storage stability (136, 137). Short aggregation prone regions (APR's) in antibody CDR's have also been recognized as contributors to antibody aggregation (139, 140).

Surface hydrophobicity is a major component of the variable domains, particularly the CDR's that form the antigen recognition site. Often antigen binding residues and APRs are coupled or overlapping (139, 140). For instance, in a recently reported case study of mAb solubility engineering, a large hydrophobic patch in the CDR's forms CDR–CDR and CDR–FR interactions in the crystal structures of the Fab fragments of IgG1 and IgG2 subtypes (124). Some of the antibody variants containing mutations of less hydrophobic residues in this patch or glycans in the C_{H1} domain or isotype switching to interfere with such interactions exhibited significantly improved solubility (124). This study emphasized the contribution of antibody CDR surface hydrophobicity to protein aggregation.

While most hydrophobic interactions that contribute to protein aggregation are considered nonspecific, specific protein–protein interactions can also play a significant role in protein/mAb aggregation. A recent study of IgG of CNTO607 was an example (141). CNTO607, a mAb of IgG1 isotype derived from a phage library, was poorly soluble (~13 mg/mL in PBS). The Fab fragment is very soluble to more than 100 mg/mL in PBS. This very large discrepancy was initially difficult to reconcile. The crystal structure of the Fab fragment, however, revealed a tetrameric arrangement of the Fab. In the Fab tetramer, the CDR-H3, which is very important for antigen binding and could not tolerate mutations without loss of activity, contacts the elbow regions of the other Fabs within the tetramer (142). The four Fabs thus form a closed ring system that is very stable and soluble (see Fig. 5). It was postulated that the two Fab moieties of a mAb can engage in similar CDR3–elbow

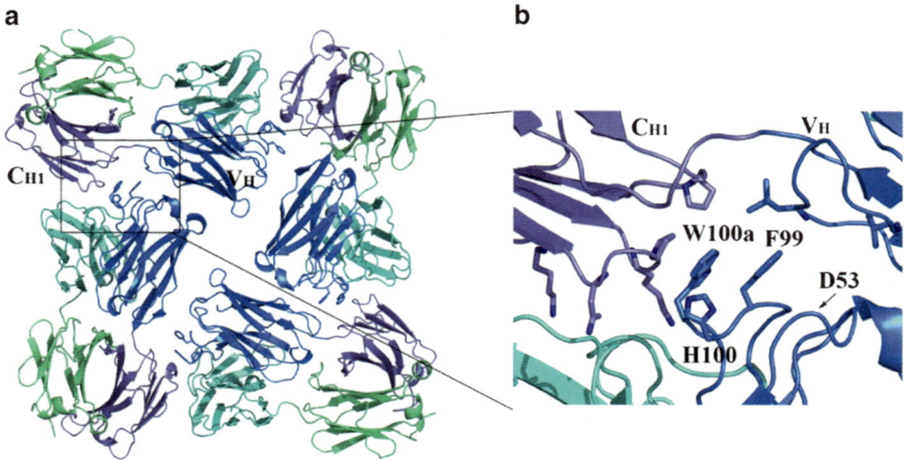

Fig. 5. (**a**) The crystal packing arrangement of CNTO607 Fabs (142). (**b**) Close-up of the region of the antigen-binding site highlighting the positions of residues W100a, H100, F99 and others involved in the interaction with elbow region of symmetry related Fab. D53 is the position selected for mutation to N53 for glycosylation, which sterically prevents this tetrameric CDR–elbow interaction.

interactions independently with other mAb molecules at high concentrations, leading to the formation of large clusters and eventually insoluble aggregates. This potential mechanism of mAb aggregation was validated in a further experiment. When a glycan was strategically placed in the CDR-H2 that was designed to sterically interfere with this presumed mechanism, the authors were able to obtain an IgG variant with solubility well over 100 mg/mL in PBS (141).

These studies suggest that antibody aggregation is a very complex issue. It often involves surface hydrophobicity in both the constant and variable regions in nonspecific interactions. Under certain circumstances, specific antibody–antibody interactions can also play a role due to the bivalent nature of the antibody molecule. Therefore, a successful engineering effort should include careful studies, including structure determinations, to understand the molecular mechanism of aggregation to rationally improve solubility of therapeutic mAbs.

7. Conclusions/ Future Directions

In conclusion, the last decade of antibody V region engineering has seen a diversification and refinement of humanization methods guided in good part by our knowledge of the structure of the antibody molecule. For instance, currently CDR grafted antibodies made not only of fully human germ-line genes but also of nonhuman CDRs are being engineered for increased human germ-line gene content based on sequence and structural analysis. This further engineering process makes humanized antibodies indistinguishable from the so-called fully human antibodies, which should in principle minimize even further immunogenic reactions than the current generation of humanized antibodies. Also, as more information on how to engineer antibodies for increased potency via residues in contact with the antigen and/or higher stability and solubility based on analysis of the type of liable amino acids and their position in the structure (structural restraints and degree of exposure to solvent), it is expected that the probability of success in formulation and clinical trials increases. In the Fc space, the last 10 years have yielded distinct amino acid variants that singularly or in combination(s) can increase or reduce capacities to bind FcγRs and thereby selectively enhance or eliminate effector functions (ADCC, ADCP, CDC) and/or extend antibody half-life. The unique applications of such variants in the context of therapeutic antibodies and their potential benefits await clinical validation, which may contribute to differentiated drugs with desired characteristics, including increased therapeutic windows as defined by improved clinical safety and efficacy profiles.

References

1. Reichert JM. (2008) Monoclonal antibodies as innovative therapeutics. *Curr Pharm. Biotechnol.* **9**, 423–430.
2. Aires da Silva F, Corte-Real S, Goncalves J. (2008) Recombinant antibodies as therapeutic agents: pathways for modeling new biodrugs. *BioDrugs.* **22**, 301–314.
3. Beck A, Wurch T, Bailly C, Corvaia N. (2010) Strategies and challenges for the next generation of therapeutic antibodies. *Nat. Rev. Immunol.* **10**, 345–352.
4. Kohler G, Milstein C. (1975) Continuous cultures of fused cells secreting antibody of predefined specificity. *Nature.* **256**, 495–497.
5. Smith GP. (1985) Filamentous fusion phage: novel expression vectors that display cloned antigens on the virion surface. *Science.* **228**, 1315–1317.
6. McCafferty J, Griffiths AD, Winter G, Chiswell DJ. (1990) Phage antibodies: filamentous phage displaying antibody variable domains. *Nature.* **348**, 552–554.
7. Marks JD, Hoogenboom HR, Bonnert TP, McCafferty J, Griffiths AD, Winter G. (1991) By-passing immunization. Human antibodies from V-gene libraries displayed on phage. *J. Mol. Biol.* **222**, 581–597.
8. Kelley B. (2009) Industrialization of mAb production technology: the bioprocessing industry at a crossroads. *MAbs.* **1**, 443–452.
9. Jefferis R. (2007) Antibody therapeutics: isotype and glycoform selection. *Expert Opin. Biol. Ther.* **7**, 1401–1413.
10. Presta LG. (2008) Molecular engineering and design of therapeutic antibodies. *Curr. Opin. Immunol.* **20**, 460–470.
11. Jefferis R. (2009) Recombinant antibody therapeutics: the impact of glycosylation on mechanisms of action. *Trends Pharmacol. Sci.* **30**, 356–362.
12. Holliger P, Hudson PJ. (2005) Engineered antibody fragments and the rise of single domains. *Nat. Biotechnol.* **23**, 1126–1136.
13. Skerra A. (2007) Alternative non-antibody scaffolds for molecular recognition. *Curr Opin Biotechnol.* **18**, 295–304.
14. Padlan EA. (1977) Structural basis for the specificity of antibody-antigen reactions and structural mechanisms for the diversification of antigen-binding specificities. *Q. Rev. Biophys.* **10**, 35–65.
15. Amzel LM, Poljak RJ. (1979) Three-dimensional structure of immunoglobulins. *Annu. Rev. Biochem.* **48**, 961–997.
16. Davies DR, Metzger H. (1983) Structural basis of antibody function. *Annu. Rev. Immunol.* **1**, 87–117.
17. Wilson IA, Stanfield RL. (1994) Antibody-antigen interactions: new structures and new conformational changes. *Curr. Opin. Struct. Biol.* **4**, 857–867.
18. Stanfield RL, Wilson, I.A. Antibody Molecular Structure. In: An Z, ed. Therapeutic Monoclonal Antibodies: From Bench to Clinic: John Wiley & Sons, Inc.; 2010.
19. Porter RR. (1959) The hydrolysis of rabbit y-globulin and antibodies with crystalline papain. *Biochem. J.* **73**, 119–126.
20. Berman HM, Westbrook J, Feng Z, *et al.* (2000) The Protein Data Bank. *Nucleic Acids Res.* **28**, 235–242.
21. Schiffer M, Girling RL, Ely KR, Edmundson AB. (1973) Structure of a lambda-type Bence-Jones protein at 3.5-A resolution. *Biochemistry.* **12**, 4620–4631.
22. Poljak RJ, Amzel LM, Chen BL, Phizackerley RP, Saul F. (1974) The three-dimensional structure of the fab' fragment of a human myeloma immunoglobulin at 2.0-angstrom resolution. *Proc. Natl. Acad. Sci. USA.* **71**, 3440–3444.
23. Stanfield RL, Zemla A, Wilson IA, Rupp B. (2006) Antibody elbow angles are influenced by their light chain class. *J. Mol. Biol.* **357**, 1566–1574.
24. Lesk AM, Chothia C. (1988) Elbow motion in the immunoglobulins involves a molecular ball-and-socket joint. *Nature.* **335**, 188–190.
25. Wu TT, Kabat EA. (1970) An analysis of the sequences of the variable regions of Bence Jones proteins and myeloma light chains and their implications for antibody complementarity. *J. Exp. Med.* **132**, 211–250.
26. Kabat EA, Wu TT. (1971) Attempts to locate complementarity-determining residues in the variable positions of light and heavy chains. *Ann. N Y Acad. Sci.* **190**, 382–393.
27. Chothia C, Lesk AM. (1987) Canonical structures for the hypervariable regions of immunoglobulins. *J. Mol. Biol.* **196**, 901–917.
28. Chothia C, Lesk AM, Tramontano A, et al. (1989) Conformations of immunoglobulin hypervariable regions. *Nature.* **342**, 877–883.
29. Chothia C, Lesk AM, Gherardi E, et al. (1992) Structural repertoire of the human VH segments. *J. Mol. Biol.* **227**, 799–817.
30. Tomlinson IM, Cox JP, Gherardi E, Lesk AM, Chothia C. (1995) The structural repertoire

of the human V kappa domain. *EMBO J.* **14**, 4628–4638.

31. Vargas-Madrazo E, Lara-Ochoa F, Almagro JC. (1995) Canonical structure repertoire of the antigen-binding site of immunoglobulins suggests strong geometrical restrictions associated to the mechanism of immune recognition. *J. Mol. Biol.* **254**, 497–504.
32. Shirai H, Kidera A, Nakamura H. (1996) Structural classification of CDR-H3 in antibodies. *FEBS Lett.* **399**, 1–8.
33. Morea V, Tramontano A, Rustici M, Chothia C, Lesk AM. (1998) Conformations of the third hypervariable region in the VH domain of immunoglobulins. *J Mol Biol.* **275**, 269–294.
34. Collis AV, Brouwer AP, Martin AC. (2003) Analysis of the antigen combining site: correlations between length and sequence composition of the hypervariable loops and the nature of the antigen. *J. Mol. Biol.* **325**, 337–354.
35. Al-Lazikani B, Lesk AM, Chothia C. (1997) Standard conformations for the canonical structures of immunoglobulins. *J. Mol. Biol.* **273**, 927–948.
36. North B, Lehmann A, Dunbrack RL, Jr. (2010) A new clustering of antibody CDR loop conformations. *J. Mol. Biol.* **406**, 228–256.
37. Martin AC, Thornton JM. (1996) Structural families in loops of homologous proteins: automatic classification, modelling and application to antibodies. *J. Mol. Biol.* **263**, 800–815.
38. Sivasubramanian A, Sircar A, Chaudhury S, Gray JJ. (2009) Toward high-resolution homology modeling of antibody Fv regions and application to antibody-antigen docking. *Proteins.* **74**, 497–514.
39. Sircar A, Kim ET, Gray JJ. (2009) RosettaAntibody: antibody variable region homology modeling server. *Nucleic Acids Res.* **37**, W474–479.
40. Marcatili P, Rosi A, Tramontano A. (2008) PIGS: automatic prediction of antibody structures. *Bioinformatics.* **24**, 1953–1954.
41. Almagro JC, Beavers MP, Hernandez-Guzman F, Shaulsky J, Butenhof K, Maier J, Kelly K, Labute P, Thorsteinson N, Teplyakov A, Luo J, Sweet R, Gilliland GL (2011) Antibody Modeling Assessment. *Proteins.* **79**, 3050–3066.
42. Marquart M, Deisenhofer J, Huber R, Palm W. (1980) Crystallographic refinement and atomic models of the intact immunoglobulin molecule Kol and its antigen-binding fragment at 3.0 A and 1.0 A resolution. *J. Mol. Biol.* **141**, 369–391.
43. Saphire EO, Stanfield RL, Crispin MD, *et al.* (2002) Contrasting IgG structures reveal extreme asymmetry and flexibility. *J. Mol. Biol.* **319**, 9–18.
44. Sondermann P, Huber R, Oosthuizen V, Jacob U. (2000) The 3.2-A crystal structure of the human IgG1 Fc fragment-Fc gammaRIII complex. *Nature.* **406**, 267–273.
45. Deisenhofer J. (1981) Crystallographic refinement and atomic models of a human Fc fragment and its complex with fragment B of protein A from Staphylococcus aureus at 2.9- and 2.8-A resolution. *Biochemistry.* **20**, 2361–2370.
46. Matsumiya S, Yamaguchi Y, Saito J, *et al.* (2007) Structural comparison of fucosylated and nonfucosylated Fc fragments of human immunoglobulin G1. *J. Mol. Biol.* **368**, 767–779.
47. Jones PT, Dear PH, Foote J, Neuberger MS, Winter G. (1986) Replacing the complementarity-determining regions in a human antibody with those from a mouse. *Nature.* **321**, 522–525.
48. Reichert JM. (2010) Metrics for antibody therapeutics development. *MAbs.* **2**, 695–700.
49. Almagro JC, Strohl WR. Antibody Engineering: Humanization, Affinity Maturation and Selection Methods. In: An Z, ed. Therapeutic Monoclonal Antibodies: From Bench to Clinic: John Wiley & Sons, Inc.; 2009:302–327.
50. De Groot AS, Goldberg M, Moise L, Martin W. (2006) Evolutionary deimmunization: an ancillary mechanism for self-tolerance? *Cell Immunol.* **244**, 148–153.
51. Tamura M, Milenic DE, Iwahashi M, Padlan E, Schlom J, Kashmiri SV. (2000) Structural correlates of an anticarcinoma antibody: identification of specificity-determining residues (SDRs) and development of a minimally immunogenic antibody variant by retention of SDRs only. *J. Immunol.* **164**, 1432–1441.
52. Tan P, Mitchell DA, Buss TN, Holmes MA, Anasetti C, Foote J. (2002) "Superhumanized" antibodies: reduction of immunogenic potential by complementarity-determining region grafting with human germline sequences: application to an anti-CD28. *J. Immunol.* **169**, 1119–1125.
53. Lazar GA, Desjarlais JR, Jacinto J, Karki S, Hammond PW. (2007) A molecular immunology approach to antibody humanization and functional optimization. *Mol. Immunol.* **44**, 1986–1998.

54. Pelat T, Bedouelle H, Rees AR, Crennell SJ, Lefranc MP, Thullier P. (2008) Germline humanization of a non-human primate antibody that neutralizes the anthrax toxin, by in vitro and in silico engineering. *J. Mol. Biol.* **384**, 1400–1407.

55. Almagro JC, Fransson J. (2008) Humanization of antibodies. *Front Biosci.* **13**, 1619–1633.

56. Osbourn J, Groves M, Vaughan T. (2005) From rodent reagents to human therapeutics using antibody guided selection. *Methods.* **36**, 61–68.

57. Dall'Acqua WF, Damschroder MM, Zhang J, et al. (2005) Antibody humanization by framework shuffling. *Methods.* **36**, 43–60.

58. Padlan EA, Abergel C, Tipper JP. (1995) Identification of specificity-determining residues in antibodies. *FASEB J.* **9**, 133–139.

59. Matsuda F, Ishii K, Bourvagnet P, et al. (1998) The complete nucleotide sequence of the human immunoglobulin heavy chain variable region locus. *J. Exp. Med.* **188**, 2151–2162.

60. Schroeder HW, Jr., Cavacini L. (2010) Structure and function of immunoglobulins. *J. Allergy Clin.. Immunol.* **125**, S41–52.

61. Lefranc MP, Giudicelli V, Kaas Q, *et al.* (2005) IMGT, the international ImMunoGeneTics information system. *Nucleic Acids Res.* **33**, D593–597.

62. Queen C, Schneider WP, Selick HE, *et al.* (1989) A humanized antibody that binds to the interleukin 2 receptor. *Proc. Natl. Acad. Sci. USA.* **86**, 10029–10033.

63. Brezinschek HP, Foster SJ, Dorner T, Brezinschek RI, Lipsky PE. (1998) Pairing of variable heavy and variable kappa chains in individual naive and memory B cells. *J. Immunol.* **160**, 4762–4767.

64. de Wildt RM, Hoet RM, van Venrooij WJ, Tomlinson IM, Winter G. (1999) Analysis of heavy and light chain pairings indicates that receptor editing shapes the human antibody repertoire. *J. Mol. Biol.* **285**, 895–901.

65. Shi L, Wheeler JC, Sweet RW, *et al.* (2010) De novo selection of high-affinity antibodies from synthetic fab libraries displayed on phage as pIX fusion proteins. *J. Mol. Biol.* **397**, 385–396.

66. Fransson J, Teplyakov A, Raghunathan G, *et al.* (2010) Human framework adaptation of a mouse anti-human IL-13 antibody. *J. Mol. Biol.* **398**, 214–231.

67. Getts DR, Getts MT, McCarthy DP, Chastain EM, Miller SD. (2010) Have we overestimated the benefit of human(ized) antibodies? *MAbs.* **2**, 682–694.

68. Bernett MJ, Karki S, Moore GL, *et al.* (2010) Engineering fully human monoclonal antibodies from murine variable regions. *J. Mol. Biol.* **396**, 1474–1490.

69. Harding FA, Stickler MM, Razo J, DuBridge RB. (2010) The immunogenicity of humanized and fully human antibodies: residual immunogenicity resides in the CDR regions. *MAbs.* **2**, 256–265.

70. Hwang WY, Almagro JC, Buss TN, Tan P, Foote J. (2005) Use of human germline genes in a CDR homology-based approach to antibody humanization. *Methods.* **36**, 35–42.

71. Hoet RM, Cohen EH, Kent RB, et al. (2005) Generation of high-affinity human antibodies by combining donor-derived and synthetic complementarity-determining-region diversity. *Nat. Biotechnol.* **23**, 344–348.

72. Rothe C, Urlinger S, Lohning C, et al. (2008) The human combinatorial antibody library HuCAL GOLD combines diversification of all six CDRs according to the natural immune system with a novel display method for efficient selection of high-affinity antibodies. *J. Mol. Biol.* **376**, 1182–1200.

73. Barbas CF, III, Hu D, Dunlop N, et al. (1994) In vitro evolution of a neutralizing human antibody to human immunodeficiency virus type 1 to enhance affinity and broaden strain cross-reactivity. *Proc. Natl. Acad. Sci. USA.* **91**, 3809–3813.

74. Hanes J, Jermutus L, Weber-Bornhauser S, Bosshard HR, Pluckthun A. (1998) Ribosome display efficiently selects and evolves high-affinity antibodies in vitro from immune libraries. *Proc. Natl. Acad. Sci. USA.* **95**, 14130–14135.

75. Midelfort KS, Hernandez HH, Lippow SM, Tidor B, Drennan CL, Wittrup KD. (2004) Substantial energetic improvement with minimal structural perturbation in a high affinity mutant antibody. *J. Mol. Biol.* **343**, 685-701.

76. Hawkins RE, Russell SJ, Winter G. (1992) Selection of phage antibodies by binding affinity. Mimicking affinity maturation. *J. Mol. Biol.* **226**, 889–896.

77. Hawkins RE, Russell SJ, Baier M, Winter G. (1993) The contribution of contact and non-contact residues of antibody in the affinity of binding to antigen. The interaction of mutant D1.3 antibodies with lysozyme. *J. Mol. Biol.* **234**, 958–964.

78. Lowman HB, Bass SH, Simpson N, Wells JA. (1991) Selecting high-affinity binding proteins by monovalent phage display. *Biochemistry.* **30**, 10832–10838.

79. Yang WP, Green K, Pinz-Sweeney S, Briones AT, Burton DR, Barbas CF, III. (1995) CDR

walking mutagenesis for the affinity maturation of a potent human anti-HIV-1 antibody into the picomolar range. *J. Mol. Biol.* **254**, 392–403.

80. Bruhns P, Iannascoli B, England P, et al. (2009) Specificity and affinity of human Fcgamma receptors and their polymorphic variants for human IgG subclasses. *Blood.* **113**, 3716–3725.
81. Duncan AR, Woof JM, Partridge LJ, Burton DR, Winter G. (1988) Localization of the binding site for the human high-affinity Fc receptor on IgG. *Nature.* **332**, 563–564.
82. Alegre ML, Collins AM, Pulito VL, et al. (1992) Effect of a single amino acid mutation on the activating and immunosuppressive properties of a "humanized" OKT3 monoclonal antibody. *J. Immunol.* **148**, 3461–3468.
83. Shields RL, Namenuk AK, Hong K, et al. (2001) High resolution mapping of the binding site on human IgG1 for Fc gamma RI, Fc gamma RII, Fc gamma RIII, and FcRn and design of IgG1 variants with improved binding to the Fc gamma R. *J. Biol. Chem.* **276**, 6591–6604.
84. Bredius RG, Driedijk PC, Schouten MF, Weening RS, Out TA. (1992) Complement activation by polyclonal immunoglobulin G1 and G2 antibodies against Staphylococcus aureus, Haemophilus influenzae type b, and tetanus toxoid. *Infect. Immun.* **60**, 4838–4847.
85. Saeland E, Vidarsson G, Leusen JH, et al. (2003) Central role of complement in passive protection by human IgG1 and IgG2 anti-pneumococcal antibodies in mice. *J. Immunol.* **170**, 6158-6164.
86. Tawara T, Hasegawa K, Sugiura Y, et al. (2008) Complement activation plays a key role in antibody-induced infusion toxicity in monkeys and rats. *J. Immunol.* **180**, 2294–2298.
87. Armour KL, Clark MR, Hadley AG, Williamson LM. (1999) Recombinant human IgG molecules lacking Fcgamma receptor I binding and monocyte triggering activities. *Eur. J. Immunol.* 29, 2613–2624.
88. Armour KL, van de Winkel JG, Williamson LM, Clark MR. (2003) Differential binding to human FcgammaRIIa and FcgammaRIIb receptors by human IgG wildtype and mutant antibodies. *Mol. Immunol.* **40**, 585–593.
89. Lazar GA, Dang W, Karki S, *et al.* (2006) Engineered antibody Fc variants with enhanced effector function. *Proc. Natl. Acad. Sci. USA.* **103**, 4005–4010.
90. Tao MH, Morrison SL. (1989) Studies of aglycosylated chimeric mouse-human IgG. Role of carbohydrate in the structure and effector functions mediated by the human IgG constant region. *J. Immunol.* **143**, 2595–2601.
91. Raju TS. (2008) Terminal sugars of Fc glycans influence antibody effector functions of IgGs. *Curr. Opin. Immunol.* **20**, 471–478.
92. Moore GL, Chen H, Karki S, Lazar GA. (2010) Engineered Fc variant antibodies with enhanced ability to recruit complement and mediate effector functions. *MAbs.* **2**, 320–328.
93. Labrijn AF, Aalberse RC, Schuurman J. (2008) When binding is enough: nonactivating antibody formats. *Curr. Opin. Immunol.* **20**, 479–485.
94. An Z, Forrest G, Moore R, et al. (2009) IgG2m4, an engineered antibody isotype with reduced Fc function. *MAbs.* **1**, 572–579.
95. Angal S, King DJ, Bodmer MW, et al. (1993) A single amino acid substitution abolishes the heterogeneity of chimeric mouse/human (IgG4) antibody. *Mol. Immunol.* **30**, 105–108.
96. Brambell FW, Hemmings WA, Morris IG. (1964) A Theoretical Model of Gamma-Globulin Catabolism. *Nature.* **203**, 1352–1354.
97. Roopenian DC, Akilesh S. (2007) FcRn: the neonatal Fc receptor comes of age. *Nat. Rev. Immunol.* 7, 715–725.
98. Dall'Acqua WF, Woods RM, Ward ES, et al. (2002) Increasing the affinity of a human IgG1 for the neonatal Fc receptor: biological consequences. *J. Immunol.* **169**, 5171–5180.
99. Zalevsky J, Chamberlain AK, Horton HM, et al. (2010) Enhanced antibody half-life improves in vivo activity. *Nat. Biotechnol.* **28**, 157–159.
100. Qiao SW, Kobayashi K, Johansen FE, et al. (2008) Dependence of antibody-mediated presentation of antigen on FcRn. *Proc. Natl. Acad. Sci. USA.* **105**, 9337–9342.
101. Abel CA, Spiegelberg HL, Grey HM. (1968) The carbohydrate contents of fragments and polypeptide chains of human gamma-G-myeloma proteins of different heavy-chain subclasses. *Biochemistry.* 7, 1271–1278.
102. Spiegelberg HL, Abel CA, Fishkin BG, Grey HM. (1970) Localization of the carbohydrate within the variable region of light and heavy chains of human gamma g myeloma proteins. *Biochemistry.* **9**, 4217–4223.
103. Wright A, Morrison SL. (1993) Antibody variable region glycosylation: biochemical and clinical effects. *Springer Semin. Immunopathol.* **15**, 259–273.

104. Co MS, Scheinberg DA, Avdalovic NM, et al. (1993) Genetically engineered deglycosylation of the variable domain increases the affinity of an anti-CD33 monoclonal antibody. *Mol. Immunol.* **30**, 1361–1367.

105. Khurana S, Raghunathan V, Salunke DM. (1997) The variable domain glycosylation in a monoclonal antibody specific to GnRH modulates antigen binding. *Biochem. Biophys. Res. Commun.* **234**, 465–469.

106. Leibiger H, Wustner D, Stigler RD, Marx U. (1999) Variable domain-linked oligosaccharides of a human monoclonal IgG: structure and influence on antigen binding. *Biochem. J.* **338** (Pt 2), 529–538.

107. Wallick SC, Kabat EA, Morrison SL. (1988) Glycosylation of a VH residue of a monoclonal antibody against alpha (1----6) dextran increases its affinity for antigen. *J. Exp. Med.* **168**, 1099–1109.

108. Wright A, Tao MH, Kabat EA, Morrison SL. (1991) Antibody variable region glycosylation: position effects on antigen binding and carbohydrate structure. *EMBO J.* **10**, 2717–2723.

109. Coloma MJ, Trinh RK, Martinez AR, Morrison SL. (1999) Position effects of variable region carbohydrate on the affinity and in vivo behavior of an anti-(1-->6) dextran antibody. *J. Immunol.* **162**, 2162–2170.

110. Rudd PM, Wormald MR, Harvey DJ, et al. (1999) Oligosaccharide analysis and molecular modeling of soluble forms of glycoproteins belonging to the Ly-6, scavenger receptor, and immunoglobulin superfamilies expressed in Chinese hamster ovary cells. *Glycobiology.* **9**, 443–458.

111. Qian J, Liu T, Yang L, Daus A, Crowley R, Zhou Q. (2007) Structural characterization of N-linked oligosaccharides on monoclonal antibody cetuximab by the combination of orthogonal matrix-assisted laser desorption/ionization hybrid quadrupole-quadrupole time-of-flight tandem mass spectrometry and sequential enzymatic digestion. *Anal. Biochem.* **364**, 8–18.

112. Imclone Systems, Inc Biologic License Application 125084, Erbitux (Cetuximab). In; 2004.

113. Frazer J, Capra J. Fundamental immunology. In: Immunoglobulins: Structure and function. 4th ed: Lippincott-Raven; 1999:37–74.

114. Wang W, Singh S, Zeng DL, King K, Nema S. (2007) Antibody structure, instability, and formulation. *J. Pharm. Sci.* **96**, 1–26.

115. Robinson N. (2002) Protein deamidation. *Proc. Natl. Acad. Sci. USA.* **99**, 5283–5288.

116. Geiger T, Clarke S. (1987) Deamidation, isomerization, and racemization at asparaginyl and aspartyl residues in peptides. Succinimide-linked reactions that contribute to protein degradation. *J. Biol. Chem.* **262**, 785–794.

117. Clarke S. (1987) Propensity for spontaneous succinimide formation from aspartyl and asparaginyl residues in cellular proteins. *Int. J. Pept. Protein Res.* **30**, 808–821.

118. Robinson NE RZ, Robinson BR, Robinson AL, Robinson JA, Robinson ML, Robinson AB. (2004) Structure-dependent nonenzymatic deamidation of glutaminyl and asparaginyl pentapeptides. *J. Peptide Res.* **63**, 426–436.

119. Ji JA, Zhang B, Cheng W, Wang YJ. (2009) Methionine, tryptophan, and histidine oxidation in a model protein, PTH: mechanisms and stabilization. *J. Pharm. Sci.* **98**, 4485–4500.

120. Pan H, Chen K, Chu L, Kinderman F, Apostol I, Huang G. (2009) Methionine oxidation in human IgG2 Fc decreases binding affinities to protein A and FcRn. *Protein Sci.* **18**, 424–433.

121. Tomlinson IM, Walter G, Marks JD, Llewelyn MB, Winter G. (1992) The repertoire of human germline VH sequences reveals about fifty groups of VH segments with different hypervariable loops. *J. Mol. Biol.* **227**, 776–798.

122. Remmele RL, Gombotz WR. (2000) Differential scanning calorimetry: A practical tool for elucidating stability of liquid biopharmaceuticals. *Biopharm.* **13**, 36–46.

123. Garber E, Demarest SJ. (2007) A broad range of Fab stabilities within a host of therapeutic IgGs. *Biochem Biophys. Res. Commun.* **355**, 751–757.

124. Pepinsky RB, Silvian L, Berkowitz SA, et al. (2010) Improving the solubility of anti-LINGO-1 monoclonal antibody Li33 by isotype switching and targeted mutagenesis. *Protein Science.* **19**, 954–966.

125. Ewert S, Honegger A, Pluckthun A. (2004) Stability improvement of antibodies for extra-cellular and intracellular applications: CDR grafting to stable frameworks and structure-based framework engineering. *Methods.* **34**, 184–199.

126. Ewert S, Huber T, Honegger A, Pluckthun A. (2003) Biophysical properties of human antibody variable domains. *J. Mol. Biol.* **325**, 531–553.

127. Worn A, Pluckthun A. (1998) Mutual stabilization of VL and VH in single-chain antibody fragments, investigated with mutants

engineered for stability. *Biochemistry.* **37**, 13120–13127.

128. Spada S, Honegger A, Pluckthun A. (1998) Reproducing the natural evolution of protein structural features with the selectively infective phage (SIP) technology. The kink in the first strand of antibody kappa domains. *J. Mol. Biol.* **283**, 395–407.
129. Worn A, Pluckthun A. (2001) Stability engineering of antibody single-chain Fv fragments. *J. Mol. Biol.* **305**, 989–1010.
130. Glockshuber R, Malia M, Pfitzinger I, Pluckthun A. (1990) A comparison of strategies to stabilize immunoglobulin Fv-fragments. *Biochemistry.* **29**, 1362–1367.
131. Forrer P, Jung S, Pluckthun A. (1999) Beyond binding: using phage display to select for structure, folding and enzymatic activity in proteins. *Curr. Opin. Struct. Biol.* **9**, 514–520.
132. Luo J, Obmolova G, Huang A, et al. (2010) Coevolution of antibody stability and Vkappa CDR-L3 canonical structure. *J. Mol. Biol.* **402**, 708–719.
133. Wiens GD, Roberts VA, Whitcomb EA, O'Hare T, Stenzel-Poore MP, Rittenberg MB. (1998) Harmful somatic mutations: lessons from the dark side. *Immunol. Rev.* **162**, 197–209.
134. Wang W, Nema S, Teagarden D. (2010) Protein aggregation--pathways and influencing factors. *Int. J. Pharm.* **390**, 89–99.
135. Trevino SR, Scholtz, J. M., Pace, C. N. (2008) Measuring and Increasing Protein Solubility. *Journal of Pharmaceutical Sciences.* **97**, 4155–4166.
136. Chennamsetty N, Helk B, Voynov V, Kayser V, Trout BL. (2009) Aggregation-prone motifs in human immunoglobulin G. *J. Mol. Biol.* **391**, 404–413.
137. Chennamsetty N, Voynov V, Kayser V, Helk B, Trout BL. (2009) Design of therapeutic proteins with enhanced stability. *Proc. Natl. Acad. Sci. USA.* **106**, 11937–11942.
138. Chennamsetty N, Voynov V, Kayser V, Helk B, Trout BL. (2010) Prediction of aggregation prone regions of therapeutic proteins. *J. Phys. Chem. B.* **114**, 6614–6624.
139. Wang X, Singh SK, Kumar S. (2010) Potential aggregation-prone regions in complementarity-determining regions of antibodies and their contribution towards antigen recognition: a computational analysis. *Pharm. Res.* **27**, 1512–1529.
140. Wang X, Das TK, Singh SK, Kumar S. (2009) Potential aggregation prone regions in biotherapeutics: A survey of commercial monoclonal antibodies. *MAbs.* **1**, 254–267.
141. Wu S-J, Luo J, O'Neil KT, Kang J, Lacy ER, Canziani G, Baker A, Huang M, Tang Q, Raju TS, Jacobs SA, Teplyakov A, Gilliland G, Feng Y. (2010) Structure-based engineering of a monoclonal antibody for improved solubility. *Pro. Eng. Des. Sel.* **23**, 643–651.
142. Teplyakov A, Obmolova G, Wu SJ, et al. (2009) Epitope mapping of anti-interleukin-13 neutralizing antibody CNTO607. *J. Mol. Biol.* **389**, 115–123.
143. Harris LJ, Larson SB, Hasel KW, Day J, Greenwood A, McPherson A. (1992) The three-dimensional structure of an intact monoclonal antibody for canine lymphoma. *Nature.* **360**, 369–372.
144. Bhat TN, Bentley GA, Boulot G, et al. (1994) Bound water molecules and conformational stabilization help mediate an antigen-antibody association. *Proc. Natl. Acad. Sci. USA.* **91**, 1089–1093.

Chapter 15

A Medicinal Chemistry Perspective on Structure-Based Drug Design and Development

Shawn P. Maddaford

Abstract

The application of X-ray crystallography and molecular modeling can provide valuable insight into the optimization of the molecular interactions of a drug–protein complex to achieve potency and selectivity of a drug candidate. For the successful application of SBDD in a drug development program, the impact of these structural modifications required to improve potency and selectivity must be considered in the context of balancing of a multitude of drug properties and other considerations that include solubility, bioavailability, metabolism, distribution, toxicology, chemical stability, and intellectual property space. The utility of structure-based design from the medicinal chemist's perspective is described in this chapter.

Key words: Structure-based drug design, Lipophilicity, Multi-parameter optimization, Drug-like properties, Lead optimization

1. Introduction

As discussed in Chapter 1, the drug discovery and development process is a complex multidisciplinary process that is driven by economic and regulatory factors. The ability to reduce development time has an obvious impact on a drug development programs likelihood of success and the development of technologies such as high-resolution X-ray crystallography, computational screening and docking methods, combinatorial chemistry, functional genomics, and other technologies have the potential to streamline the development process. The application of high-throughput screening and combinatorial chemistry in the 1990s in the pharmaceutical industry dramatically increased the numbers of compounds screened against a promising molecular target (1). Despite the development of these innovative approaches to drug discovery,

Leslie W. Tari (ed.), *Structure-Based Drug Discovery*, Methods in Molecular Biology, vol. 841,
DOI 10.1007/978-1-61779-520-6_15,

there has not been a corresponding increase in the number of new chemical entities reaching approval. Interestingly these new approaches, while innovations in themselves, actually moved the pharmaceutical industry away from innovation and pushed it towards a volume approach with corresponding high levels of drug attrition. Further pressures on declining pipelines might be explained in part by a decrease in R&D productivity (2), increasing regulatory challenges, or a misguided consolidation of the pharmaceutical industry in attempts to stave off losses of product revenues. Despite the current challenges in the pharmaceutical industry, the use of technology such as structure-based design combined with intelligent medicinal chemistry can be used to drive innovation and creative thinking, and expedite the drug discovery and development process if one keeps in mind that the ultimate goal is to produce a safe, effective drug.

2. Challenges in Drug Discovery and Development

Drug discovery is a multivariate process that requires optimization of a drug's potency, efficacy, and oral bioavailability. Layered on this is the minimization of off target binding and toxicological effects, and, depending on the therapeutic area, the requirement of therapeutic advantages over existing therapies. This complex process in general results in high attrition rates for development candidates on the order of 90% (3). During the period from 1991 to 2000, only 11% of all drugs tested in Phase I trials completed registration approval. Some of the main reasons for failure include poor pharmacokinetics related to poor absorption or excessive metabolism, animal toxicity observed in GLP toxicology studies, lack of clinical efficacy, or adverse events in humans. Poor or complete lack of clinical efficacy in part stems from inadequate target validation and translational models as well as suboptimal pharmacokinetics (4).

3. Considerations for Structure-Based Design: Optimizing Drug-Like Properties

The optimization of drug binding and functional activity can be a challenging task for medicinal chemists, especially when antitargets have high homology with the primary molecular target(s). Such is generally the case when multiple enzyme or receptor subtypes exist and for which activity at one or more of the subtypes may result in untoward toxicity or side effects. While target optimization for potency and selectivity can be aided by X-ray structural information of the primary and related off targets, concomitant optimization of pharmacokinetic and pharmacodynamic parameters is generally less amenable. Thus additional strategies must be employed by the medicinal chemist in an attempt to control drug adsorption by the gut or oral mucosa, metabolism by gut and liver, excretion of the drug via the kidney and billiary systems, distribution of the drug into the appropriate tissue(s), and to minimize toxicological effects related to drug-drug interactions, reactive metabolite formation, genotoxicity, phospholipidosis, or hERG related QTc prolongation.

4. Physicochemical Parameters in Optimization of Drug Candidates

The PK properties of a drug are dependent on factors such as solubility and the rate of dissolution of API in the stomach and absorption across the intestine, cellular permeability, drug efflux and drug mediated transport, plasma protein binding, metabolism and distribution. These parameters in turn are dependent on the physicochemical properties of the drug molecule. In general, the relationship between the physicochemical properties of compounds and their in vitro potency is often better understood than the relationship of these properties with their corresponding pharmacokinetic profiles. Finding a compound with suitable PK properties will depend on finding an optimum balance of these properties including molecular weight, lipophilicity, hydrogen-bond donors and acceptors, p*Ka*, and the number of rotatable bonds. One practical approach to understanding the relationship of these properties that leads to compounds with suitable properties for oral absorption is to profile the physicochemical properties of known orally active drugs. Molecular weight was one of the first parameters investigated, and later additional parameters were profiled leading to Lipinski's "rule of 5" (5, 6). Lipinski profiled 2,245 compounds selected from the World Drug Index and with a United States Adopted Name (USAN) characterized as compounds having entered phase II clinical trials. After removing polymers, peptides and quaternary ammonium salts, approximately 90% of the remaining compounds, had physicochemical properties that now form the basis of Lipinski's rule of five: molecular weight less than 500, calculated log*P* less than 5, sum of hydrogen-bond donors less than 5 (e.g., NH and OH), and sum of hydrogen bond acceptors less than 10 (e.g., N and O). Empirically, it was found that when two or more of these parameters fell outside of these ranges that the compounds had poor permeation and oral absorption.

Following the groundbreaking work of Lipinski, many other groups have investigated the relationships of physicochemical properties of development vs. marketed drugs (7), the influence of protein target classes and functional activity on the physicochemical characteristics of preclinical compounds (8), challenges in optimizing these parameters in the design of multiple ligands (DMLs or multiple mechanism compounds) (9, 10) to developing simple rules for interpreting ADMET trends (11), and multiparameter optimization strategies for aligning drug-like properties for drugs targeting the CNS (12, 13).

4.1. Development of Orally Acting Therapeutics: Factor Xa

As mentioned previously, one of the basic tenets of drug development is the ability to find a balance between hydrophobicity-dependent potency and hydrophilicity-dependent pharmacokinetics. The development of orally acting antithrombotic therapeutics against Factor Xa exemplifies some of the difficulties associated with striking an appropriate balance of physicochemical properties of the drug candidate. Factor Xa is a trypsin-like serine protease that plays a key role in several steps of the blood coagulation or clotting cascade (14). The enzyme converts prothrombin to thrombin, a serine protease that subsequently converts soluble fibrinogen into the clot forming fibrin and is the final step in the clotting cascade (15). Unlike previous drugs such as warfarin that have been used in clinical practice for anticoagulant therapy, specific inhibitors of Factor Xa represent a significant innovation given their broad therapeutic window and by allowing physicians to use these drugs without the need for routine monitoring (16).

4.1.1. Structure and Design of Orally Acting Factor Xa Inhibitors

Factor Xa shares homology with other serine proteases including trypsin and other proteases that regulate the clotting cascade including FVIIa, FIXa, FXIa, and thrombin. The development of orally active FXa inhibitors is predicated on attaining selectivity over these and other related serine proteases. As trypsin is involved in the digestion of foods, achieving high selectivity over this enzyme is important given its disposition in the gut and the potentially high levels of drug after oral dosing (17). However, the earliest direct inhibitors of FXa that were designed entered PII trials as parenteral agents (see Fig. 1) due to poor oral bioavailability. The early and selective agent DX-9065a (FXa K_i = 41 nm, thrombin K_i > 2,000 μM, trypsin = 620 nM, see Fig. 1) was extensively studied in preclinical models of thrombosis when dosed intravenously (iv), subcutaneously (sc), and orally (po) (18) and was found to increase activated partial thromboplastin time at twice the control [$aPTT_{2\times}$] at a concentration of 0.97 μM. However, due to the high polarity of the compound and the very basic nature of the amidine group (p*K*a = 10–11), the compound suffered from low oral bioavailability in humans (*F* = 2–3%), precluding the development of an oral agent (19). Several other benzamidine-based selective FXa inhibitors such as otamixaban (FXV-673) (20, 21) and fidexaban (ZK-807834) (22, 23) and a dual acting FXa/thrombin inhibitor tanogitran (BIBT 986; FXa K_i = 26 nM, thrombin K_i = 2.7 nM) (24) were also studied in human clinical trials via parenteral administration due to poor oral availability.

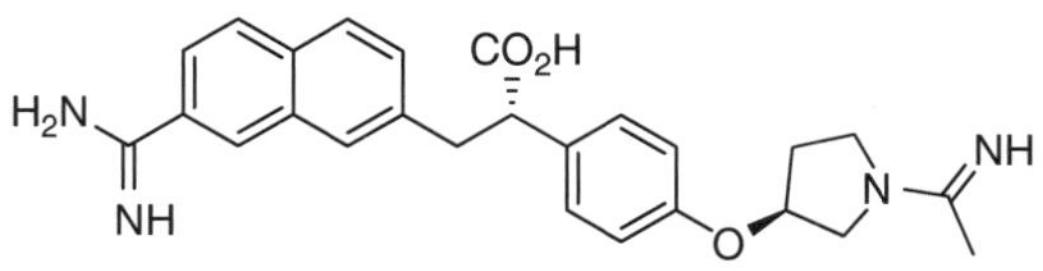

1 DX-9065a
FXa = 41 nM
thrombin>2000 μM
trypsin = 620 nM
F = 2-3%

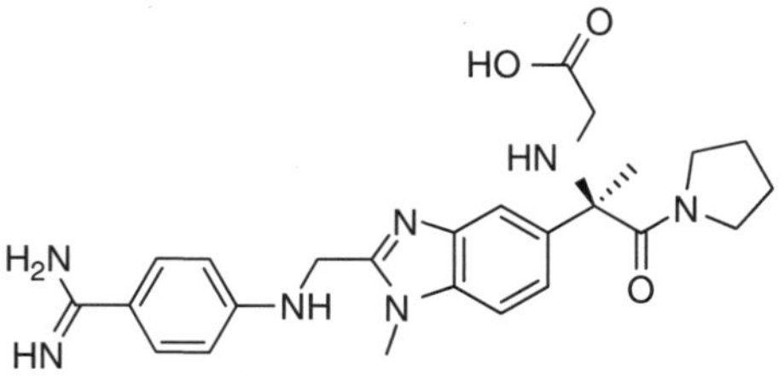

2 tanogitran (BIBT 986)
FXa = 26 nM
thrombin = 2.7 nM

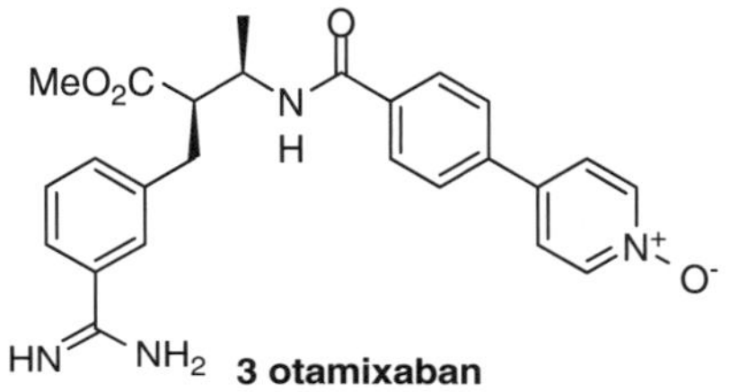

3 otamixaban

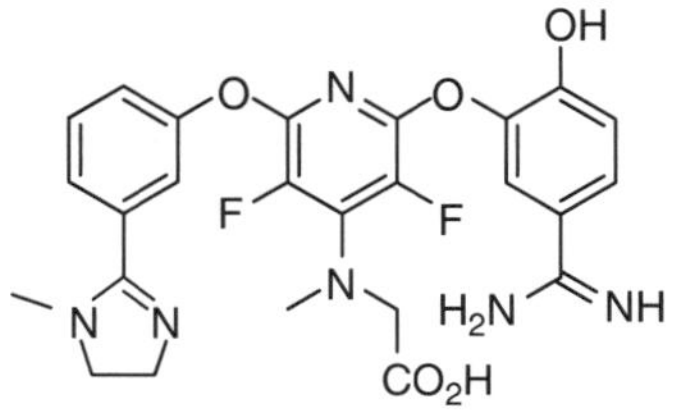

4 fidexiban

Fig. 1. Early dibasic factor Xa inhibitors investigated in human clinical trials.

The development of an orally active FXa agent was driven by information gained from structure-based drug design and an understanding of the nature of the binding pocket. The substrate binding site of the enzyme has four pockets (S1–S4). The enzyme active-site for which the majority of inhibitors interact with is primarily defined by the S1 and S4 pockets (see Fig. 2). The S1 pocket is largely a hydrophobic pocket that extends deep into the protein and terminates with Asp189 and Try228 amino acids side chains. The other principle S4 pocket is a highly hydrophobic area flanked by Try99, Phe174, and Trp215. Adjacent to S1 are the catalytic triad residues that include His57, Asp102, and Ser195. The shape of the binding pocket and the key inhibitor–amino acid interactions in the active site were delineated by the 3 Å cocrystal structure of an early prototype compound DX-9065a (FXa K_i = 41 nm, thrombin K_i > 2,000 μM, trypsin = 620 nM) complexed with FXa (see Fig. 3) (25). The naphthylamidine group occupies the S1 hydrophobic pocket formed by the backbone residues of Ala190-Gln192, Ser214-Gly218, and Gly215-Trp216, the side chains of Asp189, Val213, Tyr288, and a disulfide bridge between Cys191 and Cys220. One of the key interactions at the bottom of the S1 pocket is a salt bridge between the amidine group and the carboxylate of Asp189. The early Factor Xa inhibitors possessing low bioavailability thus developed for parenteral use all utilize this interaction. In addition to the charged amidine group, these early inhibitors also contained a second basic group that interacts in the

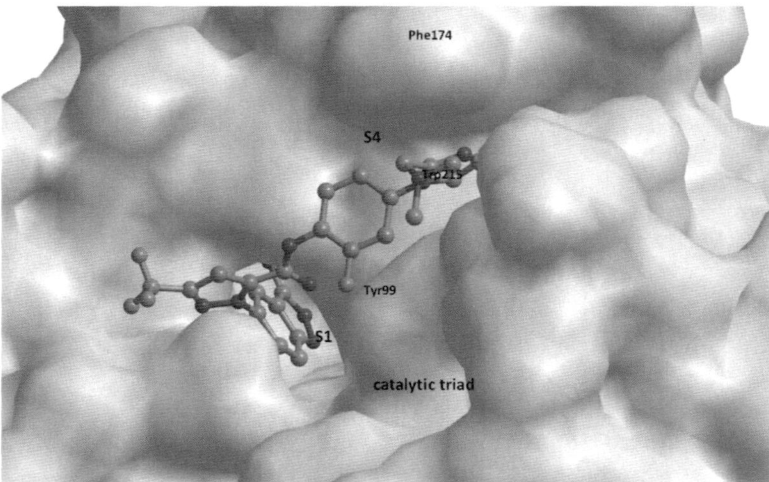

Fig. 2. The binding pocket for factor Xa inhibitors is characterized mainly by two hydrophobic pockets S1 and S4. S1 is a deep hydrophobic pocket that terminates with the side-chains of Asp189 and Tyr228, while the lipophilic S4 pocket is flanked by Tyr99, Phe174, and Trp215.

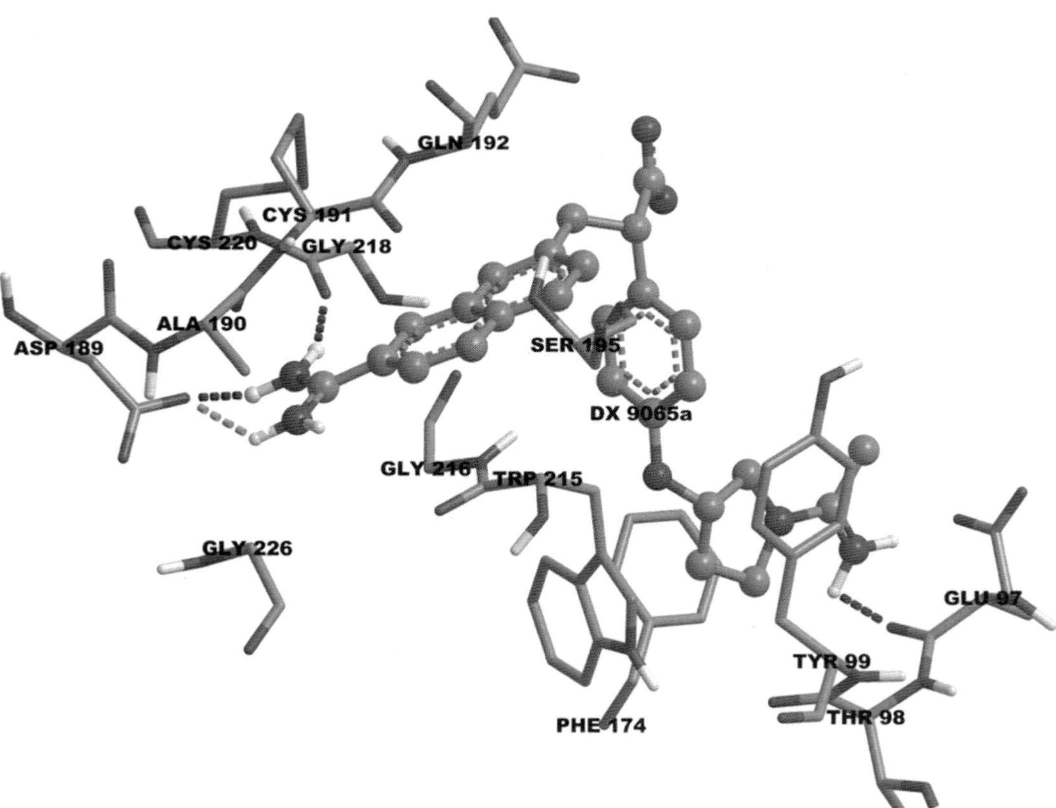

Fig. 3. X-ray cocrystal structure representing the P1 and P4 interactions with the early parenteral inhibitor DX-9065a. The P1 amidine group forms a salt-bridge with Asp189 and an H-bond with Gly218 carbonyl. In addition, the pyrrolyl-methyl-amidine group is positioned close to Glu97 and Thr98 for an electrostatic and H-bonding interaction respectively. This pyrrolyl-methylamidine group is sandwiched in a "hydrophobic enclosure" formed by Tyr99, Trp215, and Phe174 and the backbone atoms of Try99 and Thr98.

hydrophobic S4 pocket bound by residues Tyr99, Trp215, and Phe174 (see Figs. 1 and 3). The S4 pocket is a deep cleft characterized as a hydrophobic enclosure wherein water molecules with correlated hydrogen bonds induce entropic and enthalpic penalties (26). Thus, large improvements in affinity can be achieved by stabilization of the bound state relative to the hydrated state via displacement of these water molecules. In addition to hydrophobic groups, cationic species such as amidines, primary and quaternary amines such as piperidines, pyrolidines and piperazines, pyridines, and others also bind to S4 potentially through π-cationic interactions (27).

From a physicochemical property perspective, the apparent requirement for the two highly basic amidine groups in the early FXa inhibitors represents a barrier to the development of orally active agents. In order for the drug to cross the gut by passive diffusion, the drug molecule must exist to an appreciable extent as a neutral species, as pH partition theory suggests that only the neutral species can diffuse transcellularly (28, 29). While hydrophilic or highly charged compounds can diffuse through aqueous pores, the limited surface area of the pores and the size restriction imposed by tight cellular junctions limits this contribution to the absorption process. Thus it is apparent that the development of orally active FXa inhibitors would depend on finding a less basic or nonbasic replacement of the benzamidine group. The identification of isoxazoline 5 (FXa $K_i = 94$ nM,) via modification of a high-throughput screening hit provided the framework for development of the monobasic inhibitor 6 (see Fig. 4) (30, 31). Structure-based design allowed replacement of the charged amidine in 5 by a neutral *O*-phenylsulfonamide giving 6 (FXa $K_i = 6.3$ nM), which displayed improved potency and permeability. This compound was designed to fill the hydrophobic enclosure with the phenylsulfonamide ring, making an edge to face stacking interaction with Trp215 (31). To simplify the structure in 6, replacement of the chiral isoxazoline ring with a planar achiral 1,3,5-trisubstituted pyrazole resulted in 7 (SN429), a very potent inhibitor of FXa (0.013 nM) albeit with poor oral bioavailability (F=4%), short half-life ($t_{1/2} = 0.82$ h), and poor selectivity over other trypsin-like enzymes (32). Potent activity as observed for compound 7 allows latitude to adjust physicochemical properties to improve ADME properties at the expense of potency. A strategy involving replacement of the remaining P1 benzamidine group (pKa = 10.7) with less basic groups (see Fig. 4, e.g., 7–12) resulted in compounds with improved oral bioavailability (33). Thus, replacement of the benzamidine group in 7 with a benzylic primary amine in 8 resulted in a drop in the pKa from 10.7 to 8.8 with an improvement in bioavailability from 4 to 13% and a reduction in potency from 0.013 to 2.7 nM. Further improvements in potency, permeability and bioavailability could be achieved by incorporation of a 3-fluoro moiety on the biaryl group,

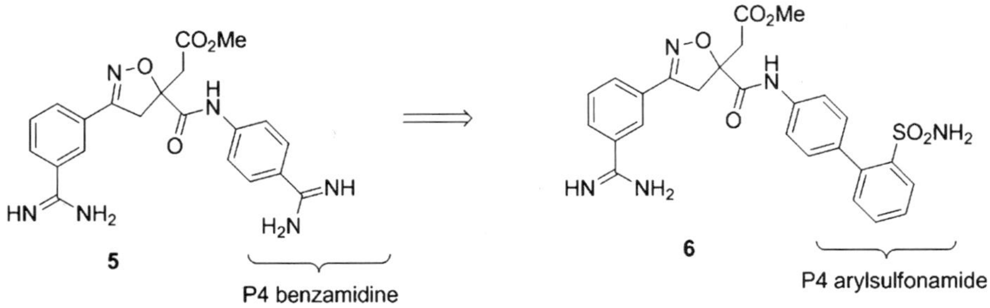

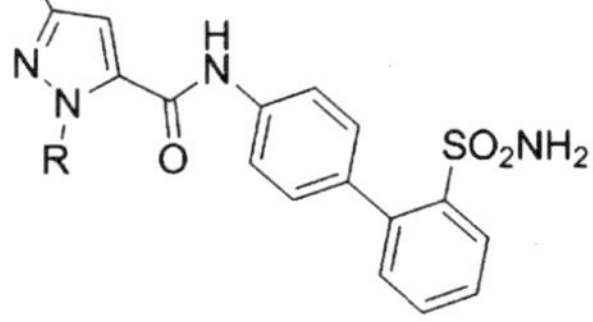

5 — P4 benzamidine ⟹ 6 — P4 arylsulfonamide

7-12

R =	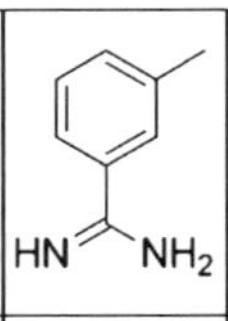 $HN{=}C{-}NH_2$	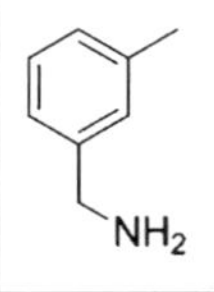 NH_2	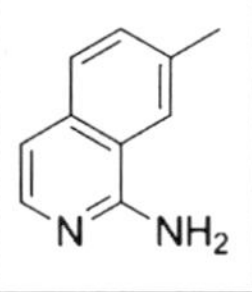 N, NH_2	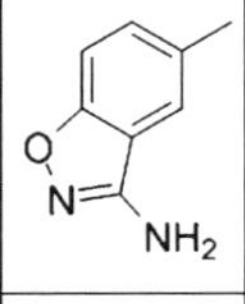 O, N, NH_2	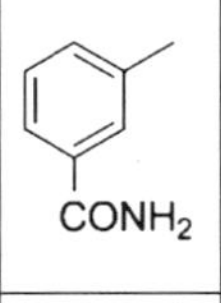 $CONH_2$	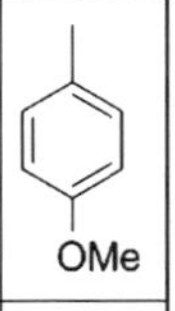 OMe
Cmpd.	**7**	**8**	**9**	**10**	**11**	**12**
pKa	10.7	8.8	6.7	2.3	< 0	< 0
FXa Ki (nM)	0.013	2.7	0.3	1.4	19	11
%F (dogs)	4	13	13	26	46	48

Fig. 4. The transition from bis-benzamidines 5 to monobasic and more neutral factor Xa inhibitors 10–12. A dramatic improvement is observed in the bioavailability in dogs as the basicity (p*K*a) of the inhibitors is reduced.

a 3-trifluoromethyl group on the pyrazole ring and methylsulfonyl (SO_2CH_3) group on the 2′-biphenyl ring. These modifications led to the discovery of the orally acting FXa inhibitor DPC423 13 ($F=57\%$; see Fig. 5) and razaxaban 14 ($F=84\%$, FXa $K_i=0.15$ nM, trypsin >5000; see Fig. 5) with excellent potency and selectivity against thrombin and trypsin (33). The crystal structure of razaxaban at 1.8 Å resolution is shown in Fig. 6. The aminoisoxazole ring system binds to the S1 pocket and forms hydrogen-bonds with Asp189 and the carbonyl of Gly218. Selectivity against trypsin for this inhibitor is attributed to the close contact with the side chain of Ala190 (3.4 Å) that is replaced by a larger Ser190 in trypsin. As such, the trypsin pocket is smaller, resulting in a steric clash between the isoxazole ring of razaxaban and Ser190 in trypsin. Additionally the trifluoromethyl group fits into a small lipophilic pocket created by the Cys191-220 disulfide bridge. Of interest is a shared water-mediated contact the imidazole N-3 nitrogen of P4 makes with Glu97 (see Fig. 6).

One important aspect of drug development is ensuring the drug reaches or maintains appropriate levels (i.e., drug half-life or $t_{1/2}$) in the appropriate compartment. With respect to FXa as an anticoagulant, the relevant compartment is the vascular system. Thus an important property for clotting therapy is achieving a low volume of

Compound	fXa k_i nM	Thrombin K_i nM	aPPT (PT) (EC_{50}) μM	PPB (h) %	Cl L/Kg/h	Vd_{ss} L/kg	$T_{1/2}$ (po) h	%F dog	tPSA	ClogP
13 DPC423	0.15	6000	4.86	89	0.24	0.9	7.5	57	104.9	2.88
14 Razaxaban	0.19		6.1 (2.1)	91	1.1	5.3	3.4	84	111.2	2.47
15	0.23	4400	(36)	>99					65.5	4.17
16 Apixaban	0.04	3100	5.1 (3.8)	87	0.02	0.2	5.8	58	108.5	1.89

aPPT (activated partial thromboplastin time); PPB (human plasma protein binding); PT (prothrombin time) Cl (clearance); Vd_{ss} (steady state volume of distribution); $T_{1/2}$ (po) (drug plasma half life after oral administration); % F (% Fraction absorbed or oral bioavailability); tPSA (total polar surface area); ClogP (calculated logP).

13 DPC 423

14 Razaxaban

15

16 Apixaban

Fig. 5. Chemical structures, pharmacokinetic and physicochemical parameters and activity in the in vitro clotting assays aPPT (activated partial thromboplastin time twice the control) and PT (prothrombin time twice the control).

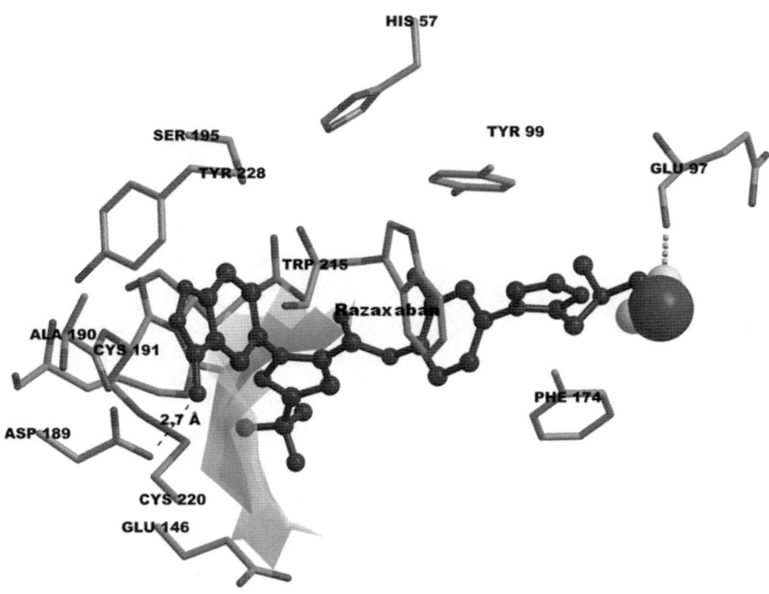

Fig. 6. X-ray crystal structure of razaxaban in the FXa binding site (PDB code 1Z6E). The aminoisoxazole group makes an electrostatic interaction with Asp189 (2.7 Å) providing potency, while the isoxazole ring makes a close contact with the side chain of Ala190. In trypsin, Ala190 is replaced by a serine resulting in steric clashes and reduced binding to this enzyme. The terminal P4 imidazole ring occupies the hydrophobic pocket created by Phe174, Tyr99 and Trp215. The 3-trifluoromethyl group attached to the pyrazole ring sits in a small hydrophobic cleft (*shaded surface*) created by the Cys191-220 disulfide bridge and the side chains of Glu146 and Arg143. Finally, a water-mediated H-bond to the N3 of the P4 imidazole ring shared with Glu97 is also present.

distribution (V_d or V_{dss}) such that the drug remains within the blood volume and does not clear rapidly. Thus, the pharmacokinetic parameters that drive half-life and the frequency of dosing are the volume of distribution and clearance, parameters that depend on the physicochemical attributes of the molecule. In the case of DPC423 13, a moderate steady state volume of distribution (V_{dss} = 0.90 L/kg), low clearance (CL = 0.24 L/h/kg), and long half-life was achieved ($t_{½}$ = 7.5 h). To obtain proof of concept, razaxaban 14 (V_{dss} = 5.3 L/kg, CL = 1.1 L/h/kg, $t_{½}$ = 3.4 h) was advanced into phase II trials but concerns over potential metabolism of the amide linkage to release 4-aminobiphenyl raised a genotoxic structural alert within this series of compounds (33). While both 13 and 14 were not susceptible to metabolic cleavage, the *ortho*-substituted analogue of 13 gave rise to a mutagenic aniline cleavage product. This potential liability prompted efforts to prevent this hydrolysis by cyclization the amide to give various bicyclic lactam analogues such as 15 with excellent in vitro potency and selectivity, but no activity in the in vitro clotting assays (see Fig. 5). The lack of activity in the clotting assay is likely related to its high plasma protein binding (PPB > 99%), high lipophilicity (Clog*P*=4.17), and low polarity (tPSA = 65.5). Modulation of the lipophilicity of 15 by replacing the trifluoromethyl group with a polar carboxamido group found in apixaban 16 resulted in a clinical compound with subnanomolar FXa binding (0.08 nM) with lower Clog*P* (1.89) and increased polar surface area (108.5) and improved potency in the clotting assay. In addition, apixaban 16 had weak activity in the hERG assay (IC_{50} > 25 μM), likely due to its relatively low Clog*P* (<3.7) value and its lack of a positive charge.

As can be seen in the case of the development of factor FXa inhibitors, appropriate modulation of the physicochemical properties such as basicity (p*K*a) and lipophilicity (Clog*P*) can have a dramatic impact on plasma protein binding and in vivo efficacy, safety (hERG), PK properties such as oral bioavailability, volume of distribution, clearance, and half-life all of which are important in obtaining suitable dosing regimens and limiting side effects. Given the relevant compartment for FXa in the clotting mechanism is in the blood, optimizing a compound by minimizing tissue distribution may be advantageous to limit side effects.

4.2. Comparing Selectivity Modes in Isoforms of Nitric Oxide Synthase: Design of Selective nNOS Inhibitors

Nitric Oxide (NO) is a ubiquitous signaling molecule that plays multiple physiological roles such as the regulation of blood pressure, neuromodulation and immune response. NO is synthesized by three structurally related isoforms of the enzyme nitric oxide synthase (NOS) with 50–60% overall sequence homology. NOS is a heterodimeric complex containing a heme oxygenase domain that catalyzes the two step oxidation of L-arginine to L-citrulline (see Fig. 7) in a process involving transfer of electrons from NADPH via the FAD and FMN groups in the C-terminal reductase

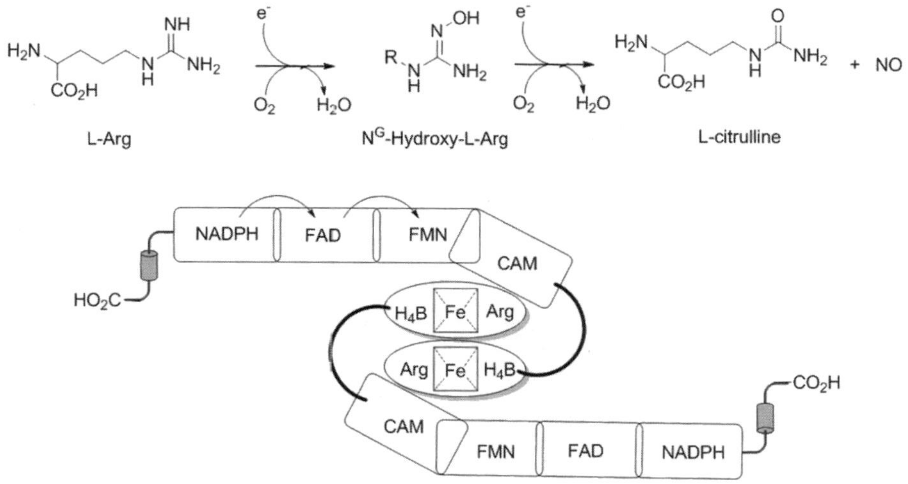

Fig. 7. The oxidation of L-arginine to L-citrulline is a two step process requiring oxygen and electrons supplied by NADPH flowing from the reductase domain to the oxygenase domain. The figure shows a model for the NOS dimer structure that accounts for the observed transfer of electrons from the reductase domain of one monomer to the oxygenase of the other monomer.

domain to the adjacent subunit N-terminal oxygenase domain that binds the arginine substrate (1′R, 2′R, 6R)-5,6,7,8-tetrahydrobiopterin (BH4) and heme (34–36). Dimer formation is important for activity of NOS and is stabilized by structural zinc binding at the oxygenase dimer interface (37). The activity of nNOS and eNOS are dependent on changes in Ca^{2+} concentration and the binding of a Ca^{2+}/calmodulin complex (CAM), while the activity of iNOS appears independent of Ca^{2+} concentration due the tight binding of the complex at the dimer interface (38). The neuronal or brain NOS (nNOS or NOS1) and the endothelial form (eNOS or NOS3) are constitutively expressed in the nervous and vascular systems, respectively. The inducible form (iNOS or NOS2) is expressed by macrophages and glia under conditions of stress or upon the release of inflammatory mediators such as TNFα, IL-1, or LPS. The overproduction of NO by these systems in various tissues is implicated in the pathogenesis of various diseases. For example, iNOS is implicated in septic shock, arthritis, Alzheimer's disease, and multiple sclerosis, while nNOS is implicated in Parkinson's disease, neuropathic pain and migraine. Given the important role of eNOS in regulating vascular tone, selective targeting of either the nNOS or iNOS isoforms is crucial in the development of a therapeutic for the treatment of disease. Unlike the development of FXa inhibitors, which were designed to reside in and specifically target the vascular system, the development of a nNOS inhibitor for the treatment of neuropathic pain requires tuning of drug properties to optimize brain penetration and activity without undesirable activity in the vasculature. Thus, the physicochemical attributes of drugs that are required to achieve the correct PK/PD relation against these targets are distinct.

With the exception of human nNOS, X-ray structures for all three isoforms have been determined, enabling structure-guided inhibitor design (37, 39–44). The oxygenase domains possess striking active-site structural conservation across the three isoforms with 16 of 18 residues within 6 Å of the substrate binding site identical (45). The high degree of structural similarity between the three isoforms accounts for the difficulties in obtaining isoform selective inhibitors. Many early NOS inhibitors were based on modifications of the natural L-arginine substrate and include compounds such as N^G-methyl-L-arginine (L-NMMA), N^G-nitro-L-arginine (L-NNA), and other amino acid-based inhibitors (46). Although potent, these inhibitors were not selective and not drug-like likely due to their similarity with the natural substrate. The nonselective NOS inhibitor L-NMMA has been shown to increase blood pressure in human clinical trials due to inhibition of eNOS (47). In order to improve selectivity for nNOS over eNOS, a series of highly basic dipeptide and reduced amide-based inhibitors were designed, despite the high similarity between the three isoform binding-pockets. Inhibitors 17–24 derived from L-NNA show selectivity for rat nNOS vs. bovine eNOS (see Fig. 8). These compounds interact with the L-arginine binding pocket, making hydrogen bonds to Glu592, Gln478 and Ser477 (rat nNOS) (48). As with many NOS inhibitors, the guanidine and aminopyridine groups make bifurcated H-bonds with the conserved glutamate (e.g., Glu363 or 592) across the three isoforms. The nitro group of 17 and 18 enhances affinity over L-arginine by providing additional hydrogen bonds and nonbonded contacts with the protein backbone (49). X-ray crystal structure analysis provides a rationale for the selectivity of 17–22 for nNOS over eNOS that appears arise from a single amino acid substitution of a neutral Asn368 (eNOS) to a charged Asp597 (nNOS) residue (see Figs. 8 and 9). Inhibitors bound to nNOS adopt a curled conformation in the active site placing the α-amino group between Glu592 and Asp597 making a direct H-bond to Glu592 (2.8 Å) and electrostatic interactions with the charged Asp597 (nNOS). By contrast, inhibitors bind to eNOS in a fully extended conformation with the α-amino in 17 reorienting to make a H-bond with Gln249 (Gln478 in rat nNOS) rather than the conserved Glu592. Similar to 17, compounds 18 and 19 also adopt a curled orientation. Mutagenesis studies (D597N) in nNOS revealed in a drop in potency for 17 (K_i wild-type = 0.13 μM, D597N = 67 μM) against the D597N mutant with an increased potency against the N368D mutant in eNOS (IC_{50}, wild-type = 107 μM, N368D = 9.5 μM), thus confirming that a single residue was responsible for the two orders of magnitude difference in potency for binding between the nNOS and eNOS isoforms (48).

In an effort to prepare new analogues, isosteric replacement of the nitroguanidine with an aminopyridine resulted in the potent and selective nNOS inhibitor 20 (see Fig. 8) (50). However, the therapeutic

Peptide, reduced amide, and aminopyridine based substrate inhibitor selectivity for **17-22**.

Cmpd[50]	Rat nNOS	Bovine eNOS	Murine iNOS	e/n	i/n	pKa*	cLogD*	HDB*	D597N nNOS	N368D eNOS
17	0.13	200	25	1538	192	8.9	-6.66	10	67	9.5
18	0.10	314	39	2617	320	9.85	-5.97	8	21	5.1
19	0.388	434.5	58.4	1114	150	9.67	-4.86	6		
20	0.015	31	9.5	2100	630	8.94	-1.12	4		
21	0.036	36	13	1000	360	7.32	-0.3	4		
22	0.08	62	52	8780	650	5.56	0.23	4		

K_i values are in μM. *Calculated with ACD/Log D version 12.5. Advanced Chemistry Development Inc., Toronto, Canada.

17 **Amide**

18 **Reduced Amide**

19

20 X_1,X_2=H
21 X_1=H, X_2=F
22 X_1=X_2=F
Aminopyridine

Fig. 8. Selective nNOS inhibitors containing: a nitroguanidine group and amide 17 and reduced amide 18 dipeptide-based inhibitors and a series of polybasic aminopyridine inhibitors 19–22. The p*Ka* values of aminopyridine analogues 20–22 were lowered by addition of fluorine in the beta position of the alkoxyaryl side chain.

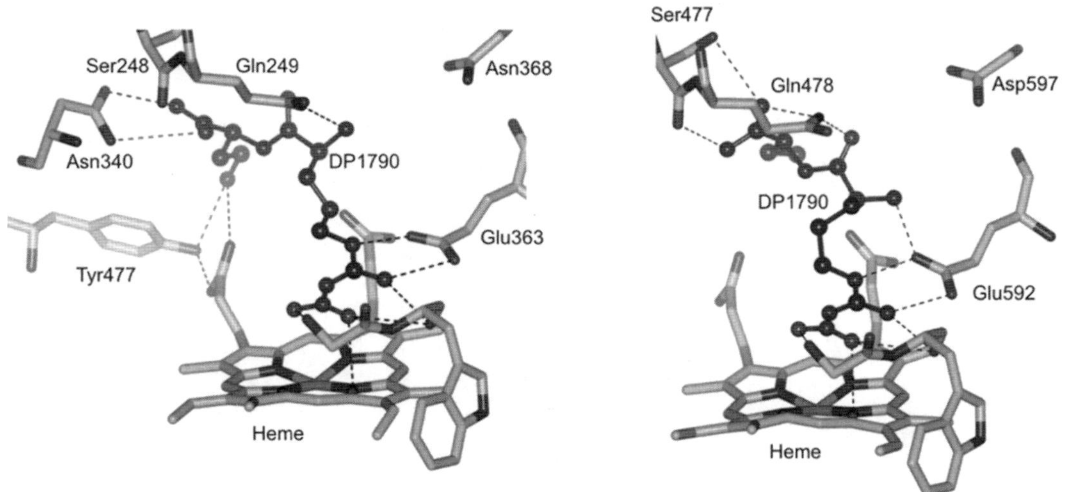

Fig. 9. Crystal structure of 17 in bovine eNOS *left* (PDB code: 1P6L) and rat nNOS *right* (PDB code: 1P6H). In the case of the eNOS complex, the inhibitor adopts a linear conformation with H-bonding interactions with Glu363, Gln249, Asn340, and Tyr477. Compound 17 in rat nNOS adopts a curled conformation that places the primary amino group of the nitroarginine residue within H-bonding distance to Glu592 and an electrostatic interaction with Asp597. In the case of eNOS, the amino group of the nitroarginine residue in 17 is H-bonded to Gln249. Substitution of Asp597 in nNOS for Asn results in loss of potency.

utility of 20 is limited by poor blood–brain barrier (BBB) permeability. It was speculated that the brain penetration of 20 was limited by the two positive charges at physiological pH. Incorporation of electron withdrawing fluorine beta to the basic nitrogen in the alkaryl side chain resulted in reduction in the calculated p*Ka* values from 8.94 to 7.32 and 5.56 for the mono and difluoro species 21 and 22 with a slight reduction in potency but potentially with an improvement in cellular permeability (50). Given the physicochemical properties of compounds 18–20, and possibly 21 and 22, it is not surprising that the compounds did not readily cross the BBB. The reported CLog*D* values for 17–20 range from −6.66 to 0.23 (Fig. 8) and have total polar surface areas (TPSA) values that range from 72 to 218, suggesting most of the compounds are very polar and exist primarily in the ionized state at physiological pH (50). In addition, all of the compounds possess 4–10 hydrogen bond donor groups (HBD) and multiple positive charges and as such do not satisfy Lipinski's rules.

A series of potent and selective nNOS inhibitors based on tetrahydroquinoline and quinolone cores with drug-like properties have been reported (see Fig. 10) (51). Quinolone 23 displayed good activity at human nNOS (0.58 μM) with 71-fold selectivity over human eNOS (IC50 = 41 μM) and good selectivity over human iNOS. In addition, compound 23 was selective against a broad panel of CNS and other targets (~80) (52). Despite the activity in the nNOS enzyme, the compound was inactive after i.p. (intraperitoneal) injection when evaluated in an animal model of neuropathic pain, potentially indicating poor brain penetration. As CLog*P* is correlated to brain penetration, it was anticipated that the reduction of the amide bond to give the more lipophilic candidate 24 would result in better brain penetration. The calculated CLog*P* for 24 has increased

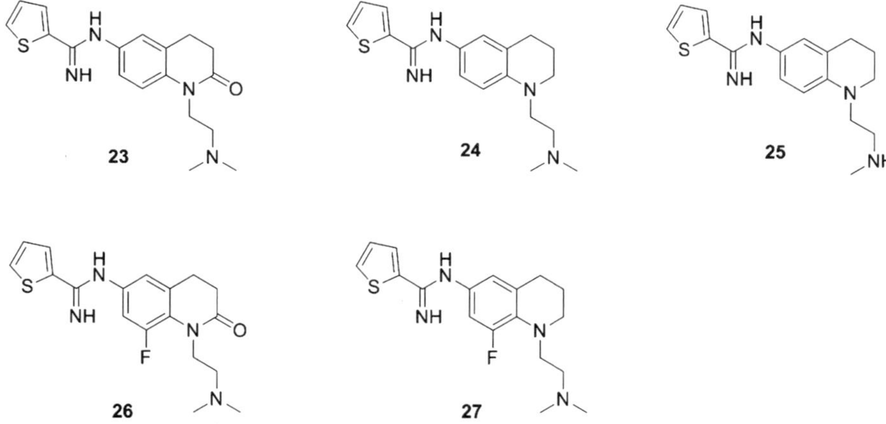

Fig. 10. Structures, NOS inhibitory data and physicochemical properties of a series of selective nNOS inhibitors based on a tetrahydroquinoline/quinolone scaffold. All compounds contain a thiophene amidine group that anchors the inhibitor by binding to the conserved glutamic acid (Glu371). A secondary or tertiary basic nitrogen containing side provides selectivity for nNOS over eNOS presumably through electrostatic interactions with Asp376 in human nNOS which is substituted by a neutral Asn residue inhuman eNOS. By decreasing the Clog*P* and increasing tPSA, it is possible to remove hERG activity while still maintaining potency and selectivity at nNOS (cf compound 25 and 24).

from 2.26 for 23 to 3.59 and the PSA decreased from 87.7 to 70.6 Å. In contrast to 23, i.p. administration of 24 at a dose of 30 mg/kg i.p. resulted in full reversal of thermal hyperalgesia in the spinal nerve ligation (SNL or Chung) model of neuropathic pain. The minimally effective dose (MED) for 24 in the SNL model was as low as 3 mg/kg. It is believed that good brain penetration is important for activity in this model but neither total brain concentration nor free brain concentration alone is a predictor of efficacy (53). However the brain–plasma ratio appears to correlate with in vivo efficacy.

As a surrogate of the drug brain–plasma concentration ratio, determination of the CSF/plasma (cerebrospinal fluid/plasma) for 23 and 24 revealed that indeed 24 crossed into the BBB more readily than 23 and had a higher CSF–plasma ratio (0.53 vs. 0.25). The absolute free drug in the CSF for 24 was higher on a dose corrected basis (28 vs. 14.9 ng/mL). Based on the data, it appears that increasing the CLog*P* and/or decreasing the PSA results in better in vivo activity. While 24 had excellent in vivo activity, the compound had a higher clearance rate of 42.9 L/h/kg and a corresponding shorter half-life of 4.8 h in comparison to 23. In addition, the compound had low oral bioavailability (13.5%) and inhibited the hERG channel with an IC_{50} value of 2.9 μM. Inhibition at the hERG channel is known to cause Torsade de pointes cardiac arrhythmias via a prolongation of cardiac QT action potential and as such, is an important safety concern (54).

The hERG channel is a potassium channel formed by four identical subunits consisting of S5, S6, a pore helix, and a selectivity filter domain. A homology model of suggests the S5–S6 pore domain forms a 6-Å pore through which K^+ ions pass. Drug access to the channel is believed to occur from the intracellular side via blockade of the activated channel. Several pharmacophore models (55, 56) and a homology model (57) are available, and the relationship between physicochemical properties, particularly lipophilicity and hERG activity (58, 59) is understood. The pharmacophore for hERG requires a basic nitrogen and lipophilic aromatic groups. Drugs containing a basic nitrogen interact via a π-cation interaction with Tyr652, while lipophilic groups interact with Phe656. H-bond donor/acceptor groups above the lipophilic region in the channel may also contribute to binding. In general, it was determined that to have a >70% chance of being less active than 10 μM against the hERG channel, compounds should have a log*D* value of <3.3 for neutral species and a <1.4 for bases (60). Thus, a reduction in lipophilicity (e.g., log*P* or Clog*P*) appears to be a simple strategy for reducing hERG binding. Other strategies include introducing an acid group to minimize intracellular access to the channel but this is not suitable for CNS drugs. A reduction in basicity of the amine group would also minimize the π-cation interaction. In the case of NOS inhibitors 23–27, both a basic nitrogen and a lipophilic group may be important for nNOS activity and selectivity. To reduce the hERG activity of 24, a reduction in the log*P* (or Clog*P*) and an increase in the PSA of 24

23	Value	Score
ClogP	2.26	1.00
ClogD	0.15	1.00
TPSA	87.7	1.00
MW	342.458	1.00
HBD	2	0.50
pKa	8.93	0.54
	CNS MPO	5.0

24	Value	Score
ClogP	3.59	0.70
ClogD	-0.47	1.00
TPSA	70.6	1.00
MW	328.47	1.00
HBD	2	0.50
pKa	9.1	0.45
	CNS MPO	4.7

25	Value	Score
ClogP	3.01	1.00
ClogD	-1.08	1.00
TPSA	79.4	1.00
MW	314.448	1.00
HBD	3	0.17
pKa	9.52	0.24
	CNS MPO	4.4

Pharmacokinetic parameters for compounds **23-25** in rats.

Cmpd	F %	Dose p.o/i.v mg/kg	Cmax ng/mL	CSF ng/mL	CSF /Plasma	Vss L/kg	CL L/hr/kg	$T_{1/2elim}$ hr	RLM $T_{1/2}$ min	hERG*
23	-	-/3	-	14.9	0.25	194	23.2	7.7	-	
24	13.5	10/15	7.5	140(28*)	0.53	262	42.9	4.77	30.7	2.9
25	60	10/3	277	86.2	0.85	83.9	5.3	18.3	>125	54

*Dose normalized to 3 mg/kg. LM (liver microsomes). Cmax (maximum drug plasma levels); CSF (cerebrospinal fluid); CSF/plasma (ratio of the concentration of drug in CSF to that in the plasma); Vss (steady state volume of distribution); CL (drug clearance rate); $T_{1/2}$(terminal half life); RLM (rat liver microsomes).

Fig. 11. Calculated CNS MPO scores for compounds 23, 24 and 25. Parameters having a score of 1 (*light shading*) suggest suitable properties for a CNS active drug while those with less than 0.5 are less than ideal (*darker gray*). Despite having the highest score, compound 23 is not active in the neuropathic pain model when administered at a dose of 30 mg/kg i.p. whereas compounds with a higher Clog*P*> 3.0 are active in the model.

was achieved by removal of a methyl group. The activity of the demethylated compound 25 retained potency and selectivity in nNOS. However a dramatic decrease in the IC_{50} value was observed in the hERG patch clamp assay (nNOS IC_{50} = 0.38 μM, eNOS IC_{50} = 32 μM, hERG = 54 μM). In addition to reducing the hERG activity, an improvement in the PK properties was observed. The oral bioavailability of 25 was improved to 60% together with an increase in the CSF–plasma ratio (CSF–plasma = 0.85) and the absolute CSF concentration (Cmax = 86 ng/mL) (see Fig. 11). A reduction in the clearance (Cl = 5.3 L/h/kg) of 25 and a corresponding increase in plasma half-life (elimination = 18 h) was observed. Relative to 24, compound 25 was also more stable in rat liver microsomes (RLM, see Fig. 11). The in vitro activity and the favorable drug-like properties of 25 were corroborated by the reversal of thermal hyperalgesia in the SNL model of neuropathic pain after i.p. dosing (52).

The effects of introduction of fluorine into compounds 23 and 24 were also investigated. Fluorine is extensively used in drug discovery and is increasingly found in drugs entering the marketplace (60). Although fluorine is larger than a hydrogen atom, its high electronegativity allows for manipulation drug properties including reduction of p*K*a, modification of lipophilicity (lowers if attached to an alkyl chain and increases if attached to an aromatic ring), improving metabolic stability and duration of action (61), enhancing potency, reducing the potential for covalent protein binding and idiosyncratic toxicity (62, 63), attenuation of biliary clearance (64), and creating amide isosteres via trifluoroethylamines (65) (e.g., cathepsin K inhibitors). To increase lipophilicity

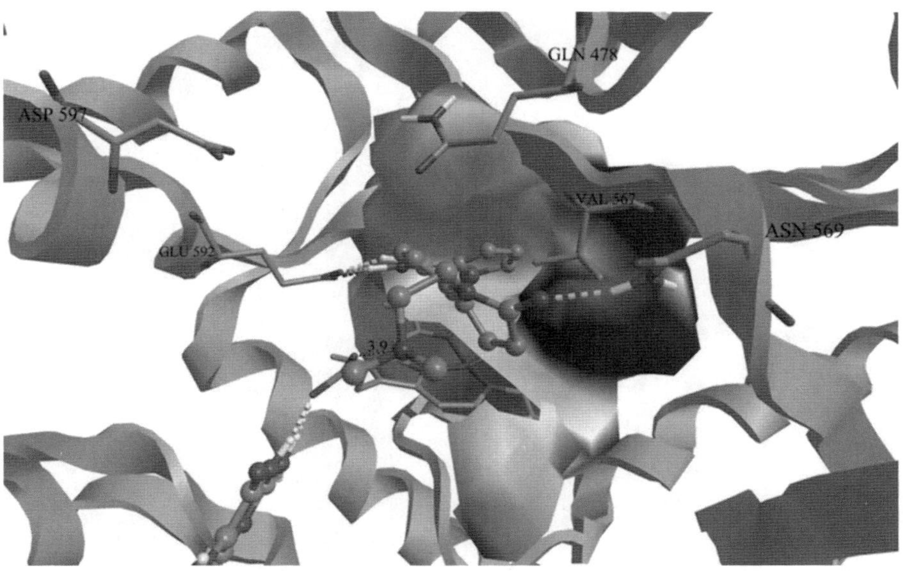

Fig. 12. The docked structure of 22 in a human nNOS homology model. The amidine group makes a bifurcated H-bond/electrostatic interaction with Glu592. The quinolone carbonyl group makes an H-bond with Asn569 while the quinolone ring system makes a hydrophobic interaction with Val567. The thiophene sulfur sits over the heme iron group and the ring engages the back of the pocket. In this model, the basic ethylamine side chain sits between the two heme propionate groups. An extended conformation of this side chain would place the positively charged nitrogen in range to interact with Asp597.

and improve brain penetration of 23, a fluorine was introduced at the 8-position of the quinolone ring. While an improvement in CLog*P* was observed for 26, a dramatic reduction in potency at nNOS was observed. However, introduction of a fluorine substituent in 24 at the same position resulted in nearly fourfold increase in potency in nNOS and an increase in selectivity over eNOS. The nature of the effects on potency reduction of combining both the amide group that introduces ring planarity together with a fluorine atom is unknown but might be attributed to steric influences in the conformation of the ethylamine side chain group (see Figs. 12 and 13). A minor reduction in potency in the hERG patch clamp activity was observed for 27 relative to 24.

Compounds 23–25 have physicochemical properties within acceptable ranges for a CNS active drug according to the CNS MPO (12, 13) (Multiparameter Optimization; see Fig. 11). The CNS MPO method is a modification of the Lipinski rules that was developed as a prospective design tool to enable the alignment of multiple drug-like properties (Clog*P*, Clog*D*, TPSA, MW, HBD, and p*K*a) and to improve the probability of finding a suitable CNS drug candidate. By comparison, Lipinski's rules employ hard cutoff values for physicochemical properties and limit the design space. Based on an analysis of the physicochemical property ranges of 119 CNS marketed drugs and a database of candidates, a property scoring function was developed (12, 13). Using an Excel based calculator, the CNS MPO property calculations revealed compound 25 had the lowest overall score of 4.4/6, however all compounds scored well.

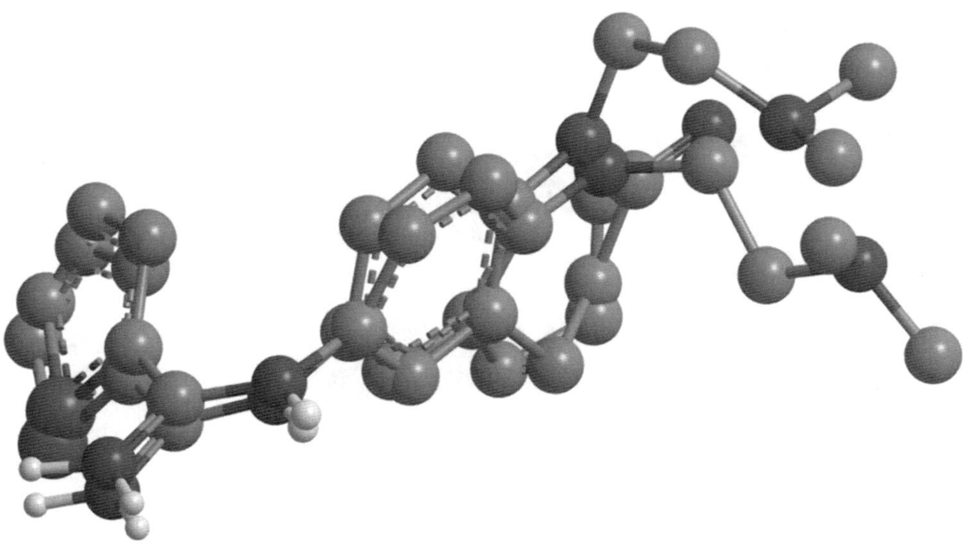

Fig. 13. Overlay of two low-energy proposed bioactive conformations of 22 and 23 calculated using AM1 semiempirical methods. The planarity of the amide group in 22 restricts the side chain closer to the plane of the quinoline ring, while the reduced tetrahydroquinoline nitrogen is bent, placing the ethylamine side chain perpendicular to the quinoline ring system.

The Clog*P* scores indicate that 24 trends on the upper end of lipophilicity. In general, higher CNS MPO score generally indicates those drugs for which there is a corresponding alignment of ADME attributes such as high permeability (Papp), low P-gp (P-glycoprotein efflux), low clearance, and low hERG susceptibility (12).

The potential binding mode of compounds such as 22 and 23 could be determined by a flexible receptor docking approach. Compound 22 was docked into the oxygenase domain of a human nNOS homology model built from the crystal structure of the rat nNOSoxy domain (PDB code 1VAG) (66). As with other NOS inhibitors, the amidine group makes a bifurcated H-bond/electrostatic interaction with the conserved Glu592. The quinolone carbonyl group makes a H-bond with Asn569, while the quinolone ring system makes a hydrophobic interaction with Val567. The thiophene sulfur is positioned over the heme iron group and the ring sits against the back of the pocket. The basis for selectivity is likely due to the basic amine side chain which sits between the two heme propionate groups. An extended conformation of this side chain would place the positively charged nitrogen near to Asp597 which substituted by an Asn residue in eNOS.

4.3. Isoform Specific Structural Plasticity in NOS: Anchored Plasticity Approach to iNOS Selective Inhibitors

Drug induced-fit is a known phenomenon in structure-based drug design, an example of which is the binding of the influenza drug Tamiflu to neuraminidase N1 (67). A recent example of the use of drug-induced fit was employed in the development of selective iNOS inhibitors. The binding of a series of bulky quinazoline and aminopyridine inhibitors to the iNOS enzyme induced isozyme-specific residue plasticity of residues distant from the active site

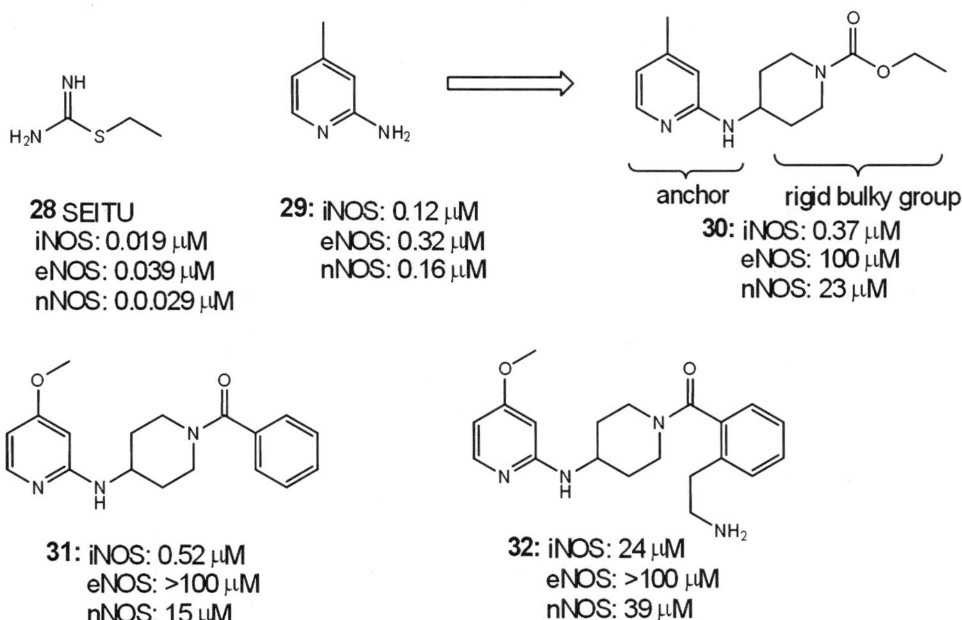

Fig. 14. Ligands binding to the oxygenase domain of the NOS enzyme. Both SEITU 28 and 4-methylaminopyridine 29 are nonselective ligands that anchor tightly to the binding pocket. In the anchored plasticity approach, attachment of a bulky side chain (e.g., 30 and 31) onto an anchor scaffold results in selectivity via isozyme-specific residue plasticity that creates a new binding pocket in the iNOS structure that is not available in the corresponding eNOS structure. Compound 32 was synthesized in an attempt to exploit an electrostatic interaction with an Asp residue within the iNOS active-site.

modulated conformational changes of conserved residues in contact with the inhibitors (44). In this approach, a core fragment is anchored into a conserved binding pocket while a rigid bulky side chain projects into remote specificity pockets that form upon induced conformational changes of "plastic residues." This contrasts with the approach to the design of nNOS selective inhibitors wherein a flexible ligand undergoes conformational adaptation to interact with variant residues within the immediate active site. The Glu377 residue in iNOS serves as an anchoring point for the binding of small nonselective ligands such as SEITU (*S*-ethylisothiourea) 28, aminopyridine 29, or the selective aminopyridines 30 and 31 (see Fig. 14) mediated by bidentate hydrogen bonding. Bulky ligands 31 and 32 induce an opening of invariant first shell residues Gln263 and Arg373 inducing a cascade-like movement of second shell residue Asn283 towards third shell residues Phe286 and Val305 (see Fig. 15). Unlike selective ligands, the binding of small ligands 28 and 29 do not induce this conformational change and thus bind to a first shell Gln-closed state. In human eNOS, the binding of bulky ligands are unable to induce conformation change of the invariant first (Gln246 and Arg249) and second shell residue (Asn266), since the final third shell residues are much bulkier isoleucine and leucine residues

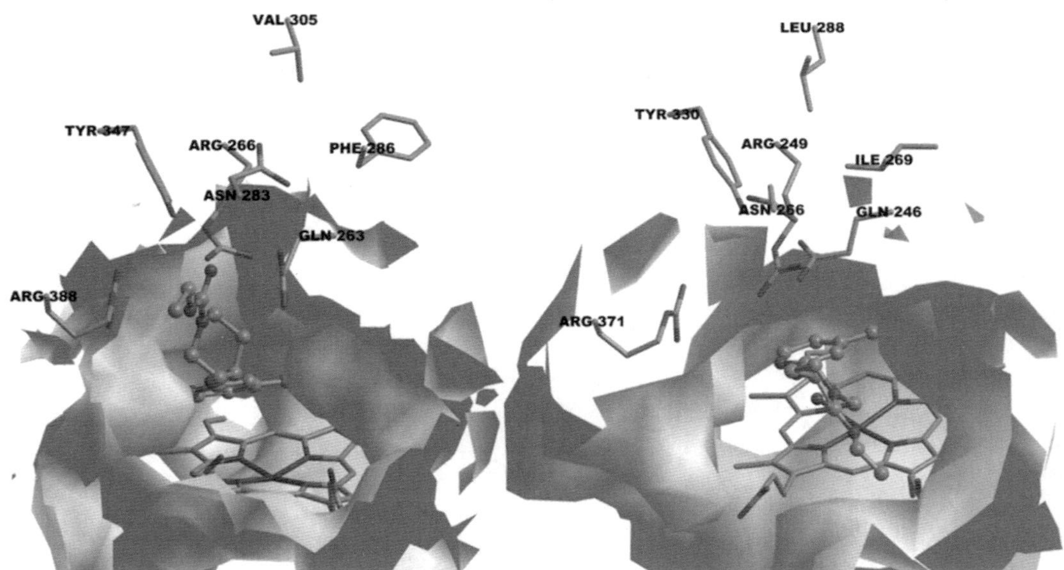

Fig. 15. *Left*: cocrystal structure of 30 (AR-C95791) in human iNOS. *Right*: cocrystal structure of 30 in bovine eNOS. In human iNOS, enzyme structural plasticity allows conversion of the Gln263 closed form to a Gln open form creating a new selectivity pocket upon binding a bulky group. A cascading motion of first shell residues Gln263 and Arg266 to the second shell residue Asn283 is ultimately accommodated by flexibility of motion of third shell residues Phe286 and Val305. Small nonselective compounds such as SEITU 28 or 4-methylaminopyridine 29 bind to the closed form of iNOS which is similar in shape to the eNOS pocket. In eNOS, rotation of the first shell Gln246 residue is not possible due to the increased steric demands of the isozyme specific third shell residues Ile269 and Leu388.

(Ile269 and Leu288). In iNOS, the third shell residues are much smaller Phe and Val residues and thus accommodate structural plasticity within the enzyme. Substitution of the simple aminopyidine 29 with the rigid and bulky 4-ethoxycarbonylpiperidino group resulted in a both decrease binding to nNOS and eNOS without affecting iNOS activity. In the case of the nNOS isoform, the third shell residues are Leu525 and Phe506. It was predicted that similar conformation changes would occur in the nNOS isoform but the slightly larger size of Leu525 of nNOS compared with the smaller Val305 residue would result in restricted motion (68, 69). This restricted conformational motion in nNOS would account for the drop in potency of 30 and 31 relative to 29.

The conformational adaptation of residues distal from the active-site opens new binding-pockets to allow for potent and selective binding of inhibitors selective for iNOS over eNOS. This anchored plasticity approach is likely suitable for the design of selective inhibitors against other targets. By identifying conserved anchor points within the pocket and locating variable protein sequences outside of the pocket, it is conceivable that one could characterize the pathways that lead to conformational plasticity by visualizing solvent accessible access channels and subsequently design an inhibitor that contains an anchoring point with attached substituents that extend towards these flexible residues (69).

Despite the potential of crystal structures in guiding a structure-based design program, caution must be exercised by medicinal chemists and modelers during the optimization stage. One potential caveat lies in the interpretation of structural data; for example, the resolution of the structure employed must be adequate to identify the orientation of the ligand and to allow accurate ligand docking. In addition, the resolution of the structure may be insufficient to determine the identity or orientation of a ligand in the active site or the positions of side chain residues. Second, enzyme plasticity as described in the above example may be hard to predict and will lead to erroneous results if the wrong enzyme structure is used in the docking study. Even the methods used in X-ray data collection can have an impact on ligand docking. For instance, the volume of the internal cavities of enzymes can shrink by several percent upon crystal freezing and low temperature data collection (68). Finally, the crystal structure provides little, if any information on the thermodynamics of ligand binding and desolvation contributions to the overall binding. Medicinal chemists and modelers therefore make several assumptions about the protein crystal structure (1) the protein structure and the ligand interactions with the protein are correct; and (2) the protein-ligand structure is relevant for drug discovery. A recent publication highlights some of the limitations of X-ray crystallography in the design of iNOS inhibitors related to structures 31 and 32 (see Fig. 14) (69). The cocrystal structure of 31 in mouse iNOS revealed a contact between the inhibitor and an amino acid that differs between iNOS and eNOS (Asp376 in iNOS substituted for Asn in eNOS). To exploit this substitution for selectivity, via an electrostatic interaction with the Asp376 residue, compound 32 was designed with a basic nitrogen group. To validate the interaction, docking studies and a low resolution X-ray structure (3.3 Å) seemed to confirm this interaction. However, the in vitro NOS assay revealed a substantial reduction in potency for iNOS (24 μM). Potentially, the low resolution of the structure suggests a weaker binding of the ligand. The reduction in potency may be due to a loss of hydrogen-bonds between the displaced water molecule with the protein or an energy penalty associated with desolvation of the primary amine group. Often, water molecules are observed H-bonded to this residue and are displaced by the carboxylate group of the native ligand. Potentially, the formation of a negative charge at this site is unfavorable for a suitable interaction with a protonated amine group. In the design of selective nNOS inhibitors, Silverman and coworkers designed an *N*-hydroxy analogue of 18 with the aim of displacing a structural water molecule hydrogen-bonded between the heme propionate groups of the nNOS enzyme (70). The water molecule was successfully displaced as evidenced by the cocrystal structure of the *N*-hydroxy analogue; however, the modification failed to improve the in vitro potency of 18. It is apparent that use of the structural information in isolation does not always guide the design in the right direction in all cases.

5. Crystal Packing, Pharmaceutical Co-crystals and Physicochemical Properties; Drug Solubility and Absorption

Once a suitable lead compound has been identified, it will become necessary to identify an appropriate crystalline form to allow formulation of a dosage form with acceptable pharmacokinetic and stability properties. The impact of crystalline state on the physicochemical properties of active pharmaceutical ingredients (APIs) is an important aspect of drug development from a regulatory and patentability perspective. Crystalline forms of APIs often form scalable products of high purity and reproducibility and the crystalline form is often thermodynamically more stable in comparison with amorphous solids. From a development perspective, a stable crystalline polymorph is often sought to establish an extended retest or expiry date for an API. Screening programs are typically undertaken early in the drug development process to find the optimum form (71). Identification of a new polymorphic form or an alternative salt form may be patentable due to the unpredictability in their preparation (72, 73), and as such fulfill the criteria of novelty, non-obviousness, and utility. In addition to the potential advantage of offering enhanced stability and patentability, the thermodynamic solubility of a particular polymorphic form of an API controls the rate of dissolution and potentially the bioavailability and pharmacokinetics of an API. As the majority of APIs belong to the class II category according to the Biopharmaceutical classification system (BCS II: low solubility and high permeability), programs aimed at screening for alternative highly soluble crystalline forms or new salt forms are important with respect to optimizing solubility and bioavailability. Alternative crystalline forms include salts, hydrates, solvates or cocrystals of the drug substance. The formation of a salt to improve solubility of a drug substance either via a basic or an acidic site represents approximately 40% of marketed drugs (74). While this approach is an important strategy for improving solubility, issues arise when a compound has no ionizable group at physiological pH and formation of a stable salt is not possible. As an example, the VR1 antagonist AMG 517 is insoluble in water and aqueous buffers at physiological pH, and has a high log*P* value (5.01) and high melting point (75, 76). While a stable hydrated polymorph could be prepared (stable for 3 years at ambient temperature and humidity), prepared crystalline salts were found to disproportionate to the free base hydrate in aqueous media and the acidic nature of the incipient solution resulted in cleavage of the ether bond. The p*K*a values for AMG 517, which range from –1.94 for the pyrimidine group to 0.28 for the aminobenzothiazole group to 12.12 for the amide group, shed light on the problems associated with formation of a suitable salt form (see Fig. 16). Given the low p*K*a values of the most basic moieties, it is understandable that any true salt would result in highly acidic solutions upon dissolution.

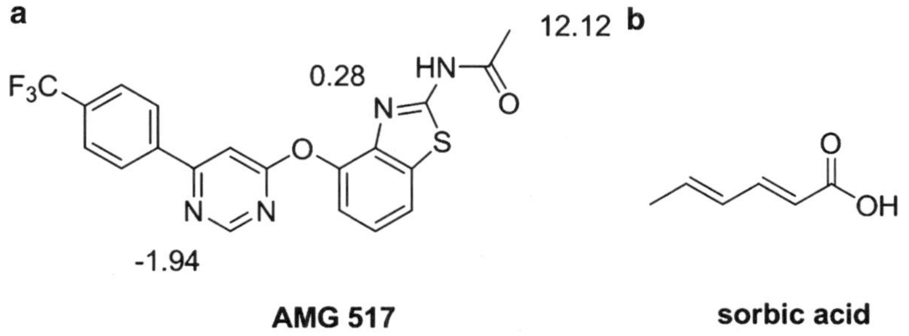

Fig. 16. Structure of the VR1 antagonist AMG 517 and its crystal coformer sorbic acid. The p*Ka* values for the pyridine group (−1.94), the aminobenzothiazole group (0.28) and the amide group (12.12) are outside the normal physiological range. The p*Ka* value of sorbic acid (4.76) is outside the required range of at least two p*Ka* units lower than the base involved in proton transfer and as such, do not form a salt pair.

Despite the insolubility of the free base, a suspension of AMG 517 in OraPlus® containing sorbic acid resulted in good exposure after low doses but poor solubility limited absorption at higher doses. This behavior was ascribed to the formation of a more water soluble 1:1 cocrystal complex between AMG 517 and sorbic acid.

Pharmaceutical cocrystals represent an alternative strategy to altering the physicochemical properties of a drug (77). In general, pharmaceutical cocrystals are defined as a multiple component crystal containing at least one molecular component that is a solid at room temperature, of which at least two are held together by weak interaction where one of the compounds is a cocrystal former (78). The principle goal is to optimize the physicochemical properties of the API while maintaining the intrinsic activity of the drug substance. For instance, APIs that are poorly soluble, have stability issues or are unable to undergo salt formation (see above) can be targeted for cocrystal formation and more than 90 compounds have been studied in this context (79).

In general, physicochemical properties such as melting point correlate with solubility and dissolution and thus oral bioavailability. In some cases, rational design of co-crystals with select physicochemical properties and predictable properties such as melting points is possible. In a study of AM 517 with ten commercially available acids, a 78% correlation was observed between the melting point of the cocrystal formers and the melting point of the cocrystal and a 55% correlation between the melting point and the Log solubility of the cocrystals was observed (80). Following from this work, 15 cocrystals of AMG 517 and several related compounds with various acids and amides were investigated (see Fig. 17) (78). However, a low correlation between the melting point of the cocrystal former and the melting point of the cocrystal was observed and no correlation was observed between the cocrystal melting point and the Log solubility. Thus, when a diverse array of coformers and APIs are employed containing multiple hydrogen bond donors and/or

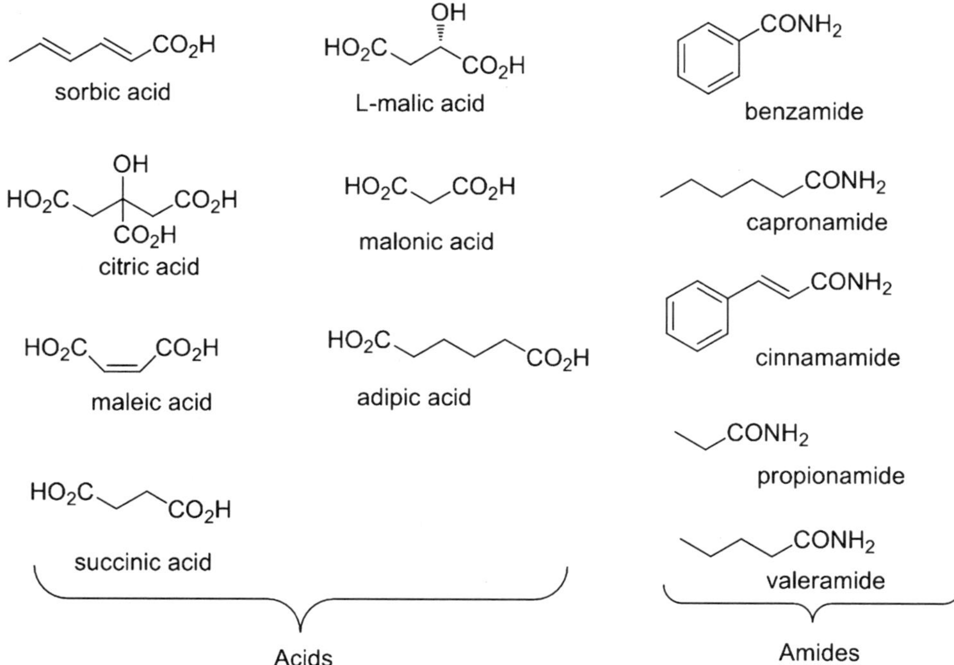

Fig. 17. Cocrystal formers crystallized with AMG 517 and three related TRPV1 antagonists.

Fig. 18. Cocrystal hydrogen-bonding motifs observed between AMG 517 and three related TRPV1 antagonists and cocrystal former acids and amides.

acceptors (see Fig. 17), the ability to find cocrystals with desirable properties is unpredictable and success depends on experimentation rather than prediction. In general, the minimum requirement for cocrystal formation is the ability of the API to form hydrogen bonds with the coformer although the presence ionic interactions may come into play. In the above examples, all the cocrystal formers and APIs interact through similar hydrogen-bonded heterosynthons (see Fig. 18). Thus both amide and acid coformers form H-bonds with the aminobenzothiazole moiety. The distinction between neutral cocrystal and salt was made on the basis of the difference between the two CO lengths. Thus, a carboxylate group has similar

CO bond lengths with a difference of ΔCO <0.03 Å, while the neutral CO2H group has CO bond length differences ΔCO > 0.08 Å. All of the complexes with AMG 517, AMG 831664, and AMG 678809 studied formed neutral cocrystal with the exception of AMG 670129-sorbic acid cocrystal, which had characteristics of a salt formation (ΔCO = 0.021 Å). When Δp*K*a (p*K*a base – p*K*a acid) values are <0, neutral cocrystal formation results almost exclusively. Examination of the aminoquinoline core of AMG 670129 reveals a p*K*a of 5.63 with a Δp*K*a of 1.04 and thus the cocrystal has greater salt characteristics. Of interest within this series of cocrystals is the varying degree of solubility in FASIF (Fasted State Simulated Intestinal Fluid) and the dissolution profiles observed, with solubility ranges from 2.1 μg/mL for the 2:1 AMG 517-adipic acid complex to 24 μg/mL for the corresponding 1:1 malonic acid complex. In comparison the solubility of the free base is 5 μg/mL. Solubility of these cocrystals did not appear to depend on the particle size as the solubility of the citric acid cocrystal with the largest particle size (512 μM) was similar to the malonic acid cocrystal with the smallest particle size (31 μM). The maximum compound solubility was reached within 1–2 h with the solubility of all but two compounds decreasing over time by conversion to the free base. Despite this, the initial solubility increase may increase drug exposure and improve bioavailability.

As mentioned previously, the formation of a salt is general strategy for improving solubility of a drug substance with hydrochloride salt being the most common form an API (81). The presence of a chloride anion in a salt complex of a cationic amine drug offers significant potential for cocrystal engineering that has not been widely exploited (82). The cocrystallization of the hydrochloride salt of the amine API fluoxetine (Prozac) with various acid co-formers was reported by Childs et al. (82). Their approach relied on the exceptional ability of chloride ion to serve as a hydrogen bond acceptor. While a N-H...Cl bond is the key interaction which contributes greatest to the stability of the crystal structure for amine compounds with hydrogen bond donor groups, other neutral chloride-assisted hydrogen bonds also participate within the coordination sphere of chloride. In the case of the fluoxetine hydrochloride salt, these include weaker alpha C-H donors (see Fig. 19). Three cocrystals of fluoxetine hydrochloride with benzoic acid (1:1), succinic acid (2:1), and fumaric acid (2:1) were prepared by slow evaporation from acetonitrile or ethanol solutions. In these three structures, bifurcated H-bonds form between the protonated amine NH-groups and chloride ion and additional H-bond of the carbonyl of neutral carboxylic acid with the chloride ion (see Fig. 20) for a total of three H-bonds to chloride ion. Additionally, charge assisted H-bonds between an aromatic C–H···Cl^- group and one activated –CH3···Cl^- together with van der Waals interactions stabilize the crystal structure.

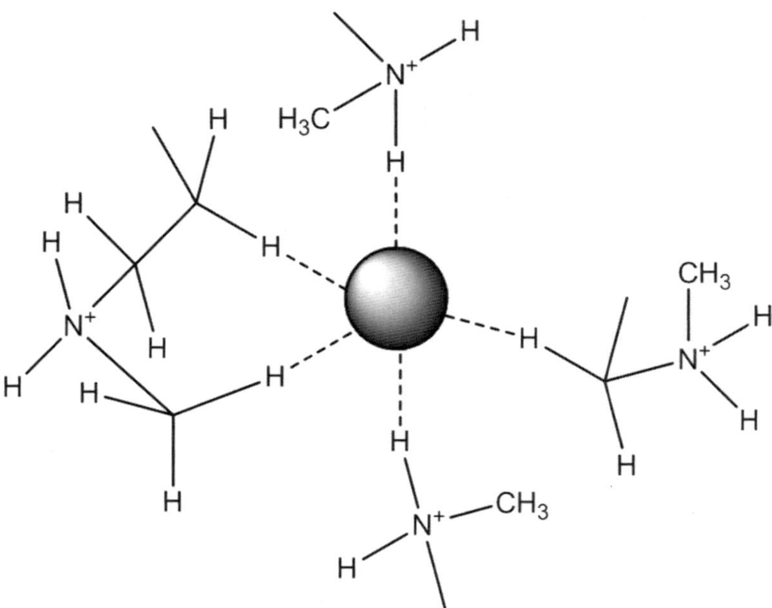

Fig. 19. Coordination sphere of chloride observed in the crystal structure of fluoxetine hydrochloride. Both N–H···Cl⁻ H-bonds and weaker CH···Cl⁻ bonds participate in the stabilization of the crystal structure. In general, the N–H···Cl⁻ H-bond lengths are approximately 2.1 Å, while that of the weaker C–H···Cl⁻ bonds range from 2.69 to 2.77 Å.

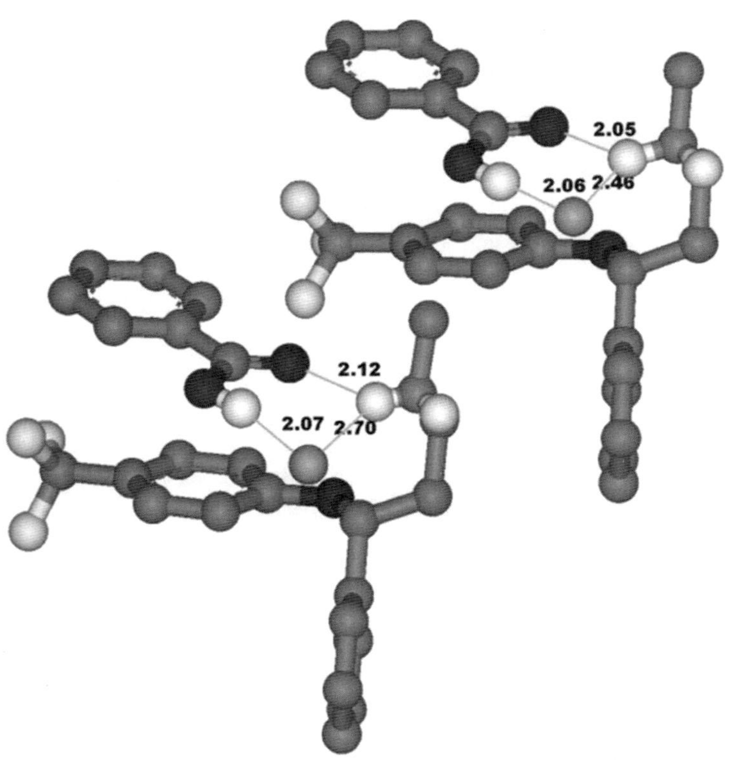

Fig. 20. X-ray structure of a cocrystal containing fluoxetine hydrochloride and benzoic acid. The asymmetric unit contains two chloride, two fluoxetine, and two benzoic acid moieties with the aggregate being held together by OH···Cl⁻, N–H···Cl⁻, and N–H···O hydrogen bonds. H-bond distances differ slightly for the various groups within the asymmetric unit as follows: OH···Cl⁻ (2.06, 2.07 Å), between and NH···Cl and for benzoic acid and fluoxetine NH···O are 2.5, 2.7 Å and 2.05, 2.12 Å, respectively. Additional van der Waals and aromatic C–H···Cl⁻ and $-CH_3$···Cl⁻ interactions are also observed.

Of interest for the three cocrystal complexes are the rates of dissolution and solubilities. The equilibrium solubilities and intrinsic dissolution rates for the fluoxetine HCl–benzoic acid (5.6 mg/mL; $k_{rel} = 0.5$) cocrystal decreased with respect to the parent fluoxetine HCl API (11.4 mg/mL; $k_{rel} = 1$), while fluoxetine HCl–succinnic acid (20.2 mg/mL; $k_{rel} > 2.9$) and the equilibrium solubility of fluoxetine HCl–fumaric acid (14.8 mg/mL; $k_{rel} = 0.9$) cocrystals increased (82). Importantly, the ordering of the intrinsic dissolution rates of the cocrystals appears to follow the solubilities of the corresponding guest acids thus allowing a crystal engineering approach to improving the solubility of the parent API.

6. Concluding Remarks

Structure-based design continues to play an important role in the drug discovery and development process. In the ideal case, the structure of ligand–protein complexes allows for rational design of potent and selective compounds against a validated molecular target. However, the development of a drug goes well beyond that of simply optimizing these parameters. As has been highlighted in this chapter, compounds must be designed with suitable physicochemical properties to aid bioavailability, reduce toxicity, and minimize metabolism, to ensure that the drug reaches the appropriate target tissue and that the API has suitable properties to aid in dissolution and maintenance of chemical stability. In some cases the use of structure-based design could potentially hinder drug development as attempts to increase potency by inflating molecular weight and lipophilicity may lead to poor absorption, increased metabolic liability, or increase risk of toxicity such as hERG or phospholipidosis (83). However, in the ideal situation, the use of structure-based design could be used to guide the rational modifications of drug candidates' physicochemical properties by allowing placement of solubilizing groups into solvent exposed areas (e.g., a basic site) or identifying allowed sites for fluorine substitutions to enhance permeability or to minimize metabolism. A further step in the application of X-ray crystallography is the understanding it can provide to help guide the design of pharmaceutical cocrystals to improve the physicochemical properties of APIs. Ultimately, the use of molecular structure to guide drug design and development must be practiced in conjunction with careful control of the physicochemical properties to maximize the probability of getting a drug approved.

References

1. Abou-Gharbia, M. (2009) Discovery of innovative small molecule therapeutics. *J. Med. Chem.* **52**, 2–9.
2. Ullman, F.; Boutellier, R. (2008) Drug discovery: are productivity metrics inhibiting motivation and creativity? *Drug Disc. Today*, **13**(21/22), 997–1001.
3. Prentis, R. A., Lis, Y., Walker, S.R. Pharmaceutical innovation by the seven UK-owned pharmaceutical companies (1964-1985). *Br. J. Clin. Pharm.* (1988) **25**, 387–396.
4. Wahling, M. Assessing the translatability of drug projects: what needs to be scored to predict success? *Nat. Rev. Drug. Disc.* (2009) **8**, 541–546.
5. Pidgeon, C., Ong, S., Liu, H., Qiu, X., Pidgeon, M., Dantzig, A.H., Munroe, J., Hornback, W.J., Kasher, J.S., Glunz, L., Szczerba, T. (1995) IAM chromatography: An *in vitro* screen for predicting drug membrane permeability. *J. Med. Chem.* **38**, 590–594.
6. Lipinski,C.A., Lombardo, F., Dominy, B.W., Feeny, P.J., Experimental and computational approaches to estimate solubility and permeability in drug discovery and development settings. *Adv. Drug Delivery Rev.* (1997) **23**, 3–25.
7. Wenlock, M.C. Rupert, P.A., Barton, P., Davis, A.M.; Leeson, P.D. (2003) A Comparison of Physicochemical Profiles of Development and Marketed Oral Drugs. *J. Med. Chem.* **46**, 1250–1256.
8. Morphy, R. (2006) The influence of target family and functional activityon the physicochemical properties of pre-clinical compounds. *J. Med. Chem.* **49**, 2969–2978.
9. Morhpy, R.; Rankovic, Z. (2006) The physicochemical challenges of designing multiple ligands. *J. Med. Chem.* **49**(16), 4961–70.
10. Morphy, R.; Rankovic, Z. (2009) Designing multiple ligands - medicinal chemistry strategies and challenges. *Curr. Pharm. Des.* **15**(6), 587–600.
11. Gleeson, M.P. (2008) Generation of a set of simple, interpretable ADMET rules of thumb. *J. Med. Chem.* **51**, 817–834.
12. Wager, T.T., Hou, X., Verhoest, P.R.; Villalobos, A. (2010) Moving beyond rules: The development of a central nervous system multiparameter optimization (CNS MPO) approach to enable alignment of druglike properties. *ACS Chem. Neurosci.* 1(6), 435–449.
13. Wager, T.T., Ramalakshmi Y. Chandrasekaran, Hou, X., Troutman, M.D., Verhoest, P.R., Villalobos, V., Will, Y. (2010) Defining Desirable Central Nervous System Drug Space through the Alignment of Molecular Properties, *in Vitro* ADME, and Safety Attributes . *ACS Chem. Neurosci.* 1(6), 420–434.
14. Hoffman, M., Monroe, D.M. (2007) Coagulation 2006: a modern view of hemostasis. *Hematol. Oncol. Clin. North Am.* **21**(1), 1–11.
15. Ansell, J. (2007) Factor Xa or thrombin: is factor Xa a better target? *J. Thromb. Haemostasis*, **5**(Suppl. 1), 60–64.
16. Galanis, T.; Thomson, L.; Palladino, M.; Merli, G.J. (2011) New oral anticoagulants. *J. Thromb. Thrombolysis* **31**, 310–320.
17. Liener, I.E. (1996) Trypsin inhibitors: concern for human nutrition or not? *J. Nutr.* **116**, 920–923.
18. Herbert, J.M.; Bernat, A.; Dol. F.; Hérault, J.P.; Crépon, B.; Lormeau, J.C. (1996) DX 9065A a novel, synthetic, selective and orally active inhibitor of factor Xa: *in vitro* and *in vivo* studies. *JPET*, **276**(3), 1030–1038.
19. Kaiser, B. (2003) DX-9065a, a direct inhibitor of factor Xa. *Cardiovasc. Drug Rev.* **21**, 91–104.
20. Hinder, M.; Frick, A.; Jordaan, P.; Hesse, G.; Gebauer, A.; Maas, J.; Paccaly, A. (2006) Direct and rapid inhibition of factor Xa by otamixaban: a pharmacokinetic and pharmacodynamic investigation in patients with coronary artery disease. *Clin. Pharmacol. Ther.* **80**(6), 691–702.
21. Cohen, M., Bhatt, D.L., Alexander, JH, Montalescot, G., Bode, C, Henry, T, Tamby, J.F., Saaiman, J, Simek, S., De Swart, J. (2007) Randomized, double-blind, dose-ranging study of otamixaban, a novel, parenteral, short-acting direct factor Xa inhibitor in percutaneous coronary intervention: the SEPIA-PCI trial. *Circulation.* **115**(20), 2642–2651.
22. Sinha, U. (2002) ZK-807834. Berlex. *Curr. Opin. Invest. Drugs.* **3**, 1736–1741.
23. Diaz-Ricart, M.; Castaner, J. (2002) ZK-807834: Anticoagulant factor Xa inhibitor. *Drugs Future.* **27**, 748–752.
24. Leitner, J.M.; Jilma, B.; Mayr, F.B.; Cardona, F.; Spiel, A.O.; Firbas, C.; Rathgen, K.; Stähl, H.; Schühly, U; Graefe-Mody, E.U. (2007) Pharmacokinetics and pharmacodynamics of the dual FII/FX inhibitor BIBT 986 in endotoxin-induced coagulation. *Clin. Pharmacol. Ther.* **81**, 858–866.
25. Brandstetter, H.; Kühne, A.; Bode, W.; Huber, R.; von der Saal, W.; Wirthensohn, K.; Engh, R.A. (1996) X-ray structure of active site-inhibited clotting factor Xa. Implications for

drug design and substrate recognition. *J. Biol. Chem.* **271**, 29988–29992.

26. Young, T., Abel, R., Byungchan, K., Berne, B.J., Friesner, R.A. (2007) Motifs for molecular recognition exploiting hydrophobic enclosure in protein-ligand binding. *Proc. Natl. Acad. Sci.* **104**(3), 808–813.
27. Pinto, D.J.P.; Smallheer, J.M.; Cheny, D.L.; Knabb, R.M.; Wexler, R.R. (2010) Factor Xa Inhibitors, Next-Generation Antithrombotic Agents. *J. Med. Chem.* **53**, 6243–6274.
28. Hogben, C.A.M., Tocco, D.J., Brodie, B.B., Schanker, L.S. (1959) On the mechanism of intestinal absorption of drugs. *J. Pharmacol. Exp. Ther.* **125**, 275–282.
29. Palm, K.; Luthman, K.; Ros, J.; Gråsjö, J.; Artursson, P. (1999) Effect of molecular charge on intestinal epithelial drug transport: pH-dependent transport of cationic drugs. *J. Pharmacol. Exp. Ther.* **291**, 435–443.
30. Quan, M.L., Pruitt, J.R., Ellis, C.D., Liauw, A.Y., Galemmo, R.A., Jr., Stouten, P.F.W., Wityak, J., Knabb, R.M., Thoolen, M.J.; Wong, P.C., Wexler, R.R. (1997) Bisbenzamidine isoxazoline derivatives as factor Xa inhibitors. *Bioorg. Med. Chem. Lett.* 7, 2813–2818.
31. Quan, M.L., Liauw, A.Y., Ellis, C.D., Pruitt, J.R., Carini, D.J.; Bostrom, L.L., Huang, P.P., Harrison, K., Knabb, R.M., Thoolen, M.J., Wong, P.C., Wexler, R.R. (1999) Design and synthesis of isoxazoline derivatives as factor Xa inhibitors. *J. Med. Chem.* **42**, 2752–2759.
32. Fevig, J.M., Pinto, D.J., Han, Q., Quan, M.L., Pruitt, J.R., Jacobson, I.C., Galemo, R.A., Wang, S., Orwat, M.J., Bostrom, L.L., Knabb, R.M., Wong, P.C., Lam, P.Y.S., Wexler, R.R. (2001) Synthesis and SAR of benzamidine factor Xa inhibitors containing a vicinally-substituted heterocyclic core. *Bioorg. Med. Chem. Lett.* **11**, 641–645.
33. Lam, P.Y.S., Clark, C.G., Li, R., Pinto, D.J.P., Orwat, M.J., Galemmo, R.A., Fevig, J.M., Teleha, C.A., Alexander, R.S., Smallwood, A.M., Rossi, K.A., Wright, M.R., Bai, S.A., He, K., Leuttgen, J.M., Wong, P.C., Knabb, R.M., Wexler, R.R. (2003) Structure-based design of novel guanidine/benzamidine mimics: potent and orally bioavailable factor Xa inhibitors as novel anticoagulants. *J. Med.Chem.* **46**, 4405–4418.
34. Bredt, D. S., Hwang, P. M., EGlatt, C. E., Lowenstein, C., Reed, R. R., Snyder, S. H. (1991) Cloned and expressed nitric oxide synthase structurally resembles cytochrome P-450 reductase. *Nature.* **351**, 714.
35. Stuehr, D. J. (1997) Structure-function aspects in the nitric oxide synthases. *Annu. Rev. Pharmacol. Toxicol.* **37**, 339.
36. Siddhanta, U., Presta, A., Fan, B., Wolan, D.D., Rousseau, D.L., Stuehr, D.J. (1998) Domain swapping in inducible nitric-oxide synthase. Electron transfer occurs between flavin and heme groups located on adjacent subunits in the dimer. *J.Biol.Chem.* **273**(30), 18950–8.
37. Fischmann, T. H., Hruza, A., Niu, X. D., Fossetta, J. D., Lunn, C. A.; Dolphin, E., Prongay, A. J., Reichert, P., Lundell, D. J., Narula, S.K., Weber, P. C. (1999) Structural characterization of nitric oxide synthase isoforms reveals striking active-site conservation. *Nat. Struct. Biol.* **6**, 233.
38. Cho, H. J., Xie, Q. W., Calaycay, J., Mumford, R. A., Swiderek, K. M., Lee, T. D., Nathan, C. (1992) Calmodulin is a subunit of nitric oxide synthase from macrophages. *J. Exp. Med.* **176**, 599.
39. Crane, B. R., Arvai, A. S., Gachhui, R., Wu, C., Ghosh, D. K., Getzoff, E. D., Stuehr, D. J., Tainer, J. A. (1997) The structure of nitric oxide synthase oxygenase domain and inhibitor complexes. *Science.* **278**, 425.
40. Crane, B. R., Arvai, A. S., Ghosh, D. K., Wu, C., Getzoff, E. D., Stuehr, D. J., Tainer. J. A. (1998) Structure of nitric oxide synthase oxygenase dimer with pterin and substrate. *Science.* **279**, 2121.
41. Raman, C. S., Li, H., Martásek, P., Král, V., Masters, B. S., Poulos, T. L. (1998) Crystal structure of constitutive endothelial nitric oxide synthase: a paradigm for pterin function involving a novel metal center. *Cell.* **95**, 939.
42. Li, H., Raman, C. S., Glaser, C. B., Blasko, E., Young, T. A., Parkinson, J. F., Whitlow, M., Poulos, T. L. (1999) Crystal structures of zinc-free and -bound heme domain of human inducible nitric-oxide synthase. Implications for dimer stability and comparison with endothelial nitric-oxide Synthase. *J. Biol. Chem.* **274**, 21276.
43. Li, H., Shimizu, H., Flinspach, M., Jamal, J., Yang, W., Xian, M., Cai, T., Wen, E. Z., Jia, Q., Wang, P. G., Poulos, T. L. (2002) The novel binding mode of N-alkyl-N′-hydroxyguanidine to neuronal nitric oxide synthase provides mechanistic insights into NO biosynthesis. *Biochemistry.* **41**, 13868.
44. Garcin, E. D., Arvai, A. S., Rosenfeld, R. J., Kroger, M. D., Crane, B. R., Andersson, G., Andrews, G., Hamley, P. J., Mallinder, P. R., Nicholls, D. J., St-Gallay, S. A., Tinker, A. C. Gensmantel, N. P., Mete, A., Cheshire, D. R., Connolly, S., Stuehr, D. J., Aberg, A., Wallace, A. V., Tainer, J. A., Getzoff, E. D. (2008) Anchored plasticity opens doors for selective inhibitor design in nitric oxide Synthase. *Nat. Chem. Biol.* **4**, 700.

45. Haitao, J., Huiying, L., Martásek, P., Roman, L. J., Poulos, T. L., Silverman, R. B. (2009) Discovery of highly potent and selective inhibitors of neuronal nitric oxide synthase by fragment hopping. *J. Med. Chem.*, **52**, 779.
46. Erdal, E.P.; Litzinger, A.; Seo, J.; Zhu, Y.; Ji, H; Silverman, R.B. (2005) Selective neuronal nitric oxide synthase inhibitors. *Curr. Top. Med. Chem.* **5**, 603.
47. Ashina, M.; Lassen, L.H.; Bendtsen, L.; Jensen, R.; Olesen, J. (1999) Effect of inhibition of nitric oxide Synthase on chronic tension-type headache: a randomized crossover trial. *The Lancet* **353**, 287–289.
48. Flinspach, H. Li, J. Jamal, W. Yang, H. Huang, J-M. Hah, J.A. Gomez-Vidal, E. A. Litzinger, R. B. Silverman and T. L. Poulos. (2004) Structural basis for dipeptide amide isoform-selective inhibition of neuronal nitric oxide synthase. *Nat. Struct. Mol. Biol.* **11**(1), 54–9.
49. Raman, C.S., Li, H., Martásek, P., Southan, G., Masters, B.S., Poulos. T.L. (2001) Crystal structure of nitric oxide synthase bound to nitro indazole reveals a novel inactivation mechanism. *Biochemistry.* **40**(45), 13448–55.
50. Xue, F.; Fang, J. Lewis, W.W.; Martásek, P. Roman, L.J. Silverman, R.B. (2010) Potent and selective neuronal nitric oxide Synthase inhibitors with improved cellular permeability. *Bioorg. Med. Chem. Lett.* **20**, 554–557.
51. Maddaford, S.; Ramnauth, J.; Rakhit, S., Patman, J., Annedi, S. C., Andrews, J., Dove, P., Silverman, S., Renton, P. (2008) Preparation of tetrahydroquinolines and related compounds having NOS inhibitory activity. US 20080234237 A1.
52. Unpublished results.
53. Zheng, G.Z., Bhatia, P., Kolasa, T., Patel, M., El Kouhen, O.F., Chang, R., Uchic, M. E., Miller, L., Baker, S., Lehto, S.G., Honore, P., Wetter, J.M., Marsh, K.C., Moreland, R.B., Brioni, J.D., Stewart, A.O. (2006) Correlation between brain/plasma ratios and efficacy in neuropathic pain models of selective metabotropic glutamate receptor 1 antagonists. *Bioorg. Med. Chem. Lett.* **16**, 4936–4940.
54. Finlayson, K., Witchel, H.J., McCulloch, J., Sharkey, J. (2004) Acquired QT interval prolongation and HERG: implications for drug discovery and development. *Eur. J. Pharmacol.* **500** (1-3), 129–42.
55. Cavalli, A., Poluzzi, E., De Ponti, F., Recanatini, M. (2002) Toward a pharmacophore for drugs inducing the long QT syndrome: Insights from a CoMFA study of hERG K+ channel blockers. *J. Med. Chem.* 2002, **45**, 3844–3853.
56. Ekins, S., Crumb, W.J., Sarazan, R.D., Wikel, J.H., Wrighton, S.A. (2002) Three-dimensional quantitative structure-activity relationship for inhibition of human ether-a-go-go-related gene potassium channel. *J. Pharmacol. Exp. Ther.* **301**(2), 427–34.
57. Pearlstein, R.A., Vaz, R.J., Kang, J., Chen, X.L., Preobrazhenskaya, M., Shchekotikhin, A.E., Korolev, A.M., Lysenkova, L.N., Miroshnikova, O.V., Hendrix,J., Rampe, D. (2003) Characterization of hERG potassium channel inhibition using CoMSiA 3D QSAR and homology modeling approaches. *Bioorg. Med. Chem. Lett.* **13**(10),1829–35.
58. Waring, M.J., Johnstone, C.A. (2007) A quantitative assessment of hERG liability as a function of lipophilicity. *Bioorg. Med. Chem. Lett.* **17**, 1759–64.
59. Wang, M.J. (2010) Lipophilicity in drug discovery. *Expert Opin. Drug Discov.* **5**(3), 235–48.
60. Hagmann, W.K. (2008) The many roles of fluorine in medicinal chemistry. *J. Med. Chem.* **51**(15), 4359–4369.
61. Hale, J.J., Mills, S.G., MacCoss, M., Finke, P.E., Cascieri, M.A., Sadowski, S., Ber, E., Chicchi, G.G., Kurtz, M., Metzger, J., Eiermann, G., Tsou, N.N., Tattersall, F.D., Rupniak, N.M., Williams, A.R., Rycroft, W., Hargreaves, R., MacIntyre, D.E. (1998) Structural optimization affording 2-(R)-(1-(R)-3, 5-bis(trifluoromethyl)phenylethoxy)-3-(S)-(4-fluoro)phenyl-4- (3-oxo-1,2,4-triazol-5-yl) methylmorpholine, a potent, orally active, long-acting morpholine acetal human NK-1 receptor antagonist. *J. Med. Chem.* **41**(23), 4607–14.
62. Uetrecht, J. (2007) Idiosyncratic drug reactions: current understanding. *Annu. Rev. Pharmacol. Toxicol.* **47**, 513–539.
63. Lin, L.S., Lanza, T.J. Jr, Jewell, J.P., Liu, P., Shah, S.K., Qi, H., Tong, X., Wang, J., Xu, S.S., Fong, T.M., Shen, C.P., Lao, J., Xiao, J.C., Shearman, L.P., Stribling, D.S., Rosko, K., Strack, A., Marsh, D.J., Feng, Y., Kumar, S., Samuel, K., Yin, W., Van der Ploeg, L.H., Goulet, M.T., Hagmann, W.K. (2006) Discovery of N-[(1S,2S)-3-(4-Chlorophenyl)-2-(3-cyanophenyl)-1-methylpropyl]-2-methyl-2-{[5-(trifluoromethyl)pyridin-2-yl]oxy} propanamide (MK-0364), a novel, acyclic cannabinoid-1 receptor inverse agonist for the treatment of obesity. *J. Med.Chem.* **49**(26), 7584–7.
64. Sturino, C.F., O'Neill,G., Lachance, N., Boyd, M., Berthelette, C., Labelle, M., Li, L., Roy, B., Scheigetz, J., Tsou, N., Aubin, Y., Bateman, K.P., Chauret, N., Day S.H., Lévesque, J.F., Seto, C.,

Silva, J.H., Trimble, L.A., Carriere, M.C., Denis, D., Greig, G., Kargman, S., Lamontagne, S., Mathieu, M.C., Sawyer, N., Slipetz, D., Abraham, W.M., Jones, T., McAuliffe, M., Piechuta, H., Nicoll-Griffith, D.A., Wang, Z., Zamboni, R., Young, R.N., Metters, K.M. (2007) Discovery of a potent and selective prostaglandin D2 receptor antagonist, [(3R)-4-(4-chloro-benzyl)-7-fluoro-5-(methylsulfonyl)-1,2,3,4-tetrahydrocyclopenta[b]indol-3-yl]-acetic acid (MK-0524). *J. Med. Chem.* **50**(4),794–806.

65. Zanda, M. (2004) Trifluoromethyl group: an effective xenobiotic function for peptide backbone modifications. *New J. Chem.* **28**, 1401–1411.
66. Fedorov, R., Vasan, R., Ghosh, D.K., Schlichting, L. (2004) Structures of nitric oxide synthase isoforms complex with the inhibitor AR-R17477 suggest a rational basis for specificity and inhibitor design. *Proc. Natl. Acad. Sci.* **101**(16), 5892–97.
67. Russell, R.J., Haire, L.F., Stevens, D.J., Collins, P.J., Lin, Y.P., Blackburn, G.M., Hay, A.J., Gamblin, S.J., Skehel, J.J. (2006) The structure of H5N1 avian influenza neuraminidase suggests new opportunities for drug design. *Nature.* **443**(7107), 45–9.
68. Skrzypcazk-Jankun, E., Borbulevych, O.Y., Zavodsky, M.I., Baranski, M.R., Padmanabhan, K., Petricek, V., Jankun, J. (2006) Effect of crystal freezing and small-molecule binding on internal cavity size in a large protein: X-ray and docking studies of lipoxygenase at ambient and low temperature at 2.0 Å resolution. *Acta Crystalogr. D Biol. Crystallogr.* **62**(Pt 7), 766–75.
69. Davis, A.M., St-Gallay, S.A., Kleywegt, G.J. (2008) Limitations and lessons in the use of X-ray structural information in drug design. *Drug. Disc.Today.* **13**, 831–41.
70. Seo, J., Igarashi, J., Li, H., Martásek, P., Roman, L.J., Poulos, T.L., Silverman, R.B. (2007) Structure-based design and synthesis of N(omega)-nitro-L-arginine-containing peptidomimetics as selective inhibitors of neuronal nitric oxide synthase. Displacement of the heme structural water. *J. Med. Chem.* **50**, 2089–99.
71. Gardner, C. R.; Walsh, C. T.; Almarsson, O. (2004) Drugs as materials: valuing physical form in drug discovery. *Nat. Rev. Drug Discov.* **3**(11), 926–34.
72. Maddox, J. (1988) Crystals from first principles. *Nature.* 335, 201.
73. Ball, P. (1996) Scandal of crystal design. *Nature.* 381, **20**, 648–650.
74. Lipinski, C. (2002) Poor Aqueous Solubility – an Industry Wide Problem in Drug Discovery. *Am. Pharm. Rev.* **5**, 82–85.
75. Bak, A.; Gore, A.; Yanez, E.; Stanton, M.; Tufekcic, S.; Syed, R.; Akrami, A.; Rose, M.; Surpaneni, S.; Bostick, T.; King, A.; Neervannan, S.; Ostovic, D.; Koparkar, A. (2008) The co-crystal approach to improve the exposure of a water-insoluble compound: AMG 517 sorbic acid co-crystal characterization and pharmacokinetics. *J. Pharm. Sci.* **97**(9), 3942–3956.
76. Stanton, M.K., Tufekcic, S., Morgan, C., Bak, A. (2009) Drug Substance and Former Structure Property Relationships in 15 Diverse Pharmaceutical Co-Crystals *Cryst. Growth Des.* **9**(3), 1344–52.
77. Shan, N.; Zaworotko, M. J. (2008) The role of cocrystals in pharmaceutical science. *Drug Disc. Today.* **13** (9-10), 440–6.
78. Zaworotko, M.J. (2007) Molecules to Crystals, Crystals to Molecules ... and Back Again? *Cryst. Growth Des.* 7(1), 4–9.
79. Cheney, M. L., Shan, N., Healey, E.R., Hanna, M., Wojtas, L., Zaworotko, M. J., Sava, V., Song, S., Sanchez-Ramos, J. R. (2010) Effects of Crystal Form on Solubility and Pharmacokinetics: A Crystal Engineering Case Study of Lamotrigine. *Cryst. Growth Des.* **10**, 394–405.
80. Stanton, M., Bak, A. (2008) Physicochemical Properties of Pharmaceutical Co-Crystals: A Case Study of Ten AMG 517 Co-Crystals. *Cryst. Growth Des.* **8**(10), 3856–62.
81. Stahl, P.H.; Wermuth, C.G.; Eds. (2002) Handbook of Pharmacetuical Salts: Properties, Selection and Use; Wiley: Weinheim, Germany.
82. Childs, S. L., Chyall, L. J., Dunlap, J. T., Smolenskaya, V. N., Stahly, B. C. Stahly, G. P. (2004) Crystal engineering approach to forming cocrystals of amine hydrochlorides with organic acids. Molecular complexes of fluoxetine hydrochloride with benzoic, succinic, and fumaric acids. *J. Am. Chem. Soc.* **126** (41), 13335–42.
83. Waring, M.J. (2010) Lipophilicity in drug discovery. *Expert Opin. Drug Discov.* **5**(3), 235–48.

INDEX

Leslie W. Tari (ed.), *Structure-Based Drug Discovery*, Methods in Molecular Biology, vol. 841,
DOI 10.1007/978-1-61779-520-6, © Springer Science+Business Media, LLC 2012

Q

R

S

T

U

V

W

X

Z